Microelectronic Materi

T0258961

GRADUATE STUDENT SERIES IN
MATERIALS SCIENCE AND ENGINEERING

Series Editor: B Cantor
Department of Metallurgy and Science of Materials,
University of Oxford

Microelectronic Materials

C R M Grovenor

Department of Metallurgy and Science
of Materials, University of Oxford

Taylor & Francis
Taylor & Francis Group
New York   London

Published in 1998 by
Taylor & Francis Group
270 Madison Avenue
New York, NY 10016

Published in Great Britain by
Taylor & Francis Group
2 Park Square
Milton Park, Abingdon
Oxon OX14 4RN

International Standard Book Number-10: 0-85274-270-3 (Softcover)
International Standard Book Number-13: 978-0-85274-270-9 (Softcover)

Library of Congress Cataloging-in-Publication Data

Catalog record is available from the Library of Congress

Taylor & Francis Group
is the Academic Division of Informa plc.

Visit the Taylor & Francis Web site at
http://www.taylorandfrancis.com

Contents

Preface

The materials used in the fabrication of microelectronic devices and systems often play an important part in limiting the performance of components. As the size of individual devices is reduced, problems with the properties and stability of these materials become very much more severe. Appreciation of this fact has led to considerable interest in understanding the factors which determine the electrical and mechanical properties of thin-film materials in particular and the mechanisms by which they degrade. Lattice defects often have an especially significant role in controlling the properties of all the materials in a microelectronic component.

This book seeks to introduce how a simple materials science approach can be used to explain the properties needed of materials in microelectronic components and then to present the ways in which materials can be chosen to optimise the performance of the component. The materials which are considered include semiconductors themselves, but also the metallic and insulating materials which are often ignored as the 'inactive' partners in the devices. A central theme of the book is that successful devices are only prepared when the properties and interactions of all the materials are taken into account. I have chosen to describe the selection of materials for the production of integrated circuits, light-emitting diodes and photovoltaic devices, while being aware that there are many other kinds of microelectronic devices in which materials problems are equally severe.

The book is intended for senior undergraduate or graduate students and assumes that the reader has a grasp of basic semiconductor device properties and an understanding of materials science to the level of being able to read a phase diagram. References have been used sparingly throughout the text, with the intention being to indicate sources where the student can find a more detailed description of a particular phenomenon or process rather than give a comprehensive list of all the significant work in a field.

No volume of this size is ever written without important contributions from many other people. I have benefited over the last few years from discussions with friends and colleagues in many countries, and the ideas and arguments which are included in the book owe much to their guidance and patience. Mistakes and contentious opinions can be blamed wholly on my imperfect understanding. Numerous colleagues have provided me with the figures and photographs needed to give clear illustrations of complex structures and processes, and I am very grateful to all those whom I have persuaded to delve into their 'micrograph drawer'. Similar thanks are due to those whom I have cajoled into reading my draft manuscripts, including in particular Drs A Cerezo and S R Ortner who have done more than their fair share of this drudgery.

C R M Grovenor
Oxford, February 1988

1

Semiconducting Materials

1.1 INTRODUCTION

The aim of this first chapter is to introduce the basic properties of a wide range of semiconducting elements and compounds. The crystal structures and phase chemistries of semiconductors will be described first and then the structure and properties of point defects (of both intrinsic and extrinsic character) and dislocations will be presented. The role that point defects play in determining the dominant mechanism and the kinetics of the diffusion processes in semiconductors will be considered in a separate section. The emphasis of this chapter will be on presenting how the structure of, and characteristic defect concentration in, a semiconductor crystal control its electrical properties.

Here, and throughout this book, it will be assumed that the reader has a working knowledge of the simple theory of semiconductor behaviour and is familiar with the use of a one-electron diagram in describing the properties of semiconductors, defects and devices. Excellent introductions to these properties can be found in the texts by Frazer (1983) and Sze (1981).

Finally, short sections on amorphous and polymeric semiconductors are included at the end of the chapter. Some of the properties of the metallic and insulating materials used in microelectronic devices will be described in Chapters 4–7.

1.2 SEMICONDUCTORS — ELEMENTS, COMPOUNDS AND ALLOYS

Many semiconductor materials have crystal structures which are related to the simple diamond cubic lattice, and a unit cell of this structure is sketched in figure 1.1(a). Each atom is tetrahedrally coordinated, but the local atomic environments are not identical for all atoms. The structure can also be thought of as consisting of two interpenetrating FCC lattices, one with origin at $(0, 0, 0)$ and the other at $(\frac{1}{4}a, \frac{1}{4}a, \frac{1}{4}a)$. A centre of symmetry lies at the point equidistant between the origins of the two FCC lattices — at $(\frac{1}{8}a, \frac{1}{8}a, \frac{1}{8}a)$. Both silicon and germanium crystallise in this structure, and since these elements have two s and two p valence electrons it is easy to see that a full set of eight valence electrons can be gained around each atom by the promotion of one s electron to a p orbital and the formation of four covalent sp^3 hybrid bonds. We shall see that most semiconductor compounds and alloys obey the principle of keeping the average electron-to-atom ratio equal to four.

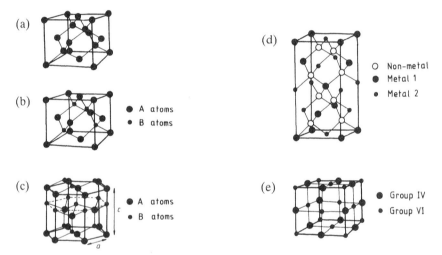

Figure 1.1 Sketches of the unit cells of semiconductor elements and compounds: (a) the diamond cubic lattice; (b) the sphalerite lattice characteristic of most III–V compounds; (c) the wurtzite lattice which can be found in some II–VI compounds; (d) the sphalerite superlattice taken up by the chalcopyrite semiconductor compounds; (e) the rocksalt crystal structure assumed by the IV–VI compounds.

The simplest illustration of this principle is given by the range of semiconductor materials AB formed between a group III element (A = B, Al, Ga and In) and one from group V (B = N, P, As and Sb). Clearly the average electron to atom ratio will be equal to four in an equiatomic compound AB. Figure 1.1(b) shows the crystal structure of one of these so-called III–V semiconductors. Let us assume that A is gallium and B is arsenic so that the compound is gallium arsenide,

GaAs. The gallium and arsenic atoms occupy alternate sites in the diamond cubic lattice so that each gallium atom is surrounded by four arsenic atoms and vice versa. Crystals of this kind are often said to have the 'zincblende' structure because their structure is identical to that of the compound zincblende, ZnS. However, it is also called the sphalerite structure and consists of two interpenetrating FCC lattices with gallium atoms occupying one sublattice and arsenic atoms the other. All the important III–V compounds crystallise naturally with the sphalerite structure.

An interesting feature of this structure is that it has a lower symmetry than the diamond cubic lattice because there is no longer a centre of symmetry between the origins of the two FCC sublattices. This means that directions and planes with indices 111 and $\bar{1}\bar{1}1$ are not identical and each {111} surface will consist of only one kind of atom. By convention, the {111} surfaces in an AB compound are defined as containing only A atoms, while the {$\bar{1}\bar{1}\bar{1}$} surfaces contain only B atoms. The chemical properties of these two surfaces are quite different and the processes of chemical dissolution and etching proceed at different rates depending on which surface is being attacked. All crystals with the sphalerite structure show this anisotropic behaviour. In addition, these crystals show preferential cleavage on {110} planes, while diamond cubic crystals of silicon or germanium cleave more readily on {111} planes.

A second important class of compound semiconductors can be prepared by combination of a group II element (Zn, Cd and Hg) with a group VI element (S, Se and Te) to form an AB compound — the II–VI semiconductor family. Most of these compounds will crystallise with either the sphalerite or wurtzite lattice structure depending on the conditions under which the crystal is grown. A unit cell of the hexagonal wurtzite structure is sketched in figure 1.1(c). The sphalerite structure is normally the equilibrium structure at room temperature. The semiconducting properties of a II–VI semiconductor are almost independent of the crystal structure, and this can be understood when we notice that the atoms in the wurtzite lattice are tetrahedrally coordinated just as in the sphalerite lattice and the separation of nearest-neighbour atoms in the two structures is almost identical.

Another group of semiconducting materials can be formed with chemical formulae of the kind I–III–VI$_2$ or II–IV–V$_2$ and these compounds crystallise in a superlattice of the sphalerite structure. Some examples of the first of these kinds of compounds are AgInTe$_2$, CuInSe$_2$ and CuGaSe$_2$, while ZnSiP$_2$ and CdSiP$_2$ are clearly II–IV–V$_2$ compounds. The metallic sublattice in the sphalerite crystal structure is now ordered to give a unit cell like that sketched in figure 1.1(d). These ternary compound semiconductors are normally referred to as the chalcopyrites.

The III–V, II–VI and chalcopyrite semiconductor families obey the principle of having on average four valence electrons/atom, and these materials are often called $A^N B^{8-N}$ compounds, where the superscripts are the valencies of the A and B atoms. However it is also possible to form compounds between some of the heavier group IV elements, such as Sn and Pb, and group VI chalcogenide elements like S, Se and Te. These are called the $A^N B^{10-N}$ compounds and have an average of five valence electrons/atom. This class of materials crystallise with the rocksalt structure which is sketched in figure 1.1(e). Here each metal atom is surrounded by not four but six chalcogenide atoms. The characteristic sp^3 hybridisation which is important in the $A^N B^{8-N}$ compounds is now replaced by s^{2}p^3 hybridisation with a large ionic component in the bonds which stabilises the six-fold coordinated structure (see Phillips 1973).

These few kinds of elements and compounds are the most important semiconducting materials for practical microelectronic devices, although a number of other materials show characteristic semiconducting properties. Table 1.1 gives the values of basic physical parameters for some of the semiconductors which will be discussed in this book: the crystal structure and lattice constants; and also the width of the forbidden band gap and its type (direct or indirect). These last two parameters are especially important in determining the use to which a semiconducting material can be put in devices and circuits, as we will see in Chapters 2, 8 and 9.

There are two interesting features of the data shown in table 1.1 which deserve special mention. Firstly, the lattice constants of these materials are generally rather larger than those of metallic elements and compounds and lie in the range 0.5– 0.65 nm. This simple fact has some significant implications for the quality of the information which we can expect to obtain on the detailed structure of crystalline defects with the latest generation of electron microscopy techniques. An example of this kind of study will be given in §1.6. Secondly, we can see that binary semiconductor compounds exist with band gaps which span the range from approximately 0 to 3 eV. This might lead us to believe that at least one of these materials will be suitable for any conceivable application in microelectronic devices. Unfortunately, it has proved hard to prepare a number of important optoelectronic devices using only this rather limited range of semiconductors (see Chapters 8 and 9). This is a result both of difficulties in preparing suitably pure and defect-free material of some compounds (see Chapter 3 for descriptions of the problems encountered in crystal growth) and also the requirement for semiconductor materials with very precisely controlled band gaps to optimise the performance of particular devices. As a result, considerable advantage has been taken of the ability to create alloys by the combination of

Table 1.1 Crystal structure and basic electrical properties of important semiconductor elements and compounds.

	Crystal structure†	Lattice spacing (nm)	Band-gap width and type (at 300 K) (eV)	
Element				
Si	D	0.5431	1.12	Indirect
Ge	D	0.5646	0.66	Indirect
III–V compounds				
GaAs	S	0.5653	1.42	Direct
GaP	S	0.5451	2.26	Indirect
GaSb	S	0.6096	0.72	Direct
InAs	S	0.6058	0.36	Direct
InP	S	0.5869	1.35	Direct
InSb	S	0.6479	0.17	Direct
AlAs	S	0.5661	2.16	Indirect
AlSb	S	0.6136	1.58	Indirect
II–VI compounds				
CdS	S/W	$0.5832/a = 0.416$ $c = 0.6756$	2.42	Direct
CdSe	S	0.605	1.7	Direct
CdTe	S	0.6482	1.56	Direct
ZnS	S/W	$0.542/a = 0.382$ $c = 0.626$	3.68	Direct
ZnSe	S	0.5669	2.7	Direct
ZnTe	S	0.6089	2.2	Direct
HgTe	S	0.644	0	—
Chalcopyrite				
CuInSe$_2$	S	$a = 0.5782$ $c = 1.1564$	1.04	Direct
IV–VI compounds				
PbS	N	0.594	0.41	Direct
PbSe	N	0.612	0.27	Direct
PbTe	N	0.646	0.31	Direct
SnTe	N	0.632	0.18	Direct

†D is diamond cubic, S is sphalerite, W is wurtzite and N is the rocksalt lattice.

binary semiconductor compounds to give ternary and even quaternary semiconductors. Pairs of binary compounds, GaAs and InAs or AlAs and GaAs for example, form a series of solid solution alloys over the whole of the composition range and the band-gap and crystal lattice constants usually vary smoothly with composition. This allows the preparation of materials with exactly the properties required for a particular application, and it is clear that the valence electron/atom ratio is constant in this kind of 'pseudobinary' alloy. Alloys of the III–V semiconductors will be discussed extensively in Chapter 8, but it is worth considering here the effect of forming alloys between two II–IV compounds — cadmium telluride, CdTe, and mercury telluride, HgTe — because the resulting material is one of the most important semiconductors currently used in the fabrication of microelectronic devices.

Properly speaking, mercury telluride, HgTe, should not be included in table 1.1 at all since it is not really a semiconductor. At room temperature HgTe shows no separation between the conduction and valence bands and even has a small band overlap — a negative band gap! We would normally expect a material with a partly filled conduction band to behave as a metal, but because the density of free carriers is very low in the regions of band overlap, materials of this kind (or with very small positive band gaps) are called semimetals. By contrast, CdTe is a more conventional semiconductor compound with a direct band gap of 1.56 eV. Both compounds have the sphalerite crystal structure, which is one reason why ternary alloys can readily be prepared. When CdTe is added to HgTe to form an alloy the conduction and valence bands move apart and at a composition of about $Hg_{85}Cd_{15}Te$ the band gap is zero. (This is a convenient point at which to introduce the convention which will be used throughout this book, namely that a ternary alloy $A_{1-x}B_xC$ consists of x moles of BC combined with $1 - x$ moles of AC.) The addition of further CdTe to the alloy produces a material with a very small direct band gap — 0.1 eV at a composition of $Hg_{80}Cd_{20}Te$ and at 77 K. A semiconductor material with a direct band gap of this width is precisely suited for the detection of light in the wavelength range 8–12 μm, the so-called 'atmospheric window' where optical absorption is minimised. For this reason, HgCdTe alloys are extremely important materials in military communications systems (Kruse 1981).

A similar way of producing a semiconductor with a very small band gap is by the alloying of two IV–VI semiconductors, SnTe and PbTe for example. Both of these compounds have very small band gaps and the same rocksalt crystal structure, but have rather different band structures. The addition of SnTe to PbTe to form $Pb_{1-x}Sn_xTe$ results in a decrease in the band gap, E_g, according to the empirical equation $E_g = 0.19 - 0.543x$ eV. E_g reaches zero at a composition of approximately $Pb_{60}Sn_{40}Te$ at 77 K. Surprisingly, further addition of SnTe again results in an increase in the band gap, and an explanation for this kind of behaviour must be sought in the detailed band structure of the two compounds (see for instance Lovatt (1977) and Harman and Melngailis (1974)). An alloy of composition $Pb_{88}Sn_{12}Te$ has a band gap of about 0.14 eV at 77 K and this material has been used for the production of efficient long-wavelength lasers (see Chapter 8). Ternary IV–VI alloys of this kind are extremely interesting, since the width of the band gap is easily modified not only by altering the composition of the alloy but also by changing its temperature. The lead chalcogenide compounds show a positive coefficient of band-gap variation with temperature (an unusual phenomenon in a semiconductor material) and increasing the operating temperature of a PbSnTe laser diode decreases the emission wavelength.

1.3 PHASE EQUILIBRIA IN COMPOUND SEMICONDUCTORS

This section will concentrate on describing the basic features of the equilibrium phase diagrams of the common binary compound semiconductors. These phase diagrams are very similar for all III–V, II–VI and IV–VI compounds, and so only a single representative compound will be considered — cadmium telluride. Included at the end of the section is a brief discussion on the form of the more complex phase diagrams which contain the ternary chalcopyrite compounds.

Figure 1.2 shows the binary phase diagram formed between cadmium and tellurium as it appears in a standard reference text like Zanio (1978). There is only one stable compound phase, cadmium telluride (CdTe), but there are eutectic reactions at compositions close to pure Cd and pure Te. CdTe is seen to be a congruently melting compound, and this is very important if large crystals are to be successfully grown from the melt (Chapter 3). This kind of temperature–composition (T–x) phase diagram illustrates the equilibrium established between the solid and liquid phases in this binary system, but in many cases it is also necessary to consider the equilibrium established between the solid and vapour phases. This is particularly true when one or more of the elements forming the semiconductor have high vapour pressures, arsenic, phosphorus, cadmium or mercury for instance. The pressure–temperature (p–T) phase diagram for CdTe is shown in figure 1.3 and we can clearly see that the equilibrium partial pressure of cadmium above the compound is generally much higher than that of tellurium. (The form of the p–T phase diagram has a significant influence on the ease with which the processes of compound synthesis and crystal growth can be carried out, as we will see in §3.2.) We will now show that changing the partial pressures of the component elements in the vapour phase can result in small but significant changes in the equilibrium stoichiometry of the solid compound semiconductor.

Returning to the phase diagram in figure 1.2, it is tempting to think of the CdTe compound phase as being perfectly stoichiometric at all temperatures, since it is represented as having a phase field of negligible

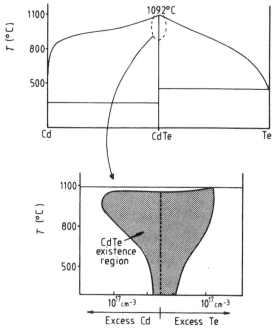

Figure 1.2 The Cd–Te binary T–x phase diagram showing that the compound CdTe is congruently melting and that no other phases are formed between these two elements. The expanded view of the central region of the phase diagram shows the extent of the CdTe existence region.

width. However, there is a narrow existence region for CdTe at high temperatures and this is of enormous significance for the electrical properties of this semiconductor. The lower part of figure 1.2 shows an expansion of the central region of the CdTe phase diagram to illustrate the extent of this existence region at temperatures close to the melting point of the compound. It is clear that the maximum melting point is found in material with composition slightly to the Te-rich side of the diagram and that at temperatures above about 500 °C the compound phase can exist over a range of compositions from the slightly Te-rich to the slightly Cd-rich. The extent of this non-stoichiometry is extremely limited, with a maximum excess of Te of about 10^{-3} at.%, and a similar deviation from exact stoichiometry on the Cd–rich side of the existence region. To use units more commonly applied to semiconductor materials, this corresponds to a stoichiometry range from about 5×10^{17} cm^{-3} excess Te atoms to the same concentration of excess Cd atoms (Zanio 1978). By combining the T–x and p–T phase diagrams shown in figures 1.2 and 1.3 it is possible to predict how we can prepare CdTe samples which are either Te- or Cd-rich simply by heating them in an atmosphere of carefully controlled composition. The composition of the vapour phase which is in equilibrium with solid CdTe can be read off the $p-T$ phase diagram. Thus at 900 °C, the equilibrium partial pressures of Te_2 and Cd over a Te-saturated crystal are 0.2 and 2×10^{-3} atmospheres respectively, while the Cd-saturated crystal will be in equilibrium with a vapour phase containing a partial pressure of about 3 atmospheres of cadmium (the equilibrium tellurium partial pressure is negligible). Adjusting the vapour phase composition from Te–rich to a large overpressure of Cd, and heating the CdTe sample to allow equilibrium to be established, will force the composition of the solid compound phase across to the far left-hand side of the existence region. The importance of this will become clear when we consider that an excess of Cd in a CdTe crystal will create point defects and that these point defects can be electrically active and control the conductivity of the crystals.

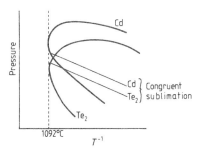

Figure 1.3 The $p-T$ phase diagram for the Cd–Te binary system showing that the vapour pressure of cadmium in equilibrium with solid CdTe is usually much higher than the equilibrium vapour pressure of tellurium dimers.

The $T-x$ phase diagrams of most common binary semiconductors are very similar in form to that shown for CdTe, although the shapes of the existence regions at high temperatures can be different. The existence region of gallium arsenide is very similar to that of CdTe (see figure 3.8), but the phase diagram of zinc telluride shows an existence region which lies almost completely on the Te–rich side of the equiatomic composition. As a result of this, the dominant native point defects in ZnTe at all temperatures are those characteristic of an excess of tellurium. By contrast, the GaP existence region is almost completely on the Ga–rich side of the equiatomic composition. The $p-T$ phase diagrams vary considerably for different compound semiconductors because of the enormous variations in the equilibrium vapour pressures of the component elements.

Now we should consider the form of the phase diagrams characteristic of ternary alloy semiconductors formed between two wholly miscible binary compounds. Since either the metallic or non-metallic elements must be common to both compounds, and the stoichiometry of the compounds is the same, AB, it is simplest to represent the 'pseudobinary' phase field between the two compounds rather than the full ternary phase diagram. An example of a pseudobinary phase diagram is shown in figure 1.4 for the case of GaInAs formed by the combination of GaAs and InAs. The complete mutual miscibility of the two binary compounds results in a very simple phase diagram, and most useful ternary semiconductor alloys have phase diagrams of this kind. We will see in §3.3 that an alloy which has a phase diagram with a wide separation between solidus and liquidus can be very hard to prepare in the form of large homogeneous single crystals.

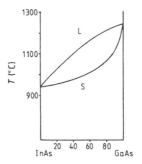

Figure 1.4 The GaAs–InAs 'pseudobinary' phase diagram showing the complete miscibility of these two compounds to form the range of ternary semiconductor alloys.

Finally, I shall describe the two basic forms of phase diagram in which the chalcopyrite semiconductor compounds are found. Some of these compounds melt congruently like the III–V and II–VI compounds and so single crystals can readily be grown from a melt. A detail from the complex pseudobinary phase diagram formed between Ag_2Se and Ga_2Se_3 is shown in figure 1.5(a) and illustrates the congruent melting of the compound $AgGaSe_2$. However, other chalcopyrite compounds are formed by peritectic reactions, for instance $CuGaSe_2$ as shown in figure 1.5(b). Large crystals of this compound are not easily prepared from a melt and are also hard to grow in a solid phase reaction process. This is one of the reasons why this group of interesting semiconductor materials has found little application in practical semiconductor devices (Shay and Wernick 1975).

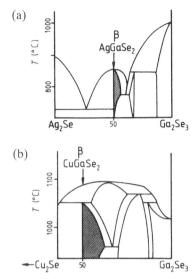

Figure 1.5 Details from the phase diagrams containing the chalcopyrite compounds $AgGaSe_2$ and $CuGaSe_2$ (after Palatnik and Belova 1967).

1.4 POINT DEFECTS IN SEMICONDUCTOR CRYSTALS

There are two kinds of point defects in which we will be interested in semiconductor crystals: native point defects — vacancies and self-interstitials for example — and impurity atoms in the semiconductor lattice. These latter defects can be further divided into two classes: dopant elements deliberately added to a semiconductor sample to modify the electronic properties; and impurity elements present in the raw materials or added involuntarily during crystal growth or device fabrication processes. We will consider native and impurity point defects separately in the following two sections and the interaction between them will be described when the electrical properties of III–V and II–VI compounds are compared in §1.4.3. This interaction between impurities and native point defects will also be important when we describe diffusion in semiconductors in §1.5.

1.4.1 Native Point Defects

Before we deal with the properties of native point defects in semiconductors, it is important to establish a nomenclature for distinguishing between the large number of different point defects which can be formed in semiconductor crystals. Here I will use a modified form of the system introduced by Kroger and Vink (1956) for II–VI materials. The three basic point defects which we must describe are the vacancy, the interstitial and the the antisite defect in compound semiconductors (an A atom on a B site or vice versa). In a semiconductor compound, AB, we can then have:

V_A^x a vacancy on the A sublattice

V_B^x a vacancy on the B sublattice

A_i^x an interstitial A atom

B_i^x an interstitial B atom

A_B^x an A atom on the B sublattice

B_A^x a B atom on the A sublattice.

The subscript shows the position of the point defect in the crystal structure and the superscript gives the 'effective' charge on the defect. The superscript x is used to signify those defects which have no effective charge, i.e. the charge is the same as that on a 'perfect' lattice site (Kroger and Vink 1956). The possibility that native point defects can be ionised is considered later in this chapter.

In an elemental semiconductor like silicon, the vacancy, V_{Si}, and the silicon self-interstitial, Si_i, are the only simple native point defects. The concentration of these point defects in a silicon crystal will depend in the normal manner on the enthalpy and entropy of formation of the vacancy or interstitial in the perfect lattice and on the temperature of the sample. For a vacancy in silicon:

$$[V_{Si}^x] \propto \exp\left(\frac{\Delta S_V}{k}\right)\exp\left(\frac{-\Delta H_V}{kT}\right). \tag{1.1}$$

(Note that throughout this chapter the symbol [X] will be used in the conventional chemical sense to mean the mole fraction of element or point defect, X, present in the crystal.) The formation enthalpy of a single vacancy in silicon has been estimated to be about 2.5 eV, and so the equilibrium concentration of vacancies will be low at all temperatures below about 1200 °C.

The thermodynamic treatment of point defect concentrations in compound semiconductors is slightly more complicated. The equilibrium concentration of any native point defect, V_A^x for example, will depend on the temperature and the enthalpy and entropy change on forming a vacancy on the A sublattice in a perfect AB crystal as above, but also on the partial pressure of A in the gas phase surrounding the crystal, PP_A:

$$[V_A^x] \propto \exp\left(\frac{\Delta S_{V_A}}{k}\right) \exp\left(\frac{-\Delta H_{V_A}}{kT}\right)(PP_A)^n. \qquad (1.2)$$

The exponent of the partial pressure of A, n, can be determined by counting the change in the number of A atoms in the crystal as a result of creating the defect. For the case of the vacancy on the A sublattice $n = -1$, but for the antisite defect A_B $n = +2$, as the crystal has not only gained one A atom but also lost a B atom, making a net gain of two A atoms. Equation (1.2) reinforces the point made in the previous section that the equilibrium concentration of native point defects, and so the overall stoichiometry of the AB crystal, can be controlled by the partial pressure of the component elements A and B in the gas phase. Values for the enthalpies of formation have been calculated for a wide range of native point defects in semiconductors (Van Vechten 1975) and a few of these results are given in table 1.2. A general feature of these values is that the formation enthalpy seems to depend on the size of the atom removed or in the interstitial site. For instance, ΔH_{V_A} increases with the covalent tetrahedral radius of the A atom. As a result, in some compound semiconductors there are significant differences in the formation enthalpies for simple vacancies on the two sublattices. In GaP, for example, $\Delta H_{V_{Ga}}$ is calculated to be 2.98 eV, while ΔH_{V_P} is 2.64 eV. This implies that at all temperatures the concentration of neutral phosphorus vacancies should exceed that of the neutral gallium vacancies. This trend is even more marked in cadmium sulphide, in which the equivalent values are $\Delta H_{V_{Cd}} = 3.56$ eV and $\Delta H_{V_S} = 2.69$ eV. Table 1.2 shows that if the ratio of covalent radii, r_A/r_B, is greater than 1, then at equilibrium we expect $[V_A]$ to be less than $[V_B]$. More generally, we do not expect the total number of defects on the A sublattice to equal the total number on the B sublattice and so the stoichiometry of the crystal will be modified, $[A] \neq [B]$. The manner in which the stoichiometry deviates from that of the perfect AB compound will depend on the relative values of the formation enthalpy of *all* the possible native point defects, and it is important to remember that many compound semiconductors are intrinsically non-stoichiometric. This is the same point as has already been made in the previous section when we considered the form of the existence regions of these compounds in the phase diagrams. The existence region of the compound ZnTe in the Zn–Te binary phase diagram lies wholly on the Te-rich side of the stoichiometric composition. This can be understood when we see in table 1.2 that $\Delta H_{V_{Te}}$ is more than 1 eV greater than $\Delta H_{V_{Zn}}$. The tellurium is thus expected always to be in excess in a ZnTe crystal, even when it is annealed in zinc vapour. By contrast, cadmium telluride, where r_A/r_B is approximately unity, has $\Delta H_{V_{Cd}} = \Delta H_{V_{Te}}$ and the existence region for this compound lies roughly symmetrically around the stoichiometric composition (figure 1.2).

Table 1.2 Calculated formation enthalpies for vacancy and antisite defects in III–V and II–VI semiconductors (after Van Vechten 1975).

Semiconductor	V_A(eV)	V_B(eV)	B_A(eV)	A_B(eV)	r_A/r_B†
GaAs	2.59	2.59	0.35	0.35	1.00
GaP	2.98	2.64	0.38	0.68	1.09
GaSb	2.03	2.56	0.08	0.32	0.87
InP	3.04	2.17	0.42	0.89	1.25
CdS	3.56	2.69	1.48	1.86	1.25
CdSe	3.18	2.65	1.25	1.42	1.15
ZnS	3.47	3.13	1.45	1.74	1.09
CdTe	2.75	2.75	0.94	0.94	1.00
ZnTe	2.54	3.60	0.96	0.78	0.87

†Covalent radius ratio for atoms on A and B sublattices.

Now we should move on to consider the electrical properties of these native point defects. I shall assume that all readers are familiar with the basic principles of doping semiconductors with impurities so as to create new electronic states lying in the forbidden band gap. The lattice distortion around point defects in semiconductors also results in the creation of new electronic states, some of which may lie in the band gap. If these states are sufficiently close to the valence or conduction band edges, then the point defect can be thermally ionised with the release of free carriers. Clearly this means that point defects can contribute significantly to the conductivity of semiconductors. If we restrict ourselves for the moment to considering only vacancies, it is possible to describe the ionisation process in terms of simple chemical reactions

$$V_{Si}^x \quad \rightleftharpoons \quad V_{Si}' \quad + \quad h^+ \qquad\qquad (1.3a)$$

neutral silicon   singly ionised   free hole

vacancy   vacancy

$$V_{S}^x \quad \rightleftharpoons \quad V_S^{\cdot\cdot} \quad + \quad 2e^-. \qquad\qquad (1.3b)$$

neutral sulphur   doubly ionised   two free

vacancy   vacancy   electrons

The notation used here for the effective charge on the point defect is that a superscript ˙ indicates a single positive charge and ′ a negative charge (Kroger and Vink 1956). We expect metallic vacancies in III–V semiconductors to have an effective single negative charge when ionised, while group II vacancies in II–VI materials are more likely to be doubly ionised, e.g. V_{Zn}'' in ZnTe. Extending this argument to other native point defects leads us to the general conclusion that we can divide these defects in III–V and II–VI materials into two classes depending on whether they act as donors or acceptors:

A_i, A_B and V_B are single or double donors

B_i, B_A and V_A are single or double acceptors

Vacancies in silicon are found to exist in the neutral, single and double acceptor and single donor states, V_{Si}^x, $V_{Si}^{\cdot}$, V_{Si}', and V_{Si}''.

It is important to be able to predict the concentration of ionised native point defects in a semiconductor crystal, and this value obviously depends on the total formation enthalpy of the defect. For a zinc vacancy in ZnTe, the value of $\Delta H_{V_{Zn}}$ has been calculated to be 2.54 eV, but the formation enthalpy of the ionised vacancy must include a contribution given by the separation of the neutral point defect state from the top of the valence band. These ionisation enthalpies can be quite large, especially in semiconductor materials with wide band gaps where the point defect state can lie far from either conduction or valence band. As an example of the importance of this second term, the ionisation enthalpy of the sulphur vacancy in CdS is very small, about 0.05 eV, while the equivalent enthalphy for the cadmium vacancy is much larger, at least 0.7 eV. From table 1.2 we can estimate the total enthalpy change on creation of ionised vacancies in CdS as $V_{Cd}'' \sim 4.3$ eV and $V_S^{\cdot\cdot} \sim 2.75$ eV. In the absence of a high concentration of another ionised native point defect to compensate for the sulphur vacancies, it is not surprising that CdS crystals are normally found to contain an excess of free electrons — native n-type behaviour. Quite the reverse is found in ZnTe crystals, where ionised zinc vacancies and free holes are present in excess giving native p-type behaviour. In other semiconductors, the native point defects are not easily ionised, e.g. GaP where the phosphorus vacancies appear to be electrically inactive.

It is convenient to be able to describe the relationship between the composition of the gas phase with which a semiconductor crystal is in equilibrium and the electrical properties of the crystal. From equation (1.2) we can predict that the concentration of neutral vacancies on the A sublattice is given by

$$[V_A^x] = K_1(PP_A)^{-1} \qquad (1.4a)$$

where K_1 is the equilibrium constant for the reaction which takes an A atom from the centre of the AB crystal and places it on the surface in equilibrium with the gas phase. If AB is taken to be a II–VI compound, then the ionisation process can be described by the equation

$$V_A^x \xrightarrow{K_2} V_A'' + 2h^+ \qquad (1.4b)$$

from which, using the law of mass action,

$$K_2 = \frac{[V_A'']p^2}{[V_A^x]} \qquad (1.4c)$$

where p is the free hole concentration. (The use of the law of mass action implies that we have made the assumptions that the concentrations of all the species in the reaction are negligibly low when compared with the total number of lattice sites in the crystal and that Henry's law is obeyed. This allows the activity of a defect to be replaced by its concentration.) Combining equations (1.4a) and (1.4c) gives

$$[V_A'']p^2 = K_3(PP_A)^{-1}. \qquad (1.5a)$$

It is simple to show that the equilibrium concentration of ionised vacancies on the B sublattice, $[V_B^{\cdot\cdot}]$, is given by a very similar equation with a different equilibrium constant:

$$[V_B^{\cdot\cdot}]n^2 = K_4(PP_A) \qquad (1.5b)$$

where n is the free electron concentration as usual. It is clear that these two equations are not independent because the values of n and p in semiconductors must obey the expression $pn = n_i^2$, where n_i is the intrinsic carrier density or the concentration of free carriers in the semiconductor crystal. The concentration of A vacancies thus depends on the concentration of B vacancies and on the concentration of all other electrically active defects in the crystal as well. Equations like (1.5a) and (1.5b) can be generated for all the possible native point defects in both II–VI (see for instance Albers (1967)) and III–V semiconductor compounds. If the values of the equilibrium constants are known, or can be estimated reliably, the total free carrier concentration can be calculated at any temperature and with the crystal in equilibrium with any chosen composition of the gas phase. This will be shown to be very important in §1.4.3 where the influence of point defects on the electrical properties of compound semiconductor materials will be described.

Equations (1.5a) and (1.5b) can also be used to determine which of the possible ionised native point defects is present in the highest concentration. If it is suspected that the most important acceptor native defect in a II–VI crystal is the doubly ionised metal vacancy, V_A'', then when the partial pressure of the metallic component in the vapour phase is low, so that an excess of metal vacancies will be produced, we can assume that the free hole concentration, p, is directly proportional to $[V_A'']$. From equation (1.5a) it follows that p will vary with $(PP_A)^{-1/3}$. This can be tested experimentally and has proved to be one of the most important methods of identifying the character of the dominant native point defects in semiconductor materials. Three examples of the electri-

cally active native point defects thought to have the controlling influence
on the electrical properties of undoped semiconductors are

Ga''_{Sb} in GaSb leading to native p-type behaviour

V''_{Zn} in ZnTe leading to native p-type behaviour

$Cd_i^{··} \; or \; V_S^{··}$ in CdS leading to native n-type behaviour.

In cadmium suphide it is not clear whether cadmium interstitials or
sulphur vacancies are the dominant defect, and in most other compound
semiconductors even less is known about the relative concentrations of
electrically active native point defects. It is frequently conjectured that a
wide range of defect types, including vacancy–impurity complexes, are
important (Kroger 1977, Zanio 1978). Part of the reason for this lack of
information on the character of the dominant native point defects lies in
the difficulties experienced in obtaining compound semiconductor samples which are sufficiently pure for the residual levels of electrically
active impurities to be insignificant when compared with the concentrations of the native point defects.

1.4.2 Dopants and Impurities

Most impurity atoms in substitutional sites in semiconductor crystals,
and some in interstitial sites as well, create new electronic states in the
forbidden band gap. The only real distinction between the behaviour of
an impurity element, whose presence is usually undesirable, and a
dopant, which is deliberately added to the semiconductor, lies in the
position of these new electronic states. Impurity elements which are
used as dopants create donor or acceptor states which lie close to the
conduction or valence band edges. These dopant atoms are thus easily
ionised and we are normally able to assume that the concentration of
free carriers in a doped semiconductor crystal is approximately equal to
the concentration of dopant atoms. When the dopant atoms are present
in concentrations close to the solubility limit of that particular dopant in
the semiconductor, some of the dopant atoms are electrically inactive.
These solubility limits are generally between 10^{19} and 2×10^{21} cm^{-3} for
most common dopant elements in silicon at 1100 °C for example, but
ionised dopant concentrations above 10^{19} cm^{-3} are rarely achieved
before saturation occurs. It is worth making a brief digression to see
how this saturation phenomenon occurs.

One popular model suggests that when dopant elements are present in
high concentrations clustering can occur and the existence of groups of
two, three, four or more dopant atoms has been postulated. These
clusters may be electrically active to some extent, but are not thought to

release as many free carriers as would the same number of dopant atoms acting independently. It has been suggested that arsenic atoms in silicon for example form As_3^+ clusters at high temperatures, but neutral As_3 clusters at room temperature (Tsai *et al* 1980). It is normally expected that each arsenic atom added to a silicon crystal will contribute one free electron to the conduction band, but if clustering occurs then the electron concentration will begin to saturate as more and more inactive arsenic clusters are formed. A second mechanism by which dopant atoms are rendered electrically inactive in a semiconductor crystal is the phenomenon of self-compensation by native point defects. This is a very serious problem in many II–VI semiconductor compounds, as we will see in the next section.

Table 1.3 lists some of the common elements used as dopants in semiconducting materials. We can see that the replacement of a host semiconductor atom by a dopant atom with one fewer valence electron will result in the formation of a free hole — a p-type or acceptor dopant — while the introduction of a dopant atom with one more valence electron creates a free electron — an n-type or donor dopant. We are assuming that the electronic states created by the dopant atoms lie sufficiently close to a band edge for complete ionisation. The ionisation enthalpies given in table 1.3 are all below about 150 meV and the characteristic values of only a few meV for many dopants in GaAs are particularly noteworthy. Thus a group II metal substituting for gallium in GaAs will act as a p-type dopant, and halogen atoms substituting for sulphur in CdS are donor dopants. The position that a dopant atom takes up in a compound semiconductor crystal will be dictated by the site at which the chemical potential is minimised, and it is generally observed that metallic impurities segregate to the metallic (A) sublattice and non-metallic impurities to the B sublattice. We can see that a compound semiconductor can be efficiently doped by substituting atoms on either the A or B sublattice. GaAs, for instance, can be doped n-type by the introduction of tellurium atoms (group VII) on the arsenic sublattice, or tin atoms (group IV) on the gallium sublattice. This allows considerable flexibility in the choice of which particular dopant element to use when carrying out a doping operation during device fabrication.

Another interesting kind of doping phenomenon is that which results from the inclusion of elements like silicon in III–V compound semiconductors. Table 1.3 shows that silicon can act as an n–type dopant when substituting for gallium on the A sublattice and as a p-type dopant when substituting for arsenic on the B sublattice. Impurities of this kind are called 'amphoteric' dopants. The preferred position of the silicon atoms in the GaAs lattice is controlled by the chemical potential of the impurity atoms. It is observed at room temperature that silicon prefers to substitute for gallium atoms and is thus an n-type dopant. However, decreasing the chemical potential of silicon atoms on arsenic sites below that for silicon on gallium sites will result in a reversal of the sublattice occupancy. This can be achieved by increasing the concentration of

Table 1.3 Dopant elements in semiconductors with their doping character and ionisation enthalpies (after Zanio 1978 and Sze 1981).

Semiconductor	Dopant	Atomic position	Doping character	Ionisation enthalpies (meV)
Silicon				
	Phosphorus	—	n-type	45
	Arsenic	—	n-type	54
	Antimony	—	n-type	39
	Boron	—	p-type	45
	Aluminium	—	p-type	67
Gallium arsenide				
	Tellurium	As	n-type	30
	Sulphur	As	n-type	6
	Tin	Ga	n-type	6
	Silicon	Ga	n-type	6
	Germanium	Ga	n-type	6
	Zinc	Ga	p-type	3
	Chromium	Ga	p-type	0.85
	Silicon	As	p-type	35
	Germanium	As	p-type	~40
Gallium phosphide				
	Tellurium	P	n-type	90
	Selenium	P	n-type	102
	Sulphur	P	n-type	104
	Zinc	Ga	p-type	64
	Magnesium	Ga	p-type	53
	Tin	P	p-type	65
Cadmium telluride				
	Indium	Cd	n-type	14
	Aluminium	Cd	n-type	14
	Chlorine	Te	n-type	14
	Phosphorus	Te	p-type	30
	Lithium	Cd	p-type	30
	Sodium	Cd	p-type	30
Cadmium sulphide				
	Gallium	Cd	n-type	33
	Iodine	S	n-type	34
	Fluorine	S	n-type	35
	Lithium	Cd	p-type	165
	Sodium	Cd	p-type	169

arsenic vacancies (by annealing the crystal in a vapour containing a low arsenic partial pressure) and will convert the GaAs into a p-type conductor by the following reaction:

$$Si'_{Ga} + V^x_{As} \overset{K_a}{\rightleftharpoons} Si'_{As} + V^x_{Ga} + 2h^+. \tag{1.6}$$

Above about 800 °C the silicon atoms also lie preferentially on arsenic sites, and we will see in Chapter 8 that the switch in doping character from n-type to p-type at this temperature can be very useful in the fabrication of light-emitting diodes. Doping the GaAs with 'normal' dopants like tellurium or zinc will also alter the preferred position of the silicon atoms. An increase in the hole concentration as a result of the addition of zinc substituting on gallium sites will force equation (1.6) in the reverse direction and the silicon will act as a compensating n-type dopant on the gallium sublattice. Doping with tellurium will have the reverse effect, with the silicon atoms moving to the arsenic sublattice. The extent to which an amphoteric dopant like silicon will compensate for a deliberately added dopant can be estimated from the law of mass action if a value is known for K_a, the equilibrium constant for the reaction in equation (1.6). Amphoteric dopants exist in other compound semiconductors as well, but the equilibrium lattice position of the dopant can rarely be changed as readily as for silicon in GaAs. If the chemical potential of the impurity atoms is always much lower on one sublattice than the other then the amphoteric nature of the dopant will never be observed.

A final class of dopant elements which deserve a brief mention are the so-called 'isoelectronic' dopants. These are atoms with the same number of valence electrons as those for which they substitute and so we would not expect them to create new electronic states in the forbidden band gap of the semiconductor. However, if the core electronic structure of the dopant atom differs substantially from that of the host atom, the local crystal potential in the semiconductor crystal is sufficiently distorted to create shallow electronic states. A particularly important example of an isoelectronic dopant is nitrogen substituting for phosphorus in gallium phosphide. The nitrogen atoms create a new electronic state close to the conduction band edge and this state can have a very marked effect on the optoelectronic properties of GaP crystals, as we will see in §8.2. However, not all isoelectronic impurities have this effect, and the substitution of indium for gallium in GaAs for example does not result in the formation of any shallow states. A reasonably effective way of predicting whether an isoelectronic impurity will be electrically active in a semiconductor crystal is to consider the relative size of host and impurity atoms. If the impurity is much larger or smaller than the atom it replaces in the lattice, then it is likely that new doping states in the band gap will be formed.

Now we should briefly turn our attention to the effect of other impurity elements in semiconductor crystals, i.e. those which do not have beneficial doping effects. These elements generally create new electronic states which lie deep in the semiconductor band gap and which are not easily ionised. Some impurities create several deep states spread throughout the band gap. Figure 1.6 illustrates in a simple one-electron diagram the position of some impurity states in silicon and gallium arsenide. Transition metal impurities in both of these semiconductors create several deep states, and these elements are recognised as some of the most damaging impurities in silicon wafers intended for the fabrication of integrated devices. The five acceptor states characteristic of copper in GaAs will be shown to be of particular importance in controlling the performance of luminescent devices (see §8.2). The influence of these deep states on the properties of semiconducting materials lies not in their intrinsic electrical properties, since they are not usually ionised, but on the efficient trapping of free carriers. The recombination of free carriers from the conduction and valence bands occurs especially rapidly in crystals which have deep states lying approximately at the centre of the band gap, and so the lifetime of a free carrier can be significantly decreased if the concentration of transition metal impurities is high. This a very important mechanism for the degradation of the performance of microelectronic devices in which the lifetime of minority carriers is required to be as long as possible. The solubilities of most transition metals in silicon and GaAs are very low,

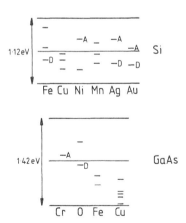

Figure 1.6 One-electron diagrams illustrating the electronic states created by the presence of some impurity elements in silicon and gallium arsenide. States in the upper half of the band gap can be assumed to be donor states unless marked A, while states in the lower half of the band gap are acceptors unless marked D.

often less than 10^{16} cm^{-3} even at high temperatures. Unfortunately, this is a sufficiently high concentration to have a significant effect on the efficiency of minority carrier devices. This fact gives a very good indication of how pure the starting material needs to be for the fabrication of microelectronic devices and systems. Parts per million levels of impurity are often too high to be tolerated in device-quality material. Some of the problems associated with achieving and preserving these high purities in semiconductor materials will be discussed in the next two chapters. The phenomenon of 'gettering' which is used to remove residual transition metal impurities from the active regions of silicon wafers will be described in §2.4.

1.4.3 Electronic and Atomic Disorder

We have considered the properties of both native and impurity point defects in semiconductors and have seen that all dopant, some impurity atoms and many native defects are electrically active since they are easily ionised to release free carriers. At this point we can ask the important question — are the electrical properties of semiconductor materials controlled by native point defects or deliberately added dopant elements? (Here I will assume that the residual impurity concentrations are too low to have any significant influence on the bulk electrical properties.)

Let us take the example of zinc telluride and see what happens when iodine is added as an n-type dopant substituting on the tellurium sublattice. The iodine atoms create an electronic state lying close to the bottom of the conduction band and this state is easily ionised to release a free electron. However the energy of the crystal as a whole can be reduced if there is a mechanism for these free electrons to drop down to the valence band. In a semiconductor material we do not normally expect this to be possible because there will be very few vacant states in the valence band. In §1.4.1 we saw that ZnTe crystals are expected always to contain an excess of zinc vacancies, and the formation enthalpy of these defects has been given in table 1.2. The ionisation of each zinc vacancy creates two free holes in the valence band and these holes are available for recombination with the free electrons introduced by the iodine dopant. The enthalpy balance associated with this process is that two electrons fall the whole width of the band gap for every one ionised zinc vacancy which is produced. The band gap of ZnTe is about 2.2 eV and the formation enthalpy of a ionised zinc vacancy is $(2.54 + x)$ eV, where x is the ionisation enthalpy and is negligibly small in this case. Thus the creation of a zinc vacancy with the expenditure of about 2.6 eV results in a decrease in electron energy of over 4 eV. It is thus energetically favourable for zinc vacancies to form spontaneously when iodine is added as a dopant to a ZnTe crystal and to remove all the free electrons introduced by the dopant. This is the mechanism of 'self-compensation' and in ZnTe we expect the zinc vacancies to compensate completely for any concentration of n-type dopant. Not only is ZnTe a native p-type semiconductor, but also it *cannot* be doped to give an n-type material. Similarly, in cadmium sulphide, the sulphur vacancy has a formation enthalpy of 2.69 eV, a negligible ionisation enthalpy and can compensate for two free holes in the valence band with a total decrease in enthalpy of about (4.8–2.7) eV. (The band gap of CdS is about 2.4 eV.) Thus CdS is a native n-type semiconductor and cannot be prepared with p-type conduction character.

We should not assume that this favourable energy balance for self-compensation always exists. The formation enthalpy of the cadmium vacancy in cadmium telluride is 2.75 eV and the band gap 1.56 eV. It is easy to show that compensation for deliberately added donor dopants will only occur if the electronic state created by the cadmium vacancy lies within 0.37 eV of the valence band. However, the cadmium vacancy state lies approximately at the centre of the band gap, about 0.96 eV above the valence band, and so we only expect self-compensation to occur if a vacancy–impurity complex state is created lying much closer to the valence band edge (Zanio 1978). In fact, CdTe is usually found to have a conductivity dominated by residual impurities in the undoped state and can be readily doped both n- and p-type with only a small amount of self-compensation. The enthalpy balance for self-compensation of n-type dopants in CdS by cadmium vacancies is also unfavourable because of the large ionisation enthalpy of this native defect. The free electron concentration in this semiconductor is thus controlled by the concentration of dopants added to the crystal.

It is clear that self-compensation will occur when the formation enthalpies of the native point defects are small, the electronic states created by these defects lie close to the valence or conduction band edges, and the band gap is wide as well. In general, the II–VI semiconductors satisfy these requirements quite well, with the important exception of CdTe, and show self-compensating behaviour. These semiconductors are said to have electrical properties controlled by native atomic disorder. We should also remember that the equilibrium concentrations of native point defects are determined by the composition of the gas phase surrounding the crystal. For some II–VI compounds the phenomenon of self-compensation can be suppressed by choosing a combination of temperature and gas phase composition which limits the equilibrium concentration of the native point defect required to compensate for the dopant elements. We are able to assume that atomic disorder will always dominate the electrical properties of ZnTe because the equilibrium phase diagram shows that tellurium is always present in excess in this compound and so it is not possible to suppress the formation of zinc vacancies. This is not true for compound semiconductors where the existence region lies on both sides of the stoichiometric composition.

It is convenient to be able to define when we expect self-compensation to be possible in a semiconductor compound. Brouwer (1954) has suggested a simple graphical construction to illustrate how the concentrations of the native point defects change with gas phase composition at a given temperature. This construction requires the measurement, or estimation, of values of the equilibrium constants for the reactions by which all the possible point defects are formed in the crystal (see equations (1.3)–(1.5)). It is then possible to plot the equilibrium point defect concentration against the gas phase composition in equilibrium with the crystal across the whole existence range of the semiconductor compound. A particularly clear description of the construction of Brouwer diagrams has been given by Tuck (1974). Two examples of calculated Brouwer diagrams are shown in figure 1.7: one for CdTe at 900 °C and the second for CdS. The plot for CdTe shows that over a wide range of low cadmium pressures no ionised point defect is present in a higher concentration than that of the native free carriers. These free carriers are produced by residual impurities and the thermal excitation of electrons across the relatively narrow band gap. However, at very high cadmium pressures the concentration of doubly ionised cadmium interstitials dominates the electrical properties of CdTe. The Brouwer diagram for CdS shows that the influence of the sulphur vacancies (or cadmium interstitials) is dominant at all gas phase compositions.

The effect of adding a dopant element to the semiconductor crystal can be included in a Brouwer diagram but will result in changes in the equilibrium constants of all the equations governing the formation of the native point defects. In the case of CdTe at 900 °C, we expect impurity doping to control the electrical properties when the cadmium overpressure is less than about 3×10^{-1} atmospheres, but it may be difficult to prepare p-type CdTe at higher cadmium pressures because of the high concentration of compensating cadmium interstitials. The free electron concentration in n-doped CdS will depend on the concentration of the n-type dopant, but sulphur vacancies will compensate for p-type dopants over the whole composition range. A more detailed description of the use of Brouwer diagrams to describe the electrical properties of doped II–VI compounds can be found in Albers (1967).

III–V compound semiconductors, in contrast to the II–VI compounds considered so far in this section, rarely have electrical properties which are controlled by atomic disorder. This is not because non-stoichiometry cannot occur in these materials, as we have already seen that GaAs has

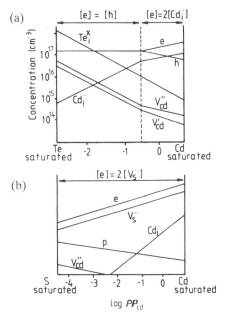

Figure 1.7 Two examples of Brouwer diagrams: (a) for CdTe at 900 °C; and (b) for CdS. These diagrams allow the dominant point defects to be identified in a crystal in equilibrium with any given composition of the gas phase.

an extended existence region and that GaP is normally deficient in phosphorus. We would thus expect GaP to show n-type conduction if the phosphorus vacancies are easily ionised and perhaps self-compensation of p-type dopants by these native point defects as well. In fact, GaP is readily doped both n- and p-type. The explanation of this behaviour lies in the important role played by the antisite defects in III–V compounds, a native defect which we have been able to ignore in II–VI compounds. Table 1.2 shows that the formation enthalpies of these defects in III–V materials can be very small indeed and so they should be present in high concentrations. Van Vechten (1975) has proposed a model for the equilibrium native point defect concentration in GaP, assuming that the only important defects are neutral vacancies and charged antisite defects: V_{Ga}^x, V_P^x, Ga_P'', and $P_{Ga}^{\cdot\cdot}$. In this model, any singly ionised vacancies are immediately compensated by the following reactions:

$$A_A^x + V_B^{\cdot} \rightleftharpoons A_B'' + V_A' + 4h^+$$

$$V_A' + B_B^x \rightleftharpoons B_A^{\cdot\cdot} + V_B^{\cdot} + 4e^-$$

(1.7)

in which only a single diffusive jump has to occur in each case. This kind of autocompensation of ionised native point defects is very effective because of the low enthalpies of formation of the antisite defects. In addition, no mechanism exists for the self-compensation of dopants in III–V materials. The energy balance which is favourable in the case of the zinc vacancy in ZnTe trapping two electrons from the conduction band is very unfavourable for a gallium vacancy in GaP which can trap only a single electron. Thus we have the situation where ionised native point defects are rapidly compensated for by other native point defects, yet free carriers introduced by dopant and impurity elements are not. III–V compound semiconductors are thus examples of materials whose electrical properties are dominated by 'electronic disorder', and the conductivity of undoped crystals is normally controlled by the concentration of residual impurities with a small contribution from direct thermal excitation of free carriers across the band gap. The addition of dopant elements to III–V compounds allows the preparation of both n- and p-type material, with almost no compensation from native point defects. The only common exception to this rule is GaSb where the Ga''_{Sb} native defect is easily formed. There is thus a significant contribution to the electrical properties from atomic disorder and this compound is naturally p-type.

There is clearly a major difference between III–V and II–VI materials; the first class of compounds can readily be doped to any desired level and with both n- or p-type character, but the II–VI compounds are generally much less tractable. We will find many examples in the following chapters where the atomic disorder in II–VI compounds makes them awkward to fabricate into useful microelectronic devices. CdTe is an important exception and can be amphoterically doped. Some amphoteric II–VI and IV–VI materials can be conveniently prepared as p-type material by annealing in a metal-deficient atmosphere to create excess metal vacancies, and as an n-type material by annealing in a metal-rich atmosphere. Both of the important ternary alloy semiconductors $Hg_{1-x}Cd_xTe$ and $Pb_{1-x}Sn_xTe$ can be treated in this way to give n- or p-type material for instance.

Finally, let us consider the role of native point defects in controlling the electrical properties of elemental semiconductors, using the convenient example of arsenic-doped silicon. We have seen that silicon vacancies can be ionised according to the reaction

$$V^x_{Si} \stackrel{K_5}{\rightleftharpoons} V'_{Si} + h^+ \qquad (1.8a)$$

where the diffusion constant K_5 is given by the law of mass action as

$$K_5 = \frac{[V'_{Si}]p}{V^x_{Si}}. \qquad (1.8b)$$

Using the fact that $np = n_i^2$, and assuming that all the free electrons in the silicon crystal are contributed by the arsenic dopant, $[As_{Si}^{\cdot}] = n$, then it is easy to see that $[V'_{Si}] = [As_{Si}^{\cdot}]$. Thus an increase in the concentration of the ionised dopant results in an increase in the concentration of compensating ionised vacancies. The extent of self-compensation is limited because of the relatively low equilibrium concentrations of vacancies in silicon, but does put an upper bound on the free electron concentration which can be achieved by doping silicon with arsenic which is well below the solubility limit of arsenic in silicon. In addition, the dependence of the concentration of charged vacancy acceptor defects on the the doping level has important consequences for the diffusion kinetics in silicon, as we will see in the following section.

1.5 DIFFUSION IN SEMICONDUCTORS

The phenomenon of diffusion in semiconductors is much more complex than in metals. It is possible to assume that solute diffusion in a bulk metal sample occurs either by a simple vacancy mechanism, if the solute substitutes for atoms on the host lattice sites, or by interstitial diffusion. Bulk self-diffusion is usually observed to occur by a vacancy mechanism. We are thus able to describe the kinetics of diffusion in metals in terms of Fick's first and second laws, where we can define a diffusion coefficient, D, for each diffusing species in terms of a single activation energy, Q, and a pre-exponential factor which takes into account the frequency with which each atom can attempt a diffusive jump, D_0:

$$D = D_0 \exp\left(\frac{-Q}{kT}\right). \tag{1.9}$$

If we are considering substitutional diffusion, then Q contains a contribution from the enthalpy change on creating a vacancy in the perfect metallic sublattice. While the effective diffusion coefficient is not necessarily independent of composition, it is possible to assume that the diffusion mechanism for any given solute is either vacancy or interstitial and that the point defects are not charged. (Readers unfamiliar with the standard treatments of diffusion in metallic elements and alloys are referred to Shewmon (1963).)

In the case of semiconductors, however, it is clear from §1.4 that a wide variety of point defects can exist in significant concentrations, with vacancies, antisite defects, impurities and dopant atoms all being found in several charge states. In addition, the equilibrium concentration of any particular charged point defect has been shown to depend directly on the concentration of all the others (cf equations (1.6)–(1.8)) and on the free carrier concentration (which we can think of in terms of the position of the Fermi level in the forbidden band gap). A further complication is introduced by the fact that the diamond cubic related crystal structures are not close packed and so many solute elements with reasonably small covalent radii can occupy both substitutional and interstitial sites. It seems likely that both substitutional and interstitial diffusion mechanisms will be important in semiconductor materials, and we might expect impurities which can lie on substitutional and interstitial sites in the same semiconductor lattice to diffuse by both mechanisms *at the same time*. (Only a few impurities are sufficiently small to occupy interstitial sites in the more closely packed metallic lattices, and the occupation of both kinds of sites in the same crystal is very rare.) These considerations imply that there will only rarely be a single mechanism of diffusion operating for a particular impurity species in a given semiconductor, and we must expect the measured diffusion coefficient, D, to be made up of several contributions, each of which will have a characteristic activation enthalpy, Q. The values of D will in many cases depend on the dopant concentration as well.

A number of electrical techniques have been developed for measuring the diffusivity of dopant elements in semiconductors and these have been reviewed by Yeh (1973) and Tsai (1983). The techniques of secondary ion mass spectrometry (SIMS) and Rutherford backscattering (RBS) can also be used to obtain concentration profiles of impurity elements diffused into a semiconductor surface; these techniques are described in Chapter 10.

We will now consider the phenomenon of diffusion in semiconductors in a little more detail, beginning with the relatively straightforward case of diffusion in silicon.

1.5.1 Diffusion in Silicon Crystals

I will start this section by describing how the processes of self-diffusion in silicon and germanium are observed to differ. The diffusion of germanium atoms in germanium crystals seems to occur by a simple vacancy mechanism and is thus a rather similar process to self-diffusion in metals. However, self-diffusion in silicon is observed to be very slow indeed, with a measured activation enthalpy, Q, well in excess of $4 \, \text{eV/atom}$. This high value is difficult to explain because the formation enthalpy of a vacancy is only $2.5 \, \text{eV}$ and the estimated enthalpy of migration of a simple vacancy is about $1 \, \text{eV}$ — a total of $3.5 \, \text{eV}$ for a substitutional diffusion process. Fair (1981) has suggested that the explanation for this discrepancy lies in the fact that vacancies in silicon can exist in neutral, donor or acceptor charge states (V_{Si}^{x}, $V_{Si}^{\cdot}$, $V_{Si}^{\prime}$ and $V_{Si}^{\prime\prime}$) and that self-diffusion can occur by the migration of all these defects but with a different activation enthalpy and pre-exponential factor in each case. Since the concentration of the charged vacancies depends on the free carrier concentration, so will the contribution that each kind of defect makes to the total self-diffusion coefficient, D_{Si}:

$$D_{Si} = D^{V_{Si}^x} + D^{V_{Si}'}\left(\frac{n}{n_i}\right) + D^{V_{Si}''}\left(\frac{n}{n_i}\right)^2 + D^{V_{Si}^{\cdot}}\left(\frac{n_i}{n}\right). \tag{1.10}$$

The estimated activation enthalpies for self-diffusion via these native point defects range from 3.89 eV (V^x) to 5.1 eV ($V^{\cdot}$ and V''), but the pre-exponential factor, $D_0^{V^x}$, is rather small so that self-diffusion via neutral vacancies is not the dominant diffusion mechanism at most temperatures. This is why the measured value of the activation enthalpy is greater than 4 eV. It is clear from this brief description that even self-diffusion in silicon is a complicated phenomenon and a model which is fully consistent with the experimental data at all temperatures has yet to be developed.

Now let us turn our attention to the diffusion of the most commonly used dopant elements in silicon devices — boron, phosphorus, arsenic and antimony. All of these elements occupy substitutional sites in the silicon lattice and so we expect vacancy diffusion to be the dominant transport mechanism. Arsenic atoms in a silicon lattice are easily ionised so that the normal charge state of the dopant is $As_{Si}^{\cdot}$. It seems reasonable to assume that this defect will interact preferentially with negatively charged silicon vacancies, V_{Si}', and that the diffusion of arsenic in a silicon crystal will occur primarily via this defect. The inclusion of ionised arsenic atoms in the silicon will modify the equilibrium concentration of singly charged silicon vacancies. From equation (1.8) it can be shown that $[V_{Si}']$ increases linearly with the ratio (n/n_i), the ratio of the free electron concentration to the intrinsic free carrier concentration. At low arsenic concentrations, the diffusion coefficient of arsenic in silicon, D_{As}, is observed to vary as

$$D_{As} = D_{As}^i\left(\frac{n}{n_i}\right) \tag{1.11}$$

where D_{As}^i is the diffusion coefficient for arsenic in undoped silicon, i.e. at infinite dilution. This observation is very strong evidence for the model that charged arsenic atoms diffuse preferentially via the V_{Si}' native point defects.

Similarly, boron is observed to diffuse via donor-type silicon vacancies, $V_{Si}^{\cdot}$, and the diffusion coefficient, D_B, depends on the boron concentration and hence on the concentration of free holes. More surprising is the behaviour of phosphorus and antimony, which both appear to diffuse preferentially via uncharged silicon vacancies (see for instance Tsai (1983)). The activation enthalpies for all these vacancy diffusion processes lie in the range 3.4–4.2 eV/atom and so dopant diffusion is quite slow in silicon crystals even at high temperatures. Typical values of diffusion coefficient at 1200 °C are 10^{-12} cm^2 s^{-1} for boron and phosphorus in intrinsic silicon, and rather similar values for arsenic and antimony can be calculated from the data given in table 1.4.

Table 1.4 Self- and impurity diffusion data in silicon (after Tsai 1983).

Diffusant	Dominant diffusion mechanism	Q (eV)	D_0 (cm^2 s^{-1})
Si	Complex vacancy mechanism	4–5	—
As	V'_{Si}	4.1	~40
B	V'_{Si}	3.46	0.76
P	V^x_{Si}	3.66	3.85
Sb	V^x_{Si}	3.65	0.2
Na	Interstitial	0.76	2×10^{-3}
K	Interstitial	0.76	1×10^{-3}
O	Interstitial	2.5	1×10^{-1}
Cu	Interstitial	0.43	5×10^{-3}
Au	{ Interstitial	0.39	2×10^{-4}
	{ Substitutional	2.04	3×10^{-3}
Ag		1.6	2×10^{-3}
Fe	Probably interstitial–substitutional	0.87	6×10^{-3}
Ni		1.9	1×10^{-1}
Cr		1.0	1×10^{-2}

If the dopant species are present in high concentrations, then clustering and the formation of dopant–point defect complexes can modify the observed diffusion kinetics. We have already discussed the formation of arsenic clusters in silicon which is heavily doped and can assume that these clusters are immobile. The measured rate of diffusion of arsenic in heavily doped samples ($> 10^{20}$ cm^{-3}) falls well below that predicted by equation (1.11), as the arsenic atoms in the clusters are neither ionised nor mobile. The measured diffusivity of boron in silicon also decreases rapidly at high doping levels. The diffusivity of phosphorus varies in a more complicated manner as the doping level is increased, with a $(P_{Si}V_{Si})'$ complex playing an important part in the overall diffusion process (Fair 1981). In general, it is impossible to define a single value of the diffusion coefficient when the dopant concentrations are high, and it is important to remember this when attempting to model how a dopant profile will develop during diffusion processing in the fabrication of silicon devices (see §2.4).

Now let us consider the diffusion of some common impurity elements in silicon. The alkali metals have small atomic radii and so occupy interstitial sites in the silicon lattice. They are thus expected to diffuse by an interstitial mechanism with a low activation enthalpy, and the measured value for sodium and potassium is 0.76 eV. Both these elements create deep states in silicon and so are very undesirable impurities in device-quality wafers. At temperatures below 700 °C, copper is also an interstitial solute in silicon and has an activation enthalpy for diffusion of only 0.43 eV, which means that it is extremely mobile in a silicon crystal even at very low temperatures. Gold shows a more complex behaviour, since it can occupy both interstitial and substitutional sites. As a result, gold can diffuse by both interstitial and vacancy mechanisms, but the activation enthalpy for interstitial diffusion is much lower than that for vacancy diffusion (0.39 eV compared with about 2 eV). We would therefore expect more of the diffusive transport of gold in silicon to occur by an interstitial mechanism. However, the equilibrium concentration of gold interstitials in silicon is very low and is controlled by the concentration of silicon vacancies:

$$Au_i + V_{Si} \rightleftharpoons Au_{Si}. \tag{1.12}$$

If we assume that interstitial diffusion is so rapid that the whole of a silicon crystal in contact with a source of gold atoms always contains a saturated concentration of gold interstitials, then the concentration of substitutional gold atoms will depend directly on the concentration of silicon vacancies. At the surface of the wafer, where the rate of formation of vacancies is high, rapid interstitial diffusion will provide the gold atoms which will preferentially populate the silicon vacancies, giving a high concentration of substitutional gold atoms. At the centre of the wafer, where the vacancy formation rate is low, the interstitial gold atoms will be unable to take up substitutional sites once the pre-existing vacancies are filled and a saturated interstitial solution will be formed. Diffusion of gold into a silicon wafer will thus result in a concentration profile like that shown in figure 1.8(a). The precise form of this profile is controlled by the diffusion of both interstitial gold and silicon vacancies. This mechanism of diffusion is called the 'interstitial–substitutional' model (Frank and Turnbull 1956) and will prove to be of great importance in interpreting diffusion data in compound semiconductors as well.

Table 1.4 gives values of activation enthalpies and pre-exponential factors for several transition metal impurities in silicon and it is clear that these impurities, which create deep electronic states in silicon (see figure 1.5), are also rapid diffusers in the silicon lattice, although some of them seem to diffuse via a substitutional mechanism. This is an extremely unfortunate combination of properties since the performance of a wide range of silicon devices is severely degraded by the presence

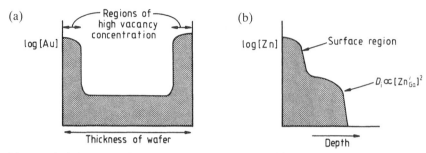

Figure 1.8 Schematic plots of the concentrations of impurities in semiconductor materials as a result of diffusion processes: (a) the concentration profile of gold in a silicon wafer after operation of the interstitial–substitutional diffusion mechanism; (b) the zinc concentration profile in a gallium arsenide wafer formed as a result of the dependence of the interstitial diffusion coefficient, D_i, on the square of the substitutional zinc concentration.

of even very low concentrations of these elements, and yet they can penetrate into a silicon wafer at low temperatures from any accidental contamination site on the wafer surface. The disastrous consequences on device yield of handling wafers with metallic tweezers at any stage in the fabrication process is a particularly clear indication of the care which has to be exercised to keep transition metal impurities away from silicon wafers.

1.5.2 Diffusion in Compound Semiconductors

We have already seen that substitutional diffusion rates in semiconductors are determined by the concentrations of charged vacancies and thus on the position of the Fermi level in the semiconductor band gap. In compound semiconductor materials we must also take into account the effect of the external gas phase composition on the equilibrium concentration of native point defects (equation (1.2)). We should not assume, however, that the point defect which has the most important effect on the electrical properties of the crystal, as predicted from the relevant Brouwer diagram, will also have the dominant influence on the mechanism of substitutional diffusion. We must also remember that an atom on the metallic (A) sublattice must jump directly to a vacancy on a second-nearest-neighbour atomic site unless it briefly forms an antisite defect, A_B, which requires that a divacancy ($V_A V_B$) be present. In fact, a divacancy diffusion mechanism has been postulated to be important in self-diffusion and the diffusion of impurity elements in some III–V compounds. In addition, it may be possible for some impurities to diffuse by an interstitial mechanism, even though they are normally considered to occupy substitutional lattice sites.

I shall now describe the comprehensively studied phenomenon of zinc diffusion in gallium arsenide because of the importance of zinc indiffusion as a method of preparing p–n junctions in GaAs devices. Experimental measurements have demonstrated that the diffusion coefficient of zinc in GaAs, D_{Zn}, depends on the square of the zinc concentration, $[Zn_{Ga}]^2$, as long as the doping level does not approach the solubility limit. This is hard to explain if we assume that the zinc species diffuse by a substitutional mechanism, but if we allow the formation of ionised zinc interstitials, $Zn_i^{\cdot}$, then diffusion can occur by the interstitial–substitutional mechanism. The equilibrium concentration of ionised zinc interstitials is controlled by the reaction

$$Zn_i^{\cdot} + V_{Ga}^{x} \rightleftharpoons Zn_{Ga}' + 2h^+. \tag{1.13}$$

We will assume that the rate of diffusion of the zinc interstitials is very fast and that of the substitutional zinc atoms in negligibly slow, and that in regions which contain a high concentration of gallium vacancies the interstitials will preferentially occupy substitutional sites where they will act as ordinary ionised dopant atoms. We are thus able to put $D_{Zn} \propto$ $[Zn_i^{\cdot}]$ and it can be shown that an effective diffusion coefficient for zinc can be derived which depends on $[Zn_{Ga}']^2$ (see for instance Tuck (1988; Chapter 4)).

The diffusivity of zinc in GaAs thus falls off very rapidly as the concentration of substitutional zinc species decreases, leading to some unusual diffusion profiles in zinc-diffused GaAs wafers. Close to the wafer surface, where the concentration of gallium vacancies is always high, zinc atoms diffusing by an interstitial mechanism from the zinc source on the surface will occupy these vacant sites, leading to an increased interstitial diffusivity. Further from the surface, pre-existing gallium vacancies are quickly saturated by the indiffusing zinc interstitials and no further increase in $[Zn_{Ga}']$ or the interstitial diffusivity can occur. A typical zinc concentration profile in a GaAs wafer is sketched in figure 1.8(b). It is clearly impossible to define a single value of the diffusion coefficient of zinc in GaAs since it varies with position in the sample. Zinc diffusion in GaP and InP seems to occur by the same interstitial–substitutional mechanism although the details of the diffusion process in InP are quite different to those in GaAs (Tuck 1988).

Much less is known about the mechanisms of diffusion of other elements in III–V compounds, although we can separate the diffusion processes into the relatively simple 'Fickian' behaviour showed by a few dopant elements in III–V compounds and the constituent elements of the compounds as well, and more complex interstitial–substitutional diffusion shown by most other impurity elements. We are unable to define a single activation enthalpy for the interstitial–substitutional diffusion processes of course, although it is sometimes possible to obtain separate values for the interstitial and the substitutional mechanism.

It seems generally to be true that self-diffusion in III–V materials is slower than the diffusion of impurities, gallium and arsenic both diffusing extremely slowly in GaAs for example. Some measured values of activation enthalpies and pre-exponential factors for elements which diffuse in a Fickian manner in GaAs and InP are given in table 1.5. Donor dopant elements like sulphur and tellurium diffuse rather slowly in GaAs and so it is not usually practicable to create a p–n junction by the diffusion of either of these dopants into a p-type GaAs wafer. Many important impurities in the III–V compounds diffuse by the interstitial–substitutional mechanism. The interstitial diffusion process can be very fast indeed, resulting in the formation of diffusion profiles similar to those shown for gold in silicon in figure 1.8(a). Table 1.5 includes a list of a few of these impurities in GaAs and InP, chosen because they have particular significance in the preparation of microelectronic devices in these semiconductors. Copper creates several unwanted deep states in

Table 1.5 Some experimental data on diffusion in compound semiconductors (after the comprehensive compilation of data given by Tuck (1988)).

Semiconductor	Diffusant	Q (eV)	D_0 ($cm^2 s^{-1}$)
(a) Fickian diffusing elements			
GaAs	Ga	5.6 and 2.6	1×10^7 4×10^{-5}
GaAs	As	3.2	7×10^{-1}
GaAs	S	1.6–9.4	10^{-5}–10^6
InP	In	3.85	1×10^{-5}
InP	P	5.65	7×10^{10}
(b) Impurities which diffuse by an interstitial–substitutional mechanism			
GaAs	Cr		
	Fe		
	Cu		
InP	Fe		
	Cu		
	Ag		
(c) CdS (after Kumar and Kroger 1971)†			
High Cd pressure	Cd ($Cd_i^{.}$)	1.2	7×10^{-5}
	S ($V_S^{..}$)	1.9	3×10^{-6}
High S pressure	Cd (V_{Cd}'')	—	—
	S (S_i^x)	2.1	1×10^{-2}

†The dominant diffusing species is identified in parentheses.

GaAs (figure 1.6) and is also an extremely rapid interstitial diffuser, making it very difficult to avoid ruining a whole GaAs sample if copper is present during a heat-treatment stage in device fabrication. Chromium and iron are often used to prepare semi-insulating GaAs and InP respectively, which are then used as substrate materials for the growth of epitaxial layers (Chapter 3). However, rapid diffusion of these metals from the substrate into the epitaxial layer can modify the electrical properties of the material in which the devices are to be fabricated. Gold and silver are often used as contact materials to GaAs and InP devices, but rapid indiffusion of the metal can alter the operating characteristics of devices in the underlying semiconductor.

Further details on the complexities of diffusion processes in III–V compounds can be found in Casey (1973) and Tuck (1988), who also describe how important it is to control the composition of both the source of the diffusing element and of the surrounding gas phase if reliable measurements are to be made of the diffusion parameters. The exceedingly wide range of pre-exponential factors and activation enthalpies quoted for the slowly diffusing elements in table 1.5 illustrates how easy it is to obtain quite different results on the same diffusing system under slightly altered experimental conditions.

We have seen in §1.4 that II–VI and IV–VI semiconductor compounds have electrical properties which are dominated by the concentrations of charged native point defects — atomic disorder. It is thus possible to prepare material with the desired properties by annealing in a gas of a carefully chosen composition to create the required concentration of the relevant native point defects. Indiffusion of these defects is a necessary part of this process and so both self-diffusion and dopant diffusion in these compounds will be of interest. Stevenson (1973) has pointed out that self-diffusion in II-VI semiconductors does not necessarily take place via the native point defects which dominate the electrical properties of the material. In cadmium telluride for example, the cadmium interstitial, $Cd_i^{..}$, has the dominant effect on the conductivity, but self-diffusion seems to take place via cadmium vacancies and neutral tellurium interstitials. The equilibration concentration of any particular native point defect also depends on the composition of the gas phase surrounding the crystal. After equilibrium in a cadmium-rich vapour, the self-diffusion of cadmium in CdS occurs by the diffusion of ionised cadmium interstitials, but if the gas phase is sulphur rich then the transport occurs preferentially via cadmium vacancies, V_{Cd}''. In both II–VI and IV–VI compounds the diffusivities of the non-metallic elements are generally found to be lower than of the metallic species, and this means that it is convenient to form a buried p–n junction by the in- or outdiffusion of metallic species to create metal interstitials (donors) or metal vacancies (acceptors). The surface regions of a p-type HgCdTe wafer are readily type converted to n-type material by annealing in a mercury overpressure. II–VI compounds can also be doped by the introduction of impurity elements and the halogens are effective n–type dopants in many of these semiconductors. These dopants diffuse by an interstitial mechanism and with a very low activation enthalpy, and so can be used to prepare deep p–n junctions.

Table 1.5 includes some data on diffusion parameters in CdS. There is very little reliable information on the diffusion mechanisms or kinetics in binary II–VI and IV–VI compounds and even less in ternary alloys like HgCdTe and PbSnTe. For instance, values for the activation energy of indium diffusion in $Hg_{60}Cd_{40}Te$ have been measured ranging from 0.37 to 1.6 eV. This is in part because setting up the experimental conditions so that the diffusion source is in equilibrium with the semiconductor crystal becomes increasingly difficult the larger the number of elements in the crystal.

1.6 DISLOCATIONS IN SEMICONDUCTORS

The structure and electronic properties of dislocations in semiconductors have been extensively investigated because of the influence that these defects can have on the performance of microelectronic devices. Detailed descriptions of the particular effects which can be attributed to the presence of dislocations in semiconductor materials will be deferred to Chapters 2 and 8. However, we will also see that dislocations in semiconductor wafers can be useful in attracting the unwanted heavy metal impurities away from the active device regions on the wafer surface, the phenomenon of 'gettering'. In this section, we will consider first the structure of dislocations in diamond cubic related lattices and then see how dislocations can alter the local band structure and so have an effect on the electronic properties of the whole semiconductor crystal.

1.6.1 The Structure of Dislocations and Stacking Faults in Semiconductors

The structures of the diamound cubic and sphalerite lattices have already been described and can be thought of in terms of two inter-penetrating FCC lattices, as we have seen in figures 1.1(a) and (b). We are thus able to predict that the Burger's vectors of perfect lattice dislocations in both these crystal structures will be the same as in the FCC lattice, $\frac{1}{2}a\langle 110 \rangle$, and that these dislocations will glide on {111} planes. It is commonly observed that dislocations in crystals with diamond cubic lattice structures lie preferentially along ⟨110⟩ directions as well, which means that there are only two kinds of perfect dislocations which we expect to find: the screw dislocation where the Burger's vector is aligned along the dislocation; and 60° dislocations where this is the angle between the Burger's vector and the dislocation line.

If we look at a projection of the diamond cubic lattice along a ⟨110⟩ direction (figure 1.9(a)) then it is clear that there are six separate (111) planes corresponding to the ABC stacking sequence on both A and B sublattices. (111) planes of the two sublattices are normally distinguished by calling them ABC and abc respectively. Looking down the [111] direction, the atoms on a and A planes are projected on top of one another. The closest plane separation is between planes of different

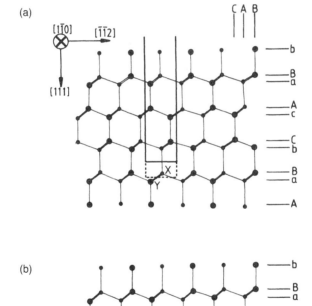

Figure 1.9 A sketch of the atomic structure of: (a) the perfect diamond cubic lattice projected along [1 1̄0]; and (b) the same crystal with a C and a c plane removed to form an intrinsic stacking fault. The large and small circles represent atoms in the plane of the paper, and in the {110} plane immediately above the paper, respectively.

stacking character, B and a, and A and c, etc. The simplest way of forming a low-energy stacking fault in this structure is by removing a pair of (111) planes, for instance the central C and c planes in figure 1.9(a). This results in the formation of the structure shown in figure 1.9(b). This is an intrinsic stacking fault and we can see that the tetrahedral bonding configuration around each atom is preserved, but the shape of the characteristic structural units in this projection is modified in the planes surrounding the stacking fault.

We can create a 60° dislocation by removing a pair of half planes as indicated by the outlined region in figure 1.9(a). However, there are two ways in which we can choose to terminate these half planes — before the atom marked X, or by removing an extra line of atoms before the atom marked Y. It was originally thought that the first of these possibilities, where the extra half plane of the dislocation ends between two planes of the same stacking character, e.g. B and b, would occur, but more recent observations indicate that it is the second kind of dislocation which dominates in semiconductor materials. For instance, Pirouz *et al* (1983) have shown that most 60° dislocations in silicon are dissociated into two partials by a reaction of the kind

$$\tfrac{1}{2}a[011] \rightarrow \tfrac{1}{6}a[121] + \tfrac{1}{6}a[\bar{1}12] \tag{1.14}$$

where the two partials are separated by a region of stacking fault. (See Hirth and Lothe (1978) for some excellent diagrams illustrating the structure of dissociated 60° dislocations.) This dissociation process can only occur freely for 60° dislocations where the extra half plane terminates between planes of dissimilar stacking position, B and a for example. This is because the stacking fault can only lie at this position, as is shown in figure 1.9(b). The observation that the dissociated 60° dislocation can glide easily is an additional indication that the extra half planes of these defects terminate on Y atoms (figure 1.9(a)), since the dissociation products of an X-terminated 60° dislocation are not glissile. High-resolution electron microscopy (HREM) images provide the final evidence that the atomic structure of the core region of 60° dislocations in silicon is that characteristic of Y-terminated extra half planes. Figure 1.10 shows a high-resolution image of a 60° dislocation in silicon and from images of this kind the structure of the dislocation core can be determined. These Y-terminated dislocations are called the 'glide' set, because they are mobile in the diamond cubic lattice.

A further complication arises if we consider the structure of the same 60° dislocations in crystals with the sphalerite lattice. Now the extra half planes can be terminated on a line of either metallic or non-metallic atoms, and in the glide dislocation set the convention is to call non-metal-terminated defects α dislocations and metal-terminated defects β dislocations. It has now been clearly established that the mobilities of these two kinds of dislocations can be quite different and depend on the doping level and doping character in the semiconductor crystal. This is very strong evidence that the dislocations themselves are electrically charged. One model suggests that the variation in the mobility of dislocations in semiconductors can be understood in terms of the motion of charged kinks (Hirsch 1979). (A kink is a point where the end of the half plane jumps from between the B and a planes in figure 1.9(a) to between the C and b planes for example, putting a step on the dislocation line.) The importance of understanding the mechanisms of dislocation migration in semiconductor crystals lies in the desire to control the number of these defects which are found in the active device regions of semiconductor wafers. Dislocations can be introduced at many stages during crystal growth and device fabrication procedures, as we will see in later chapters.

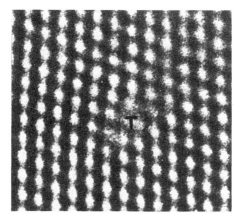

Figure 1.10 A high-resolution transmission electron micrograph showing the core structure of an undissociated 60° dislocation in silicon in a ⟨110⟩ projection. In this image the dark spots are columns of pairs of atoms in the diamond cubic lattice. (Courtesy of Dr J Hutchison.)

1.6.2 The Electrical Properties of Dislocations

If we construct a 60° glide dislocation in a crystal with the diamond cubic lattice by removing two half planes as shown in figure 1.9(a) and rewelding the crystal together, we obtain a structure like that shown in figure 1.11 where a row of dangling bonds is left along the core of the dislocation. Dangling bonds can create electronic states with both donor and acceptor character and some of these states will lie in the band gap. It was at first thought that these dangling bonds would control the electronic properties of dislocations in semiconductors. However, it is possible for dislocation cores to reconstruct, resulting in the removal of the dangling bonds. It has been shown that this reconstruction is energetically favourable and so the electronic properties of dislocations may not be dominated by effects due to dangling bonds. Experimental observations on the electrical activity of dislocations in silicon clearly indicate that bands of electronic states lying deep in the band gap are created, but that the density of these states is relatively low — one deep state for every 1000 atoms on the end of the extra half plane of the dislocation. There are two possible explanations for this observation: firstly that kinks on the dislocation lines are responsible for these deep states (in agreement with the model that kinks are charged defects in semiconductors); and secondly that they are created by impurity atoms segregated to the dislocation line (Cottrell atmospheres). It is not known at present which of these two explanations is correct and the evidence on either side of the argument has been discussed by Petroff (1985). We will see in Chapter 4 that the question of whether defect states are produced as a result of structural disorder or because of impurity decoration remains unanswered for grain boundaries in semiconductors as well.

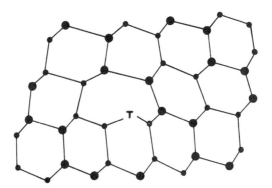

Figure 1.11 A sketch of the core structure of a 60° dislocation in a diamond cubic lattice in the same projection as figure 1.8. In a model of this kind, a continuous line of dangling bonds is formed along the dislocation core.

Even though we do not know exactly the cause of the deep states created by dislocations, we can make some attempt to predict how they will influence the properties of semiconductor crystals. Any deep states have the potential to trap free carriers and so act as recombination centres, just as has already been described for impurities like copper in GaAs and the transition metals in silicon. Direct evidence for enhanced recombination rates at dislocations in semiconductors can be obtained in techniques like cathodoluminescence and EBIC (see Chapter 10). Enhanced recombination at defects can have a particularly severe effect on the operating efficiency of minority carrier devices. Even in majority carrier devices, where recombination phenomena are less important, dislocations can act as short-circuit conduction paths across p–n junctions. A dislocation which is heavily decorated with metallic impurities can have a high one-dimensional conductivity, leading to 'leaky' transistors and a low yield of satisfactory integrated circuits. We will see in Chapters 2, 3 and 8 that considerable effort has been expended to ensure that the densities of dislocations in the active regions of semiconductor devices are as low as possible. Even after taking all possible precautions, the performance of many kinds of microelectronic components is still limited by the effect of residual dislocations.

1.7 AMORPHOUS SEMICONDUCTORS

A wide range of elements and compounds can be prepared in a metastable amorphous form and some of these materials show characteristic semiconducting properties. (The techniques used for the deposition of amorphous semiconductors in thin film form will be described in Chapters 4 and 9.) Here I will concentrate on the properties of amorphous silicon, α-Si, and merely note that a number of oxide glasses and chalcogenide compounds (e.g. As_2Te_3, GeTe and numerous related ternary alloys) are also observed to behave as amorphous semiconductors (see LeComber and Mort (1973), Tauc (1974), Mott and Davis (1979)).

The fundamental semiconducting behaviour of crystalline silicon is controlled by the tetrahedral bonding character of the diamond cubic lattice and it is not immediately clear that similar electronic properties are to be expected in amorphous silicon where all the long-range order is lost. However, Ioffe and Regel (1960) argued that many of the semiconducting properties would be retained in an amorphous material if the short-range-order characteristic of the silicon crystal was retained in the amorphous structure. There are several models for the structure of amorphous silicon, but at present the best fit to the experimental data is given by the 'continuous random network' model where the tetrahedral coordination of individual silicon atoms is preserved and the amorphous character is introduced by distorting the tetrahedral structural units so that the bond lengths and angles deviate randomly from their values in the crystal lattice (Polk 1971). This kind of structure may also contain some dangling bonds, where the local tetrahedral coordination is lost. In addition, thin films of α-Si and α-Ge often contain numerous microvoids and dangling bonds will be common on the internal surfaces of these small defects.

The band structure of amorphous silicon is rather similar to that of crystalline semiconductors, with a gap between valence and conduction bands where electronic states can be created by defects and impurities in the amorphous network. In addition, the random distortions of the tetrahedral bonding units create a set of states whose density is greatest at the valence and conduction band edges but also penetrates some way into the band gap, i.e. 'tailing states'. Electrons in these tailing states are trapped by the local nature of the bond distortion which creates the state, and the mobility of these electrons is very low. Figure 1.12 is a sketch of one of the proposed band structures for amorphous semiconductors, and Mott and Davis (1979) have discussed the relative virtues of a number of these models. Most vapour-deposited α-Si, the material in which we will be most interested for the preparation of solar cells, has an apparent band gap (sometimes also referred to as the mobility gap) of about 1.55 eV and so has a dull orange colour.

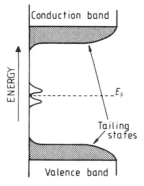

Figure 1.12 A sketch of one of the proposed band structures for amorphous silicon showing the tailing states from the conduction and valence band edges, and the dangling bond states at the centre of the band gap which pin the Fermi level (after Mott and Davis 1979).

It is not easy to dope a sample of amorphous silicon which has this structure because the addition of an impurity element from groups III or V will result only in a local atomic rearrangement such that all the bonds on the impurity atom are satisfied (Mott 1967). The dopant atom will thus not be able to create a shallow electronic state, and α-Si thin films are normally found to have a very high resistivity, $10^8 \, \Omega \, cm$ at room temperature being a typical figure. Only by saturating all the dangling bonds in the α-Si network, which suppresses the ability of the network to reconstruct around dopant atoms, can doping be used to move the Fermi level from the centre of the band gap. The use of atomic hydrogen to saturate dangling bonds in α-Si will be described in §9.4 Hydrogen-saturated α-Si can be readily doped both n- and p-type and is used in the preparation of some important microelectronic devices. A comprehensive description of the properties of α-Si can be found in Chapter 7 of Mott and Davis (1979).

1.8 CONDUCTING POLYMERS

There is at present considerable interest in the use of polymer materials as components of microelectronic systems. The particular features of polymers which make them attractive for these applications are their mouldability, conformability and extreme ease of deposition in thin film form. These features are exploited in the use of polymers in bulk packaging technologies and in the deposition of thin films which are excellent interlayer dielectrics in multilayer metallisation structures. A wide range of these more familiar applications of insulating polymers in microelectronics will be described in Chapters 2, 6 and 7. However, it has become clear in the last 15 years that some polymers can be processed to have semiconducting or even metallic behaviour, and it may even be possible to use these novel polymeric materials as contacting or interconnecting layers in microelectronic circuitry. These conducting polymers are also being investigated for a number of other applications of course and their potential as components in lightweight batteries, photochromic displays and conductive adhesives appears very promising (Blythe 1981).

1.8.1 Polymer Structure and Electrical Properties

All polymers consist of long chain macromolecules which are bonded together into 'bundles' or fibres which are usually 10–20 nm in diameter. The strength of the bonding between individual macromolecules in a fibre depends on the character of the particular molecular units which are found along the chain. Even when the interchain bonding is relatively strong however, and the macromolecules are arranged in a fairly regular array (a so-called crystalline polymer), the degree of disorder is very high when compared with a crystalline semiconductor. In particular, the fibres themselves are not very closely packed, so that about 60% of a typical polymer material is free volume. The electrical properties of conducting polymers are thus characteristic of a highly disordered system, just as we have already seen for the case of α-Si. Macroscopic conduction through a polymer must take place not only by charge hopping along the polymer chains, but also between the macromolecules that make up individual fibres and between the fibres themselves. Only the first one of these conduction steps has been understood in any detail.

If we introduce a free charge onto a polymer macromolecule, there are a large number of sites at which the charge can attach itself firmly to a molecular subunit. In fact, the strong polarisation of the molecular units by the presence of the charge itself results in the formation of deep states in which the charge will then be trapped. Stable and stationary 'molecular ions' are thus readily formed, and this is the mechanism by which the insulating properties of most polymer materials are developed. The trapped charges are rather unlikely to 'hop' from a molecular ion trapping site to an adjacent site, or to put it in more formal terms — the hopping integral between two sites on the polymer chains is vanishingly low. In a recent review Duke (1985) has given a typical value of the trapping energy of charges in molecular ions as about 1 eV, while the hopping integrals in most conventional polymers are less than 0.01 eV. It is clear that in these materials the charges are essentially localised. This state is characterised as the Fermi-glass limit and although materials of this kind have semiconductor-like properties their resistivities are generally very high (at least 10^6 Ω cm). The band structure of these polymers is most conveniently thought of as a continuous distribution of localised states with no real band gap, although there is a region of the energy spectrum where the density of states does seem to drop to a minimum. This region is often referred to as the 'band gap' of the polymer.

There are, however, some polymers where the hopping probabilities along the macromolecular chains, but not between them, are rather larger and charge migration along the chains is possible. This condition can be described as 'macromolecular localisation' of the charges and is observed in polymers like polyacetylene and polyethylene. The resistivity of these materials is rather lower than of the Fermi glasses. A few polymers have sufficiently large hopping probabilities for partial delocalisation of the charges to occur and a metallic-type conduction results. More properly, the band structure of these materials resembles that of narrow band-gap semiconductors or semimetals.

Here we can introduce an interesting analogy between more conventional semiconductor materials and conducting polymers — that both of them can be doped to give material of lower resistivity. The addition of iodine or AsF_5 to polyacetylene creates states in the polymer band gap, in a manner rather similar to the doping effect of phosphorus atoms in a silicon crystal. In a doped polymer the dopant species form molecular ions between which the hopping probability is relatively large. We can use the example of polyacetylene to describe the phenomenon of doping in a polymer in a little more detail. Figure 1.13(a) shows a perfect section of a polyacetylene molecule containing alternating single and double carbon–carbon bonds. The region of minimum electronic state density in this polymer is a result of the energy difference between the filled π bonds (double bonds) and unfilled π^* bonds. A charge which is propagating along the macromolecular chain must hop from π to π^* bonds. Let us now introduce an acceptor dopant — iodine — into this region of the molecule. The dopant atom will remove one bonding electron from a π bond, creating a carbonium ion on the polymer chain (figure 1.13(b)). Calculations on the lowest energy structure of the carbonium ion indicate that they are partially delocalised, spreading out along the polymer chain to form a larger molecular ion (figure 1.13(c)). This can now be considered a 'molecular hole'. Careful inspection of figure 1.13(c) will reveal that the bonding arrangement, and in particular the pattern of alternation of single and double bonds, is disturbed at both ends of the molecular ion to form a central region where the bonding arrangement is inverted when compared with the rest of the polymer chain. This inverted region is bounded by two 'kinks'. Conduction in iodine-doped polyacetylene can occur by migration of the molecular holes by bond redistribution at the two kink sites, or by variable range hopping of charge between randomly spaced ions. There

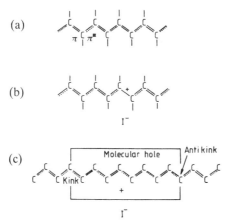

Figure 1.13 A schematic illustration of the process by which a molecular ion can be formed in a polymer chain. (a) A section of a perfect polyacetylene chain; (b) the formation of a carbonium ion by addition of an iodine ion; (c) the delocalisation of the carbonium ion to form a molecular hole bounded by two kinks.

is at present no completely general consensus as to which of these processes dominates. As the concentration of dopant species is increased, metallic conductivity may be achieved and we can think of this as the result of the molecular ions being packed so closely together that they directly interact, forming a simple conduction path along the polymer chains.

Heavy doping of polymers can result in the preparation of materials with resistivities as low as 10^{-2} Ω cm, but the mobilities remain very low with typical values of about 10^{-7} cm^2 V^{-1}s^{-1}(Duke 1985). The process of doping a polymer is remarkably simple. Using a sheet of the polymer as an electrode in an electrochemical cell containing I$^-$ ions will result in the removal of some electrons from the π bonds and the introduction of I$^-$ ions, i.e. p-type doping. Reversing the applied voltage in a solution of Na$^+$ ions results in n-type doping. The level of doping depends on the applied voltage and the doping process is completely reversible. Polymers can also be doped by indiffusion of the dopant from the gas phase. The open fibrillar structure of the polymers explains why the dopant species are so easily introduced from both gas and liquid phases. Unfortunately, heavily doped polymers are often very brittle and sensitive to chemical attack by water and oxygen.

It may also prove possible to increase the range of conductivities available in polymer materials by 'engineering' molecules with particularly small band gaps. The most favourable chain structure for reducing the band gap seems to be when there are alternating double and single carbon–carbon bonds *with equal bond lengths*. In some cases, this kind of structure has been created, for example by adding a benzene ring to thiophene. A few basic properties of some conducting polymers are listed in table 1.6.

Table 1.6 Properties of some conducting polymers.

Material	Band gap (eV)	Resistivity range
Polyacetylene	1.4	10^{10}–10^{-2} Ω cm
Thiophene	2.0	10^{10}–10^{-2} Ω cm
Benzene-modified thiophene	1.0	10^8 –10^{-3} Ω cm

	Dopants	
p-type	I, AsF$_5$	
n-type	Na, Li	

An alternative way of preparing a conducting polymer is by introducing a fine distribution of a conducting phase into an insulating polymer matrix. Carbon and silver particles are popularly used for this application and the process is called 'particulate doping'. The conduction takes place by tunnelling between adjacent filler particles. A more subtle way of achieving the same kind of microstructure is to anneal a polymer-like polyimide (see Chapter 6) above 650 °C, when small islands of conducting carbonaceous material are formed. The conductivity of the polyimide increases by more than 10 orders of magnitude to about 10^{-2} Ω cm.

Most commercially exploited conducting polymers are of the particulately doped kind at present, but molecularly doped materials have some very attractive properties and are being tested for many applications in microelectronic systems. The use of organic materials in integrated circuits is still regarded with some suspicion, but the substantial cost advantages offered by polymeric materials and processing techniques may soon result in the introduction of organic conductors into some components.

REFERENCES

Albers W 1967 in *Physics and Chemistry of II–VI Compounds* ed. M Aven and J S Prener (Amsterdam: North-Holland) p166

Blythe A R 1981 *Electrical Properties of Polymers* (Cambridge: Cambridge University Press)

Brouwer G 1954 *Phillips Res. Rep.* **9** 366

Casey H C 1973 in *Atomic Diffusion in Semiconductors* ed. D Shaw (London: Plenum) p351

Duke C B 1985 *J. Vac. Sci. Technol.* **A3** 732

Fair R B 1981 in *Impurity Doping Processes in Silicon* ed. F F Y Wang (New York: North-Holland) ch. 7

Frank F C and Turnbull D 1956 *Phys. Rev.* **104** 617

Frazer D A 1983 *The Physics of Semiconductor Devices* (Oxford: Oxford University Press)

Harman T C and Melngailis I 1974 in *Applied Solid State Science* ed. R Wolfe (New York: Academic) p1

Hirsch P B 1979 *J. Physique* **40** C6 27

Hirth J P and Lothe J 1978 *Theory of Dislocations* (New York: McGraw-Hill) ch. 11

Ioffe A F and Regel A R 1960 in *Progress in Semiconductors* vol. 4 ed. A F Gibson (London: Heywood) p237

Kroger F A 1977 *Ann. Rev. Mater. Sci.* **7** 449

Kroger F A and Vink H J 1956 *Solid State Phys.* **3** ed. F Seitz and D Turnbull (New York: Academic) p307

Kruse P W 1981 in *Semiconductors and Semimetals* **18** ed. R K Willardson and A C Beer (New York: Academic) p1

Kumar V and Kroger F A 1971 *J. Solid State Chem.* **3** 387

LeComber P G and Mort J 1973 *Electronic and Structural Properties of Amorphous Semiconductors* (New York: Academic)

Lovatt D R 1977 *Semimetals and Narrow Bandgap Semiconductors* (London: Pion)

Mott N F 1967 *Adv. Phys.* **16** 49

Mott N F and Davis E A 1979 *Electronic Processes in Non-Crystalline Materials* (London: Oxford University Press)

Palatnik L S and Belova E K 1967 *Izv. Acad. Nauk Neoyan Mater.* **3** 2194

Petroff P M 1985 in *Semiconductors and Semimetals* **22A** ed. W T Tsang (New York: Academic) p379

Phillips J C 1973 *Bands and Bonds in Semiconductors* (New York: Academic)

Pirouz P, Cockayne D J H, Sumida N, Hirsch P B and Lang A R 1983 *Proc. R. Soc.* A **386** 241

Polk D E 1971 *J. Non-Cryst. Solids.* **5** 365

Shay J L and Wernick J H 1975 *Ternary Chalcopyrite Semiconductors* (Oxford: Pergamon)

Shewmon P G 1963 *Diffusion in Solids* (New York: McGraw-Hill)

Stevenson D A 1973 in *Atomic Diffusion in Semiconductors* ed. D Shaw (London: Plenum) p351

Sze S M 1981 *Physics of Semiconductor Devices* (New York: Wiley)

Tauc J 1974 *Amorphous and Liquid Semiconductors* (London: Plenum)

Tsai J C C 1983 in *VLSI Technology* ed. S M Sze (New York: McGraw-Hill) p169

Tsai M Y, Morehead F F and Baglin J E E 1980 *J. Appl. Phys.* **51** 3230

Tuck B 1974 *Introduction to Diffusion in Semiconductors* (Stevenage: Peter Peregrinus)

—— 1988 *Atomic Diffusion in III—V Semiconductors* (Bristol: Adam Hilger)

Van Vechten J A 1975 *J. Electrochem. Soc.* **122** 419

Yeh T H 1973 in *Atomic Diffusion in Semiconductors* ed. D Shaw (London: Plenum) p155

Zanio K 1978 *Cadmium Telluride* (*Semiconductors and Semimetals* **13**) (New York: Academic)

2

Contacts, Devices and Integrated Circuits

2.1 INTRODUCTION

In this chapter we will see how microelectronic devices can be fabricated in the semiconductor materials introduced in the previous chapter, starting with a discussion of a component which is found in all devices—the metal–semiconductor contact. The structure and properties of these contacts will be considered in some detail because it is now becoming clear that the chemistry of, and density of crystalline defects in, the interface regions play an important part in determining their electrical properties. It will then be shown that simple transistor devices can be built up from combinations of a few basic components: metal–semiconductor contacts, p–n junctions in the semiconductor material itself and oxide–semiconductor interfaces. The electrical properties of neither p–n junctions nor the complete devices will be described in any detail here as they are the subject of several excellent textbooks. The author has benefited greatly from a study of the sections on semiconductor devices in Sze (1981) and Frazer (1983) in particular. The concept of device integration will also be introduced, with a brief consideration of how the rapidly shrinking dimensions of modern microelectronic devices may lead to significant problems with the choice and long-term stability of the materials used to fabricated the devices.

The final section of this chapter contains a brief description of several of the most important processes involved in the production of a semiconductor device or circuit. (The emphasis here will be on the fabrication of devices on silicon wafers, and the more specialised methods used in the fabrication of optoelectronic and photovoltaic devices in compound semiconductors will be described in Chapters 8 and 9.) We will consider the preparation of semiconductor wafers, diffusion and implantation techniques for controlled doping of the semiconductor material and the lithography and etching stages needed to define small device features on the wafer surface. Throughout this section it will be the materials aspects of the fabrication processes which will be stressed and references to publications which offer more detailed information on the technological aspects will be cited wherever relevant. Several stages in the device fabrication process are of sufficient importance to be treated in greater detail elsewhere and topics like single-crystal growth, the deposition of polycrystalline thin films, the stability and performance of metallisation systems, and the growth of insulating layers will be described in Chapters 3–6 respectively.

2.2 METAL–SEMICONDUCTOR CONTACTS

The structure and electronic properties of the interfaces produced by the deposition of a thin metal film on a semiconductor surface have been intensively studied since the discovery of transistor properties in semiconductors. A vast body of empirical evidence has been collected on the behaviour of these interfaces in practical device applications and more recent detailed studies on the chemistry and electronic state density at these interfaces have begun to reveal the complex interactions which occur between metal and semiconductor. However, there is as yet no generally accepted model of the mechanism by which the electrical properties are developed, as we will see below. We will start this section by defining the two basic kinds of metal–semiconductor contact—the Schottky and the ohmic contact—and will then see how a simple model for predicting which kind of interface behaviour we expect for any particular combination of metal and semiconductor fails to explain many of the experimental results. I will then show that a model based on the presence of defect states at the interface can give a better, but still not complete, explanation of the behaviour of many Schottky contacts. The defects responsible for these characteristic interface states will be discussed in a separate section. The preparation of ohmic contacts will be the subject of §2.2.3, where we will see that an ideal ohmic contact is rarely achieved in practice. Finally, §2.2.4 will contain a brief compilation of data on the properties of some metal–semiconductor contacts commonly used in microelectronic devices.

2.2.1 The Schottky and Bardeen Models of Metal–Semiconductor Contacts

The simplest way of modelling the electrical properties of a metal–semiconductor contact is to assume that no electronic states lying in the semiconductor band gap are created when a semiconductor crystal is cleaved or polished to produce a free surface. (It is known that a freshly cleaved gallium arsenide (110) surface has no surface states for instance.) Then when a metal is brought into contact with this surface, charge will flow to equalise the Fermi levels in the two materials. Let us consider two particular cases of metal contacts on an n-type semiconductor with an electron affinity χ_s: one where the workfunction of the metal, Φ_m, is greater than χ_s (figure 2.1(a)); and the second where Φ_m is less than χ_s (figure 2.1(b). χ_s and Φ_m are parameters which define the energy difference between an electron at the vacuum level and a free electron in semiconductor and metal respectively. The effect of bringing the metal and semiconductor into contact is shown in figure 2.1. In both cases, charge will flow to equalise the position of the Fermi level, but the one-electron potential energy diagrams are quite different in form. In figure 2.1(a), where Φ_m is greater than χ_s, a depletion region is established in the semiconductor and a potential barrier is set up at the metal–semiconductor interface. This is called a Schottky barrier and current transport across such an interface will show strong rectifying characteristics. The height of the potential barrier, Φ_{Bn}, on an n-type semiconductor is given by the simple equation

$$\Phi_{Bn} = \Phi_m - \chi_s. \tag{2.1}$$

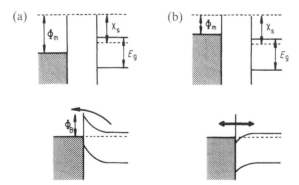

Figure 2.1 One-dimensional potential energy diagrams to illustrate the formation of (a) a Schottky and (b) an ohmic contact at interfaces between metals and an n-type semiconductor.

At this point it is appropriate to consider briefly the mechanisms of carrier transport across a Schottky barrier of the kind shown in figure 2.1(a). If a positive voltage is applied to the semiconductor, the two most important mechanisms by which electrons can move to the metal contact are: (i) thermionic emission over the top of the potential barrier; and (ii) quantum mechanical tunnelling of electrons through the barrier. Mechanism (i) dominates when the semiconductor is relatively lightly doped, i.e. a donor doping level, N_D, of less than about 10^{17} cm^{-3}, while mechanism (ii) will become important when the semiconductor is heavily doped. Crowell and Sze (1966) derived an expression for the current passing by thermionic emission across a Schottky barrier of height Φ_{Bn} when the applied voltage is V:

$$ J \propto \exp\left(\frac{-q\Phi_{Bn}}{kT}\right)\left[\exp\left(\frac{qV}{kT}\right) - 1\right]. \tag{2.2}$$

Tunnelling transport across a Schottky contact can be modelled by an approximate expression of the form

$$ J \sim \exp\left(\frac{-q\Phi_{Bn}}{E_{00}}\right) \tag{2.3}$$

where E_{00} is a parameter which varies with the root of the doping concentration, $N_D^{1/2}$. The tunnelling current will increase exponentially with $N_D^{1/2}$. A more detailed discussion of the theories of carrier transport across Schottky barriers can be found in Rhoderick (1978) and Sze (1981). Here it is only important for us to appreciate that when the doping level in the semiconductor is lower than about 10^{17} cm^{-3} we expect thermionic emission to be dominant, but as the doping level is increased above 10^{19} cm^{-3} there will be a rapid increase in the tunnelling current.

Returning to figure 2.1(b), we can see that when Φ_m is less than χ_s no barrier is created at the interface when the Fermi levels are equalised and the free carriers in the semiconductor (electrons in this case) can flow freely in both directions across the contact. This is called an ohmic contact and will have perfectly linear current–voltage (J–V) characteristics. Construction of similar figures for the case of a p-type semiconductor will lead to the prediction that when Φ_m is greater than χ_s an ohmic contact would be produced, but when Φ_m is less than χ_s a Schottky barrier is formed with height Φ_{Bp} given by the difference between χ_s and Φ_m. This is the Schottky model for the behaviour of metal–semiconductor contacts (Schottky 1939). It predicts that on n-type semiconductors most metal contacts should have rectifying properties since the workfunctions of most common metals exceed the typical values of χ_s. The height of the Schottky barriers at these contacts is predicted by equation (2.1) to depend linearly on Φ_m.

However, experimental measurements on the electrical properties of contacts to a range of important semiconductor materials have shown that the Schottky barrier heights are generally rather lower than predicted by this simple model and that there is, in many cases, little sign of a linear dependence of barrier height on the workfunction of the metal. This observation led Bardeen (1947) to propose that localised

states are formed at free semiconductor surfaces and that at a metal–
semiconductor contact these states can pin the Fermi level in the centre
of the band gap. It was also necessary to propose that a very thin oxide
layer is present on the surface of the semiconductor before deposition of
the metal, as is shown in figure 2.4(a), because this oxide can shield the
surface states from the metal. Crowell and Sze (1966) suggested that at
such an interface the potential barrier height can be estimated from the
expression

$$\Phi_{Bn} = \gamma(\Phi_m - \chi_s) + (1 - \gamma)(E_g - \Phi_I) \qquad (2.4)$$

where $\gamma = \varepsilon_i/(\varepsilon_i + q\delta N_I)$ and ε_i is the permittivity of the oxide layer, δ
its thickness, N_I the density of interfacial trapping states and Φ_I the
position of the Fermi level at the semiconductor surface when the metal
is not present. This expression reduces to the Schottky equation (2.1)
when $N_I = 0$, but when N_I is large predicts a fixed value of Φ_{Bn} since
the Fermi level is pinned by the surface states—the Bardeen limit. In
both the Bardeen and the Schottky models the sum of the barrier
heights to p- and n-type material should be equal to the band gap of the
semiconductor, i.e. $\Phi_{Bn} + \Phi_{Bp} = E_g$.

We should now consider whether the Bardeen model can be used to
predict the behaviour of metal–semiconductor contacts. In order to do
this I will present a few sets of experimental data on the electrical
properties of some of these interfaces. These data are chosen because
they give a very clear indication of the fact that not all contacts show
the same kind of properties. Figure 2.2 shows the result of measuring
the barrier height formed when gold contacts are deposited on a wide
range of semiconducting materials. At most of these contacts the Fermi
level is pinned at a position approximately two-thirds of the width of the
semiconductor band gap below the conduction band edge, the so-called
'$\frac{2}{3}E_g$ rule'. (The Au–InP contact is a notable exception to this rule, with
the Fermi level pinned only 0.3 eV below the conduction band edge.)
The observation of the two-thirds rule strongly implies that the properties of these interfaces can be explained by the Bardeen model, with
the assumption that a high density of interfacial pinning states is located
slightly below the centre of the band gap. However, figure 2.3 gives
another set of experimental data on the measured Schottky barrier
heights for a variety of metal contacts on two semiconductors: covalently
bonded gallium arsenide; and zinc sulphide where the bonding has a
strong ionic component. While the barrier height is almost invariant at
about 0.9 eV for all the metal contacts to n-type GaAs, as we would
expect from figure 2.2 and the two-thirds E_g rule, the barrier heights of
the contacts to ZnS increase sharply with the electronegativity of the
metal, X_m. (We should remember that the workfunction of a metal, Φ_m,

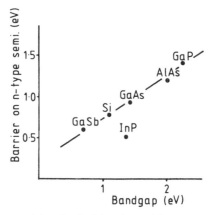

Figure 2.2 Measured barrier heights for gold contacts on a range of semiconductor compounds, illustrating that the Fermi level is pinned approximately two-thirds of the width of the band gap below the conduction band edge for all these semiconductors except InP (after Mead 1966).

is directly related to its electronegativity.) This second set of data seems to indicate that contacts on GaAs follow the behaviour expected at the Bardeen limit, but that contacts to ZnS obey the Schottky model. Kurtin *et al* (1969) made a compilation of data on the properties of metal–semiconductor contacts on a much wider range of semiconductor materials and plotted a parameter $S = \delta\Phi_B/\delta X_m$ against the heat of formation of the semiconductor compound. (Highly ionic compounds have in general much higher heats of formation than covalently bonded semiconductors.) This plot showed that S was approximately unity for ionic semiconductor compounds, which means that the Schottky model,

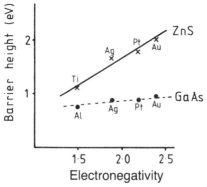

Figure 2.3 Measured barrier heights for contacts formed between a variety of metals and GaAs and ZnS (after Mead 1966).

where Φ_B is linearly dependent on X_m (or Φ_m), is obeyed. For the covalently bonded semiconductors, S is approximately equal to zero, and the Bardeen limit is obeyed, with the Fermi level at the interface firmly pinned by interface states.

While there is some debate as to the precise interpretation of the values of S (see Schluter (1978) for example), these observations lead us to the interesting idea that high densities of interface states are produced at contacts between metals and covalent semiconductors, but that the interface state densities are much lower if the semiconductor has a strongly ionic bonding character. Several models have been proposed to explain the density and nature of interfacial states at metal–semiconductor contacts and it is appropriate to consider some of these in a little more detail.

2.2.2 Interface States

The Bardeen model for metal–semiconductor contacts introduced above assumes that the states formed on semiconductor surfaces are preserved under a metal contact by the screening effect of a thin native oxide layer. However, it is now possible to prepare intimate metal–semiconductor interfaces by ensuring that no native oxide layer can grow on the semiconductor before deposition of the metal contact. Characteristic interface states are still found at these interfaces. Further evidence that we can no longer identify interface states with intrinsic surface states comes from the observation that the deposition of a thin metal film on a semiconductor surface on which there are *no* intrinsic states, freshly cleaved (110) surfaces of GaAs, InP and GaSb for example, still results in the formation of interface states (Spicer *et al* 1979). We have to assume that the presence of a metal on the surface of a covalently bonded semiconductor creates new states lying in the semiconductor band gap.

The first explanation of this effect was offered by Heine (1965) who suggested that the decay of the metallic wavefunctions into the semiconductor would create interface states (figure 2.4(b). These states were called metal induced gap states, or MIGS, and their position in the semiconductor band gap was expected to depend on the properties of the metal forming the contact as well as the electronic properties of the semiconductor. This may be seen as a problem with Heine's model when we remember that the barrier heights of contacts formed by many metals on semiconductors like GaAs are all very similar (figure 2.2). However, Tersoff (1985) has argued that the position of the interface states will depend primarily on the band structure of the semiconductor and has demonstrated that the barrier heights at gold contacts to silicon and some III–V semiconductors can be accurately predicted with this assumption.

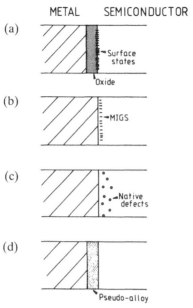

Figure 2.4 (a) The model of the metal–semiconductor interface proposed by Bardeen to explain the pinning of the Fermi level. Illustrations of a few of the more recent theories are: (b) the MIGS model; (c) the unified defect model; (d) the 'pseudoalloy' model.

A more serious limitation of the MIGS model is that it seems improbable that this kind of interface state can be formed until a sufficient thickness of metal has been deposited to allow the true 'bulk' metallic band structure to be developed. An estimate of the critical thickness might be a single complete unit cell of the metal, or about 0.5 nm. Spicer *et al* (1979, 1986) have demonstrated in a very extensive set of experiments that the Fermi level on the surface of a covalently bonded semiconductor is pinned by a metal coverage of only a tenth of a monolayer, far less than that needed to give a bulk metallic band structure. Even more interestingly, a similar coverage of oxygen atoms pins the Fermi level in the same position in the semiconductor band gap. The Fermi level at a contact to GaAs is pinned roughly at the centre of the band gap, while at a contact on InP the Fermi level is pinned closer to the conduction band edge. Both of these observations are in excellent agreement with the Schottky barrier heights measured on thick metal contacts to these semiconductors. These experiments led to the suggestion that the release of the heat of condensation by metal, or oxygen, atoms arriving on the surface of the semiconductor creates native point defects in the semiconductor crystal itself, the 'unified defect model' (Spicer *et al* 1980). In this model, it is the electronic states characteristic of the native point defects which pin the Fermi level at the metal–semiconductor interface (figure 2.4(c)). Clearly we expect the barrier heights of metallic contacts to any covalent semiconductor to be very similar irrespective of the choice of metal since the native point defects will be the same in each case. Strong additional support for this model has come from the calculations of Allen *et al* (1986), who have shown that the position of the electronic states characteristic of antisite defects near the surface of many III–V compounds are very close to where the interface pinning states must lie to explain the observed Schottky barrier heights. We should remember that the formation enthalpies of antisite point defects in III–V semiconductors are very low (see table 1.2). The observed Schottky barrier heights of metal contacts to silicon are better explained by the presence of dangling bonds on the silicon surface. Within the unified defect model, we can also explain why Fermi-level pinning does not occur at metal contacts to ionic semiconductors by remembering that these materials are very strongly self-compensating (see §1.4.3). Any interface state created by the condensation of the contacting metal layer will immediately be removed by the spontaneous formation of compensating native point defects.

At present, the unified defect model seems to offer a very attractive way of explaining many of the experimental observations on the properties of metal–semiconductor contacts. However, it has also become clear that there are often diffusion and chemical reaction processes occurring at these same interfaces, and we must ask to what extent these processes will effect the electrical properties of the contacts. There is abundant evidence that the presence of a metal layer on the surface of many semiconductors promotes the rapid dissociation of the semiconductor, even when there is no direct chemical reaction between metal and semiconductor. Deposition of a thin gold film onto a silicon crystal, for instance, results in a weakening of the silicon bonds close to the interface, leading to mixing of the gold and silicon and the formation of an interfacial 'phase' or 'pseudoalloy' of poorly defined stoichiometry. Similarly, at gold contacts to GaAs, InP and CdS, dissociation of the compound semiconductor occurs, with rapid outdiffusion of one or both of the semiconductor components into the gold film. Some indiffusion of gold may also occur and this has been suggested to be of particular importance in determining the electrical properties of the contact.

Even more complex interactions are found when there is a chemical reaction between the metal and one component of the compound semiconductor, for example between an aluminium contact and arsenic in GaAs. In this case, the semiconductor dissociates under the contact and the gallium diffuses away from the interface to be replaced by aluminium with the formation of AlAs as an interfacial compound.

Brillson (1979) proposed that the electrical properties of metal–semiconductor contacts can be predicted from a knowledge of whether a true chemical reaction occurs at the interface (a reactive interface) and whether the semiconductor is ionic or covalent in bonding character. These two parameters determine the extent of charge redistribution at the interface and so control the interfacial dipole. A rather similar model proposed by Freeouf and Woodall (1981) suggests that the properties of contacts are controlled primarily by the workfunctions of the 'pseudoalloys' formed by diffusional mixing at the metal–semiconductor interface—the effective workfunction model. It is important not to forget when faced with this plethora of models that the precise nature of the interfacial chemistry will also control the nature and concentration of native defects in the semiconductor material under the contact. It is possible therefore that models based on interfacial chemistry may not differ very much from the basic principles of the unified defect model. One important additional feature of the 'chemical' models is that they allow for the possbility of indiffusing metal atoms *doping* the top layer of the semiconductor. This concept will be especially important when we come to consider the formation of practical ohmic contacts in the next section.

One further complication has been added to the debate on the behaviour of metal–semiconductor interfaces by the observation of Tung (1984) that merely changing the atomic structure *but not the chemistry* of $NiSi_2$–Si interfaces resulted in a change in the barrier height. Brugger *et al* (1984) also observed that a monolayer of germanium on GaAs gave an interface where the Fermi level was strongly pinned, while deposition of a thicker layer of germanium, presumably forming a more ordered interfacial structure, produced an ohmic contact. Once again, this implies that surface disorder is an important contributory factor in determining the position and density of surface states. It is not completely obvious that data on the position of Fermi-level pinning obtained in experiments where only a submonolayer coverage of metal is used on the semiconductor surface will tell us very much about the properties of practical metal–semiconductor contacts where the metal films will be at least 100 nm thick. There is, however, some evidence that the pinning positions are very nearly identical for both thin and thick metal layers on the same semiconductor surface (e.g Spicer *et al* 1986).

In summary, it is not clear whether interface states are created by interfacial disorder, native point defects in the semiconductor or dopant indiffusion from the contact, or by charge redistribution at reacted or unreacted metal–semiconductor interfaces. This unsatisfactory state of affairs arises primarily because it is very hard to characterise fully the nature of the point defect which creates a particular electronic state in semiconductor materials; we will find many other examples of this problem in Chapters 3, 4 and 8. We can be certain that the structure and electronic properties of metal–semiconductor interfaces will continue to excite a great deal of intensive fundamental research, since the problems that remain to be solved are both scientifically profound and of great technological importance. An excellent set of chapters on various aspects of contact properties can be found in Einspruch and Bauer (1985) and are recommended to the interested reader.

2.2.3 Ohmic Contacts

In §2.2.1 we saw that the Schottky model of metal–semiconductor contacts predicts that an ohmic contact will be formed on an n-type semiconductor crystal when a metal is deposited whose workfunction is smaller than the electron affinity of the semiconductor, i.e. $\Phi_m < \chi_s$. However, metal contacts to covalent semiconductors, including technologically important materials like silicon and gallium arsenide, show properties which are dominated by the pinning of the Fermi level by interface states and are usually rectifying. Even on ionic semiconductors, where the interface state density is low, it is often hard to find a metal with a workfunction which is sufficiently low to create an ideal ohmic contact. In general, it is not possible to prepare contacts which have perfectly linear $J–V$ characteristics and negligible specific contact resistances. Values of this last parameter lower than $10^{-4} \; \Omega \, cm^2$ are required for the efficient operation of many devices, although in solar cells rather higher values can be tolerated. One of the very few ideal ohmic interfaces is that between metallic indium and n-type cadmium sulphide, and this contact can be used on the CdS/Cu_xS solar cells described in Chapter 9.

There are two ways in which the properties of a metal–semiconductor contact can be forced to approximate to those of an ideal ohmic contact: (i) by reducing the Schottky barrier height as much as possible to increase the thermionic emission current across the contact (see equation (2.2); or (ii) by increasing the doping level, which narrows the potential barrier and encourages tunnelling through the contact (equation (2.3)). It is possible to choose combinations of metal and semiconductor where the Schottky barrier height is relatively low, but even if Φ_B is only 0.25 eV the specific contact resistance will usually be of the order of $10^{-3} \; \Omega \, cm^2$. As a result, the use of a heavily doped semiconductor layer directly under the metallic contact to increase the tunnelling current is a common way of creating a 'quasi-ohmic' contact. The measured values of the barrier height are observed to fall sharply as the doping level under the contact exceeds about $10^{19} \; cm^{-3}$.

We can distinguish two ways in which this heavily doped layer can be formed: firstly by deliberate doping of the semiconductor surface before the metal contact is deposited and secondly by stimulating a chemical reaction between the contact material and the semiconductor. Heavily doped layers can be prepared at the surface of a semiconductor wafer by the epitaxial growth processes described in Chapter 3, or by the diffusion or implantation techniques presented later in this chapter. The metal film is then a passive component of the ohmic contact and is normally chosen to minimise the value of Φ_B. The preparation of quasi-ohmic contacts by placing a passive metal film on a heavily doped semiconductor is especially common in the laser and solar cell structures described in Chapters 8 and 9.

It is also possible to choose a metal for the contact which will itself diffuse into and dope the semiconductor surface; an aluminium contact on a p-type silicon wafer is a good example of this phenomenon. Aluminium is a p-type dopant in silicon and a heavily doped p-type layer can be formed directly under the contact when the aluminium diffuses into the silicon. It is a general feature of the preparation of these 'reactive' ohmic contacts that quite severe heat treatments are needed after the deposition of the metal layer to encourage the indiffusion process. More complex alloying chemistries are used to produce ohmic contacts to III–V semiconductor materials, and these contacts are so reactive that relatively massive particles of a variety of compound phases are often found after annealing. It is quite common for the contact to be at least partly molten during annealing, due to the formation of low-melting-point compounds by chemical reaction between the contact and the semiconductor. Gold-germanium alloy contacts to GaAs have been particularly comprehensively studied because they form some of the most reliable low-resistance ohmic contacts. It is believed that the germanium, and possibly the gold as well, diffuse into and dope the GaAs surface layers during annealing at temperatures between 400 and 600 °C. Germanium is an amphoteric dopant in GaAs and so quasi-ohmic contacts to both n- and p-type material can be prepared with the same alloy. At the same time as the germanium is diffusing into the GaAs, a number of intermetallic phases are formed at the interface, and the severity of this chemical reaction is well illustrated by the cross-sectional electron micrograph shown in figure 2.5. Whether the presence of the intermetallic phases plays any role in determining the electrical properties of the contacts is not known, although it is clear that extensive interfacial reaction occurs at temperatures well below those needed to achieve the lowest specific contact resistances.

It should be apparent from the above that the preparation of satisfactory ohmic contacts depends not only on the choice of a suitable dopant element to include in the contact-forming alloy, but also on discovering the optimum annealing conditions. This is the reason why methods for forming ohmic contacts with particularly low specific

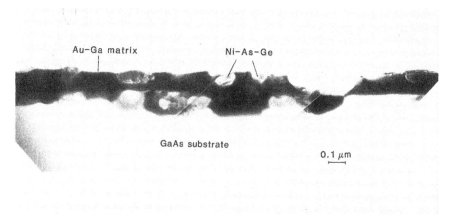

Figure 2.5 A cross-sectional transmission electron micrograph showing the inter-metallic phases formed by chemical reaction at a gold-germanium contact on GaAs. (Courtesy of Drs A Staton-Bevan and X Zhang.)

contact resistances have been passed from person to person in the microelectronics industry in a manner rather reminiscent of the way in which a successful recipe circulates in a small community. Serious problems like the 'balling-up' of Au–Ge alloy contacts when annealed on GaAs were solved by the empirical observation that the addition of a thin layer of nickel to the contact improved the wetting behaviour, and this trick soon became very widely known. The inclusion of a third element in the contact of course allows the formation of yet another set of intermetallic compounds during annealing. Details will be given in the following section of some contacting metals and alloys used to prepare low-resistance quasi-ohmic contacts to a number of semiconductor materials. Further details on the electrical properties of these contacts, and a more comprehensive list of preparation 'recipes', can be found in Schwartz (1969), Rideout (1975) and Piotrowska *et al* (1983).

2.2.4 Properties of Practical Schottky and Ohmic Contacts

This section will present some of the basic properties of Schottky and ohmic contacts in tables 2.1 and 2.2 respectively. The barrier height is the single most important parameter needed to predict the electrical properties of a Schottky contact, and table 2.1 contains values of Φ_B for a number of metals on silicon and a few compound semiconductors. These values are taken mostly from the compilation given by Sze (1981), but the precise value of the barrier height for a particular metal–semiconductor contact will always depend on the method chosen to make the measurement, the doping level in the semiconductor, and

Table 2.1 Schottky barrier heights (in eV) measured at some metal–semiconductor contacts.

		Al	Au	Cu	Pt	Ti
Si	n-type	0.72	0.8	0.58	0.9	0.5
	p-type	0.42	0.34	0.46		0.61
GaAs	n-type	0.8	0.9	0.82	0.84	
	p-type	0.5	0.42			
GaP	n-type	1.07	1.3		1.45	1.12
	p-type		0.72			
InP	n-type	0.33	0.42			
	p-type		0.76			
CdS	n-type	ohmic	0.78	0.5	1.1	
CdTe	n-type	0.76	0.71			

perhaps most importantly the cleanliness of the semiconductor surface before deposition of the contact. We have seen that the chemistry, defect concentration and even the atomic structure of the interfacial region all effect Φ_B. The values in table 2.1 should therefore only be taken as an indication of the Schottky barrier height for a given combination of metal and semiconductor. Even with these qualifications, these data clearly show the pinning of the barrier height of many metal contacts on n-type GaAs at about 0.9 eV (the $\frac{2}{3}E_g$ rule), while the barrier height of both aluminium and gold contacts on InP are very low. This fact has important consequences for the design of devices in these two semiconductors. High-quality rectifying contacts can be formed to GaAs devices but not to InP devices, where oxide–semiconductor contacts are more commonly used. The fact that the sum of Φ_{Bn} and Φ_{Bp} for any metal–semiconductor contact is approximately equal to the width of the semiconductor band gap can also be seen in table 2.1.

For successful use in microelectronic devices, the barrier height at a Schottky contact must be very stable. In Chapter 5 we will see that there are some Schottky contacts which are not ideally stable in service, and the particularly important case of the Al–Si contact will be discussed in some detail. Later in this chapter I will mention that metal–semiconductor contacts are often subjected to annealing treatments during the fabrication of integrated circuits (especially during the deposition of insulating films; see Chapter 6). It is thus important to select contact metals which do not react with the semiconductor, since the formation of interfacial phases often leads to a change in barrier height. The Au–GaAs interface is very susceptible to this kind of annealing-induced change in electrical properties and so is not a popular Schottky contact to GaAs devices.

Table 2.2 Details on the preparation of ohmic contacts.

Semiconductor	Contact alloy	Annealing treatment (°C)	Specific contact resistance (Ω cm^{-2})
Si n-type	Al (passive) Au–Sb	300	10^{-5}
Si p-type	Al In Au–In	300	10^{-6}
GaAs n-type	Au–Mn Au–Te Au–Ge–Ni	400–600	10^{-6} to 10^{-7} if correctly annealed
GaAs p-type	Au–Zn Au–Ge–Ni Au–In		
GaAs n or p	epitaxial Ge		10^{-7}
InP n-type	In–Te Au–Sn Cr	Must be annealed	Not usually better than 10^{-4}
InP p-type	Au–Zn In		
II–VI compounds n-type	Au In (ideal ohmic contact)		Not usually better than 10^{-4}
p-type	Au In Pt		

Table 2.2 lists some popular alloy compositions and suggested annealing temperatures for the preparation of ohmic contacts, and gives approximate values of the specific contact resistances which might be expected for these contacts. The contact resistance is especially sensitive to the conditions under which the metal contact alloy is deposited and the details of the annealing process, and can vary by several orders of

magnitude even when the preparation conditions are apparently identical. As the packing density in integrated devices increases it becomes ever more important to minimise the contact resistances since the area of each ohmic contact is steadily shrinking. This is why an enormous amount of attention has been paid to improving the quality of ohmic contacts to silicon and GaAs—the only two semiconductors which are currently used in integrated circuits.

2.3 DEVICE STRUCTURES AND INTEGRATED CIRCUITS

In this section I will describe the basic structure of a number of simple microelectronic devices. We will see that all these devices can be made by the combination of only four components: the two metal–semiconductor contacts described in the previous section, oxide–semiconductor interfaces, and p–n junctions in the semiconductor material itself. The preparation and properties of oxide–semiconductor interfaces are discussed in Chapter 6, while the formation of p–n junctions will be the subject of the sections on diffusion and implantation later in this chapter. The electronic properties of these junctions are described in great detail in the references cited in §2.1 and will not be discussed here. The emphasis in this section will be to illustrate how the devices can be built up by adding each of the required components in a sequence of simple fabrication processes.

It is convenient to divide semiconductor devices into categories which depend on the number of p–n junctions in the device, and whether there is one dominant carrier type involved in the operation of the device (unipolar) or the transport of both electrons and holes is important (bipolar). Table 2.3 shows how many of the most common semiconductor device types can be divided into these broad categories. Solar cells and light-emitting diodes, which are some of the most important single-junction devices, are described in Chapters 8 and 9, and so here we will concentrate on how some devices with two p–n junctions can be prepared: a bipolar transitor, various kinds of field effect transistor (FET) and a simple memory device. Figure 2.6 shows the structure of these devices in a highly schematic form which is intended to indicate how our four basic device components are combined in each case.

Figure 2.6(a) shows the structure of an npn bipolar transistor identifying the collector, emitter and base contacts. The emitter and base contacts are required to act as simple electrical connections and so should have ohmic character. This is normally achieved by heavy doping of the semiconductor material under the contacts. In some designs, the base can be a Schottky contact, and in these cases the base regions do

Table 2.3 Categories of simple semiconductor devices.

Devices with one p–n junction:
        Rectifiers, voltage regulators, tunnel diodes and avalanche diodes
        Light-emitting diodes and lasers
        Solar cells

Devices with two or more p–n junctions:
  Bipolar
        Transistors—microwave, power and logic
        Thyristors

  Unipolar
        Field effect transistors (FETS)
            metal–semiconductor (MESFETS)
            metal–oxide–semiconductor (MOSFETS)
            silicon-on-insulator FETS
        Charge coupled devices (CCD)
        Memory devices

not need to be so heavily doped. Clearly the introduction of the p- and n-type doped regions into the n-type semiconductor wafer and the patterning of the metallic contacts are important stages in fabricating a device of this kind.

Figure 2.6(b) shows the structure of an n-channel FET. Here the source and drain contacts to the two n-type regions in the p-type wafer

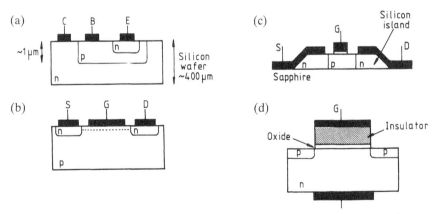

Figure 2.6 Schematic illustrations of the structure of some simple semiconductor devices: (a) an npn bipolar transistor in a silicon wafer; (b) an n-channel field effect transistor; (c) a silicon-on-insulator field effect transistor; (d) a MIOS memory device.

must be ohmic contacts, while the gate contact should have a rectifying character. FETS can be produced in both silicon and gallium arsenide and the choice of the material with which to form the Schottky contact will depend on the relative stability of metal and oxide contacts to the two semiconductors. We have seen that many metals can be used to prepare stable Schottky contacts to GaAs and in Chapter 6 it will be shown that the oxide–GaAs interface has rather poor electrical properties. Metal–semiconductor contacts are thus normally used to form the gates in GaAs FETS and so these devices are called metal semiconductor FETS or MESFETS. By contrast, the oxide–silicon interface can be prepared with a very low interface state density and metal–silicon contacts can be very unstable due to chemical reactions at the interface (see Chapter 5). For this reason, metal–oxide–semiconductor (MOS) gate contacts are used on silicon FETS and these are called MOSFETS. The Schottky barrier heights of metal–InP contacts are rather low in general and so MOSFET devices are more promising in InP than MESFETS. A more detailed description of the choice of materials for forming contacts to silicon and GaAs FETS can be found at the end of Chapter 5.

Figure 2.6(c) shows the structure of a different kind of FET where a thin single-crystalline silicon film has been grown over a sapphire substrate (see §3.5). The source, gate and drain contacts must behave in the same way as for the more conventional FET in figure 2.6(b). Bipolar transistors can also be produced in the silicon layer of course, and one of the most crucial stages in the fabrication of these devices is the patterning of the epitaxial film to give isolated islands of silicon containing the two p–n junctions. Finally, figure 2.6(d) illustrates the structure of one kind of simple memory device, the metal–insulator–oxide–semiconductor (MIOS) memory, where charge can be stored at the oxide–insulator interface. Silicon nitride, Si_3N_4, is a popular choice for the insulator layer, but the gate contact is formed by the same oxide–silicon interface found in silicon MOSFETS. The deposition or growth of SiO_2 and Si_3N_4 layers for MOSFET and memory devices is discussed in Chapter 6.

Before moving on to present the concept of device integration, it is appropriate at this point to describe the sequence of events required to fabricate one of the devices illustrated in figure 2.6. I have chosen to show how the FET structure in figure 2.6(b) can be built up, assuming that the gate electrode is an SiO_2–polysilicon contact (polysilicon is the common abbreviation of polycrystalline silicon; see Chapter 4), while metal contacts must be used for the source and drain electrodes. We will start with a p-type silicon wafer (figure 2.7(a)) and proceed through two separate oxidations, four patterning (or lithography and etching processes), the deposition from the vapour phase of polysilicon, an oxide dielectric and a metal layer, and a doping operation to create the n-type source and drain regions in the wafer (figures 2.7(b–l)). The lithography and etching stages each contain several separate processes of course and these will be described in §2.4. We can see that even to prepare this very simple device, many different processing steps have to be successfully carried out, and the number of these steps rapidly increases with the complexity of the device. Figure 2.7 also illustrates the important point that conventional lithography operations result in severe roughening of the surface of the device. The deposition of each subsequent layer of metal or dielectric must therefore be a 'conformal' process, where the new layer takes on the shape of the underlying surface.

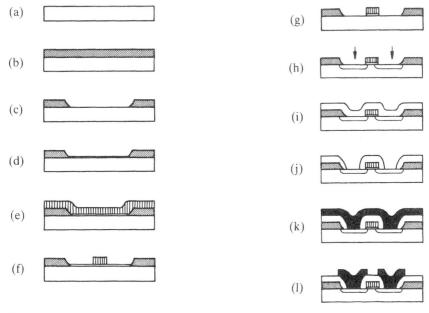

Figure 2.7 A sketch of the fabrication stages needed to prepare a simple field effect transistor structure: (a) the starting material, a p-type silicon wafer; (b) the growth of a thick oxide layer which will be used to isolate the devices; (c) the oxide layer is patterned to define the area of the wafer in which the device is to be fabricated; (d) a thin gate oxide is grown over the exposed area of the silicon wafer; (e) polysilicon is deposited over the whole wafer; (f) the gate contacts are patterned; (g) the gate oxide is removed in areas not protected by the polysilicon; (h) n-type contact regions are introduced by implantation or diffusion; (i) a second dielectric layer is deposited over the whole wafer surface, and is patterned in (j); (k) the contacting metal film is deposited; and patterned in (l).

Bipolar transistor structures generally require more patterning stages than MOS devices and consequently can be more difficult and expensive to fabricate. Figure 2.8 shows a sketch of a relatively complicated device structure, an oxide isolated bipolar transistor. This device is prepared on a p-type silicon wafer over which an n-type epitaxial layer has been grown. Several oxidation and doping operations, each requiring a lithographic and etching process, can then be used to build up the device, and oxide barriers are used to isolate adjacent transistors electrically. (We will see the importance of this isolation in the next section.) These oxide barriers are not produced by growing a thick oxide layer over the whole wafer surface and then patterning the layer, but by constraining the oxide to grow only in chosen places on the wafer by use of an oxidation mask. Silicon nitride films are commonly used as oxidation masks. The number of individual processing steps required in the fabrication of this structure is very large and each individual process must be controlled very precisely if the yield of correctly functioning devices is to be high.

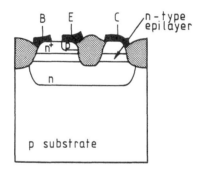

Figure 2.8 A sketch showing the structure of an oxide isolated bipolar transistor device prepared on a p-type substrate on which an n-type epitaxial layer has been grown (after Labuda and Clemens 1980).

2.3.1 Integration of Devices

The concept of device integration is founded on the simple observation that the computing power (or density of data storage) available on the surface of a chip increases greatly as the size of the individual device features is reduced and the devices packed very closely together. The operating speed of a microelectronic circuit is improved when the lengths of the interconnecting paths between devices are shortened, and the power dissipated in devices also decreases with their size. The power consumption of a complete system can thus be reduced by miniaturisation of the component devices. At the time of writing (1987), the current minimum size of device features in the most densely packed integrated circuits is about $1\ \mu m$ and it is possible to pack together about 1 million individual devices onto the surface of a silicon chip of area only $1\ cm^2$. Many of these densely populated chips contain mostly MOS devices like the n-channel MOSFET sketched in figure 2.6(b), since these devices have proved relatively easy to prepare in highly integrated arrays by extended versions of the fabrication process outlined in figure 2.7. Instead of patterning a single FET in the lithography processes, a complex array of interconnected devices is defined. A slightly more complicated MOS device is commonly used in integrated circuits, the complementary MOS or CMOS device, which combines n-channel and p-channel devices. This kind of composite structure has a very low power consumption, which makes it particularly suitable for high-level integration. However, bipolar transistors, memory devices, and even resistor and capacitive components can be included in an integrated circuit, so a very wide range of data storage and processing operations can be combined on the same chip.

The advantages of device integration are almost self-evident, but we should briefly consider a few of the problems which arise when integrated circuits are being fabricated. These problems fall into two catagories: those associated with the practical aspects of fabricating these very small devices; and those which occur because the requirements for the performance of components in the rapidly shrinking devices are so stringent that we are unable to find suitable materials to act as reliable conductors and insulators. Fabrication problems are primarily associated with the need to achieve very high levels of control over processes like the alignment of lithographic masks and the etching of metallic and insulating layers, as well as the maintenance of scrupulous cleanliness at all stages. Miracles of micromanipulation are necessary to align masks to tolerances of less than $0.1\ \mu m$ over a whole 6 inch wafer. It is also very important to avoid the creation of lattice defects near the top surface of the semiconductor wafer. A dislocation or stacking fault can act as a segregation site for heavy metal impurities and degrade the performance of nearby devices (see §2.4.1(a)).

More fundamental problems have arisen with the design of semiconductor devices themselves. When devices are very closely packed together on the top surface of a semiconductor wafer, it is hard to prevent the leakage of current between adjacent devices. This current leakage reduces the efficiency of operation of the whole circuit, but can almost be eliminated by placing reverse biased p–n junctions between each device. Unfortunately, this can result in the generation of large parasitic capacitances in the circuit. A better solution to the leakage problem is to use the deep oxide isolation structure sketched in figure 2.8, even though these devices are significantly harder to fabricate than more 'planar' structures. Semiconductor-on-insulator (SOI) structures are also effective at decreasing the communication between devices, since here each device is placed on a separate semiconductor island (figure 2.6(c)). Several methods for the preparation of soi structures will be described in §2.4. Another problem in the design of highly integrated devices lies in the need to connect each device into the complete circuit with a network of thin film metallic conductor tracks. Some semiconductor chips are still fabricated with only a single layer of metallic conductors, but it is becoming much more common to use two or even three planes of metal tracks, separated of course by thin insulating films, to achieve the desired complexity of interconnection. Figure 2.9 is an optical micrograph of the top surface of a relatively simple silicon chip, and it is clear that two layers of metal tracks have been used to make the interconnections between individual devices. Here the insulating layer is transparent and so both layers of metal tracks can be seen. These three-dimensional arrays of metal and insulator which form the 'wiring' on the chip surface are called the 'metallisation', and the materials used in practical metallisation systems are described in Chapters 5 and 6. We will see that selection of these materials is made difficult by the requirement of extreme stability in service. Chemical reaction and corrosion phenomena are clearly very dangerous when the metal conductors are prepared by the patterning of thin films. The individual tracks are commonly a few tenths of a micrometre thick and perhaps only $2\,\mu m$ wide. Conductors of this size are almost small enough for quantum mechanical effects to begin to decrease the conductivity of the metal (Ruoff and Chan 1982).

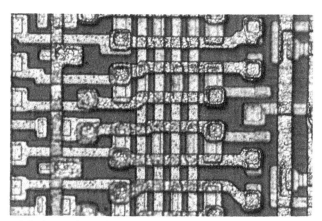

Figure 2.9 An optical micrograph of the top surface of an integrated circuit chip showing the two layers of thin film interconnection tracks. The individual tracks in this circuit are about $5\,\mu m$ wide.

The miniaturisation of devices also increases the amount of waste heat generated per unit area of the chip, even though the heat lost by each device is reduced. The packaging of integrated circuits to dissipate this heat is very important, and in Chapter 7 I will describe some of the package types designed specifically to provide good thermal contact between the chip surface and external heat sinks. It is normally considered that the upper limit on the operating temperature of a chip is about 175 °C. All the chemical reactions which degrade the performance of metallisation systems proceed much more rapidly as the operating temperature is increased, and I will describe several examples of these deleterious reactions in Chapter 5.

2.4 FABRICATION PROCESSES FOR MICROELECTRONIC DEVICES

2.4.1 Wafer Preparation

In Chapter 3 the growth of large single crystals of semiconductor materials will be discussed in detail. In this section we can start our consideration of the preparation of wafers suitable for the fabrication of integrated circuits on the assumption that a large single-crystal boule has already been grown. Figure 2.10 shows an example of a large boule of gallium arsenide grown by the liquid encapsulated Czochralski technique. From boules of this shape we wish to produce as many semiconductor wafers as possible. The basic requirements for these wafers are: (i) that the wafer be very flat to allow device features to be accurately defined by lithographic methods; (ii) the wafer surfaces should not be contaminated with heavy metal or alkali metal impurities; and (iii) the top surface of the wafers should have a low residual density of lattice defects. I shall now outline the stages used to prepare wafers from a large single-crystal boule of silicon. (Very similar processes are used for II–V and II–VI semiconductors, except that these materials are generally softer than silicon so that the introduction of mechanical damage is harder to avoid and different polishing and etching solutions need to be selected for each material.)

Firstly the top and bottom of the boule, sometimes called the 'seed' and 'tang' ends, are sawn off with a diamond-impregnated blade. These offcuts may be returned to the crystal grower for remelting. The outer diameter of the rough boule is then ground to a smooth cylinder and surface flats ground to allow easy identification of the orientation of the single-crystal wafers. These grinding operations introduce considerable lattice damage at the circumference of the boule, which can be removed by chemical dissolution of the outer few micrometres of silicon. X-ray diffraction techniques are then used to orient the shaped boule correctly with respect to a diamond-coated wafering saw; (001) wafers are usually cut precisely on orientation, while (111) wafers are cut close to, but not

Figure 2.10 An as-grown boule of gallium arsenide ready for the wafer preparation process. (Courtesy of Cambridge Instruments Ltd.)

precisely on, the (111) plane. This (111) 'vicinal' orientation is needed to improve the quality of epitaxial layers grown by vapour phase techniques (see §3.7). The diamond-covered saw blade is normally about $300 \, \mu m$ thick and so this thickness of expensive single crystal is lost as sawdust every time a wafer is cut—'kerf loss'. As much as one-half of the total volume of the boule can be lost in this way.

The as-cut wafers are fairly rough and so are first ground flat in a mechanical process using particles of SiC or Al_2O_3 as the grinding medium. The sharp edges of the wafer are then ground to a circular shape, or 'contoured', since this reduces the danger of chipping during handling. (Chips of silicon on the wafer surface can interfere very severely with the lithographic process.) Both of these grinding operations introduce lattice damage into the top surfaces of the wafers and a layer of high dislocation density around $10 \, \mu m$ deep is commonly found. This damaged layer can be removed with a chemical etch; a mixture of hydrofluoric, nitric and acetic acids is a suitable etch for a silicon wafer (see §2.4.3). About $20 \, \mu m$ of the wafer surface would typically be removed in this etching step. Finally, the surfaces are polished in a process which combines mechanical abrasion with extremely fine silica particles ($10 \, \mu m$ in radius) and chemical polishing with an aqueous solution of sodium hydroxide. This polishing stage gives the front surfaces of the wafers a mirror finish suitable for the lithography processes described below.

Further details on this sequence of preparation processes can be found in Pearce (1983) and Horne (1986). The stringent quality-control checks which are used to ensure that the finish of each wafer is of an acceptably high standard are also described in these references.

2.4.1(a) Impurity gettering in wafers

Semiconductor crystals grown by the techniques described in Chapter 3 contain significant concentrations of a number of impurity elements. Impurities can also be introduced into wafers during the sawing, grinding and polishing operations described above, and in many of the fabrication processes which are described below. It is important to try and limit the deleterious effects which some of these elements can have on the performance of individual devices in integrated circuits. The impurities which have been identified as having the most severe effects on the properties of devices in silicon wafers are the fast-diffusing elements described in §1.5, the transition metals (Fe, Ni, Cu, Cr and Co in particular) and the alkali metals like sodium and potassium. The transition metals have very low solubilities in silicon and so segregate strongly to lattice imperfections like dislocations and stacking faults, where they can form precipitates of silicide compounds. Many of these silicide compounds have the calcium fluoride lattice structure, which is very similar to that of silicon. Coherent silicide precipitates can thus nucleate very easily on lattice defects in the silicon crystal, and the intersection of these highly conductive precipitates with a p–n junction will result in unacceptably high leakage currents. An example of a nickel disilicide, $NiSi_2$, precipitate in a silicon wafer is shown in figure 2.11. In addition, carrier recombination at these impurities is very fast and so the minority carrier lifetime will be greatly reduced near impurity-decorated lattice defects. The minority carrier lifetime is also degraded severely by other metallic impurities like tungsten and tantalum which can diffuse into the wafer from silicide contacts in the metallisation system. Dislocations and stacking faults are present in the as-grown semiconductor boules, but can also be introduced during wafer preparation, the oxidation and doping stages of device fabrication, or simply as a result of careless handling. Lawrence and Huff (1982) have considered the effect of the presence of grown-in and process-induced lattice defects on the performance of silicon integrated circuits and concluded that only the precipitate-decorated defects cause serious degradation. Alkali metal impurities have a different effect on device performance. If present near the top surface of the silicon wafer they can be incorporated into a gate oxide, where they have a serious effect on the stability of the oxide–semiconductor interface (see §6.2).

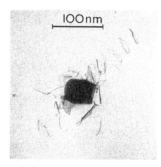

Figure 2.11 A transmission electron micrograph showing a small NiSi$_2$ precipitate in a silicon wafer. The prismatic dislocation loops punched out around the precipitate are also clearly visible. (Courtesy of Mr P Augustus and Plessey Research (Caswell) Ltd.)

It is clearly important to remove all these impurity elements from the region of the wafer where the devices are to be fabricated—the top 1–2 μm of a wafer. This can be achieved in two ways—back-surface and intrinsic 'gettering'. The principle of any gettering process is to collect together all the harmful impurity elements well away from the top surface of a semiconductor wafer. The deliberate introduction of a high concentration of dislocations or stacking faults is a convenient way to do this, as we have already seen that heavy metal impurities segregate strongly to these defects. Back-surface damage can be introduced by grinding or sandblasting the silicon wafer, but an intense laser beam can also be used to give a well controlled defect concentration. The transition and alkali metals diffuse so rapidly in silicon (see table 1.4) that an annealing treatment at 1000 °C for a few minutes is sufficient to allow the impurities to diffuse all the way through a silicon wafer and be trapped in the region of high defect concentration. This is the phenomenon of back-surface gettering.

Intrinsic gettering makes use of the fact that a high oxygen content is always present in silicon crystals grown by the Czochralski process (see §3.3.1). Annealing the shaped and polished silicon wafer in a nitrogen atmosphere at temperatures above 1050 °C encourages the evaporation of oxygen from the wafer surfaces, producing denuded zones at both front and back surfaces. A subsequent heat treatment at between 700 and 900 °C promotes the formation of oxide precipitates in the centre of the wafer where the oxygen content is still high, but not at the wafer surfaces. The oxide precipitates fit very poorly into the silicon lattice and so generate very high local stress levels. Arrays of prismatic dislocation loops are thus formed around these misfitting oxide particles and can act as gettering sites for impurities in the wafer, attracting them away from the active device region on the top surface.

Both of these gettering processes have proved extremely effective at reducing the deleterious effects of metallic impurities on the performance of integrated circuits in silicon wafers. However, annealing treatments have to be used either to create the dislocation arrays or to allow diffusion of the impurities to the dislocations. Pulling the wafers rapidly out of a furnace can set up severe stresses due to differential cooling rates at the edge and centre, and these stresses may result in significant bowing of the wafer. This is highly undesirable because it makes many of the lithographic operations harder to control accurately and must be avoided by lowering the cooling rates as much as possible.

2.4.1(b) Growth of epitaxial layers

The final stage of preparing a silicon wafer for the fabrication of devices may be the growth of an epitaxial layer of silicon on the top surface, although by no means all wafers require this epitaxial layer. An example of a device design for which an n-type epitaxial layer must be grown over a p-type silicon wafer before any of the other fabrication steps can be carried out is shown in figure 2.8. The techniques used for the growth of epitaxial layers on semiconductor substrates are described in some detail in §3.7, so here it is important only to point out the advantages of preparing devices in epitaxial layers instead of directly in the top surface of the wafer itself. Firstly, epitaxial silicon layers with very low dopant concentrations and very high purities can be deposited. This material consequently has a high resistivity and is ideal for the fabrication of densely packed devices in a highly integrated array. Cross-talk and parasitic capacitance effects are minimised if the silicon separating the individual devices has a high resistivity. Secondly, more complex devices structures can be fabricated when the opportunity exists to grow one or more heavily doped epitaxial layers which will form the active regions of devices (figure 2.8). Many of the most complex arrays of devices are thus prepared on silicon wafers which have an epitaxial silicon layer on the top surface.

We have seen in figure 2.6(c) that an alternative way of avoiding unwanted communication between adjacent devices is to grow a semiconductor-on-insulator structure where each device is fabricated in a separate semiconductor island. One way of producing this kind of structure is by the epitaxial growth of silicon layers on an insulating substrate like sapphire (see §3.5). However, it is also possible to isolate the top surface of a silicon wafer by implanting oxygen ions deep under the surface (see §2.4.4) and then annealing the sample to produce an amorphous oxide layer supporting a silicon layer which still preserves a high degree of lattice perfection. An example of the excellent quality of the silicon–oxide–silicon structures which can be prepared in this way is given in figure 2.12. The silicon layer over the oxide can then be patterned into islands just as in the case of silicon-on-sapphire structures.

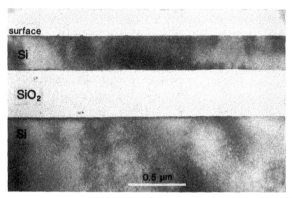

Figure 2.12 A cross-sectional transmission electron micrograph of an oxygen-implanted silicon wafer after annealing to promote epitaxial regrowth of the top silicon layer. The crystalline perfection of the regrown silicon is very high, and a device fabricated in this layer is electrically isolated from the bulk of the wafer by the oxide layer. (Courtesy of Dr C Marsh.)

2.4.2 Principles of Lithography

We have seen that the fabrication of discrete semiconductor devices and integrated circuits requires the patterning of the device features both in the semiconductor material itself (in silicon devices this means primarily the doped regions) and in the insulating and conducting films which make up the metallisation structures. In this section I will introduce the basic principles of lithographic patterning and then describe the relative advantages of the photolithographic, electron beam, ion beam and x-ray lithography processes. The preparation of suitable masks is an important stage in all lithographic processes and will be discussed at the end of the section.

As an introduction to the photolithographic process let us see how a pattern can be transferred from a mask into a layer of polymer 'resist' on a wafer surface. (The word 'resist' illustrates the role that these polymer layers are required to play—they must protect underlying surfaces during fabrication stages like etching by 'resisting' the etch process.) Figure 2.13 shows (a) resist deposition, (b) exposure of the resist to light through a mask, and (c) development of the resist to create the required pattern. I have shown this process for two different kinds of resist: positive resist, which on exposure to light becomes more

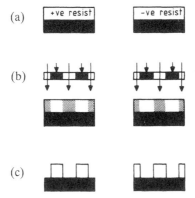

Figure 2.13 A sequence of sketches showing the patterning of positive and negative resists (a) in a two-stage process: (b) exposure and (c) development.

soluble in the developing solution; and negative resists, which become less soluble on exposure. In order to see how photolithographic patterning of resists can be used in practical device fabrication processes, figure 2.14 illustrates the separate stages required to open up a contact hole through an oxide layer on a silicon wafer surface, a 'via', and then to pattern a metal contact over this via.

The first operation is the selective etching of the via hole down to the silicon wafer, leaving the oxide undisturbed in other regions, and can be achieved by placing an etch-resistant polymer layer over the oxide surface (figure 2.14(b)). In this case let us assume that a positive resist is used. The resist is then exposed to light through an appropriate mask (figure 2.14(c)) and the resist is developed to reveal selected areas of the oxide surface. These exposed areas of oxide can then be etched away (e) and removal of the remaining resist (f) reveals the via hole in the oxide layer. A metal contact can now be patterned in a similar way. Figure 2.14(g) shows the deposition of a metal film over the whole surface, followed by the deposition of a layer of negative photoresist (h). The photoresist is then exposed (i) and developed, and the aluminium revealed under the resist can then be etched away (k). The final stage is the removal of the remaining photoresist (l). The same result could be achieved by depositing, exposing and developing a positive resist layer using the same mask as shown in (i) *before* deposition of the metal contact film. The unwanted metal surrounding the contact area is then removed along with the remaining resist — the 'lift-off' process. The use of lift-off techniques reduces the total number of processes required to pattern a metal contact and so is finding wide application in the fabrication of integrated circuits. For those readers who wish to follow this photolithographic process in more detail, Elliott

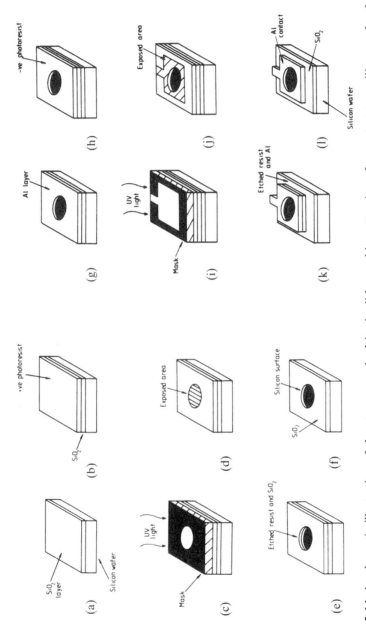

Figure 2.14 A schematic illustration of the stages required in the lithographic patterning of a contact to a silicon wafer through a via defined in an oxide layer. The separate stages in this sequence are described in the text.

(1982) has given an excellent description of all the stages in the production of integrated MOS devices.

In this outline of a photolithographic process I have used the terms expose and develop to describe the processing of resist layers. This illustrates the fact that photolithography has its roots in simple photographic techniques and the possibility of reducing the projected size of mask features by the use of an optical demagnification system has proved particularly important in the definition of small device features. It is also clear from the above that photoresist polymers must be highly sensitive to light, easily developed, readily removed once the patterning of the underlying material is complete, and yet must not be degraded by etching or implantation processes. The properties of resist polymers are described in Chapter 6, as is the process by which highly conformal polymer layers are 'spun' onto wafer surfaces. The etching of semiconductors, metals and insulators will be considered in the following section.

The three fundamental requirements of a lithography process are: throughput (the number of wafers processed in a given time); the layer-to-layer registration (or the accuracy with which the device features patterned in different lithography steps can be aligned with one another); and finally the resolution of individual device features. So far I have concentrated on describing a lithography process where the exposing radiation is in the visible or near-ultraviolet wavelength range, 200–500 nm. Photolithography has been the basic tool of the semiconductor industry for many years and techniques have been developed to allow high throughput and adequate registration with device feature spacing of about 1.5 μm. The use of positive resists can give rather better resolution than negative resists since they do not expand during development. However, diffraction of the exposing light around the edge of the mask puts a fundamental limit on the resolution of photolithography at about 1 μm. Current integrated circuit technology is seeking to produce devices where the feature size and separation is less than 1 μm, forcing the development of techiques which use exposing radiations with shorter wavelengths (λ) to overcome the diffraction limit on resolution. Electron beams ($\lambda \sim 10^{-12}$ m), x-rays ($\lambda \sim 10^{-9}$ m) and ion beams with very short wavelengths are all possible candidates, and so it has been necessary to discover new resist polymers which are sensitive to these radiations (see Chapter 6).

Table 2.4 compares the efficiency of these various illumination types for lithography applications. We can see that the resolutions available in electron beam, ion beam and x-ray lithography techniques are all substantially better than 1 μm, but there are particular problems associated with the use of each of these exposing radiations. The factors which determine the resolution limits in these lithography techniques have recently been considered in some detail by Howard and Prober (1982) and McGillis (1983) and will be briefly covered here.

Table 2.4 A comparison of lithography processes (after McGillis 1983 and Stengl et al 1986).

	Photo-lithography	Electron beam lithography	Ion beam lithography	X-ray lithography
Resolution limit	Diffraction limited to about 1 μm	0.15 μm for positive resist Limited by scattering	Better than 0.2 μm	About 0.5 μm Limited by blurring because of large source
Throughput	About 40 wafers/h	5 wafers/h at most	Low exposure time, so can be very high	Similar to photolitho-graphy
Comments	Full mask set required which is expensive Dominant technique for integrated circuits	No masks required, so very flexible circuit design Excellent registration Used for mask fabrication	Stencil masks hard to make Technique still under development	Masks hard to make Technique still under development

Large-area sources of electrons are currently not available and so a finely focused electron beam must be 'written' across the wafer surface to define each device feature individually. This has the advantage that no masks are required in electron beam lithography, but the processing of a single wafer takes a very long time so that throughput is low. However, registration is excellent because the electron beam can be used to recognise previously patterned features on the wafer surface while defining a pattern in the overlying resist. The major limitation on the resolution available in electron beam lithography arises from the scattering of the electrons in the resist and back from the underlying material, but if the resist layer is thin, feature separations as small as 0.2 μm can be achieved.

Ion beam lithography can be carried out in two ways: (i) with a focused ion beam written over the wafer surface as in electron beam lithography; and (ii) with a broad beam source which can project a demagnified ion image of a 'stencil' mask onto a wafer (Stengl et al 1986). Ion beam damage can be very effective in removing resist layers by sputtering, which removes the need for an exposing process. Some inorganic materials like SiO_2 and Si_3N_4 can also be directly patterned with a finely focused ion beam, with the exposed regions showing chemical dissolution rates up to three times faster than unexposed regions. In some cases, therefore, ion beam lithography can be carried out without any resist layers. This combination of properties makes ion beam lithography a very powerful technique and resolutions well under $1\,\mu m$ have been demonstrated with high throughputs when the broad beam source is used. The demagnification of the ion image after the mask by an ion focusing system means that the mask itself can be as much as 10 times larger than the real device features. Even so, mask making is particularly difficult since the device features usually have to be patterned in completely open holes in a metallic film. (It is hard to prepare ion-transparent materials which will not decompose rapidly under the ion beam.)

X-ray lithography appears an attractive technique because of the availability of intense broad beam sources operating in the wavelength range where diffraction effects will be negligible. This should allow a high throughput and excellent resolution. However, it is not possible to project a demagnified image of an x-ray mask because reliable x-ray optical systems have not yet been developed. This means that the device features must be defined in the mask at the same size as they are required on the wafer surface. We will see that the preparation of masks for x-ray lithography is as hard as for ion beam techniques.

These advanced imaging techniques have allowed the fabrication of devices with features as small as $0.1\,\mu m$ and single metallic conductor tracks only 10 nm wide can also be prepared. Much of the effort in the area of microfabrication has been aimed at understanding the properties of materials when reduced to dimensions where quantum mechanical effects become important, although the preparation of some novel device structures also requires reliable lithography processes with exceptionally high resolutions. At present, photolithography is still very widely used for the fabrication of semiconductor devices and integrated circuits because of the combination of rapid throughput and reasonable resolution. Electron beam lithography is used when very fine device features are required and for mask making, but is rather slow. X-ray and ion beam lithography techniques are still being developed, but may become very important in the future. In all these processes, absolute cleanliness is essential, as any dust particle on a mask or wafer will disturb the pattern generated in the resist on the wafer surface. Contamination or damage of a photomask is especially dangerous, as the defective pattern can be reproduced onto very many wafers before the source of the error is detected. Here x-ray sources could be especially valuable, since most dust particles are transparent to x-rays. Further details on lithography processes can be found in Elliott (1982) and McGillis (1983), and the equipment currently available for all kinds of lithographic operations is described by Horne (1986).

2.4.2(a) Mask fabrication

We have seen that the patterns that are to be transferred to a layer of resist on a wafer surface by photolithography or x-ray lithography processes must first be produced on a mask. In a photolithographic or ion beam process the mask features can be much larger than those required on the chip surface because of the possibility of projecting a demagnified image of the mask onto the resist. However, x-ray masks must be made with features exactly the same size as are needed on the wafer. (Electron and focused ion beam lithography systems do not need masks, as an exposing beam can be written across the resist surface to define the chosen feature pattern.) Lithography masks have to be used a large number of times and so must be very strong to allow for repeated handling.

The masks for optical lithography are made on very flat quartz, soda-lime or borosilicate glass plates, although quartz sheets have to be used if the illumination is in the far UV to reduce absorption. Opaque coating layers are deposited by sputtering or evaporation techniques (see Chapter 4) and these layers are commonly made from hard materials like chromium, silicon or iron oxide. The desired device patterns are then defined in a resist layer over this coating layer, usually with an electron beam system. The slowness of electron beam lithography is no disadvantage when making masks, and the high resolution inherent in this technique is important if high-quality masks are to be prepared. With a thin resist layer to limit beam-spreading effects, resolutions of at least $0.5 \mu m$ can be achieved. Horne (1986) has given a detailed description of the design and operation of an electron beam lithography system for mask fabrication. The coating layers under the developed resist are then patterned, usually with a carefully controlled wet chemical etching treatment (see table 2.5). The mask now has the desired device features defined in the opaque coating layer. For applications where resolution is less important, photolithography masks can be prepared by exposing emulsion-coated plates to a demagnified image of design pattern called the artwork. Developing the emulsion produces opaque regions where the mask was exposed and which act in the same way as the metallic coatings in high-resolution masks.

Masks for x-ray lithography are much harder to prepare, since the substrate has to be transparent to x-rays. Most glasses, for instance, absorb x-rays very efficiently. In addition, the individual mask features have to be of the order of $1-10 \mu m$ in size to satisfy the current design rules for integrated devices. One successful structure is a thin substrate of boron nitride and polyimide, with device features patterned in a gold masking layer with the ubiquitous electron beam lithography (Maydan 1980). Thin titanium sheet can also be used as a semi-transparent support for a gold mask pattern. The supporting layers for ion beam masks have to be transparent to high-energy ions, or the mask features must be defined in holes etched out of a thin metal sheet. The technology of mask fabrication for these last two kinds of lithography is still being developed and currently limits the application of x-ray and ion beam processes in the production of semiconductor devices and circuits.

Table 2.5 Selective chemical etches used in the fabrication of semiconductor devices.

System	Material removed	Etching solution
SiO_2/Si	SiO_2	Buffered HF
Polysilicon gate	Polysilicon	HF/HNO_3
Metallisation	Al alloys	$HPO_3/HNO_3/CH_3COOH$
Cr on glass mask	Cr	$Ce(NH_4)_2(NO_3)_6/HClO_4$
Iron oxide on mask	Iron oxide	$HCl/FeCl_2$
		H_2PO_4
$GaAs/Ga_{1-x}Al_xAs$	GaAs	H_2O_2/NH_4OH
		KI/I_2 (also etches GaP)
GaInAsP on InP	InP	HCl/H_3PO_4
		$HCl/HClO_4/C_3H_8O_3$
GaInAsP on InP	GaInAsP	$KOH/K_3Fe(CN)_6$
		H_2SO_4/H_2O_2

2.4.3 Etching Processes for Microfabrication

In the previous section the etching of materials exposed under lithographically defined windows was shown to be a fundamental process in the fabrication of semiconductor devices. Until recently, most device fabrication was carried out using wet chemical etching, but new gas phase etching processes have been developed and have become known by the generic term of 'dry etching'. In this section I shall describe the basic requirements of an etching process suitable for device fabrication and then show why dry etching is becoming more widely used than wet chemical dissolution for the preparation of devices which contain very fine features.

The fundamental etching process is the transferral of a pattern lithographically defined in a resist layer to the underlying material. There are two etching parameters which control how accurately the pattern in the resist will be reproduced: the anisotropy of the etch and its selectivity. Consider a patterned resist over a silica layer on a silicon substrate, as in figure 2.15(a). If the etching process is directionally isotropic, then the edge of the resist mask will be undercut, as is shown in figure 2.15(b). Clearly the exact form of the pattern in the resist will not be reproduced in the SiO_2 layer. However, if the etching process is anisotropic such that etching only occurs in a direction normal to the wafer surface, then a perfect replica of the resist pattern will be produced in the oxide layer (figure 2.15(c)). These straight-sided etched via holes are not ideal for the subsequent deposition of a metal contact layer, since the evaporated or sputtered metal films will not be continuous over the sharp edges of the via. A sloping via wall like that produced in an isotropic etching process will have a much more suitable surface morphology over which to deposit a metal conductor film. For some other applications the anisotropic etch morphology is preferred, so the character of the etch must be chosen to give the desired shape to the etched layer.

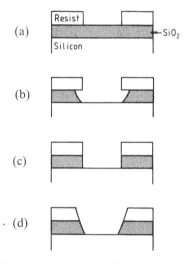

Figure 2.15 A schematic illustration of etching profiles under a resist mask (a). (b) Isotropic etching showing undercutting; (c) anisotropic etching with straight sidewalls; (d) the edge profile produced when the resist is also etched at a finite rate.

Etch selectivity is defined as the ratio of the etch rates of the material which we wish to remove, and the resist and underlying materials which should remain unattacked as much as possible. Any etching of the resist will degrade the fidelity of the pattern transfer and produce a morphology rather similar to isotropic etching (figure 2.15(d)). Unselective etching of other materials in the device structure can result in considerable damage to semiconductor surfaces and lower levels of a metallisation structure. In most practical circumstances, the selectivity of an etch is required to be as high as possible, and table 2.5 gives a few examples of successful selective etches used in the fabrication of several kinds of semiconductor devices. Hydrofluoric acid solutions are particularly effective at removing SiO_2 layers from silicon surfaces while hardly attacking the silicon at all and so are used to pattern vias to silicon devices. A mixture of hydrofluoric and nitric acids can be used to etch single-crystal and polycrystalline silicon. A variety of etching solutions have also been developed for the selective removal of III–V compounds and the patterning of GaInAsP films on InP substrates is an especially important step in the fabrication of some laser diodes (see Chapter 8). All of these chemical etches work in roughly the same way, with one component forming an oxide layer on the semiconductor surface which a second component then dissolves. On a silicon surface for instance, the nitric acid forms an oxide which is dissolved by the hydrofluoric acid. Table 2.5 also contains etches which can be used for the selective removal of the materials which form the opaque features in photomasks—Cr and iron oxide—and for the etching of the aluminium alloys used in metallisation structures.

It has also been found that altering the precise composition of these etches can change the etching characteristics from isotropic to anisotropic. In the latter case, etches have been found which delineate particular planes of a semiconductor crystal, so that very sharply defined crystallographic features can be prepared. The sloping edge profile of the silicon-on-sapphire mesa sketched in figure 2.6(c) is an example of this kind of crystallographic etching and is needed to allow continuous metal conductor tracks to be deposited running from the device surface over the edge of the mesa and across the sapphire substrate.

However, the resolution achieved with wet etching techniques is usually not sufficiently good for the fabrication of devices with the smallest design rules. In addition, no reliable chemical etch for silicon nitride layers has been found and this material is used for several important applications in integrated circuit manufacture. For these reasons, wet etching is not used much in the production of the latest generation of integrated circuits, and dry etching techniques have been developed to pattern semiconductor, metal and insulator layers in these structures. Wet etching techniques continue to be used in the production of devices where resolution is less important.

2.4.3(a) *Dry etching processes*

The simplest dry etching process is sputtering, or the bombardment of a surface by high-energy ions or atoms. Energy is transferred from the incident particles to the surface atoms and when this energy exceeds the binding energy, then atoms will be displaced out into the vapour phase. This etching process has a rather poor selectivity and may also create significant amounts of lattice damage in a semiconductor material, including the formation of dislocations. A particularly severe problem arises with the sputter etching of InP with inert gas ions. The InP is dissociated by the incident ion beam and the phosphorus escapes into the vapour phase but the indium remains on the surface in the form of small islands. These islands act as miniature sputter masks and, as sputtering is continued, lead to the formation of a characteristic rough surface morphology like that shown in figure 2.16. This is by no means an extreme case of this highly undesirable effect and other compound semiconductors are also prone to form rough surfaces when sputter etched. In §10.3.3, we will see that this roughening behaviour has important consequences for the depth resolutions which can be achieved in some common microanalytical methods. Sputtering is thus not a useful technique for the etching of microelectronic materials, but the very high anisotropy which arises from the directional nature of the sputtering process proves important in the development of other dry etching techniques.

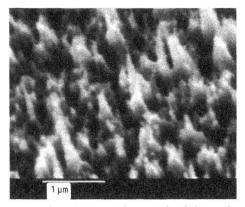

Figure 2.16 A scanning electron micrograph of the surface of an InP wafer after sputter etching with 5 keV argon ions showing the characteristic rough surface structure. (Courtesy of Dr A Webb and Plessey Research (Caswell) Ltd.)

A second method of dry etching relies on the formation of highly reactive chemical species in the gas phase above the sample, which can then react with the sample surface to form volatile reaction products. Striking a plasma discharge in a gas causes some ionisation and dissociation of the gas molecules when they are hit by high-energy electrons. For example, a plasma struck in carbon tetrafluoride, CF_4, results in the production of CF_3^+ and F^+ ions and free fluorine atoms, while a plasma in molecular oxygen forms O_2^- and O^- ions and free oxygen atoms. These species are extremely reactive and even at room temperature can form a complex mixture of volatile products on reaction with exposed surfaces of a semiconductor wafer. The species produced in CF_4 plasmas can be used to etch silicon surfaces, while oxygen radicals react with polymers to form volatile products like CO, CO_2 and H_2O. A mixture of CF_4 and oxygen can be used to etch silicon nitride films very effectively. This kind of etching reaction is called plasma etching, or PE.

The selectivity of a plasma etching process depends strongly on the composition of the gas phase. Carbon tetrafluoride etches both silicon and SiO_2 but adding a small partial pressure of oxygen to the CF_4 markedly increases the etching rate of silicon relative to that of SiO_2 (Mogab *et al* 1978). By contrast, the addition of hydrogen enormously increases the etch rate of the oxide relative to that of the silicon (Ephrath 1979). It is thus possible to tailor the gas composition to achieve the selectivity required in any particular etching process.

Plasma etching is rather isotropic because it depends on simple chemical reaction processes, and so we expect that a certain amount of undercutting of the resist pattern will always occur (figure 2.15(b)). This is convenient when etching vias through oxide layers on silicon wafers, but may not be desirable in other etching processes. An improved etch anisotropy can be achieved by combining plasma and sputter etching by forming reactive free radicals in the vapour phase as before, but by then bombarding the sample surface with the ionised species in the plasma. The directional nature of the sputtering process will then control the degree of anisotropy, but both the simple chemical etching and sputtering will contribute to the overall rate of material removal. In fact, etching rates seem to be enhanced when both processes are operating (Coburn and Winters 1979). The energetic ions may be assisting the free radicals to overcome the activation energy barrier of the surface reactions. Detailed models of this symbiotic process are difficult to formulate because the surface reaction paths are complex and poorly understood. The combined plasma etching and sputtering process is called reactive ion etching, or RIE, and is really a portmanteau term given to all the processing conditions which lie between simple sputter etching and plasma etching. The parameters which control the rate and nature of the etching process are the gas phase composition, plasma power and excitation frequency, and the potential on and the temperature of the sample. The values of these parameters have to be carefully selected to allow successful etching of each particular combination of materials.

Table 2.6 lists a few of the gases which are commonly used to etch semiconductors and the materials in metallisation systems. The selectivities in dry etching processes are sometimes rather low, and the etching of silicon nitride over silicon is especially inadvisable because silicon is rapidly removed by the same radicals which etch the nitride. Problems can also arise when there is a component of the layer to be etched which does not form a volatile reaction product, for example copper in aluminium alloy conductor tracks. The copper is left on the surface after dry etching and must usually be removed selectively with a wet chemical etch. III–V compound semiconductors can be difficult to etch with fluorine-containing gases for a similar reason—the group III elements do not form volatile fluoride compounds. Chlorine-containing etching gases are found to be more effective for III–V compounds.

Table 2.6 Some gases used in the dry etching of microelectronic materials.

Materials to be etched	Gas mixtures
Polysilicon or silicides	CF_4
	CCl_4
	$CF_4 + O_2$
	Cl_2
Silica	$CF_4 + H_2$
	$CHF_3 + O_2$
Silicon nitride	$CF_4 + O_2$
Aluminium alloys	CCl_4
	$BCl_3 + Cl_2$
	$SiCl_4$
Titanium–tungsten alloys	as for aluminium
Gallium arsenide	$Cl_2 + H_2$
	CCl_2F_2
Indium phosphide	$CH_4 + H_2$
	$CH_3I + O_2$

Three problems with dry etching processes are the formation of polymeric films on the sample, radiation damage from the plasma and residual contamination effects. The sidewalls of deeply etched features are preferential sites for the formation of polymer films by reaction of the molecular fragments present in the plasma, CCl_2 radicals for example. This can lead to an enhanced anisotropy of etching, but can also cause problems with the resistances of metal–semiconductor contacts. Gate oxides are particularly sensitive to radiation damage, with the formation of charged defects in the SiO_2 network affecting the electronic properties of the gate contact. Direct ion impact can also displace atoms from their lattice sites in the top 10 nm of a semiconductor wafer, producing electrically active point defects and even dislocations. An annealing treatment at 1000 °C may be required to remove this lattice damage, which is of course in the region of the semiconductor where the devices are to be fabricated. Contaminant elements can be left on the sample surface after dry etching, including halogens from the etching gases and heavy metals sputtered from the walls of the etching chamber. We have seen that many metallic impurities degrade device performance by forming leakage paths across p–n junctions and lowering minority carrier lifetimes, while halogen elements can create problems with the adhesion of metallisation layers and can also lead to the rapid corrosion of metallic conductors (see §5.5).

Plasma etching and RIE processes are currently used for numerous etching stages in the fabrication of integrated circuits. However, vital parameters like the selectivity, anisotropy and surface cleanliness can all be improved, and we are far from having a complete understanding of the complex chemical reactions which occur on the sample surface. Some more details on these technologically important processes can be found in Downey *et al* (1981) and Mogab (1983). Figure 2.17 shows examples of the kind of etch profiles which can be achieved with dry etching techniques in quite different materials: (a) a via through an oxide layer on a silicon wafer surface etched with $CHF_3 + O_2$; (b) a mesa etched in GaAs with an $Ar + Cl_2$ mixture; and (c) a trilayer metal conductor track, Ti–W/Al–Cu/Ti–W, etched with $BCl_3 + Cl_2$ in a single etching process.

2.4.4 Doping Processes

Figure 2.7(h) shows the stage in the preparation of a semiconductor device where p–n junctions are produced by the introduction of carefully controlled concentrations of a doping element into chosen regions of a semiconductor wafer. In this particular case, it is the formation of n-type regions at the surface of a p-type wafer. The fabrication of all the simple device structures sketched in figure 2.6 requires at least one such process and there are two techniques which have been developed for the doping of semiconductor wafers—diffusion and implantation. Diffusion doping was developed much earlier than implantation and so will be described first.

(a)

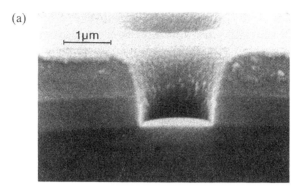

(b)

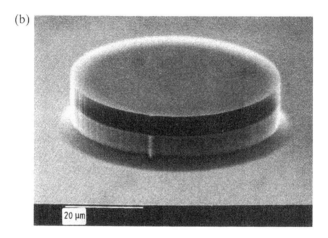

(c)

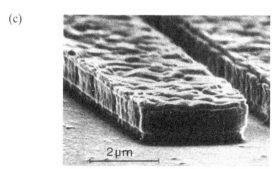

Figure 2.17 Examples of the kind of feature definition which can be achieved by dry etching: (a) a via etched through an oxide layer down to a silicon wafer surface (courtesy of Plasma Technology Ltd); (b) a mesa structure etched in a GaAs wafer (courtesy of Dr A Webb and Plessey Research (Caswell) Ltd); (c) a narrow metal track etched from a continuous trilayer metal film (courtesy of Drs P May and A Spiers and GEC Research Ltd).

2.4.4(a) Diffusion doping

The phenomenon of diffusion in semiconductors was introduced in §1.5, where it was shown that the mechanisms of diffusion are complex and in compound semiconductors often not fully understood. A particularly important feature of many diffusion processes in semiconductors is the dependence of the dopant diffusion coefficient, D, on the dopant concentration. The diffusion coefficients of arsenic and phosphorus in silicon, and of zinc in gallium arsenide, are very sensitive to the local dopant concentration, especially when this quantity exceeds 10^{18} cm^{-3}. In order to allow the prediction of the dopant diffusion rate in practical doping processes it is assumed that the diffusion coefficient for a dopant has a fixed value at any given temperature. We should not forget, however, that this is a major simplification of a complicated phenomenon. Tables 1.4 and 1.5 contain data on the diffusion parameters of several dopant elements in silicon and gallium arsenide and illustrate the point that most dopant diffusion processes are extremely slow. Extended high-temperature-annealing treatments are thus needed to carry out diffusion doping.

The first stage in a diffusion doping process for the preparation of integrated devices on silicon or GaAs wafers is the lithographic patterning of a mask on the semiconductor surface which exposes the regions to be doped. The dopant species can then be brought into contact with the wafer either by surrounding the whole sample in a gas containing the dopant element, or by coating the wafer with a solid or liquid dopant source. Volatile compounds like B_2H_6, BCl_3 or AsH_3 are used in gas phase doping and fine particles of P_2O_5 can be transported to the wafer surface by an inert carrier gas. Metallic dopants like zinc, which have high vapour pressures, are also introduced from the gas phase. Solutions containing dopant elements can be spun onto the wafer just like a polymer precursor, and thin dopant-rich inorganic films are deposited by CVD techniques (see §3.7) or mechanically 'painted' onto the wafer. Chemical reactions between these dopant sources and the semiconductor surface often create thin heavily doped oxide layers which act as the immediate source of dopant for indiffusion, but which have to be removed before the next stage in the fabrication process. This reacted layer on the semiconductor acts to set up 'constant source conditions', i.e. the concentration of dopant on the semiconductor surface, C_s, remains invariant during the diffusion process.

We should remember that many II–VI and IV–VI semiconductor compounds have electronic properties which are dominated by the concentrations of native point defects (see §1.4). p–n junctions in these materials can easily be prepared by annealing samples in an ambient which introduces a high equilibrium concentration of compensating native point defects. This is in principle a diffusion doping process as well, but here the diffusion rates, and so the depth of the p–n junction, are controlled by the self-diffusion coefficients.

Under constant source conditions it is simple to predict the dopant concentration at any depth, x, under the wafer surface from an appropriate solution of Fick's second law:

$$C(x, t) = C_s \, \mathrm{erfc}\left(\frac{x}{2\sqrt{Dt}}\right) \qquad (2.5)$$

where t is the diffusion time. The value of C_s will of course depend on the particular way chosen to bring the dopant element into contact with the semiconductor surface and D is usually assumed to have a fixed value at the diffusion temperature. The depth of the p–n junction created by diffusion of a compensating dopant into a wafer can now be calculated by finding the value of x where $C(x, t) = C_i$, the concentration of dopant in the bulk of the wafer. At this depth the indiffused dopant exactly compensates for the bulk dopant and this is called the 'metallurgical junction'. Figure 2.18 shows how the free carrier concentration varies with depth into a p-type wafer after a diffusion doping process. It is clear from this figure that diffusion doping does not give a very uniform profile and that the concentration of the dopant near the semiconductor surface can approach the solubility limit at the temperature of the diffusion treatment. This can lead to the formation of dopant-rich precipitate particles when the wafer is cooled. To limit the total amount of dopant which diffuses into the wafer, some devices are fabricated with a two-stage diffusion process—a short anneal to introduce some dopant into the semiconductor surface, followed by removal of the dopant source and a 'drive-in' anneal to form the p–n junction. (A better mathematical model for this kind of diffusion process is the 'constant total dopant' condition described in the following section.)

There are two serious problems which can arise during dopant diffusion through a mask: (i) penetration of the dopant under the edge of the mask by lateral diffusion; and (ii) diffusion through the mask itself. Both of these effects can lead to the distortion of the device structures. Lateral diffusion will always occur because diffusion is a roughly isotropic process in semiconductor materials. Clearly a metallurgical p–n junction will be formed at approximately the same distance laterally under the mask as it is deep in the semiconductor wafer (figure 2.19). This effect limits the packing density of devices which can be achieved, and in particular diffusion doping cannot be used to fabricate devices with very narrow gates. If the diffusion mask is made of silicon nitride, then the semiconductor material under the mask can be under considerable strain (see Chapter 6) and this strain can result in a considerable enhancement in dopant diffusion rates and even worse lateral penetration (Lin *et al* 1979).

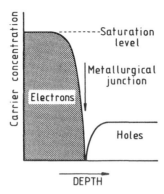

Figure 2.18 A schematic illustration of the carrier concentration profile after a diffusion doping process, showing the carrier saturation near the wafer surface and the formation of a metallurgical junction where the electron concentration compensates for the holes already present.

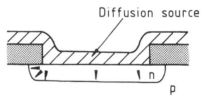

Figure 2.19 A schematic illustration of the phenomenon of lateral diffusion under a mask, which increases the size of the doped region.

The problem of diffusion through oxide masks arises primarily because group III and V dopant elements lower the melting point of silica. The diffusion coefficients of these dopants in the oxide vary with their concentration and also depend on the exact conditions under which the oxide layer was grown, i.e. wet or dry oxidation, or vapour phase deposition. The diffusion of boron and phosphorus in SiO_2 is very slow if they are present in concentrations of less than 1%, with average penetration depths of only a few nanometres after an anneal at 900 °C for 1 h. An oxide masking layer 1 μm thick should easily prevent any penetration of these dopants into the semiconductor wafer. However, gallium and aluminium diffuse rapidly in SiO_2 and are therefore not used in the diffusion doping of silicon. Further information on the practical details of dopant diffusion processes for the production of semiconductor devices can be found in Tuck (1988), Tsai (1983) and Till and Luxon (1982).

A final point which should be made when describing dopant diffusion is that lattice damage can be introduced when a misfitting dopant element is present in high concentrations. We have already seen that the formation of dislocations near p–n junctions is most undesirable and can be avoided during dopant diffusion by choosing dopant elements which do not distort the semiconductor lattice too severely, or by keeping the dopant concentrations low. Phosphorus atoms fit particularly poorly into the silicon lattice and as a result can lead to extensive lattice damage during diffusion doping, as well as acting as preferential segregation sites for metallic impurities. The interaction between phosphorus and copper atoms is especially strong. This effect may make phosphorus unattractive as a diffused dopant for the fabrication of device structures where high doping concentrations are needed, but heavy phosphorus doping can be used far away from the active device regions to getter metallic impurities.

2.4.4(b) Ion implantation

Doping by implantation is a technique in which the surface of a sample is bombarded with a beam of energetic dopant ions, which results in the penetration of these ions deep below the surface of exposed semiconductor materials. Accelerating energies of between 3 and 500 keV are common, and a wide range of dopants can be readily implanted, including boron, phosphorus and arsenic into silicon, and silicon, sulphur, beryllium and magnesium into gallium arsenide. Just as in diffusion doping, oxide or silicon nitride masks can be patterned to define the regions of a semiconductor wafer which are to be doped, but resist polymers can also act as efficient implantation masks as long as the ion dose is reasonably low. The major advantages of ion implantation doping are the very precise control over the dopant concentration and dopant depth distribution which can be achieved; the lack of lateral spreading under the mask edges which allows a greater packing density of devices; and the fact that it is a 'dry' process in which contamination of the semiconductor surface can be minimised (Seidel 1983).

The range of the ions is a very important parameter in an implantation process and is controlled by the way that an incident ion loses energy as it passes through the semiconductor lattice. Collisions with both electrons and atomic nuclei occur, as is briefly described in §10.3.2(c), and a general theory of ion stopping has been developed— the Lindhard, Scharff and Schioff, or LSS, theory. A comprehensive review of the development of this theory has recently been given by Ziegler (1984), where it is also shown that it is possible to predict the range of implanted ions very accurately. It is generally found that ion implantation results in the formation of a roughly Gaussian distribution of ions and that the peak concentration lies at a depth under the sample surface which increases approximately linearly with ion energy. Light ions penetrate much further than heavy ones because they interact less strongly with the sample. The ranges of 100 keV B^+, P^+ and As^+ ions in silicon are 0.3, 0.12 and 0.06 μm respectively. Many current device structures require the fabrication of very shallow p–n junctions, and shallow boron implantations are usually made with heavy BF_2^+ ions which do not penetrate as far as B^+ions. Implantation into single-crystal material can lead to distortion of the Gaussian depth distribution by 'channelling' of the ions along atom rows in the crystal lattice. This effect adds a long tail to the distribution profile of the dopant, which is often undesirable, and can be avoided by orienting the sample so that the incident beam is aligned 6 or 7 ° away from either the $\langle 001 \rangle$ or $\langle 111 \rangle$ directions in the semiconductor crystal. A thin layer of amorphous oxide or nitride on the semiconductor surface may also help to 'randomise' the direction of the incident ions.

Thus far we have not considered the effect that the energetic incident ions have on the semiconductor material. Much of the energy of these ions is lost in direct nuclear collisions which lead to the displacement of the atoms of the semiconductor off their lattice sites. It only requires 15 eV to displace a silicon atom, and so an incident ion accelerated through 100 keV can displace many atoms and create a damage track or 'cascade'. Heavy ions in general cause more displacement damage than light ones, and this damage is confined closer to the semiconductor surface because these ions also have short ranges. As-implanted material often contains a very high concentration of vacancies and interstitials, some of which will condense to form dislocation loops. In addition, most of the implanted dopant atoms will not lie on lattice sites and so will be electrically inactive. The electrical properties of this damaged layer of semiconductor material are usually very poor, with high resistivities and a large density of states in the semiconductor band gap. When the ion flux is sufficiently high, an amorphous layer can even be formed, since most of the atoms will be displaced off their lattice sites. This amorphous layer can lie either at the surface of the semiconductor, or be buried at the site of maximum dopant concentration. Figure 2.20(a) shows an electron micrograph of an implanted silicon sample where the top 120 nm of the semiconductor is amorphised, and under this layer a band of dislocations can be seen as well. Implantation with heavy ions like As^+ can easily create amorphous layers in silicon, but very high fluxes of boron ions are required before any amorphisation is seen. The temperature of the sample during implantation is also an important parameter in determining the extent of amorphisation. At high temperatures, the semiconductor material 'self-heals' and does not become amorphous.

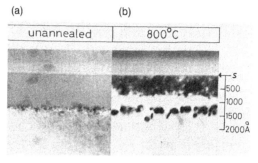

Figure 2.20 Cross-sectional transmission electron micrographs of (a) the damage created by implantation of arsenic ions into a silicon wafer. (b) The result of annealing the implanted sample, where the amorphous layer has regrown but a significant concentration of dislocations is still present. (Courtesy of the late Dr J Fletcher.)

It is clear that the as-implanted materials are quite unsuitable for operation as the active regions of semiconductor devices and the amorphous and heavily damaged layers must be removed as far as possible. The dopant atoms must also be encouraged to take up substitutional sites in the semiconductor lattice, a process called 'activation'. Some kind of annealing treatment is required to improve the structure and properties of the implanted semiconductors. It has been found that silicon crystals implanted with boron ions, and which therefore contain no amorphous regions, must be annealed above 800 °C before activation is complete and the density of trapping states is reduced to an acceptable level. However, it has also been observed that at temperatures around 500 °C amorphous layers turn into single-crystalline material by a process of solid state epitaxial regrowth from the substrate. This difference in the temperatures required for the production of perfect semiconductor material from amorphised and damaged material is presumably because the thermodynamic driving force for the amorphous-to-crystalline phase transition is much greater than for recovery in a crystal containing lattice defects. Even after expitaxial regrowth, some dislocations usually remain in the implanted regions, as shown in figure 2.20(b). These dislocations may have to be removed by a brief annealing stage at higher temperature.

While these annealing processes return the damaged semiconductor material to a high state of crystalline perfection and activate the implanted dopant species, they also allow extensive diffusion of the dopants. The as-implanted dopant profile broadens very rapidly because of the high density of lattice defects in the same region of the semiconductor, which increases the rate of diffusion. Since the total amount of diffusion dopant is fixed by the implanted dose, this is a condition of 'constant total dopant' and the diffusion process can be modelled by an expression of the form

$$C(x,\ t) = \frac{M}{(\pi Dt)^{1/2}} \exp\left(\frac{-x^2}{4Dt}\right) \tag{2.6}$$

where M is the total mass of the implanted dopant. This expression can be used to predict the concentration of dopant a distance x away from the peak of the as-implanted Gaussian distribution. Diffusion during the annealing treatments is usually undesirable. since it modifies the carefully chosen dopant profiles and can change the depth of p–n junctions and allow lateral penetration of the dopant as well. In order to limit the spreading of implanted dopant profiles, several processes have been developed for removing the implantation lattice damage in short high-temperature treatments. Techniques like heating with lasers, electron beams and intense light sources have all been used (see Gyulai 1984) and can be quite effective at producing very perfect semiconductor material without disturbing the dopant profiles too badly. The surface of the semiconductor material may be melted for very short periods in some of these processes, and it is the epitaxial regrowth from the liquid phase which produces the highly perfect semiconductor crystal.

The removal of damage and activation of dopants in compound semiconductors gives special problems since at the annealing temperatures, typically 800–900 °C, the surface of the semiconductor dissociates. At temperatures above 600 °C, for instance, gallium arsenide decomposes with the rapid loss of arsenic from the surface. These materials thus have to be encapsulated during annealing and SiO_2, Si_3N_4 and AlN layers have all been used as capping materials on GaAs substrates.

To complete this section I should mention a few more uses of ion beam implantation techniques in the fabrication of microelectronic devices. Heavy dose implantation of inert gas ions can create severe lattice damage in regions far away from the active devices which can then act as a gettering site for metallic impurities. A similar effect can be achieved by implanting phosphorus ions to getter residual copper atoms. Implantation of hydrogen ions into selected regions of GaAs wafers can be used to isolate individual devices electrically in an integrated circuit. The damaged regions become semi-insulating as a result of the high concentration of lattice defects. This process is called 'proton isolation' and is the analogue in GaAs technology of the oxide isolation process for silicon devices illustrated in figure 2.8. Finally, implantation is used to improve the crystalline quality of silicon epitaxial films grown over sapphire substrates. Amorphisation of the silicon layer with a heavy implant dose of silicon ions, and solid phase epitaxial regrowth, creates material with a lower defect density than the as-grown epitaxial layer. As a result, leakage currents and recombination rates are lowered in the devices fabricated on the silicon islands.

REFERENCES

Allen R E, Sankey O F and Dow J D 1986 *Surf. Sci.* **168** 376
Bardeen J 1947 *Phys. Rev.* **71** 717
Brillson L J 1979 *J. Vac. Sci. Technol.* **16** 1137
Brugger H, Schaffler F and Absteiter G 1984 *Phys. Rev. Lett.* **52** 141
Coburn J W and Winters H F 1979 *J. Vac. Sci. Technol.* **16** 391
Crowell C R and Sze S M 1966 *Solid State Electron.* **9** 1035
Downey D F, Boltoms W R and Hanley P R 1981 *Solid State Technol.* **26** 121
Einspruch N G and Bauer R S 1985 *VLSI Electronics Microstructure Science* **10** (New York: Academic)
Elliott D J 1982 *Integrated Circuit Fabrication Technology* (New York: McGraw-Hill)
Ephrath L M 1979 *J. Electrochem. Soc.* **126** 1419
Frazer D A 1983 *The Physics of Semiconductor Devices* (Oxford: Oxford University Press)
Freeouf J L and Woodall J M 1981 *Appl. Phys. Lett.* **39** 727
Gyulai J 1984 in *Ion Implantation, Science and Technology* ed. J F Ziegler (New York: Academic) p 139
Heine V 1965 *Phys. Rev.* A **138** 1689
Horne D F 1986 *Microcircuit Production Technology* (Bristol: Adam Hilger)
Howard R E and Prober D E 1982 in *VLSI Electronics Microstructure Science* **5** ed. N G Einspruch (New York: Academic) p 146
Kurtin S, McGill T C and Mead C A 1969 *Phys. Rev. Lett.* **22** 1433
Labuda E F and Clemens J T 1980 in *Encyclopedia of Chemical Technology* ed. R E Kirk and D F Othmer (New York: McGraw-Hill)
Lawrence J E and Huff H R 1982 in *VLSI Electronics Microstructure Science* **5** ed. N G Einspruch (New York: Academic) p 51
Lin A M, Dutton R W and Antoniadis D A 1979 *Appl. Phys. Lett.* **35** 799
McGillis D A 1983 in *VLSI Technology* ed. S M Sze (Singapore: McGraw-Hill) p 267
Maydan D 1980 *J. Vac. Sci. Technol.* **17** 1164
Mead C A 1966 *Solid State Electron.* **9** 1023
Mogab C J 1983 in *VLSI Technology* ed. S M Sze (Singapore: McGraw-Hill) p 303
Mogab C J, Adams A C and Flamm D L 1978 *J. Appl. Phys.* **49** 3769
Pearce C W 1983 in *VLSI Technology* ed. S M Sze (Singapore: McGraw-Hill) p 9
Piotrowska A, Guivarch A and Pelous G 1983 *Solid State Electron.* **26** 176
Rhoderick E H 1978 *Metal–Semiconductor Contacts* (Oxford: Clarendon)
Rideout V L 1975 *Solid State Electron.* **18** 541
Ruoff A L and Chan K-S 1982 in *VLSI Electronics Microstructure Science* **5** ed. N G Einspruch (New York: Academic) p 330
Schluter M 1978 *Phys. Rev,* B **17** 5044
Schottky W 1939 *Z. Phys.* **113** 367
Schwartz B 1969 *Ohmic Contacts to Semiconductors* (New York: Electrochemical Society)

Seidel T E 1983 in *VLSI Technology* ed. S M Sze (Singapore: McGraw-Hill) p 219

Spicer W E, Chye P W, Skeath P R, Su C Y and Lindau I 1979 *J. Vac. Sci. Technol.* **16** 1422

Spicer W E, Kendelwicz T, Newman N, Chin K K and Lindau I 1986 *Surf. Sci.* **168** 240

Spicer W E, Lindau I, Skeath P and Su C Y 1980 *J. Vac. Sci. Technol.* **17** 1019

Stengl G, Loschner H and Muray J J 1986 *Solid State Technol.* February 119

Sze S M 1981 *Physics of Semiconductor Devices* (New York: Wiley)

Tersoff J 1985 *J. Vac. Sci. Technol.* **B3** 1157

Till W C and Luxon J T 1982 *Integrated Circuit; Materials, Devices and Fabrication* (Englewood Cliffs, NJ: Prentice-Hall)

Tsai J C C 1983 in *VLSI Technology* ed. S M Sze (Singapore: McGraw-Hill) p 169

Tuck B 1988 *Atomic Diffusion in III–V Semiconductors* (Bristol: Adam Hilger)

Tung R T 1984 *Phys. Rev. Lett.* **52** 461

Ziegler J F 1984 *Ion Implantation, Science and Technology* (New York: Academic)

3

The Growth of Single Crystals of Semiconducting Materials

3.1 INTRODUCTION

In Chapter 2 I showed that many individual semiconductor devices can be integrated together on a single semiconductor chip. These devices may be prepared on the surface of single-crystal wafers of silicon or gallium arsenide, or in thin single-crystal layers of semiconductor material grown over the top of a normal single-crystal wafer, i.e. epitaxial layers. Epitaxial layers of compound semiconductor materials are needed in the fabrication of many of the optoelectronic and photovoltaic devices which are described in Chapters 8 and 9. In addition, efforts are being made to grow epitaxial silicon layers on insulating substrates for the technology referred to as semiconductor-on-insulator, or SOI. Some advantages of this kind of device structure have been described in §2.3. Epitaxial films of other semiconductor materials, most notably gallium arsenide, have also been grown on insulating substrates. Techniques have thus been developed both for the growth of bulk single crystals from which semiconductor wafers can be prepared, as described in §2.4, and for the deposition of single-crystal epitaxial layers only a few micrometres, or fractions of micrometres, thick. These thin film growth techniques allow the deposition of a very wide range of semiconductor materials, including silicon, and many III–V and II–VI compounds.

In this chapter I shall introduce some of the most commonly used techniques for the preparation of bulk and thin film crystals of silicon and compound semiconductor materials. Large crystals are normally solidified from a melt, while epitaxial layers can be grown from the vapour phase. The kinetic and thermodynamic processes involved in melt and vapour growth are quite different and so these two families of crystal growth techniques will be described separately. Section 3.3 will present the two major techniques for the growth of bulk semiconductor crystals, while methods for the deposition of epitaxial layers are described in §3.4, 3.7 and 3.8. Pamplin (1975) and Brice (1986) have both given excellent introductions to the techniques and processes of crystal growth.

Before starting to describe single-crystal growth processes, we must understand why it is so important that the semiconducting materials in which transistor and optoelectronic devices are to be made should be single crystals. Quite simply, the electrical properties of semiconductors are highly dependent on the defect concentration in the material, as I have described in §1.4. Grain boundaries are particularly effective scatterers of free carriers and so increase the resistivity of a semiconductor and reduce the carrier mobility. They also provide very high densities of recombination centres, reducing important parameters like minority carrier lifetime. We should remember that the efficiency of bipolar devices, and many kinds of optoelectronic devices as well, depends very heavily on the minority carrier lifetime. Similarly, high-speed transistors can only be made in material which shows a high carrier mobility. The presence of grain boundaries can therefore severely degrade the performance of semiconductor devices of all kinds. A more complete description of the electronic properties of grain boundaries in semiconductors is given in §4.3. Dislocations will degrade the properties of semiconductor materials by very similar mechanisms to those described for grain boundaries. We have already seen in §2.4 the ways in which a single dislocation can effect the operating efficiency of a bipolar device, and the deleterious properties of dislocations in optoelectronic and photovoltaic devices will be described in Chapters 8 and 9. It is generally recognised that both grain boundaries and dislocations must be removed from semiconductor crystals before they are of optimum device quality. This is as true for thin epitaxial layers as it is for the bulk crystals from which wafers are to be prepared. Where a device with properties that are far from the optimum is acceptable, polycrystalline material can be used. Polycrystalline solar cells are usually significantly less efficient than their counterparts fabricated in single-crystal material, but are far cheaper to produce. This brings us to an important point: single-crystal growth processes are expensive and the necessity of using perfect single crystals as the starting material for most device fabrication processes substantially increases the cost of microelectronic components.

3.2 MELT GROWTH PROCESSES

In this section I shall present an outline of the basic principles that govern the growth of bulk single crystals from a melt, and in particular the processes which control the distribution of impurity and doping elements in the crystals. Very low levels of impurity elements in an elemental or compound semiconductor have strong effects on electronic properties and it is vital that semiconductor crystals are prepared with very high purities if the expected electrical properties are to be obtained. Clearly the melt from which semiconductor crystals are to be grown must itself be formed from very pure starting materials.

Silicon crystals are grown from a melt produced from polycrystalline electronic grade silicon, EGS. This starting material is prepared by the fractional distillation of $SiHCl_3$, and the reduction of the resulting very pure compound with hydrogen (see Pearce 1983). (This reduction reaction is actually a chemical vapour deposition process of the kind discussed in §3.7.) EGS normally contains significant levels of impurities like iron, chromium, nickel and carbon, and so the melt from which silicon crystals are to be grown will contain at least 1 part per million of each of these elements. In addition, molten silicon is an extremely aggressive reagent and will attack and dissolve the crucible in which it is contained during crystal growth. The choice of crucible material has thus been dictated by the need to minimise the damage caused to electrical properties of the crystal by the unavoidable impurities from the dissolved crucible. This immediately removes refractory metals, refractory carbides and silicon carbide from consideration as crucible materials. Heavy metal impurities introduce unwanted deep states into the silicon, while excess carbon is precipitated as silicon carbide. Quartz crucibles dissolve to give considerable oxygen contamination in the melt and this will find its way into the crystals. The complex effects that oxygen can have in a silicon lattice are described in §3.3.2. Carefully cleaned quartz crucibles are almost always used for the melt growth of bulk silicon crystals, simply because the effect of oxygen in the crystals is less damaging than that of most other crucible impurities. Leipold *et al* (1980) have discussed the choice of crucible materials for containing silicon melts and concluded that in the future silicon nitride may join quartz as a suitable material. Nitrogen, like oxygen, has no very damaging electrical properties in the silicon lattice. At present it is still difficult to prepare very pure silicon nitride containing no easily dissolved grain boundary phases.

Compound semiconductor crystals can be grown from melts produced either by the direct synthesis of the compound from its component elements, or by melting presynthesised polycrystalline material. The former process is a more direct method, and so often a cheaper one, of converting the elements into a large single crystal of the compound semiconductor. The starting elements for either process must, of course, be very pure if high-quality crystals are to be grown. A wide variety of methods are used to prepare these pure precursor elements, including

zone refining, fractional distillation and chemical purification. Details of
these techniques, and the quality of the materials prepared, are given by
Grovenor (1987) and in the compilation of data by Willardson and
Goering (1962). Most of the elements needed for the synthesis and
growth of large crystals of compound semiconductors can be prepared
with total impurity levels of about 1 part per million, or approximately
10^{16} cm^{-3}, if we think of the impurities as involuntarily added dopant
elements. Once again, the compound semiconductor melts prepared
from the purified elements will be contaminated by impurities from the
crucibles. Gallium arsenide crystals are often grown from boron nitride
or quartz crucibles and so will contain significant levels of boron or
silicon contamination. II–VI semiconductors, like cadmium telluride and
zinc sulphide, can be grown from alumina and quartz crucibles respectively and so will always be contaminated with oxygen and aluminium or
silicon.

A further problem exists when melts of compound semiconductor
materials are being prepared because of the very high vapour pressures
exerted by some of the volatile elements in these compounds. Figure
3.1(a) shows a sketch of the equilibrium vapour pressure–temperature
diagram for a hypothetical binary compound semiconductor, AB, and
the AB phase diagram is sketched in figure 3.1(b). The equilibrium
vapour pressures exerted by the two elements A and B over the pure
elements are shown as broken lines in figure 3.1(a) and the full lines
trace the vapour pressure of each element in equilibrium with both solid
and liquid AB over the whole composition range. The relationship
between figures 3.1(a) and (b) is made clear by seeing that at a
temperature T_1 there are two compositions in the binary phase diagram
at which equilibrium exists between solid and liquid AB, i.e. C_1 and C_2.
At each of these compositions the equilibrium vapour pressure of A and
B can be found from the intersection of the isothermal line at T_1 in
figure 3.1(a) with the vapour pressure curves for A and B. This diagram
illustrates the case where the equilibrium vapour pressure of one of the
components, A, is much higher than that of B. It is quite common for
compound semiconductors to have one component which is much more
volatile than the others; phosphorus in GaP, arsenic in GaAs, cadmium
in CdTe and mercury in HgCdTe are a few examples. We can see that it
is not always the metalloid (or non-metallic) element which has the
highest vapour pressure in these compounds. Volatile metals like
cadmium and mercury have much higher vapour pressures than group
VI elements like tellurium for instance.

We can use figure 3.1 to follow the way in which the equilibrium
vapour pressure will vary during the formation of a compound semiconductor melt. Let us first consider the case where we are synthesising a
binary compound directly from the two component elements, A and B.

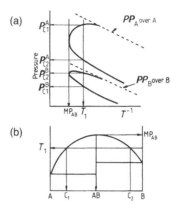

Figure 3.1 Schematic illustrations of phase diagrams for a hypothetical binary system, A–B; (a) the p–T phase diagram; (b) the conventional T–x phase diagram.

I shall assume that the vapour pressure of A is so much higher than that of B over the pure solid elements that the contribution to the total vapour pressure from B can be ignored. As we increase the temperature of the two solid elements, the vapour pressure of A increases rapidly, following the broken line in figure 3.1(a). In the absence of B, this pressure would simply continue to increase with temperature, with the slope changing as the solid A melted. However, at some temperature the two elements will react to form AB, the compound semiconductor. An example of this kind of reaction takes place between liquid gallium and solid arsenic to form gallium arsenide at about 800 °C. The equilibrium vapour pressure over solid AB will usually be lower than over the pure elements, and so the partial pressure of A in the reaction chamber now follows the full curve of A in equilibrium with solid and liquid AB. Further increase in the temperature to the melting point of the compound, MP_{AB}, causes the vapour pressure of A to decrease even further. This is because at the melting point the only composition at which equilibrium can be established between solid and liquid AB and A and B vapour is that of the equiatomic AB compound itself. Thus, the direct synthesis of a compound semiconductor melt will result in the maximum vapour pressure being developed at a temperature somewhat below the melting point of the compound. At the melting point, the vapour pressure will be significantly lower, even though the temperature is higher.

The second way of forming a compound semiconductor melt is from presynthesised material. Here, the more volatile element will evaporate from the compound during heating and a dilute melt of A in B will be formed. The vapour pressure can thus be found by adding the contributions from the upper curve of the full line of element B and the lower curve of A. Although the total vapour pressure over the compound semiconductor melt at MP_{AB} is the same as in the first case, the vapour pressure does not reach a maximum level at lower temperatures. (This higher vapour pressure has to be contended with during the presynthesis phase of the process of course.)

Table 3.1 gives the vapour pressures of the most volatile components over a few common semiconductors. VP_{max} is the peak of the full curve in figure 3.1 and VP_{MP} is the vapour pressure over the liquid and solid compound in equilibrium at the melting temperature. It is clear that compound semiconductors which contain phosphorus will have extremely high vapour pressures at all temperatures near the melting point. Arsenic and mercury also exert high vapour pressures.

Table 3.1 The vapour pressures exerted by the most volatile component in some common compound semiconductors.

	Maximum VP (atm)	VP at melting point (atm) (MP in °C)
GaAs	40 (As)	0.9 (1238)
GaP	>40 (P)	32 (1470)
InP	>40 (P)	27 (1062)
CdTe	7 (Cd)	0.5 (1092)
$Hg_{80}Cd_{20}Te$	>40 (Hg)	15 (700†)

†Solidus temperature.

There are two problems which can be created by these high vapour pressures over compound semiconductor melts. The first is the preferential loss of one component into the vapour phase. We can see in figure 3.1(a) that in the hypothetical AB binary system there is no temperature at which the vapour pressure of A is equal to that of B. This means that the compound will not evaporate or sublime congruently at any temperature. This will alter the composition of the melt and can effect the stoichiometry of the crystals. We have seen in Chapter 1 how the electrical properties of a compound semiconductor can be modified by changes in stoichiometry. Some methods by which this loss of volatile components can be avoided will be discussed in the following sections. The high vapour pressures create a more practical problem in the design of crystal-growing equipment which can safely contain very high pressures of reactive and poisonous elements like phosphorus, arsenic and mercury. The pressure of phosphorus vapour over GaP and InP melts can burst crucibles, allowing the rapid, and often explosive, reaction of phosphorus with air. All the techniques used to grow single crystals of these compounds have to be modified to include a high-pressure vessel as an integral part of the equipment.

3.2.1 The Distribution of Impurities During Melt Growth

We shall now assume that a melt has been prepared from a semiconductor material and contains involuntarily added contaminants from both starting materials and crucibles, and perhaps a deliberately included dopant element as well. While the crystal grower will wish to reduce the concentration of impurities in the crystal as much as possible, a uniform dopant distribution throughout the boule is usually required. I shall assume that crystal growth proceeds by the passage of a single solid/liquid interface through the melt, and what we need to be able to predict is how the various impurity species in the melt are partitioned between the melt and the growing crystal.

Most impurities will have different free energies in the liquid and the solid, and if equilibrium is established during growth they will partition between the melt and the solid. The free energies of the impurity atoms will be determined by parameters like atomic size, valence and bond-forming ability with the atoms of the semiconductor lattice. The equilibrium partition coefficient, k, is defined as the ratio of the equilibrium concentration of impurity in the solid, $[I]^s$, to that in the melt, $[I]^m$. A value of k greater than one thus means that the particular impurity will partition preferentially into the solid during growth.

At this point we should establish a link between the value of the equilibrium partition coefficient and the form of the phase diagram between impurity and semiconductor material. This can be conveniently done with the example of an impurity such as aluminium in an elemental semiconductor like silicon. Figure 3.2 shows a detail of the Al–Si binary phase diagram. It is clear that the result of freezing a melt of composition 10 at.% aluminium is the initial formation of a small volume of solid containing only about 1 at.% of aluminium. In this case therefore the equilibrium value of k is about 0.1. In this particular phase diagram the separation of solidus and liquidus is such that k is about 0.1 for the freezing of all liquid compositions in the range 0–10 at.% aluminium. Different combinations of impurity and semiconductor will have phase diagrams with different forms and so the impurities will have different values of equilibrium partition coefficient. Values both less than and greater than unity are possible. Table 3.2 gives the measured values of k for a number of common impurity elements in silicon, gallium arsenide, indium phosphide and gallium phosphide. We can see from this that many of these impurities have values of k that are very much less than 1 and so will partition strongly into the melt during

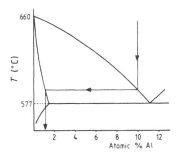

Figure 3.2 A detail of the equilibrium Si–Al binary phase diagram, showing how the equilibrium value of the partition coefficient of aluminium during the growth of a silicon crystal from the melt can be estimated to be 0.1.

crystal growth. However, a few impurities, in particular oxygen in silicon and aluminium in gallium arsenide, have values of k that are greater than 1. These impurities will partition into the growing crystals. I have also shown in table 3.2 that the measured values of k can lie over a significant range; see the values for tin in gallium arsenide and zinc in gallium phosphide for example. This is because real crystal growth processes are often insufficiently slow to allow equilibrium to be established at the solid/liquid interface. The values of k quoted in table 3.2 are thus 'effective' partition coefficients and will depend on such practical details as the rate of growth, the rate of diffusion of the impurity in the melt and in some cases the indices of the growing crystal plane. Effective partition coefficients, k_{eff}, can be estimated from:

$$k_{eff} = \left[1 - \left(\frac{1}{k_0} - 1\right)\exp\left(\frac{V\delta}{D}\right)\right]^{-1} \qquad (3.1)$$

where k_0 is the equilibrium partition coefficient, V is the growth rate, δ the boundary layer thickness and D the diffusion rate of the impurity in the melt. This is why the value of k_{eff} shown for aluminium in silicon in table 3.2 is not the same as we have just predicted from the equilibrium phase diagram. It is the values of effective partition coefficients, k_{eff}, which we must use if we are to predict impurity distributions in real crystal growth processes and which will tend to unity as the rate of growth increases. Values of k_{eff} have to be determined experimentally for each combination of dopant element and semiconductor material over a range of growth conditions.

It is often important to be able to estimate the concentration of ionised dopant atoms incorporated into a crystal grown from a melt of given composition. This depends not only on the value of k_{eff}, but also on the vacancy concentration in the crystal because a dopant atom will only be ionised when on a substitutional site in the semiconductor. For

Table 3.2 Some measured values of partition coefficients for common impurities in semiconductors.

Silicon

Impurity	Fe	Cu	Al	C	As	P	B	O
k_{eff}	0.000006	0.0004	0.002	0.07	0.3	0.35–0.56	0.8	1.25

Gallium arsenide

Impurity	Cu	Fe	Sn	In	Si	C	P	Al
k_{eff}	0.000015	0.003	0.003–0.03	0.1	0.1	0.8	2	3

Indium phosphide

Impurity	Ge	Sn	Te	Se	S
k_{eff}	0.01	0.002–0.03	0.4	0.6	0.8

Gallium phosphide

Impurity	Sn	O	Zn
k_{eff}	0.0001	0.005	0.05–0.08

the case of Zn doping of GaAs crystals, the zinc atoms will fill gallium vacancies, V_{Ga}, and we can write an equation for this incorporation reaction:

$$Zn_m + V_{Ga} \overset{K}{\rightleftharpoons} Zn_{Ga}^- + h^+ \qquad (3.2)$$

assuming that all the dopant atoms in the semiconductor lattice are ionised (Van Gool 1966). The equilibrium constant for this reaction will be given by

$$K = \frac{[Zn_{Ga}^-][h^+]}{[Zn_m][V_{Ga}]} = \frac{k_{eff}[h^+]}{[V_{Ga}]} \qquad (3.2a)$$

assuming that the activity coefficients of all these species are unity. If the dopant concentration is high in the solid then $[Zn_{Ga}^-] = [h^+]$, and from equation (3.2a) the concentration of ionised zinc in the gallium arsenide will depend on the square root of the zinc concentration in the melt $[Zn_m]$:

$$[Zn_{Ga}^-] \propto [Zn_m]^{1/2}. \qquad (3.2b)$$

At lower dopant concentrations where $[Zn_{Ga}^-]$ is much less than $[h^+]$ then the concentration of ionised zinc atoms will depend linearly on the zinc concentration in the melt:

$$[Zn_{Ga}^-] \propto [Zn_m]. \qquad (3.2c)$$

Under these conditions we must consider the hole concentration, $[h^+]$, to depend on the presence of residual acceptor impurities in the GaAs crystal.

From these equations we can see that there should be a transition from a linear dependence of the dopant concentration in the crystal, $[N_d]$, on the dopant concentration in the melt, $[N_m]$, to a parabolic dependence at some critical value of $[N_d]$. During the melt growth of doped gallium phosphide crystals this transition has been observed to occur for the dopant elements S, Se, Te and Zn. In gallium arsenide, doping with zinc shows a parabolic dependence of $[Zn_s]$ on $[Zn_m]$ under most conditions of growth, while many other dopant elements seem to obey equation (3.2c). In order to explain these observations we must consider the band structure at the crystal/melt interface. Zschauer and Vogel (1971) have suggested that when the term D_d/L_s exceeds the growth velocity by a factor of more than 10 we must consider that the bulk of the semiconductor is in equilibrium with the melt. (D_d is the solid state diffusion coefficient of the dopant in the crystal and L_s is the intrinsic Debye length). A high value of the diffusion coefficient means that dopant atoms incorporated at the growing crystal surface are able to diffuse rapidly and homogenise the dopant distribution in the whole crystal. However, if D_d/L_s is less than the growth velocity, then the melt is only in equilibrium with the surface of the semiconductor. A space-charge region and interfacial potential barrier will be set up at the semiconductor surface by the pinning of the Fermi level at the surface states. Thus we are no longer able to assume, as in equation (3.2a), that $[N_d]$ is equivalent to the carrier density, and a linear dependence of $[N_d]$ on $[N_m]$ will result. Dopants like Te, Sn and Ge in GaAs obey a linear partition law because they are slow diffusers, while Zn, which diffuses rapidly in GaAs, shows the parabolic incorporation behaviour predicted by equation (3.2b). The significance of this is that k_{eff} may not be invariant with the dopant concentration in the melt. The effect of $[N_m]$ on k_{eff} is particularly strong at high dopant concentrations. When we come to consider liquid phase epitaxial growth in §3.4 this compositional dependence of k_{eff} will be important, since liquid phase epitaxy (LPE) is often used to grow heavily doped epitaxial layers for the optoelectronic device structures described in Chapter 8.

It is tempting to think that once the value of k_{eff} for a particular impurity in a semiconductor has been determined for a given set of growth conditions, the distribution of that impurity in a crystal can be accurately predicted. However, this distribution is further influenced by the details of the solidification process. Two extreme cases can be isolated: one in which the melt is continuously stirred to ensure a homogeneous liquid phase; and the second in which the only mechanism by which the liquid can approach homogeneity is diffusion of the impurity elements. The reason why it is important to distinguish between these two situations is that an impurity with a value of k_{eff} which is less than one will be rejected into the melt ahead of the moving solid/liquid interface. The concentration of this impurity element near the interface will depend on the efficiency with which it is mixed into the bulk of the remaining liquid, and the composition of the next solid to be frozen, will also depend on the impurity concentration near the interface.

The distribution of a solute element with $k_{eff} = 0.1$ along a crystal grown from melts that are (a) completely mixed and (b) diffusionally mixed are sketched in figure 3.3. It is clear from these curves that the conditions of growth strongly affect impurity, or dopant, distributions in bulk crystals. In the case of complete mixing in the melt, the composition of the first part of the crystal, C_s, will be given by the Scheil equation

$$C_s = k_{eff}C_0(1 - X)^{(k_{eff}-1)} \qquad (3.3)$$

where C_0 is the initial composition of the melt and X is the volume fraction of the melt solidified. For the case of an unstirred diffusionally mixed melt the composition of the first two-thirds of the crystal grown is given by

$$C_s = C_0[1 - (1 - k_{eff}) \exp(-VX/D_L)] \tag{3.4}$$

where D_L is the rate of impurity diffusion in the liquid and V is the velocity of the solid/liquid interface. Two of the most important techniques for the growth of bulk crystals of semiconductor materials, Czochralski and Bridgman growth, are quite good examples of completely mixed and diffusionally mixed growth processes respectively. The practical consequences of this will be described further below.

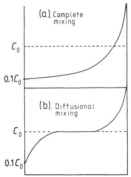

Figure 3.3 Schematic illustrations of the composition profiles expected along crystals grown from melts containing a solute element with a value of $k_{eff} = 0.1$ when: (a) the melt is completely mixed; (b) the melt is mixed by diffusion alone.

A final phenomenon which can occur during the growth of a bulk crystal from the melt is constitutional supercooling. We have seen that a build up of an impurity species can occur ahead of the moving solid/liquid interface if the value of k_{eff} is less than one and the rate of mixing of the rejected impurity into the remaining melt is too slow to maintain a homogeneous melt. This change in composition ahead of the interface will also change the equilibrium freezing temperature of the melt in these regions. By reference to figure 3.2 we can see that as the concentration of impurity species increases, approaching the solid/liquid interface, the temperature at which the liquid will freeze decreases. Thus the liquid at the interface will have an equilibrium freezing point which is lower than that of the remainder of the melt. Under these circumstances, the temperature gradient imposed by the details of the design of the furnace and crystal growth apparatus may lie below the equilibrium freezing temperature for some distance into the melt. The liquid close to the solid/liquid interface is thus supercooled and will attempt to freeze as quickly as possible. This results in the breakdown of the initially planar growth front and the production of crystals which are very inhomogeneous. This phenomenon is called 'constitutional supercooling'.

Any circumstances that encourage the preferential rejection of one component of a melt ahead of the growing crystal may lead to constitutional supercooling. Large separations between the solidus and liquidus in the phase diagram, implying values of k far from unity, and the loss of a volatile component to create a non-stoichiometric melt, can both contribute to the breakdown of the growth interface. This means that constitutional supercooling is particularly likely in the growth of crystals of indium phosphide where the loss of phosphorus to the vapour phase is very rapid. High temperature gradients at the solid/liquid interface and low impurity concentrations in the initial melt can reduce the chance of breakdown.

In this section I have introduced only very briefly the processes by which impurity and dopant elements are redistributed during the growth of semiconductor crystals. Further details on these phenomena can be found in standard references such as Chalmers (1964) and Brice (1986).

3.2.2 Stoichiometry Variations in Compound Semiconductor Crystals

We have seen how gross variations in the composition of a semiconductor melt can lead to constitutional supercooling and the production of very inhomogeneous crystals. However, the stoichiometry of compound semiconductor crystals can also be affected in more suble ways by the exact conditions under which they are grown. I shall now give examples to illustrate how the composition of the vapour surrounding the growing crystal, the composition of the melt from which the crystal is grown, and the growth temperature, can all modify the electrical properties of a compound semiconductor crystal. These effects are due to slight imbalances in the stoichiometry of the compound semiconductor itself and arise from the relative ease with which electrically active native point defects can be formed in many compound semiconductor materials and the existence of extended compound phase fields at elevated temperatures (see §1.4).

A simple example of the modification of stoichiometry during crystal growth is given by the case of cadmium telluride. If the crystal is in contact with a vapour pressure of cadmium in excess of the equilibrium vapour pressure of this element over the compound at its melting point, then an excess of cadmium is incorporated into the crystal. This will produce electrically active defects (cadmium interstitials or tellurium vacancies) and the crystal will have p-type conduction characteristics.

A second case of the non-stoichiometric growth of semiconductor crystals occurs when gallium arsenide is being grown fom melts which do not contain exactly the same amount of gallium and arsenic. The phase field of the binary compound GaAs is slightly extended at temperatures close to its melting point (see figure 3.6) and Holmes *et al* (1982) have shown that as a result of this the electrical properties of GaAs crystals can be changed very significantly if the stoichiometry of the melt is altered by only a very small amount. Material which is p-type, and has a very low resistivity, is grown from melts containing less than 48 at.% of arsenic because the crystals contain a very high concentration of shallow acceptor states. These states may be a result of the presence of excess gallium atoms in the form of antisite defects, i.e. gallium atoms on arsenic lattice sites. By contrast, n-type crystals with very high resistivities can be grown from melts containing more than 48 at.% arsenic. These crystals have been shown to contain a high density of a native defect which gives a characteristic mid-gap state, often called EL2. Gallium arsenide crystals of this second kind are very much more suitable for the production of substrate wafers on which integrated circuits may be fabricated than the low-resistivity p-type material. Most commercial gallium arsenide crystals are thus grown from slightly arsenic-rich melts, which ensures that, even with some loss of the volatile arsenic from the melt, at no stage in the growth process can a gallium-rich melt be produced.

Similar variations in conductivity are found in melt-grown ternary II–VI and IV–VI alloy materials like mercury cadmium telluride and lead tin telluride in which the concentrations of vacancies on the metallic and tellurium sublattices determine the electrical properties, as described in §1.4. PbSnTe crystals grown from a melt containing nearly equal concentrations of metal and Te are normally p-type since they contain a slight excess of Te. Lowering the growth temperature, by using a solution growth process (§3.4), can result in the growth of crystals with n-type conductivity (Harmon 1973). Once again this is a result of the form of the extended phase field of the compound at high temperatures (see figure 3.15(b)).

From these examples it is clear that careful control must be exercised over both the melt and vapour phase compositions if crystals of compound semiconductor materials with the required electrical properties are to be grown. It is also important to know the form of the compound phase field at temperatures near the melting point and, if possible, to have measured the formation energies of all native point defects in the crystals.

3.3 BULK CRYSTAL GROWTH

3.3.1 Czochralski Growth

This technique is used for almost all the commercial growth of bulk single crystals of silicon, gallium arsenide and indium phosphide. As such, Czochralski crystal growth is one of the most important stages in the production of discrete devices and integrated circuits in semiconductor wafers. Figure 3.4 shows the basic features of a Czochralski crystal-pulling facility. The molten semiconductor is contained in a large crucible in a furnace. A seed crystal is brought into contact with the melt and then slowly withdrawn. If the conditions are controlled correctly, a single crystal is grown on the retreating seed. The seed is rotated during growth to ensure that a cylindrical crystal is grown and, in most cases, the crucible is rotated as well to help mix the melt. A Czochralski growth process is usually taken to be an example of solidification from a completely mixed melt. The distribution of dopant and impurity species in the growing crystal is thus expected to follow the curve shown in figure 3.3(a). In fact, thermal convection at the crystal/melt interface is generally far from uniform and so the impurity distribution perpendicular to the growth direction is rarely completely homogeneous. The characteristic 'swirl' defects seen on sections through Czochralski-grown crystals are regions of varying dopant concentrations, produced as the growing interface is rotated through regions of melt at different temperatures. An example of swirl defects will be given in the following section.

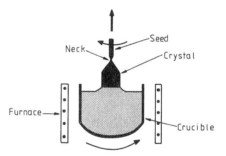

Figure 3.4 A schematic illustration of the basic features of a Czochralski crystal growth process.

The two most important parameters in the growth of single crystals by the Czochralski process are the rate at which the seed is withdrawn from the semiconductor melt and the rate at which heat can be transported away from the solid/liquid interface. The first of these depends heavily on the second. The latent heat of freezing is released at the crystal/melt interface and can be conducted away both through the melt and through the growing crystal. Assuming that the melt is at a uniform temperature above the melting point, it is much more likely that the latent heat will be carried away wholly through the crystal. At equilibrium, a simple heat balance equation can be written as

$$V\rho L = K_s \frac{dT}{dx} \tag{3.5}$$

where V is the rate of growth of the crystal, L the latent heat of freezing per unit mass, ρ the density and K_s the thermal conductivity of the crystal. This equation can be used to estimate the maximum rate of crystal growth for any particular semiconductor material since, when the latent heat released during solidification exceeds the heat which can be transported away from the solid/melt interface, the temperature at the interface increases, exceeds the melting point and remelting of the crystal will occur. For materials like silicon and gallium arsenide, equation (3.5) can easily be used to show that the maximum growth rate, V, cannot be greater than a few mm h^{-1}. This is an absolute limit on the rate at which single crystals can be grown by the Czochralski process.

Within this limited range of growth rates, the diameter of the cylindrical crystals is controlled by the rate at which the seed is withdrawn from the melt. The slower the rate of retraction, the larger the crystals. Silicon ingots with diameters in excess of 8 inches have been successfully grown, while 3 inch crystals of III–V materials are more common. This ability to control the diameter of the crystal allows the deliberate introduction of a narrow neck just below the seed, as illustrated in figure 3.4. Dislocations in the seed will be 'grown out' against the sides of the neck and so will not propagate into the bulk of the crystal. However, additional dislocations are nucleated in the crystals both by thermal strains and impurity precipitation, as will be described below.

I have already mentioned that impurity elements will always be present in a semiconductor melt, and we have seen that Czochralski growth is an example of a completely mixed growth process. We would thus expect oxygen in silicon, $k_{eff} = 1.25$, to partition into the first part of the crystal to solidify, while another crucible impurity, silicon in gallium arsenide, $k_{eff} = 0.1$, to be distributed towards the bottom end of the crystal. Dopant elements, like phosphorus in silicon and tin in gallium arsenide, will also be distributed preferentially at the bottom of the crystal. In addition, if a high concentration of any impurity is rejected ahead of the solid/liquid interface, constitutional supercooling can occur, with the breakdown of the morphology of the crystal/melt interface. Excessive concentrations of antimony in silicon melts, and the loss of volatile phosphorus from InP melts, can both lead to constitutional supercooling.

The details of the Czochralski growth processes used for the production of boules of silicon and the III–V compounds differ in some important details, and so I shall now briefly describe the growth of silicon and gallium arsenide crystals separately.

3.3.1(a) Czochralski growth of silicon

The growth of large single crystals of an elemental semiconductor like silicon uses a process like that shown in figure 3.4. The crystal growth chamber is backfilled with an inert gas, or evacuated, to stop contamination of the very reactive silicon melt. The major impurity in the melt will be oxygen from the dissolution of the quartz crucible, and the high partition coefficient of oxygen in silicon will ensure that very high levels of this impurity will be found in all Czochralski-grown silicon crystals. Concentrations of oxygen as high as 10^{18} cm^{-3} are quite common.

Oxygen in silicon crystals can have three quite distinct effects, as has been reviewed by Patel (1977). The first is that oxygen can be a donor dopant in the silicon lattice, forming SiO_4 complexes very rapidly as the crystal is cooled below about 500 °C. Most of the oxygen atoms are not complexed in this way, and are electrically inactive, but a sufficiently high density of these donor complexes is produced to distort any intentional dopant distribution in the crystals. This is, of course, especially true if a crystal containing a p-type dopant is being grown. Silicon crystals are often cooled rapidly from 600 °C to avoid the formation of these complexes. The second effect of oxygen in silicon is the solution hardening of the semiconductor lattice. The additional hardness can be advantageous in limiting the amount of plastic deformation which occurs during cooling of the crystals, so reducing the density of dislocations. Finally, during the cooling of a silicon boule which contains a supersaturation of oxygen, precipitates of an oxide phase can be formed. These precipitates are usually observed to form at temperatures below 900 °C. Arrays of prismatic dislocation loops are punched out around oxide precipitates because of the high local stresses around the misfitting particles and these dislocations play an important role in the intrinsic gettering process described in §2.4.

The inhomogeneous distribution of oxygen in the Czochralski-grown boules means that the relative importance of all three of these oxygen-related effects depends on the position in the boule from which each individual wafer is cut. The top of the boule will usually contain more oxygen than the bottom, because k_{eff} is greater than 1. The effect of this variation in oxygen concentration on the mechanical and electrical properties of silicon wafers has been reviewed by Lawrence and Huff (1982).

Large silicon crystals with almost negligible dislocation densities are routinely grown by the Czochralski process, as described by Bradshaw and Goorissen (1980). Most of the dislocations will be in the form of small loops, or the prismatic dislocations around oxide particles. Twinning occasionally occurs during growth, and twinned boules are rejected for remelting. The most important problem in the growth of large silicon crystals is probably the inhomogeneity of the dopant concentration, caused by fluctuations in temperature at the growth interface. An example of the notorious 'swirl' pattern is shown in figure 3.5. These inhomogeneities are hard to avoid, even with the most careful control over all the parameters of the growth process, although the use of strong magnetic fields around the crucible during growth can reduce the temperature variations at the crystal/melt interface.

Figure 3.5 An example of a dopant 'swirl' pattern near the edge of a silicon wafer revealed in a Lang x-ray topograph. Some dislocations are also shown propagating from the edge of the wafer. (Courtesy of Mr C Dineen and GEC Research Ltd.)

3.3.1(b) Czochralski growth of gallium arsenide
The first point to make in this section is that Czochralski growth techniques can only be used for compound semiconductors which melt congruently. Ternary compounds like mercury cadmium telluride, which form a continuous range of solid solutions and have a very wide separation between liquidus and solidus, are extremely difficult to grow as single crystals by this technique. Fortunately, all the common binary III–V compounds, GaAs, InP and GaP in particular, melt congruently and are readily grown as single crystals by simple modifications of the Czochralski process illustrated in figure 3.4. The modifications are necessary because of the very high vapour pressures of arsenic and phosphorus over the binary semiconductor melts (see table 3.1). In the case of gallium arsenide, arsenic will be lost from a melt until the equilibrium partial pressure of 0.9 atmospheres has been established in the growth chamber. The walls of the chamber must also be maintained at temperatures above 600 °C, otherwise the arsenic vapour will condense there, the pressure of arsenic will be reduced and further evaporation will occur from the melt. The technique of liquid encapsulated growth was developed to avoid the necessity of containing these high pressures of arsenic and heating the whole of the crystal-pulling chamber. This technique was originally used for the growth of PbTe crystals by Metz *et al* (1962) and applied to III–V compound semiconductors by Mullin *et al* (1965).

Liquid encapsulated Czochralski growth, LEC, uses an inert capping layer floating on the top of the semiconductor melt to stop the evaporation of a volatile component. Boric oxide, B_2O_3, is very commonly used as the capping layer as it does not react with most compound semiconductor melts, and the diffusion rates of the volatile elements through the glassy layer are very low. An overpressure of an inert gas equal to the total equilibrium vapour pressure over the molten compound is sufficient to keep the B_2O_3 firmly in contact with the melt. Since no large volumes of arsenic vapour escape to the growth chamber, the walls do not need to be heated.

Gallium arsenide wafers, on which integrated circuits or discrete devices are to be fabricated, are ideally cut from large single crystals of semi-insulating material. This is a necessary requirement if the dopant profiles introduced during device manufacture are not to be distorted by the donor or acceptor states produced by impurity elements or native defects and to reduce interference between adjacent devices in an integrated circuit. The benchmark value of resistivity which defines semi-insulating gallium arsenide is about $10^7\,\Omega$ cm, which establishes that the concentration of electrically active point defects in the crystals is very low. The mechanism by which semi-insulating gallium arsenide can be grown in a LEC process has been particularly comprehensively studied and involves compensation of the electronic states formed as a result of the presence of two important impurity elements in the crystals — silicon and carbon. I have described how impurity elements from a crucible are added to the melt, and hence to the growing crystals, in most melt growth processes. Quartz crucibles will dissolve in gallium arsenide melts to give levels of silicon around 10^{15} cm^{-3} in the crystals. Silicon is an amphoteric dopant in gallium arsenide, as described in §1.4.2, but normally creates a shallow donor state. This state can be compensated for very effectively by the deep acceptor state associated with chromium atoms in gallium arsenide, and this has led to the practice of chromium being added to gallium arsenide melts. A similar effect is obtained when iron is added as an impurity to indium phosphide crystals. Iron creates a deep acceptor state which compensates for the residual donor impurities present in the melt.

Contamination of a gallium arsenide melt with silicon can be avoided by replacing the quartz crucible with one of boron nitride. Both the crucible and the floating B_2O_3 capping layer will introduce boron contamination into the melt, but chromium need no longer be added as there will be no silicon donor states for which to compensate. Boron is a relatively harmless impurity in gallium arsenide, although it does seem to decrease the concentration of active silicon atoms on gallium sites when this element is used as a donor dopant.

Carbon can hardly ever be completely avoided as an impurity in gallium arsenide melts and will produce a shallow acceptor level in the crystals. High resistivity crystals are rarely grown unless the carbon state is compensated. There is a native defect in gallium arsenide crystals which gives a deep donor state and is normally called EL2. This native defect state very effectively compensates for the carbon impurity states and so no additional compensating impurity need be added to the gallium arsenide crystals. The lattice defect responsible for the EL2 state has not been clearly identified, but may be an antisite point defect. These defects seem to occur around, or to be produced by, dislocations in the gallium arsenide crystals.

Thus it is possible to grow high-quality semi-insulating gallium arsenide crystals by a Czochralski process from both quartz and boron nitride crucibles. In both cases the carbon impurity state is fortuitously compensated by EL2. A more detailed description of the impurities in LEC gallium arsenide, and their effect on the electrical properties of the crystals, can be found in Willardson and Beer (1984).

However, native point defects can also play an important role in LEC III–V crystals. Even when a stoichiometric melt is carefully preserved by the use of a capping layer and a high inert gas overpressure, the crystals grown may contain high densities of electrically active native point defects. For example, gallium arsenide crystals often contain a high concentration of gallium vacancies. This can be explained by examination of the detail of the Ga–As phase diagram shown in figure 3.6. The phase field of the congruently melting compound is slightly extended at high temperatures and shifted to the arsenic-rich side of the stoichiometric composition at the melting temperature. Gallium arsenide crystals may thus be grown with a slight arsenic excess and often have excellent semi-insulating properties. This may be due to the presence of arsenic antisite defects, i.e. As atoms on Ga lattice sites. In this case, the dominant native defect seems to be wholly beneficial if semi-insulating material is required and the arsenic antisite defect may be associated with the EL2 state mentioned above. A similar kind of effect is seen in melt-grown gallium phosphide crystals. Here a supersaturation of gallium vacancies will often be found, but in this case the presence of these electrically active native defects will degrade the luminescent properties of the material. These defects are very hard to avoid since they are in the thermodynamic equilibrium with the lattice at the melting temperature.

3.3.1(c) Lattice defects in LEC gallium arsenide

Now we must move on to consider the extended defects which are produced in gallium crystals during a Czochralski growth process. It was not necessary to consider the mechanisms by which dislocations may be

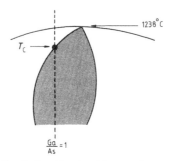

Figure 3.6 A detail of the Ga–As binary phase diagram, showing the extended phase field of the compound GaAs at high temperatures. This diagram also shows that the GaAs crystals will be arsenic rich if grown from a melt in which the composition ratio is close to Ga/As = 1.

produced in silicon crystals, because the yield stress of pure silicon is rather high (of the order of 10 MPa even at 1000 °C). Thermal contraction stresses developed during the cooling of the boules rarely exceed the yield stress if the cooling cycle is carefully controlled. However, gallium arsenide is very much softer than silicon, with a yield stress of only 1 MPa. This means that thermal stresses which would create no dislocations in a silicon crystal can cause a significant amount of plastic deformation in a gallium arsenide boule.

The stoichiometry of the melt also influences the dislocation density in the crystals, and it has been found that a minimum density is achieved when the melt composition has a gallium-to-arsenic ratio close to 1. n-type doping can reduce the dislocation density dramatically, while the reverse effect is seen when the crystal is doped p-type. These observations can be explained if we assume that the concentration of neutral gallium vacancies, $[V_{Ga}]$, is a critical factor in controlling the rate of condensation of vacancies into dislocation loops. We must also assume that condensation only occurs when the vacancies are neutral. The stoichiometry of the melt determines $[V_{Ga}]$ and the position of the Fermi level (which is of course controlled by the doping concentration and type in the crystal) determines what fraction of these vacancies is ionised. The higher the position of the Fermi level in the band gap, the larger the fraction of gallium vacancies which will be ionised, which explains why n-type doping reduces the dislocation density. This condensation mechanism provides a way in which a high density of dislocations can be created in gallium arsenide crystals, even when the thermal stresses are low.

The role that dislocations will play in modifying the properties of devices fabricated on the gallium arsenide wafers is poorly characterised, but it is generally assumed that the dislocation density should be kept to a minimum in the highest-quality material. Dislocation densities of around 10^4 cm^{-2} are common in LEC gallium arsenide crystals and are very easily revealed by etching the surface of crystal or wafer with molten KOH. The larger the diameter of the crystal, the higher the dislocation density in general, simply because the thermal stresses generated during cooling will be greater. Jordan (1980) has calculated the total stress distribution in a cylindrical crystal during cooling, assuming that the surface cools most rapidly leaving a relatively hot core region. He also assumed that dislocation slip would occur on 12 equivalent (111) $\langle 110 \rangle$ systems in the sphalerite lattice. Under these circumstances, a high tensile stress is generated at the outside of the crystal and a rather high compressive stress at the centre. Jordan predicted that a stress distribution of this kind would lead to a characteristic dislocation density profile: high in the centre of the crystal, falling to a low level in the annular ring around the centre where the stress levels are low, and finally a very high density of dislocations near the surface. This distribution of dislocations is found in most Czochralski gallium arsenide crystals, and an example of the characteristic 'W' profile is shown in figure 3.7(a). The crystallography of the slip systems in diamond cubic related lattices means that the density of dislocations will be higher along $\langle 100 \rangle$ than along $\langle 110 \rangle$ directions. The optical micrograph in figure 3.7(b) shows the dislocation arrays in a gallium arsenide crystal in greater detail; the dislocations are arranged into a cellular structure of subgrains separated by low-angle grain boundaries.

The addition of isoelectronic impurities has been used to try and reduce the dislocation densities in crystals of III–V materials, including indium in gallium arsenide and gallium in indium phosphide. These impurity elements have almost no effect on the electrical properties of the compound semiconductors. Crystals containing as much as 1 at.% of indium or gallium have been grown and the dislocation densities can be as low as 10^2 cm^{-2} — effectively dislocation-free material. The mechanism by which the generation of dislocations in these crystals is suppressed is not completely clear. Solution hardening of the soft compound semiconductor lattice seems unlikely to stop all dislocation formation, and the addition of indium to gallium arsenide has been shown to increase the yield strength at the melting point only by a factor of about four (Hobgood *et al* 1986). Some authors have suggested that Frank–Read dislocation sources are less effective in indium-doped gallium arsenide, so that the few dislocations created by the thermal stresses cannot easily multiply. However, indium has a partition coefficient of about 0.1 in gallium arsenide and so will segregate strongly to the melt. This can result in a build-up of indium in the liquid near the growing

(a)

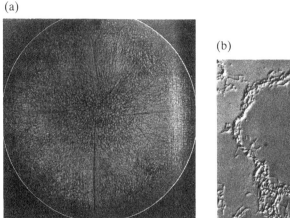

(b)

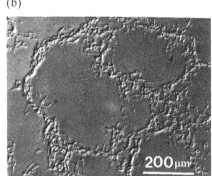

Figure 3.7 (a) A transmission x-ray topograph showing the dislocation distribution in a 2 inch GaAs wafer. A very high dislocation density can be seen at the centre and outer rim of the wafer, with a relatively dislocation-free region in between. (b) An optical micrograph showing the arrangement of dislocations into a cellular structure in a GaAs wafer. The dislocations are revealed by chemical defect etching. (Courtesy of Dr D Stirland and Plessey Research (Caswell) Ltd.)

crystal. Constitutional supercooling may then occur unless a sufficiently high growth rate is imposed to keep the temperature gradient at the interface above the equilibrium freezing temperature at all points. Figure 3.8(a) shows a detail of a section through a Czochralski crystal of indium-doped gallium arsenide where constitutional supercooling has occurred during growth. Cells of relatively pure gallium arsenide are separated by thin regions where the indium concentration is much higher. The facetted shape of the cells on the degenerated solid/liquid interface is clearly outlined by the indium-rich regions in this photograph. In addition if an indium-doped gallium arsenide crystal is grown from an undoped seed, dislocations may be nucleated at the seed/crystal interface because of the mismatch in the lattice parameters in the pure and indium-doped gallium arsenide. A high density of dislocations can be found at the centre of the gallium arsenide boule in these circumstances, as shown in figure 3.8(b). Indium doping of gallium arsenide crystals used to be the standard way of growing relatively dislocation-free material for the manufacture of integrated circuits. However, it has recently been shown that reducing the thermal gradients around a growing crystal by using a very thick layer of capping B_2O_3 will also reduce the dislocation density in the boule. This may provide a simple way of improving the quality of gallium arsenide material without the need to introduce doping elements.

(b)

(a)

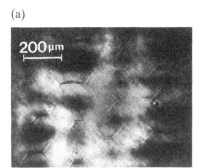

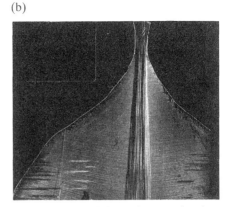

Figure 3.8 (a) A transmission infrared micrograph viewed through crossed polarisers of an indium-doped GaAs wafer in which constitutional supercooling has occurred during growth. The arrow-head features are indium-rich regions. (b) A transmission x-ray topograph of the dislocations produced at the mismatched interface between a pure GaAs seed and an indium-doped GaAs crystal. The dislocations propagate down through the centre of the growing crystal. (Courtesy of Dr D Stirland and Plessey Research (Caswell) Ltd.)

Crystals of the III–V compound semiconductors are also very prone to twin nucleation during growth. This is presumably because the energy of the twin boundary is very low indeed and a simple misstacking of the {111} planes is all that is needed to nucleate a twin. Figure 3.9 shows the plan and cross-sectional view of a gallium arsenide boule which initially grew as a single crystal, but repeated twinning created the polycrystalline material shown at one side of the plan view. Simple twin boundaries initially nucleate at the edge of the growing crystal, but interaction between intersecting twins soon produces 'random' high-angle grain boundaries like those described in §4.3.3(a) Rae *et al* (1981) have described the process by which twin boundaries can intersect to produce high-angle grain boundaries. The mechanism by which the first twin boundaries are formed is not well determined, but may be related to the presence of gallium particles on the outside of the boule. These gallium particles are formed by the preferential loss of arsenic from the hot gallium arsenide surface.

In this section, I have tried to show that LEC growth of very perfect gallium arsenide crystals, and crystals of a wide range of other compound semiconductors, is rather difficult compared to the growth of large silicon crystals. This is a result of their intrinsic softness, the high vapour pressure of their component elements and the possibility of growing non-stoichiometric crystals. II–VI compounds are rarely successfully grown by the Czochralski technique, but can readily be prepared by one of the variety of techniques related to Bridgman growth (see §3.4).

Figure 3.9 A cross-sectional and a plan view of samples cut from a 3 inch GaAs boule in which twin nucleation has occurred. The intersecting twins eventually lead to the formation of polycrystalline material. (Courtesy of Cambridge Instruments Ltd.)

3.3.2 The Growth of Shaped Crystals

Many methods have been developed for the growth of single crystals with shapes other than the cylinders produced by a simple Czochralski process. These methods have been applied to the growth of tubes, sheets and rods of materials as diverse as sapphire, silicon and barium titanate, $BaTiO_3$. Here I shall concentrate on describing a few of the methods that have been used for the growth of thin sheets of silicon from a melt. These silicon ribbons are used in the fabrication of low-cost large-area solar cells and so must be produced by inexpensive and rapid techniques. Many of the same techniques can also be used to grow sheets of sapphire, which can then be used as substrates for the epitaxial growth processes described in §3.7 and 3.8.

Most of the techniques developed for the growth of silicon sheets from the melt stem from the pioneering work of Stephanov first reported in 1959. His concept was to modify the Czochralski growth process to grow crystals of well-defined morphology by using a die to shape the meniscus at the melt/crystal interface. From this original idea, numerous related techniques have been developed, all relying on capillary action to draw the molten silicon up through an aperture in a die. Two distinct subdivisions of this kind of process have been designed, distinguished by whether a wetted or non-wetted die is used.

The Stephanov method in its original form used silica or boron nitride dies to shape the silicon melt into a sheet from which the ribbon is grown. These materials are poorly wetted by the molten silicon, and so it is the width of the slot in the die which defines the thickness of the ribbon (see figure 3.10). With a wettable die (graphite is wetted by molten silicon for example), the sheet thickness is defined by the separation of the outer edges of the flats on the top of the die and the form of the melt meniscus. This second process is normally called edge-defined film-fed growth (EFG). A second version of this kind of growth deliberately allows two silicon dendrites to form in the early stages of growth from a seed crystal, by using a supercooled melt. The growth rate is then slowed down, causing a thin sheet of silicon to solidify between the two dendrites. This process is called dendritic web growth. Ribbon-to-ribbon (RTR) sheet growth involves the zone melting and recrystallisation of a polycrystalline starting ribbon. A collection of papers on the theory and practice of growing silicon crystals in the form of ribbons from a melt can be found in a special issue of the *Journal of Crystal Growth* **50** (1980).

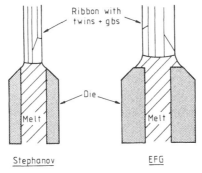

Figure 3.10 Schematic illustrations of two kinds of growth processes for the preparation of shaped crystals. Both these methods use a die to shape the melt from which the crystal is grown.

The silicon ribbons produced by any of the above techniques appear essentially the same — a few hundreds of micrometres thick with fairly smooth surfaces. The stability of the sheet growth process depends on controlling the heat transfer away from the crystal/melt interface and in EFG the meniscus shape and wetting angle at the die. Incvitably, very high thermal gradients are produced where the thin sheets are withdrawn from the melt, so that the ribbons contain a high density of dislocations, twin boundaries and low-angle grain boundaries, and may be severely buckled. However, rates of growth of several centimetres per minute can readily be achieved, so that large areas of silicon ribbon can be produced relatively rapidly. This fast growth rate is the major advantage that sheet growth processes have over the Czochralski growth of silicon boules, but means that the effective partition coefficient of most impurities in the silicon melt will be close to unity (see equation (3.1)) and almost no purification will occur during growth. The use of graphite dies introduces unacceptably high levels of carbon contamination into silicon sheets, leading to the danger of extensive precipitation of silicon carbide. Commercial graphite can also contain significant amounts of iron, titanium and chromium, which are most undesirable impurities in silicon material intended for solar cell applications. Quartz, silicon carbide and silicon nitride have all been tried as die materials (Leipold *et al* 1980) since the impurity elements they add to the ribbons — oxygen, carbon and nitrogen — modify the electrical properties of the silicon only very slightly. Metal carbides and boron nitride are generally avoided as die materials, since heavy metal impurities have deep states in silicon and dissolution of BN in the melt gives ribbons which are degenerately p-doped. Once again, we can see that the aggressive nature of the silicon melt makes it very difficult to contain. Dendritic web growth can be carried out without the need for a shaping die, so that problems associated with the dissolution of die materials can be avoided.

Finally, we should consider the microstructure of these thin silicon sheets. I have already said that they contain a high density of twins and it is these defects that control both the growth mechanism and the electrical properties of the material. A crystalline seed of a material with a diamond cubic lattice will tend, when in contact with a melt, to facet in order to increase the area of the low-energy {111} surfaces. On these perfectly flat surfaces, nucleation of a new crystal plane is difficult since the first atom of the new plane will have a relatively high energy and be prone to redissolve unless the melt is highly supercooled. Hamilton and Seidensticker (1960) showed how a seed which contains twins can nucleate new {111} planes at the re-entrant angle created by the intersection of the surfaces of the two twin facets. This mechanism of growth removes the necessity of having a high supercooling in the melt, as now the atoms from the melt can be added to the low-energy sites at the bottom of the re-entrant facets. The preferred growth directions during twin-assisted growth are of the type $\langle 211 \rangle$. Silicon sheets prepared by any of the growth methods described above almost always have a $\langle 211 \rangle$ growth direction and quite often a (011) surface texture as well. A typical ribbon also contains a very high density of twins.

Simple twin boundaries in silicon have little or no electrical activity, as will be discussed further in §4.3.3(a), but intersecting twins create shorter regions of second-, third- and higher-order twin boundaries which may be electrically active. Ast *et al* (1984) amongst others have used electron microscopy techniques to analyse the structure and recombination efficiency of grain boundaries in silicon ribbons. The EBIC technique (which is described in §10.2) has been shown to be most valuable in the identification of electrically active boundaries in this kind of material. The electrically active grain boundaries at which the recombination rate is high, appear as dark lines, while inactive twin boundaries are invisible. This kind of investigation has shown that, although the density of grain boundaries in silicon ribbon material is very high, most of these boundaries are not damaging to the electrical properties of the ribbon. This is why these ribbons can be used in the production of reasonably efficient solar cells, even though they are full of lattice defects.

3.3.3 Bridgman Crystal Growth Techniques

We have seen that the Czochralski technique can be used for the growth of large single crystals of silicon, gallium arsenide and a range of other III–V compound semiconductors. However, it is rarely used for II–VI materials because of technical difficulties in the control of the growth process. Bridgman crystal growth techniques cover a large number of processes which are linked by the basic feature that a melt is completely frozen to form a single crystal by the passage of a solid/liquid interface. Czochralski growth does not fit within this description because the mass of the initial melt usually exceeds that of the crystals which are grown.

In practice, Bridgman crystal growth is carried out by passing a semiconductor melt through a temperature gradient. Nucleation in the coolest region of the melt can be limited to the production of a single grain by using sharply pointed crucibles, a relatively shallow temperature gradient to avoid excessive undercooling, and very often seed crystals as well. Figures 3.11 and 3.12 illustrate two kinds of Bridgman crystal growth processes: the vertical growth of a compound semiconductor crystal simply by carefully cooling a crucible containing a stoichiometric melt; and the horizontal growth of a gallium arsenide crystal. In this second technique, the gallium arsenide melt is formed by reaction between a pure gallium melt and arsenic vapour. This vapour is formed by heating a source to give a pressure of 0.9 atmospheres of arsenic, the equilibrium pressure over a gallium arsenide melt.

The vertical growth of crystals in a sealed crucible can result in significant strains being set up in the crystal if the thermal expansion coefficients of crucible material and crystal are not very similar. These strains will create unwelcome high dislocation concentrations in the crystals. By contrast, horizontal growth puts little mechanical restraint on the crystals during cooling, and very low dislocation densities are

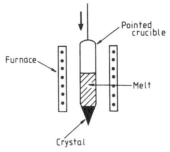

Figure 3.11 A schematic illustration of a vertical Bridgman growth process.

found even in very soft gallium arsenide crystals. Dislocation densities of 10^2 cm^{-2} are typical. However, horizontal Bridgman growth gives inconvenient 'crucible'-shaped crystals with a D-shaped cross section. It is preferable to grow cylindrical crystals if the wafers are intended for the fabrication of integrated circuits. A great deal of expensive single-crystal material is lost in the sawing operation to obtain circular wafers from D-shaped boules.

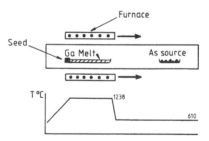

Figure 3.12 A schematic illustration of a horizontal Bridgman growth process for preparing GaAs crystals from a gallium melt and a source of arsenic vapour. The temperature profile in the reactor is also shown.

Two features of Bridgman growth processes serve further to distinguish them from Czochralski techniques. Firstly the melt is relatively static in the processes illustrated in figure 3.12, and this means that solidification by a Bridgman process must be considered as crystal growth from a diffusionally mixed melt. The distribution of impurity or dopant elements along a Bridgman crystal will therefore approximate to that shown in figure 3.3(b). Thus we expect the beginning and end of a Bridgman crystal to contain impurity concentrations that differ quite significantly from the relatively uniform distribution in the middle of the boule. (In a Czochralski crystal a much more uniform macroscopic impurity concentration is achieved by limiting the fraction of the melt that is frozen. This is not possible in a Bridgman growth process of course.) Secondly, in order to grow single crystals by a Bridgman method rather slow growth rates are needed, a few mm h^{-1} at most. Both the melt and the frozen crystal are thus in contact with the crucible for extended periods of time. Contamination from the dissolution of the crucibles and from solid state diffusion into the crystal is one of the most significant problems associated with this kind of crystal growth technique.

The horizontal Bridgman process is commonly used to grow bulk single crystals of II–VI compound semiconductors. A particularly diffi- cult material to prepare in single-crystal form is the ternary alloy $H_{0.8}Cd_{0.2}Te$. This alloy has become commercially important because of its very small band gap, which makes it an ideal choice for use in infrared photodetectors, as will be described in Chapter 8. However, the pseudobinary phase diagram of this ternary compound has the form shown in figure 3.13(a), with a very large separation between the liquidus and solidus. We can think of this as implying that the equilibrium partition coefficient of HgTe in HgCdTe is much less than unity, so that freezing a melt of composition $Hg_{0.8}Cd_{0.2}Te$ will initially produce solid containing a much higher CdTe content. The composition profile along a boule of HgCdTe grown by a Bridgman process is sketched in figure 3.13(b), where the inhomogeneity of the mercury content is very evident. If the growth rate is not kept high, even more severe macrosegregation occurs with the loss of the region in the centre of the crystal where the composition is approximately uniform. This alloy is also prone to constitutional supercooling during growth, both because of the strong partitioning of mercury into the melt and because of the rapid loss of volatile mercury into the vapour phase. The first effect produces a mercury-rich liquid just ahead of the crystal/melt interface, while the second will deplete the melt in metallic elements and create an excess of tellurium ahead of the same interface. Thus two quite separate constitutional supercooling reactions are possible during the growth of this material.

HgCdTe crystals grown at high rates are usually found to be large- grained polycrystals containing a high dislocation density. There is often evidence of constitutional breakdown in the centre of the grains with the formation of Cd-rich/Te-depleted dendrites. These crystals are annealed after growth to allow homogenisation and grain growth to occur. Only if very careful control is maintained over the partial pressure of mercury during growth do the electrical properties approximate to those required for the production of devices. This is because the concentration of metal vacancies in the alloy determines the majority carrier type and the resistivity. More details on the growth of crystals of this particularly intransigent compound can be found in the review by Micklethwaite (1981).

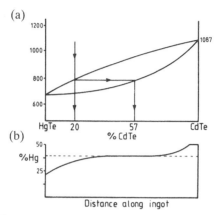

Figure 3.13 (a) A sketch of the 'pseudobinary' phase diagram for HgCdTe alloys, showing the very wide separation of the liquidus and solidus and the consequent formation of CdTe-rich solid on freezing. (b) A mercury composition profile along an HgCdTe crystal grown by a Bridgman process.

It is fortunate that only a few of the technologically important compound semiconductors have phase diagrams of the kind shown for HgCdTe. The common binary semiconductor materials, GaAs, GaP, InP, CdTe, CdS, ZnTe, etc, are all congruently melting compounds, which means that the problems described for HgCdTe are rarely encountered when growing crystals by the Bridgman technique. High vapour pressures during synthesis and growth, and the high melting points of some of these compounds, can, however, make crystal growth complicated and expensive. Many II–VI semiconductors with high melting points are more easily prepared in the form of single crystals by sublimation growth from a polycrystalline source, as described by Lorenz (1967) and Brice (1986). This is only possible when the compounds sublime congruently. Single crystals of important IV–VI alloys like $Pb_xSn_{1-x}Te$ and $Pb_xSn_{1-x}Se$, and their related binary compounds, are conveniently prepared by Bridgman techniques, and quite homogeneous material is obtained because, in the case of these ternary alloys, the liquidus and the solidus lines in the phase diagrams are only separated by about 20 °C. However, we must remember that the vapour pressure of the metallic components above crystals of these alloys during growth will control the concentration of electrically active point defects.

Table 3.3 summarises the advantages and problems associated with the Bridgman and Czochralski growth techniques for the production of large single crystals of semiconductor materials. Both these techniques are widely used in a number of forms specifically designed for particular materials and grow almost all the crystals from which wafer substrates are prepared. Czochralski crystal-pulling equipment is more expensive than most Bridgman facilities, but the greater control over the crystal growth process that it affords makes it more popular for the production of the major wafer materials, Si, GaAs and InP. For further details and discussion of these growth techniques the reader is directed to the volumes by Ray (1969), Willardson and Beer (1984) and Brice (1986).

Table 3.3 A comparison of the Bridgman and Czochralski growth techniques for the preparation of bulk semiconductor single crystals.

	Czochralski	Bridgman
Growth rate	1–20 cm h^{-1}	1 cm h^{-1}
Diameter	15 cm or more	less than 10 cm
Dislocation density	10^2–10^5 cm^{-2}	10^{-2} or better
Dopant distribution	Quite uniform	uniform in centre
Materials grown	Si, GaAs, InP	II–VI compounds, GaAs

3.3.4 Float Zone Crystal Growth

The float zone technique for crystal growth, FZ, can be used to grow bulk silicon crystals which have much higher purities than those prepared by the other techniques described in this section. In principle it is a very simple technique. A narrow molten zone is passed along a rod of polycrystalline silicon by the passage of a small furnace. The rods have a circular cross section and are about 5–10 cm in diameter. The molten zone is supported by the surface tension forces where it makes contact with the solid material on either side of the exact position of the furnace. There is thus no contact made between the melt and a crucible, and the whole apparatus can be contained in vacuum or in an inert atmosphere to reduce the possibility of pick-up of contaminant elements from the gas phase. A seed can be used to ensure nucleation of a single crystal as the molten zone is passed along the rod for the first time.

This technique is a development of the zone-refining process designed by Pfann (1966) and the impurity distribution along a rod after the passage of a single molten zone is given by

$$C_s = C_0[1 - (1 - k_{eff})\exp(-k_{eff}X/L)] \qquad (3.6)$$

where L is the length of the molten zone and X is the distance along the rod. One zone pass is sufficient to produce a single crystal from a polycrystalline rod, but if the zone is passed repeatedly all the impurity elements for which k_{eff} is less than 1 will be driven towards the end of the rod. This allows the preparation of extremely pure material, and germanium and silicon crystals with impurity levels as low as 10^{13} cm^{-3} have been obtained in this way. These high purities are important in solar cell materials (see Chapter 9) and for germanium crystals in infrared photoconducting devices like those described in Chapter 8. In both cases, the presence of impurities can lower the minority carrier lifetime and reduce the efficiency of the device. However, crystals prepared by multiple float zone processes are very expensive, and so are only used when no other techniques are available for the preparation of suitable semiconductor material. The dislocation content in FZ crystals can sometimes be rather higher than in Czochralski crystals, because of the high temperature gradients near the edge of the molten zone.

A modification of the FZ process replaces the cylindrical sample with a thin polycrystalline ribbon about 100 μm thick. A high-intensity laser beam can be used to form a narrow molten zone, which can then be scanned along the ribbon at rates as high as 10 cm min^{-1}. A heavily dislocated and twinned ribbon is produced, which is very similar in structure to EFG or dendritic web ribbon material (see§3.3.2). This process is called ribbon-to-ribbon growth (RTR) and can be a convenient way of preparing large-area silicon sheet for terrestrial solar cell applications. Matlock (1979) has compared the quality of silicon crystals grown by Czochralski and FZ techniques for a variety of microelectronic applications, concluding that Czochralski material is generally more suitable for the fabrication of integrated circuits but that FZ techniques are important where very pure crystals, or sheet materials, are required.

3.4 LIQUID PHASE EPITAXY

So far in this chapter I have been concerned with describing the growth of large single crystals of semiconductors from melts which are of approximately the same composition as the final crystal. In this section I shall introduce a process by which thin single crystals of the same semiconductor materials can be grown from dilute melts onto substrate wafers. In general, the substrates are chosen to have lattice parameters and crystal lattice types which are closely matched to those of the growing crystal. Under these circumstances, the film will grow epitaxially, with the same structure and orientation as the substrate, and so the process is called liquid phase epitaxy, LPE. This kind of crystal growth process is an example of solution growth, and I must now explain how solution growth differs from the melt growth processes described above. The kinds of defects which are produced at epitaxial interfaces, and the atomic processes by which epitaxial growth proceeds, will be discussed in the following two sections.

The common III–V binary semiconductors are congruently melting compounds and so have equilibrium phase diagrams like that illustrated for GaAs in figure 3.14(a). The freezing of a melt which contains exactly equal atomic fractions of the two elements will produce the compound phase, as described for the bulk growth processes in §3.3. However, if a melt which is highly deficient in arsenic is cooled, following the path marked in figure 3.14(a), supersaturation of the melt with arsenic will occur at about 1000 °C and GaAs will be precipitated onto any available substrate with which a reasonably low-energy interface can be formed. Nucleation of a solid phase does not occur onto the components of the LPE system itself, because these components are made from materials with which the growing crystals can form no low-energy interfaces.

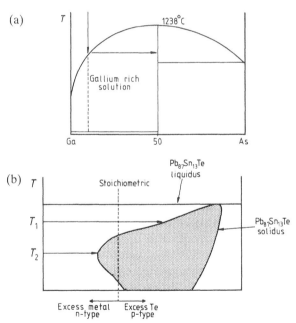

Figure 3.14 (a) A sketch of the Ga–As binary phase diagram, showing that cooling a gallium-rich melt will result in the precipitation of GaAs. (b) A sketch of the extended phase field of the alloy $Pb_{87}Sn_{13}Te$ in the Pb/Sn–Te phase diagram. If an epitaxial layer is grown from a melt at temperature T_1, the equilibrium composition is metal deficient and the material is p-type. Lowering the growth temperature to T_2 results in the deposition of Te-deficient material which is n-type.

The process of crystal growth by LPE can be described as follows: a melt of pure gallium exposed to a GaAs or GaP wafer will dissolve some of the solid to produce a dilute solution of the group V element. Cooling this solution to induce a slight supersaturation, and bringing a suitable substrate into contact with the melt surface, will result in the growth of a layer of GaAs or GaP all over the substrate surface. Care has to be taken that the substrate crystal does not dissolve too rapidly in the melt, by keeping the melt supersaturated. Even under these conditions, some 'meltback' often occurs, removing the immediate surface layer from the substrate. This effect can help to clean any contamination from the substrate, which would otherwise impede the growth of the epitaxial layer.

The thickness of the epitaxial layer is controlled by the time for which the substrate is in contact with the melt, t, the cooling rate, R, and, in situations where the melt can be considered to be stagnant (i.e. no convection or mixing is occurring), on the rate of diffusion of the slowest component element of the semiconductor through the melt, D_s. Hsieh (1974) has shown that the layer thickness, d, of solid of composition C_s can be estimated from

$$d = \frac{4Rt}{3C_s}\left(\frac{D_s t}{\pi}\right)^{1/2}\frac{\mathrm{d}[c]}{\mathrm{d}T} \qquad (3.7a)$$

where $\mathrm{d}T/\mathrm{d}[c]$ is the slope of the liquidus curve at the growth temperature. If a constant undercooling, ΔT, is used instead of a steady cooling, then

$$d = \frac{2\Delta T}{3C_s}\left(\frac{D_s t}{\pi}\right)^{1/2}\frac{\mathrm{d}[c]}{\mathrm{d}T}. \qquad (3.7b)$$

If the value of D_s is known, these equations allow values of d to be calculated for all systems for which the form of the liquidus has been accurately determined. LPE growth from solutions which are substantially supercooled before the substrate is brought into contact with the melt can give much smoother epitaxial layers than those grown by continuous cooling because of the rapid nucleation of the epitaxial layer all over the substrate (Crossley and Small 1973). These authors have also observed that constitutional supercooling rarely occurs until quite thick layers (about 15 μm) of GaAs have been grown from gallium solutions, although the growth interface for other compositions of epitaxial layer may break down at smaller thicknesses. The onset of constitutional supercooling will be controlled by D_s and the imposed temperature gradient.

Figure 3.14(a) illustrates one of the major advantages of this kind of simple solution growth, namely, that the growth temperature can be well below the melting point of the compound semiconductor which is being deposited. We have already seen that the GaAs compound phase field is slightly extended at high temperatures, as shown in figure 3.6, and so the lower the temperature of growth, the closer the crystal layer will be to the perfect GaAs stoichiometry. In figure 3.6 the only temperature at which exactly stoichiometric GaAs can be grown from a gallium-rich solution is called T_c, and it is possible to select the composition of the solution such that growth will occur at this temperature. It is also possible to control the conductivity type of some semiconductor materials by altering the growth temperature. In PbSnTe alloys for example, the temperature controls the equilibrium concentration of point defects, and lowering the growth temperature can change the character of the epitaxial layer from p-type to n-type, for instance by altering the growth temperature from T_1 to T_2 in figure 3.14(b).

From this outline of the LPE process we can see that the principles of growth are extremely simple and the equipment needed for LPE growth is also relatively modest when compared with a commercial Czochralski crystal puller. Figure 3.15 shows sketches of two kinds of LPE growth processes: a tipping arrangement first used by Nelson (1963) in (a); and a slightly more complicated sliding substrate holder in (b). In this second case, the substrate wafer can be passed under several melts of different compositions, growing multilayered epitaxial structures in a single process. These several melts can differ in doping content to allow

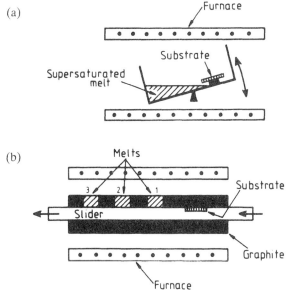

Figure 3.15 Sketches of two kinds of simple LPE growth processes: (a) a tipping arrangement; (b) a sliding substrate holder which can be passed under several melts of different compositions.

p–n junctions to be prepared, or may be solutions of completely different semiconductor compounds to deposit heteroepitaxial structures of the kind described in Chapter 8. Casey and Panish (1978) have given an excellent account of the practical details of carrying out a successful LPE growth run.

A very wide range of materials can be grown by LPE, but most attention has been paid to the growth of III–V compounds and alloys, with some experiments on growing silicon and germanium layers from gallium solutions and on the growth of IV–VI alloys from lead-rich solutions. A major problem in the production of crystals with very well defined compositions is that the effective partition coefficients of the various elements will not all be the same. Thus the composition of the relatively small volumes of each melt will rapidly change as crystal growth proceeds. This is particularly true when GaAlAs layers are grown from gallium-rich melts. Aluminium has a very high partition coefficient in GaAs (see table 3.2) and so the concentration of aluminium in the growing layer will be greatest near to the substrate, rapidly decreasing as the layer thickens and the melt is depleted in this element. Similarly, dopant concentrations in epitaxial layers can be rather inhomogeneous if the values of k_{eff} for the dopants differ significantly from 1, as they do for Te and Zn in GaP for instance. We have also seen in §3.2.1 that the values of k_{eff} depend sensitively on the dopant concentration in the melt (see equation (3.2b)) and so it is not possible to assume that k_{eff} for any particular element will be a constant even during a single growth run. Inhomogeneities in crystal composition can be avoided if very small volumes of melt are frozen onto each substrate wafer, or if the layers are annealed after deposition to allow solid state diffusion to homogenise the distribution of the elements. However, this last process can have the unfortunate effect of broadening the sharp composition changes between heteroepitaxial layers, or across p–n junctions.

A particularly demanding material to grow by LPE techniques is the quaternary III–V alloy GaInAsP (Nakajima 1982). It has proved necessary to determine the form of the liquidus and the solidus at the growth temperature of around 650 °C over a considerable region of the quaternary composition field. This is because it has been found that the values of the various partition coefficients depend heavily on the melt composition and so extrapolation from the behaviour of simple binary melts is not a satisfactory way of predicting the composition of the layers grown from the quaternary melt. The melt composition which will result in the growth of layers of the required stoichiometry and band gap can only be determined once the shape of the quaternary phase diagram is known. The form of the phase diagram for a ternary or quaternary semiconductor alloy can be determined experimentally and also predicted from thermodynamic data. The experimental method requires the measurement of the composition of layers deposited from melts of known composition, which determines the position of the solidus at any temperature. The electron microprobe has proved a convenient tool for the analysis of the composition of LPE-grown layers. The form of the liquidus can be found by the seed dissolution technique, where measurement of the weight loss of an InP seed in contact with a GaInAs melt, for instance, will allow calculation of the solubility of phosphorus in the melt. Further details on how these phase diagrams can be determined can be found in Nakajima (1982, 1985) for GaInAsP alloys and in Horikoshi (1985) for GaInAsSb and IV–VI alloys.

The calculation of ternary and quaternary phase diagrams can be attempted by using rigorous thermodynamic models for solid/liquid equilibria of the kind developed by Jordan and Ilegems (1975). These calculations are often difficult to perform accurately because of the strong non-ideality of the solutions and the lack of reliable values for the activity coefficients of the component elements in the melt. Nakajima (1982) has described how these models may be formulated for the determination of the phase diagram of GaInAsP, concluding that the match between experiment and theory is poor because the chemical interaction parameters between the elements are not well known. The match between theory and experiment in the simpler GaAlAs system is much better. Casey and Panish (1979) have given an excellent introduction to the thermodynamic models which have been used to calculate the liquidus and solidus isotherms in ternary III–V alloys.

The final result of an experimental or theoretical determination of a quaternary phase diagram is a set of plots of crystal composition against melt composition, one for each of the component elements in the melt. An example of such a plot is shown in figure 3.16 for gallium in GaInAsP for two concentrations of arsenic in the melt. From this kind of data, the melt composition can be chosen to deposit an epitaxial layer of precisely the required composition. GaInAsP alloys are usually grown from indium-rich solutions, where the arsenic, gallium and phosphorus concentrations are below 10 at.%. An additional constraint on the growth process is that the GaInAsP layer is normally required to be 'lattice matched' to a GaAs or InP wafer substrate, as will be described in the following section. Very careful control over the composition of the melt must be maintained if the layers are to have both the correct band gap and to be lattice matched to the substrate. The manner in which these compositions are chosen is described in Chapter 8.

Another important commercial application of LPE growth is the deposition of buffer layers of GaAs on Czochralski-grown GaAs semi-insulating wafers. These layers can be prepared with excellent electrical properties, ideally suited for the fabrication of integrated circuits (Stolte

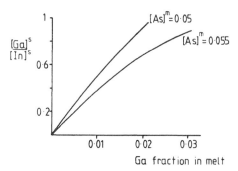

Figure 3.16 Experimental data on the variation of the composition of an LPE epitaxial layer of GaInAsP with the composition of the melt. The gallium-to-indium ratio can be seen to depend very sensitively on the gallium and arsenic concentrations in the melt (after Nakajima 1985.)

1984). LPE growth is also commonly used for the preparation of heteroepitaxial structures for laser and other optoelectronic devices. GaP light-emitting diodes with relatively high efficiencies can be produced by quite simple LPE techniques, as reviewed by Bergh and Dean (1976). There are no obvious technological reasons why a whole range of commercially important II–VI compounds cannot be deposited by LPE techniques, but rather little attention has been paid to developing these processes. One of the major disadvantages of LPE as a commercial layer deposition process is that it is not normally possible to deposit onto many substrates at once. This means that each LPE layer will be rather expensive to grow. In addition, inefficient use is made of the expensive pure melt materials; only a small fraction of the melt is frozen in any LPE growth run. Contamination from the crucible and slider materials, often made of graphite, make it inadvisable to use any one melt to grow too many layers since carbon is an electrically active element in most III–V compound materials.

3.5 THE STRUCTURE OF EPITAXIAL INTERFACES

There are two kinds of epitaxial interface which are common in semiconductor devices: those between a substrate and deposit of exactly the same material, e.g. a thin silicon layer on a perfect silicon wafer; and those between a deposit crystal and a substrate whose lattice parameter, and even crystal structure, are not matched. The first kind of interface should, if sufficient care is taken to clean the substrate surface and to design an efficient epitaxial growth process, be almost undetectable. This is because there is no structural mismatch or change in chemistry at the interface between substrate and deposit. In practice, the perfect continuation of the substrate lattice into the growing film is rarely achieved, and some examples of the kind of defects which are produced at these homoepitaxial interfaces will be described below. When the equilibrium structure of a growing deposit is poorly matched to that of the substrate, we must expect the interface to contain a high density of lattice defects.

Let us first consider the relatively simple case where both substrate and deposit have the sphalerite structure and both crystals are in exactly the same orientation with a (001) plane forming the interface between them. This arrangement is by far the most common structure for a heteroepitaxial interface produced by the epitaxial deposition of one semiconductor material on another — the growth of a (001)-oriented single-crystal layer on a (001)-oriented substrate crystal. The first few deposit atoms on the substrate surface will find their lowest-energy positions at the bottom of the potential wells created by the topmost atomic plane of the substrate crystal. (I shall assume here that the deposit atoms have a sufficiently high mobility on the substrate surface to find these equilibrium positions.) The distance between the adjacent deposit atoms adhering to the substrate in this manner will be characteristic not of the equilibrium crystal structure of the deposit, but of that of the substrate lattice. If the mismatch between the two lattice parameters is not too large, it is energetically favourable for the deposit to grow as a continuation of the substrate crystal, i.e. with a crystal lattice strained to fit exactly with that of the substrate. Under these circumstances, the deposit is under a state of plane stress, with all the interplanar spacings stretched or compressed by a constant fraction to match the equivalent spacings in the substrate. Figure 3.17 illustrates the lattice matching of a deposit crystal onto a substrate with a slightly larger lattice parameter. In the plane of the interface, all the crystal planes in the deposit are forced apart to match those in the substrate, while in the direction normal to the interface plane, the deposit planes are forced closer together — the so-called tetragonal distortion. We can estimate the excess energy of the strained deposit layer from simple elasticity theory. For a (001)-oriented substrate we can assume that the deposit crystal is distorted by equal strains along two orthogonal [100] directions: $\varepsilon_{xx} = \varepsilon_{yy} = \varepsilon$. The stresses, σ, at the interface along these two [100] directions will be equal to

$$\sigma = \frac{\varepsilon E}{1 - \nu} \tag{3.8}$$

where E is Young's modulus and ν Poisson's ratio for the deposit material. The work done per unit volume on straining a material by $d\varepsilon$,

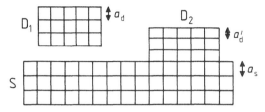

Figure 3.17 A schematic illustration of the way in which a deposit crystal, D, with equilibrium lattice parameter, a_d, can be elastically strained to match a substrate crystal with lattice parameter a_s. This matching in the interface plane results in a tetragonal distortion of the deposit crystal so that the lattice parameter normal to the interface becomes a'_d.

when the stress exerted is σ, is simply $\sigma d\varepsilon$. Therefore the total elastic energy per unit volume, E_{el}, is

$$E_{el} = 2 \int_0^\varepsilon \sigma d\varepsilon. \tag{3.9}$$

From equations (3.8) and (3.9), the elastic energy per unit area of a deposit of thickness h is

$$E_{el} = \frac{hE\varepsilon^2}{1 - v}. \tag{3.10}$$

When the misfit, f, is accommodated wholly by elastic strain at the interface, the value of ε used in equation (3.10) is obtained from the misfit between the lattice parameters of substrate, a_s, and deposit, a_d:

$$\varepsilon = \frac{a_s - a_d}{a_d}. \tag{3.11}$$

In addition, there will be a contribution to the total interfacial energy associated with the fact that there is a chemical mismatch at the substrate/deposit interface. Where a gallium phosphide layer is grown on a gallium arsenide substrate, for example, phosphorus atoms at the interface are in sites occupied by arsenic atoms in the substrate. The energy of each group V species will be increased by the presence of the other, but this energy is usually small and is generally ignored.

As a result of the parabolic relationship between the elastic energy of a heterointerface and the lattice misfit between substrate and deposit, the energy of the interfaces will be relatively low when the misfit is small. However, as the misfit increases, the elastic energy rapidly becomes very large, as shown by curve E in figure 3.18. A highly misfitting deposit layer can reduce this large elastic energy by plastic relaxation of the interface, i.e. the introduction of dislocations. In an interface formed between the (001) planes of two sphalerite crystals, the

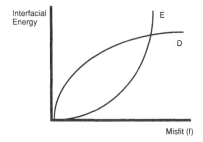

Figure 3.18 A schematic diagram showing the variation in interfacial energy with misfit between deposit and substrate lattices for the two cases in which the misfit is accommodated wholly by elastic strain (E) and by creating dislocations in the interface (D).

four-fold symmetry means that the lattice mismatch can be relaxed by the introduction of two arrays of dislocations running orthogonally. Now most of the lattice mismatch will be concentrated at the cores of the dislocations, and the regions of the interface between the dislocations can be considered to be perfect crystal. In fact, the long-range strain fields of the dislocations will extend far beyond the core regions and we must take this into account when calculating the energy of a dislocated interface. In addition, the chemical mismatch is still present at all positions in the interface. Figure 3.19 shows a sketch of the relaxation of a highly misfitting heterointerface by the introduction of an edge dislocation.

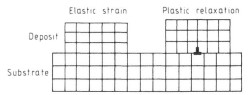

Figure 3.19 A schematic illustration of the relaxation of elastic strain in an interface by the formation of a misfit dislocation.

Van der Merwe (1963) and his co-workers (Van der Merwe and Ball 1975) have modelled the energy of this kind of dislocated interface. They treat the deposit and substrate as two interacting sinusoidal potentials and use a simple Peierls–Nabarro expression for the tangential shear stress at the interface, σ, which attempts to force these misfitting potentials into registry. The model assumes that dislocations are formed to localise the misfit between the deposit and substrate lattices —

'geometrically necessary dislocations' — and predicts that the energy of an epitaxial interface initially increases rapidly with increasing misfit, but soon levels off (see curve D in figure 3.18). This means that at low misfits it is energetically favourable to strain the deposit crystal elastically to fit the substrate lattice. However, at a critical value of the misfit, a dislocated interface becomes the lower-energy configuration, although a fraction of the total misfit will still be accommodated by elastic strain. The total misfit, f, can now be divided into two parts, $f = (\varepsilon + p)$, the elastic and plastic components respectively.

For the case of (001) heteroepitaxial systems, we can derive a much simpler expression for the energy of the interface than that given by the Van der Merwe theory. Summing the energies of all the dislocations in a plastically relaxed interface will give a rough approximation to the total interfacial energy, as has been pointed out by Kressel and Butler (1979). We obtain an expression of the form

$$E_{\text{plastic}} = \frac{\mu_{\text{eff}} b^2}{4\pi(1 - v)} \left[\ln\left(\frac{r_1}{r_0}\right) + 1 \right] 2\left(\frac{f - \varepsilon}{b}\right) \qquad (3.12)$$

where μ_{eff} is the effective shear modulus at the interface, r_1 and r_0 the separation of the dislocations and the radius of the dislocation core respectively and b the magnitude of the Burger's vector of the misfit dislocations. This equation is given for the case where all the dislocations are of pure edge character, with the Burger's vector lying in the plane of the interface, i.e. dislocations with Burger's vectors $\frac{1}{2}[1\bar{1}0]$ or $\frac{1}{2}[1\bar{1}0]$ in our model system. The first part of equation (3.12) is the ordinary expression for the energy of an edge dislocation (see for example Hirth and Lothe (1978)) and the last term gives the number of these dislocations encountered per unit length along the interface.

We can now use equations (3.10) and (3.12) to calculate the critical value of misfit at which it becomes energetically favourable to introduce the first misfit dislocations. The total energy of an interface where the misfit is accommodated by a combination of plastic and elastic deformation, E_T, is

$$E_T = \frac{E \varepsilon^2 h}{1 - v} + \frac{\mu b}{4\pi(1 - v)} \left[\ln\left(\frac{r_1}{r_0}\right) + 1 \right] 2(f - \varepsilon). \qquad (3.13)$$

For any given value of f the minimum value of this total energy can be found when $\mathrm{d}E_T/\mathrm{d}\varepsilon = 0$. Thus the value of the elastic strain, ε, for which the total energy is minimised is given by

$$\varepsilon_{\text{crit}} = \frac{1}{h} \frac{\mu b}{4\pi E} \left[\ln\left(\frac{r_1}{r_0}\right) + 1 \right]. \qquad (3.14)$$

Now the maximum misfit at which a dislocation-free interface can be formed is when $f = \varepsilon_{\text{crit}}$. Thus we have derived an approximate expression for the critical thickness to which a deposit crystal can be grown before it becomes energetically favourable to nucleate misfit dislocations:

$$h_c = \frac{1}{f} \frac{\mu b}{4\pi E} \left[\ln\left(\frac{r_1}{r_0}\right) + 1 \right]. \tag{3.15}$$

The optical properties of most devices which contain heterointerfaces are strongly degraded by the presence of misfit dislocations, as we shall see in Chapter 8. It is thus important to be able to calculate the value of h_c for any combination of deposit layer and substrate. Equation (3.15) can be used to calculate approximate values of h_c for any semiconductor materials for which the relevant parameters are known. In Table 3.4 I have given a compilation of some of these parameters for the common III–V binary compounds. Table 3.5 give calculated values of h_c for some of these semiconductor materials on substrates with which they have a variety of values of misfit. (Data on the elastic constants of ternary and quaternary III–V compounds are scarce and so calculations on the

Table 3.4 Some data on the mechanical properties of binary III–V compounds (from Adachi 1982).

Compound	Shear modulus (10^5 MPa)	Young's modulus	Poisson's ratio	Thermal expansion coefficient (10^{-6} K^{-1})	Lattice parameter (nm)
GaAs	1.2	0.6	0.31	6.63	0.565
InP	1.02	0.46	0.36	4.56	0.587
GaP	1.4	0.4	0.31	5.91	0.545
InAs	0.83	0.7	0.35	5.16	0.606

Table 3.5 Calculated values of critical epitaxial layer thicknesses before misfit dislocations will be nucleated (from equation (3.15)).

Misfit with substrate	Epitaxial layer material GaAs or GaP (μm)	InP (μm)
10^{-2}	0.005	0.005
10^{-3}	0.92	0.9
10^{-4}	1.35	1.24
10^{-5}	17.4	15.3

Note: These values are very similar for all three binary compounds because the ratio of shear modulus to Young's modulus is almost constant.

values of h_c are hard to make for these materials.) It is clear from this table that deposit layers of a useful thickness ($0.5–10\,\mu$m) must have lattice parameters very well matched with those of their substrates if dislocations are to be avoided at the heterointerface. The constraints that this can place on the choice of materials for the fabrication of thin film optoelectronic devices will be discussed in Chapter 8. If it is necessary to grow even thicker epitaxial layers, $10\,\mu$m or above, the problem in matching the lattice parameters of substrate and deposit becomes even more severe. The growth of thick undoped silicon layers on boron-doped silicon wafers, where the lattice mismatch may be as high as 10^{-4}, can result in the formation of quite dense dislocation arrays (Sugita *et al* 1969). Similarly, the preparation of p–n junctions can also create interfacial dislocation arrays, as for example in GaAs(Zn)/GaAs(Te) junctions (Abrahams and Buiocchi 1966). In both these cases, it is the modification of the lattice parameters of the host semiconductors by the misfitting dopant atoms which causes the problem.

So far, we have assumed that the misfit dislocations have Burger's vectors which lie in the interface plane. In fact, in a (001) interface between two crystals with the sphalerite structure, dislocations with $\frac{1}{2}a[101]$ or $\frac{1}{2}a[011]$ Burger's vectors are normally observed. This is because these dislocations can easily glide to the interface from the surface of the film along inclined $\{111\}$ planes. Dislocations with Burger's vectors of this kind relieve only half the interfacial misfit accommodated by dislocations with Burger's vectors of the form $\frac{1}{2}a[110]$. In Equation (3.11) we must thus double the number of dislocations needed in the interface to relieve any particular misfit, and consequently the energy of the plastically relaxed interface will be higher. We have also assumed that the glide of $\frac{1}{2}a[101]$ misfit dislocations to the substrate/deposit interface occurs immediately the critical thickness, h_c, is exceeded during film growth. However, the dislocation nucleation processes may not be sufficiently rapid to create all the geometrically necessary misfit-relieving dislocations. (A complete discussion of the various mechanisms by which these misfit dislocations can be nucleated in growing deposits can be found in Matthews (1975)). This means that it is quite common for misfit dislocations to be observed only when films have been grown which are considerably thicker than the calculated value of h_c. We can assume for semiconductor deposits with thicknesses in the range $1–5\,\mu$m that interfacial misfits which exceed 5×10^{-3} will almost always result in the formation of dislocation arrays. In systems where the deposit has a low plastic yield strength, a misfit of 10^{-4} is usually sufficient to give dislocations at the interfacial plane. Compressive strains in the deposit are less likely to create dislocations than tensile strains and for this reason substrates are normally chosen which put the deposit under compression.

A further complication in predicting the density of an interfacial dislocation network in a heterointerface arises from the fact that epitaxial growth must be carried out at elevated temperatures. We have already seen that in LPE growth processes the growth temperature is a significant fraction of the melting temperature of the deposit material. Rather lower substrate temperatures can be used in some of the vapour phase growth techniques that will be described in the following sections, but it is generally true that all epitaxial layers will be cooled by several hundreds of degrees down to room temperature immediately after growth. If the thermal expansion coefficients of the substrate and deposit, α_s and α_d respectively, are not well matched, new strains will be generated during cooling which can create additional dislocation arrays in the heterointerface. If we are attempting to create a dislocation-free interface, it is not sufficient only to match the lattice parameters of substrate and deposit at the growth temperature; it is also important to attempt to match the values of α. It is usually assumed that the α values for ternary alloys can be calculated from a weighted linear combination of the α values of the relevant binary compounds. From the data in table 3.4 it is clear that a $Ga_{50}In_{50}As$ alloy is likely to have a thermal expansion coefficient roughly the same as that of a GaP substrate, but is not well matched to the much more useful InP substrate. Very few thermal expansion coefficients have actually been measured for ternary III–V compounds, although Bisaro et al (1979) have shown that the values of α for $In_{53}Ga_{47}As$ and an InGaAsP alloy are in good agreement with the values estimated from those of the binary and ternary compounds. The lattice parameter and thermal expansion coefficient data given in table 3.4 can be used to choose substrate/deposit combinations for which the lattice mismatch is as low as possible at all temperatures between room temperature and the growth temperature.

Figure 3.20(a) shows a typical example of the two orthogonal dislocation arrays which accommodate some of the misfit between substrate and deposit in a heterointerface grown on a (001) substate. In this example, the deposit layer is GaInP grown on a GaAs substrate, and the misfit is very high, about 3×10^{-2}. Only 20% of this misfit is relaxed by these dislocation arrays, elastic strain and other dislocations accounting for the remainder. Figure 3.20(b) shows that dislocations also propagate from the misfitting interface into the growing epitaxial layer.

3.5.1 Complex Epitaxial Systems

In the previous section I restricted the kind of heterointerface under consideration to the relatively simple case of a deposit growing epitaxially on the (001) surface of the substrate and also specified that both substrate and deposit were to have the diamond cubic, or sphalerite,

(a)

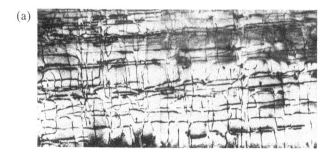

(b)

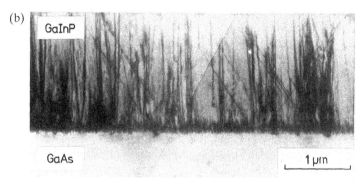

Figure 3.20 (a) A transmission electron micrograph of the dislocation array formed at the interface between a (001) GaAs substrate
and a GaInP epitaxial layer. (b) A cross-sectional transmission
electron micrograph of dislocations propagating into the epitaxial
layer from a misfitting GaInP/GaAs heterointerface. (Courtesy of
Dr M Hockly.)

structure. However, the epitaxial growth of semiconductor layers can be
used to create much more complex heterointerfaces which are important
components in a whole range of microelectronic devices. These heterointerfaces include those in which misfit is very high, but the lattice
structures of substrate and deposit are still the same, and others in
which the crystal structures are quite different. Under these circumstances of poor lattice match between substrate and deposit, we should
not be surprised to find that the density of interfacial defects is high and
we will also see that a wide range of new defects can be formed in these
interfaces.

The growth of II–VI compounds like CdTe and HgCdTe on (001)
GaAs substrates are examples of systems where the misfit between the
sphalerite lattices is rather high. Figure 3.21 compares the defect

concentration in CdTe films grown on GaAs and on an InSb substrate with which the CdTe is much better lattice matched. While the epitaxial growth process is similar in the two cases, there are many more dislocations in the CdTe layer grown on the GaAs substrate. Some of these dislocations propagate from the interface up into the CdTe layer. This example illustrates how important it is to choose the substrate material so that it matches the lattice parameter of the deposit as closely as possible, if electrically active dislocations are to be avoided in the deposit layers.

(a)   (b)

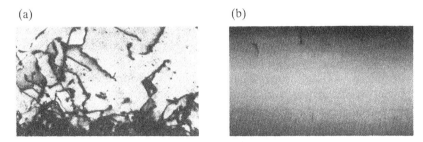

Figure 3.21 Cross-sectional transmission electron micrographs showing a comparison of the defect concentration in CdTe epitaxial layers grown on (a) GaAs and (b) InSb substrates. Both CdTe layers are approximately 1 μm thick, but a much higher dislocation density can be seen in the layer grown on GaAs. (Courtesy of Drs J Hutchison and A G Cullis.)

The growth of GaAs films on silicon substrates is a particularly important example of epitaxial growth where the mismatch between the substrate and deposit layer is large. Bulk GaAs wafers are expensive, and often contain high dislocation densities, as we have discussed earlier in this chapter. By contrast, high-quality silicon wafers are cheap and readily available. The growth of defect-free GaAs films on silicon wafers would allow the production of devices which benefit from the high electron mobility in GaAs and offer the opportunity of fabricating optoelectronic devices directly on a silicon wafer. However, the misfit between the silicon and GaAs lattices is 4×10^{-2}, so the quality of the GaAs films grown at present is rather poor, with very high densities of dislocations, twins and symmetry-related defects of the kind discussed in the following section. Figure 9.13(b) shows an example of a highly defective GaAs layer grown on a silicon substrate.

In Chapter 2 I described the philosophy which underlies the desire of device engineers to grow epitaxial layers of semiconductor materials on insulating substrates and even sandwich structures of single-crystalline semiconductor-on-insulator on semiconductor. These epitaxial samples are required for the semiconductor-on-insulator, SOI, technologies which should allow very fast and complex devices to be fabricated. I will consider here two epitaxial semiconductor/insulator systems: silicon on sapphire and the growth of crystals with the fluorite structure on silicon substrates. The epitaxial growth of silicon crystals on sapphire substrates is one of the most promising ways of producing isolated semiconductor devices on an insulating substrate.

At first sight sapphire does not appear to be a suitable material on which to grow a single crystal of silicon, since it has a rhombohedral crystal structure and provides no very obvious lattice match for the diamond cubic lattice of silicon. Figure 3.22(a) shows a sketch of the positions of the aluminium and oxygen atoms on the ($\bar{1}$012) surface of sapphire. The oxygen atoms are arranged in a pattern that is very poorly related to the distribution of atoms on any of the low index planes of the silicon lattice, but the aluminium atoms can be seen to form an approximately square array. In the figure, the atom positions in the (001) planes of silicon are shown superimposed over the aluminium sites. The correspondence between the two sets of atomic positions appears reasonably close, but there is a mismatch in atomic spacing along the [10$\bar{1}$1] directions in the sapphire surface of about 4% and along [1$\bar{2}$10] of 12%. In terms of the heteroepitaxial systems described in the last section, the match between the sapphire and silicon lattices is extremely poor. However, it has proved possible to grow epitaxial layers of silicon on carefully cut and polished sapphire ($\bar{1}$012) surfaces and in fact on a number of other planes of the sapphire lattice as well (Nolder and Cadoff 1965). A good deal of attention has been paid to developing methods for the growth and polishing of high-quality sapphire crystals. Both Czochralski and Bridgman crystal-growing techniques have been used to prepare bulk sapphire material, and the EFG technique is popular for the production of thin sapphire sheets. Cullen (1978) has reviewed the preparation of sapphire substrates for SOI devices. The high lattice mismatch means that dislocations and twins often propagate from the heterointerface up into the growing silicon film. Great efforts have been made to refine the growth processes, and to anneal out any defects after deposition, in an attempt to produce high-grade SOI material. Figure 3.22(b) shows a high-resolution transmission electron microscope image of the interface region of an epitaxial crystal of silicon grown on a ($\bar{1}$012) sapphire surface. The lattice planes in the two crystals are clearly shown in the image and no twins or dislocations can be seen in the silicon layer. This kind of image is only of tiny volume of perfect silicon, but it does illustrate that very-high-quality epitaxial silicon crystals can be grown on sapphire even though the lattice mismatch is very large indeed. However, during the growth process the sapphire substrate is attacked by the reactive silicon, resulting in the incorporation of significant levels of aluminium into the growing epita-

(a)

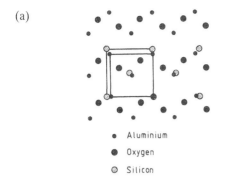

- Aluminium
- Oxygen
- Silicon

(b)

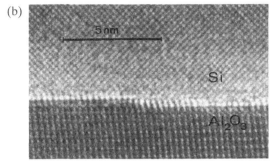

Figure 3.22 (a) A sketch of the position of the aluminium and oxygen atoms on the ($\bar{1}012$) surface of a sapphire crystal. The atom positions on a {001} plane of the silicon lattice are shown superposed over the aluminium sites. (b) A high-resolution cross-sectional transmission electron micrograph of the Si/Al_2O_3 interface, illustrating the quality of the silicon layers which can be grown. (Courtesy of Dr C Paus.)

xial film. Aluminium, of course, is a p-type dopant in silicon, and it proves very difficult to grow epitaxial silicon layers on sapphire substrates that are not heavily doped in this manner.

Now let us consider the reverse case of the growth of an epitaxial insulator layer on a semiconductor substrate. Calcium fluoride, CaF_2, has a cubic lattice called the fluorite structure, and several other fluoride compounds, including BaF_2 and SrF_2, share this same crystal structure. The lattice parameters of all these compounds lie in the range 0.5–0.65 nm, the same range as most of the common semiconductor materials. The similarity between the structures of the semiconductors and these fluorite compounds means that we might expect the epitaxial growth of one material on the other to be possible. Many of the fluoride compounds also evaporate in a congruent fashion since MF_2 molecules (where M is Ca, Ba or Sr) are extremely stable. Evaporation and molecular beam epitaxy techniques (described later in this chapter) have been successfully used to deposit films of these compounds. Numerous authors have shown that it is indeed possible to grow epitaxial layers of BaF_2, CaF_2, SrF_2 and even mixed compounds like $Ca_xSr_{1-x}F_2$ on silicon, GaAs and InP wafers with both (001) and (111) orientations, even though the misfit between the fluoride compound and semiconductor lattices may be as high as 14%. Grains of a 'twinned' orientation can be found in some films grown on (111) substrates, and under some conditions of growth the twinned structure can dominate the epitaxial relationship between the deposit and the substrate. The lattice parameters of the fluoride layers can be chosen to match those of the substrate material by growing layers with mixed compositions. $Ba_{0.17}Sr_{0.83}F_2$ has a lattice parameter that perfectly matches that of InP for example. It is convenient that BaF_2 and SrF_2, and CaF_2 and SrF_2, show complete mutual solid solubility, and that Vegard's law gives a reasonably accurate prediction of the lattice parameters of the mixed fluoride crystals.

Films of these fluorides containing very low defect densities can be grown and have an epitaxial quality very similar to that achieved in silicon films on sapphire. It is also possible to grow an epitaxial semiconductor film over the top of one of these fluoride films and it has been shown that silicon can be grown epitaxially over a CaF_2 layer on a silicon wafer (Ishiwara and Asano 1984). The dielectric properties of the fluoride layers are often excellent: resistivities in excess of $10^{12}\ \Omega$ cm and breakdown voltages of $5 \times 10^7\ V\,m^{-1}$ for example (Sullivan *et al* 1982). Further details on this fascinating and potentially extremely important, epitaxial process can be found in the above references and the review by Phillips and Gibson (1984).

Before moving on to describe the effect of crystal symmetry on interfacial structures, it is worth mentioning that the study of high-misfit epitaxial systems like $Si-Al_2O_3$ and BaF_2-Ge have shown that not all epitaxial interfaces can be thought of as two crystal lattices in intimate contact, with regions of 'good fit' being separated by dislocations which localise the misfit between the two lattices. We have assumed so far that the regions of good fit are 'coherent', with lattice planes of similar spacing in the two lattices being continuous across the interface plane. In many regions of the $Si-Al_2O_3$ sample shown in figure 3.22(b) we can trace the crystal planes across the interface, which indicates that there is some measure of coherence here. However, Phillips and Gibson have shown that the high-misfit BaF_2-Ge ($f = 9.6\%$) interface appears completely incoherent, or 'incommensurate', in a high-resolution image. The planes in the two crystals show no sign of matching across the interface. It is remarkable that high-quality epitaxial BaF_2 films can be grown from an incommensurate interface, since it is not clear how the deposit nuclei align themselves with the crystal lattice of the substrate. What is certain is that we can no longer use the argument that it is the interaction of the potential fields of substrate and deposit which dictates the orientation of the growing film. The growth of fluoride compound crystals on semiconductor surfaces is an example of an epitaxial process in which the chemical interaction between the materials at the interface is very weak, and I will describe the consequences of this in §3.6. For the particular case of BaF_2 growth on germanium, it has been suggested that the epitaxial relationship between substrate and deposit is wholly controlled by the alignment of deposit nuclei along crystallographically oriented steps on the substrate surface.

3.5.2 Bicrystallography

No discussion of the structure of epitaxial interfaces would be complete without a mention of the fact that there are characteristic crystal defects which may be present at heterointerfaces as a result of a difference in the symmetry elements of the substrate and deposit lattices. I have assumed that the misfit dislocations at epitaxial interfaces will have Burger's vectors which are characteristic of the dopant lattice. This is because these defects must glide to the interface from the deposit surface, or a dislocation source in the deposit layer (Matthews 1975). It is also possible that dislocations present in the substrate may propagate into the epitaxial interface, under which circumstances they will of course have Burger's vectors characteristic of the substrate crystal lattice. However, it is well known that grain boundaries can contain dislocations that have Burger's vectors which are characteristic of the misorientation between the two grains (see Chadwick and Smith (1976) for example). In an exactly analogous manner, dislocations with unusual Burger's vectors can be stable in heterointerfaces. In both these cases, the Burger's vectors of the interfacial dislocations will usually be much smaller than those of the lattice dislocations, and these defects are often difficult to observe even by quite sophisticated electron microscope techniques.

Pond and co-workers (1983, 1984) have presented the formal theory of interfacial defects, based on a consideration of the symmetry of heterointerfaces of all kinds. The name that has been given to the study of the symmetry-related properties of interfaces is bicrystallography. Put in a very simple form, the argument for the existence of symmetry-related defects at interfaces runs as follows.

Creating a bicrystal from any two lattices, whether it be by rotating one grain with a similar crystal structure with respect to another or by the formation of a heterointerface, will in general create an object with a lower symmetry than either of the two interacting crystals. This phenomenon is called 'desymmetrisation'.

As a necessary result of the loss of some symmetry elements, there will be more than one structurally equivalent way of arranging the atoms of the two crystals at the interface plane. This is because symmetry cannot be destroyed; if lost in the formation of the bicrystal, it must reappear in the creation of symmetry-related variants of the same interface structure. Domains of each symmetry-related morphological variant can coexist in desymmetrised epitaxial interfaces and will be separated by defects which may have dislocation character. The Burger's vectors of these dislocations can be predicted by considering the relative lattice shifts that convert one morphological variant into another, but will normally not be the same as those of the ordinary crystal lattice dislocations.

As an example of the use of the theory of bicrystallography let us consider the epitaxial growth of $NiSi_2$, which has the fluorite crystal structure on a (001) silicon substrate. The misfit between the two cubic lattices is small, about 0.4%, and so we expect only a low density of ordinary misfit-relieving lattice dislocations at the heterointerface. However, the formation of an $Si/NiSi_2$ bicrystal reduces the symmetry considerably, and there are six morphological variants in this case, all with identical interface structures. Separating these variants will be interfacial dislocations with unusual Burger's vectors of the type $\frac{1}{4}a[111]$. These dislocations have been identified experimentally by Cherns and Pond (1984). A schematic diagram of the core structure of a dislocation of this type in an interface between a crystal with the fluorite structure and one with the diamond cubic structure is shown in figure 3.23.

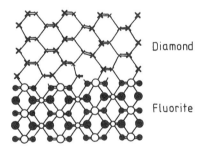

Diamond

Fluorite

Figure 3.23 A schematic illustration of a symmetry-related disloca-tion formed at the interface between a crystal with the fluorite lattice and one with the diamond cubic lattice (after Pond *et al* 1984.)

While in most heterointerfaces the density of the dislocations resulting from desymmetrisation will be low, it is important to remember that symmetry-related morphological variants will be present in almost all such interfaces. It seems unlikely that the domain structure of the deposit will affect the electrical properties of the heterointerfaces to any significant extent, since the atomic structure of the interface does not vary across a domain boundary. However, in the cases where the deposit is a polar material, GaAs for instance, antiphase domain boundaries may separate the adjacent morphological variants (Pond *et al* 1984). The presence of these boundaries may well degrade the electrical properties of an epitaxial semiconductor deposit, and yet the density of these defects can only be determined if the structure of the film is investigated in very great detail. Symmetry-related 'inversion boundaries' are frequently found in GaAs layers grown on silicon substrates and some of these defects are included in the highly defective GaAs layer shown in figure 9.13(b).

3.6 PRINCIPLES OF VAPOUR PHASE GROWTH

The homogeneous nucleation of a solid phase from a melt may require a significant undercooling — tens or even hundreds of degrees — because of the necessity of creating a large area of new solid/liquid interface. However, if a heterogeneous nucleation site is introduced into the melt, a seed crystal for instance, the undercooling required for crystal deposition is reduced. Under circumstances in which a low-energy interface can be formed between the seed and the growing crystal, the critical undercooling may be only a few degrees. The choice of a seed crystal which is pefectly, or at least closely, lattice matched to the crystal which is grown from the melt is thus an important factor in keeping the undercooling, and so the melt supersaturation, small. For a discussion of the basic features of the nucleation of a solid phase from melt the reader is directed to Porter and Easterling (1981) or Brice (1986). We have also seen that melt growth processes do not quite proceed under equilibrium conditions, since the measured partition coefficients for impurity elements are not equal to the values predicted from the equilibrium phase diagram. However, because of the low driving force for growth we can consider melt growth to be a near-equilibrium process. The composition of the crystals which are grown depends primarily on the partition coefficient and this quantity approaches the equilibrium value as the rate of growth is reduced. The rate of growth depends on the combination of thermodynamic parameters, like the latent heat of freezing of the semiconductor, and material properties like the thermal conductivity (see equation (3.3)).

The situation is quite different for the growth of crystals from the vapour phase. We can distinguish two kinds of vapour phase process: physical vapour deposition (PVD) and chemical vapour deposition (CVD). It is convenient to to describe a PVD process as one in which a highly non-equilibrium concentration of vapour phase species is produced from a source or target material by evaporation or sputtering. In this kind of deposition the supersaturation in the vapour phase is very high and we expect the rapid condensation of a deposit film onto any available surface, including the walls of the deposition chamber. Under these circumstances it is the composition of the vapour above the source which determines the composition of the deposited films; the substrate temperature will control the microstructure but not the chemistry of the films. Evaporation and sputtering are commonly used in the deposition of conducting thin films and will be discussed in more detail in Chapter 4. There I shall describe how the essentially non-equilibrium nature of PVD growth can lead to very high defect densities in deposited layers and even the growth of layers with phases and structures not found in the equilibrium phase diagram.

Chemical vapour deposition processes are distinguished from PVD techniques by the fact that film growth proceeds as a result of a series of chemical reactions above a heated substrate. The basic principles of CVD growth will be described in §3.7, but here I should make it clear that the rate of deposition and the composition of the films are controlled by parameters like:

(1) The partial pressures of both precursor compounds and reaction products in the reactor. The equilibrium position of the important chemical reactions will often be determined by the composition of the gas phase.

(2) Whether the principal chemical reaction is homogeneous or heterogeneous.

(3) The substrate and reactor temperatures, which control the rate of all the chemical reactions.

(4) The flow rates of precursor gases into the reaction chamber, which influence the rate of transport of reactant species to the substrate. The gas flow conditions (laminar or turbulent) in the reactor are also important.

It is clear from this list that chemical thermodynamics and kinetics will often play an important part in determining the composition of the deposited layers. Many CVD processes can be considered as near-equilibrium growth phenomena, while PVD is clearly very far from an equilibrium process.

Before going on to consider some of the details of the growth of epitaxial layers by CVD techniques I shall introduce some of the concepts we need to understand, such as how a single-crystal deposit is nucleated in a vapour phase growth process.

3.6.1 The Nucleation of Epitaxial Films

During our consideration of melt growth processes above, I have not introduced any discussion of the nucleation of the crystals since in most cases a seed or a substrate crystal is provided on which the melt can be deposited. Under conditions of low supersaturation in the melt, the deposit atoms can simply join onto the terminating crystal planes on the surface of the substrate or seed, the lowest-energy positions to which atoms can be added, and the growth process has begun. As long as the supersaturation of the melt is kept low, and contact is maintained between the seed and the melt, the chance of growing a perfect crystal is high. We can take this simple view of nucleation from a melt because of the near-equilibrium nature of the process, which means that the melt atoms will only condense onto a surface if they can do so in the equilbrium, or lattice, sites. The energy gain on forming the solid is so small in these near-equilibrium processes that any less energetically favourable positions for atoms on the substrate surface will be unstable and the atoms will spontaneously return to the melt.

The circumstances are quite different if the deposit species arrive on the substrate from a highly supersaturated vapour phase. Once again, the deposit atoms may take up low-energy positions at the bottom of the potential wells created by the periodic arrangement of atoms on the substrate surface. These positions may be considered as the equilibrium positions and, if occupied, will result in the growth of an epitaxial film. However, in some vapour phase growth processes, the rate of arrival of deposit atoms on the substrate surface can be very high indeed. Under these circumstances we can ask the question: do the deposit atoms have sufficient time to reach their equilibrium sites on the substrate surface before they are covered by more deposit atoms and prevented from further movement or re-evaporation? In order to try and answer this question we must think a little more about the mechanisms by which a deposit film is nucleated from a vapour. I shall follow the treatment of Venables and Price (1975) in describing the atomic processes involved in film growth.

We can picture the process of nucleating a thin film in three stages:

(1) The atoms in the vapour come into contact with the substrate, where they either form bonds with the surface atoms or re-evaporate. The atoms which do remain bonded to the substrate can be considered to have condensed from the vapour. I shall call the energy of the bonds which have been formed, i.e. the energy which binds the deposit atom to the surface, E_a.

(2) The condensed atoms may be able to migrate over the substrate surface, and I shall define the activation energy of this surface diffusion process as E_d. Obviously the higher the temperature of the substrate, the faster this surface diffusion will be and the more quickly the deposit

atoms will be able to migrate to their equilibrium sites.

(3) The deposit atoms diffusing over the substrate surface may have a mutually attractive interaction on the substrate and so join together to form clusters at an early stage of growth. I shall call the energy of the bonds between adjacent atoms in these clusters E_c. (Alternatively, the atoms can have a repulsive interaction and form an ordered two-dimensional array on the substrate. I shall not consider this second possibility any further here.)

With these parameters defined we are able to separate the condensation process into several regimes of growth, each of which will produce films with a characteristic structure. If E_d is high, and the temperature of the substrate is low, then surface diffusion is essentially frozen out. We must assume that, under these conditions, the deposit atoms remain in the positions in which they land on the substrate. Only amorphous or microcrystalline films can be produced in such a growth process; the rearrangement of the randomly distributed atoms necessary to create an epitaxial deposit cannot occur. Films with this kind of structure are discussed in the next chapter. A more interesting region of growth, if we are concerned with nucleating an epitaxial film, occurs when E_d is low or the substrate temperature is high. Now the condensed deposit atoms are able to migrate freely over the substrate and we can distinguish two kinds of growth process: the first where E_a is much larger than E_c; and the second where the reverse is true. If the binding energy of the deposit atoms to the substrate far exceeds the energy of bonds between adjacent deposit atoms, the film will tend to nucleate and grow in a plane-by-plane manner. Each arriving deposit atom will migrate over the surface until it reaches the edge of a two-dimensional deposit island (figure 3.24(a)). The atom will be bound to this position by an energy $E_a + 3E_c$ if we assume that it now has three neighbouring deposit atoms. The strong interaction between the deposit atoms and the substrate means that the formation of two-dimensional deposit islands is energetically preferred to the nucleation of any three-dimensional clusters. By contrast, if E_c is much larger than E_a, it is energetically favourable for the deposit atoms to migrate over the substrate surface and agglomerate into three-dimensional islands (figure 3.24(b)). These two nucleating processes are often referred to as Frank–Van der Merwe and Volmer–Weber growth modes respectively. A third mode of growth involves first the nucleation of two-dimensional layers, followed by the nucleation of three-dimensional islands on the first deposit layers. This is known as Stranski–Krastanov growth and is clearly controlled by the difference in the value of E_a for deposit atoms on the substrate and on the first deposit layer. The rate equations governing the migration and nucleation processes in these three modes of thin film growth have been reviewed by Venables and Price (1975) and by Stoyanov (1979), but will not be presented here. What is of more direct interest is to try to determine how these kinds of nucleation events can lead to the growth of an epitaxial film.

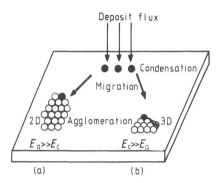

Figure 3.24 A schematic illustration of two different ways in which a deposit can be grown from the vapour phase; (a) by agglomeration into two-dimensional islands on the substrate, (b) by the formation of discrete three-dimensional islands.

Let us first consider the situation where there is a strong interaction between deposit atoms and the substrate, e.g. the growth of a III–V compound semiconductor film on an InP or GaAs wafer, or the growth of a metal film on a single-crystal metallic substrate. The large binding energy of the deposit atoms to the substrate ensures that these atoms have little choice but to fall into positions between the atoms on the substrate surface and an epitaxial film will be grown almost as a continuation of the substrate lattice. The development of the epitaxial relationship between deposit and substrate will inevitably mean that the deposit lattice is strained to some extent, unless it has the same lattice parameter as the substrate. Thus an epitaxial deposit can be assumed to nucleate as long as there is sufficient surface diffusion to allow the deposit atoms to reach their equilibrium positions after condensation. This can only occur if the substrate is heated and is also scrupulously clean. It has been shown in numerous experiments that levels of contamination, usually carbon and oxygen species at barely detectable levels, can stop the process of epitaxial nucleation completely, or change the epitaxial relationship between substrate and deposit. However, once an epitaxial layer has been nucleated, the deposit will usually continue to grow as a single crystal to whatever thickness is required.

In the situation where there is only a very weak interaction between deposit atoms and the substrate, e.g. metal or semiconductor materials grown on ionic substrates (§3.5.1) we cannot explain epitaxial growth by any strong influence of the deposit lattice over the orientation of the three-dimensional deposit islands. The thermodynamically stable nuclei may be very small indeed for these deposits, even a single atom in some cases, although these atoms quickly agglomerate into larger clusters. The orientations of small deposit nuclei have been directly observed in transmission electron microscope experiments at a very early stage of film growth. It is clear from these experiments that the nuclei are not all aligned in the same epitaxial relationship with the substrate, but a strong preference has been found for the most densely packed planes to be oriented parallel with the substrate in a wide range of materials (see for example the work of Metois et al (1972) on the growth of gold on KCl). As the deposition process continues, these nuclei will grow as they collect more condensed atoms and eventually impinge. It appears that epitaxial films are produced from these often randomly oriented nuclei by a process involving the migration, rotation and agglomeration of nuclei. The vital stage in the development of an epitaxial deposit is the recrystallisation of the polycrystalline islands during the agglomeration process. The driving forces for these complex rearrangements are not well understood, nor are some of the epitaxial relationships formed by the growth of deposits on substrates with which they interact only weakly. A common example of an epitaxial relationship of this last kind is the growth of epitaxial gold deposits on the (001) surface of rocksalt. The gold films grow in perfect epitaxy with the rocksalt although the lattice parameter of NaCl is much larger than that of gold (a misfit of 27%), and a simple rotation of the gold epitaxial deposit by 45° with respect to the rocksalt lattice, to align $\langle 110 \rangle_{Au}$ with $\langle 100 \rangle_{NaCl}$, would produce an epitaxial relationship with much reduced misfit (about 2%). The first stages of nucleation in many cubic materials on substrates with which they interact weakly seems to involve the formation of multiply twinned particles. It is the coalescence and recrystallisation phenomena in these particles which, rather surprisingly, results in the formation of a completely epitaxial deposit. A detailed discussion of the various theories of epitaxial growth in weakly interacting systems can be found in Matthews (1975) and Kern et al (1979). The role that surface steps may play in the nucleation of weakly interacting epitaxial films has been mentioned in the section on the growth of fluoride films.

In the remainder of this chapter we will be most concerned with the epitaxial growth of semiconductor materials on semiconductor substrates. Here the only requirement for epitaxial growth is that the misfit at the interface is not too large (although we have seen that reasonable epitaxial films can be grown with misfits as high as 15%) and that the substrate is extremely clean and at a high enough temperature to allow rapid surface diffusion. The precise temperature needed to ensure the growth of high-quality epitaxial films is not easily predicted and usually has to be determined empirically for each combination of substrate and deposit. Temperatures between a few hundred and 1500 °C are used in the epitaxial growth processes described in this chapter, and some specific examples of the most effective range of substrate temperatures for the growth of particular epitaxial layers are given below. Increasing the substrate temperature not only increases the rate of surface migration, but also the evaporation rate of deposit species from the substrate surface. This evaporation process can decrease the density of stable deposit nuclei formed for any given incident flux of deposit atoms and often improves the crystalline quality of the epitaxial films. The concept which I have introduced here of film growth being a process of condensation, migration and agglomeration will also prove valuable in our consideration of the growth and structure of polycrystalline thin films in the next chapter.

3.7 CHEMICAL VAPOUR DEPOSITION

The process of chemical vapour deposition, CVD, can be described most simply as the use of chemical reactions to create free product species which condense to form a thin deposit film on a substrate. These chemical reactions can be simple pyrolytic processes — the decomposition of silane, SiH_4, on substrates at 600–1000 °C to deposit silicon films for example — or a more complex interaction of two or more precursor gases to form a compound or alloy film. It is convenient to divide CVD reactions into two groups: those which take place as a result of random collisions between molecules in the gas phase (a homogeneous reaction); and those which occur predominantly on a heated substrate (heterogeneous reactions). We will see that the character of the chemical reaction can have quite a strong effect on the quality of the epitaxial film grown in a CVD process, especially if the pressure in the reactor is significantly less than one atmosphere.

Let us first consider the case where the precursor gases are introduced into an open-ended reactor tube as a dilute mixture in an inert carrier gas like H_2, and the total gas pressure is about one atmosphere. We will also assume that as the gas mixture is passed over the substrate a laminar flow condition is set up, creating a boundary layer over the substrate surface in which the gas flow is much slower than in the centre of the reactor tube. Precursor gas species must diffuse to the substrate, and excess product gases escape, across this boundary layer. The rates at which these gas phase diffusion processes occur determines the rate of growth of an epitaxial deposit and will also control the deposit composition when alloy or compound layers are being grown. For this reason, atmospheric-pressure CVD (APCVD) growth processes are often considered to be 'transport controlled'. Parameters like the substrate temperature, gas flow rates, reactor geometry and gas viscosity all effect the transport phenomena in the boundary layer and so influence the structure and composition of the deposit films.

Because the rate of film growth in a CVD reactor operating at atmospheric pressure depends strongly on the precise nature of the local gas flow, it can prove difficult to achieve uniform film growth on a large number of stacked wafers in a reactor. The wafers themselves will encourage turbulence in the gas flow and the variation in the boundary layer thicknesses will lead to quite different rates of growth in different parts of the reactor. To reduce the dependence of the growth rate and film composition on the hydrodynamics in the reactor, many CVD processes are carried out at total gas pressures well below one atmosphere. At low pressures, chemical reactions become more important in determining the character of the deposited films.

Let us consider the case where a homogeneous gas phase reaction occurs between gases A and B to form a compound C. If A and B are mixed before they enter the heated reactor then the homogeneous reaction will occur immediately they become hot. For example, the reaction between SiH_4 and O_2 is homogeneous, producing solid SiO_2 particles which then settle over all the reactor and substrate surfaces. This is obviously unsatisfactory if we are trying to grow a thin SiO_2 film over the substrate, but this pre-reaction phenomenon can be avoided by only mixing the silane and oxygen directly over the substrate surface.

By contrast, a heterogeneous reaction will only occur actually on a substrate (or on the walls of the reactor) and so will be much more likely to produce a smooth, even and conformal deposit. Using the example of the pyrolysis of silane on a hot substrate we can now look a little more closely at the details of a heterogeneous CVD deposition process. Wilkes (1984) has described how a heterogeneous CVD reaction can be considered as a series of adsorption and reaction steps. The first step in the decomposition of silane is the adsorption of the silane molecule onto a vacant surface site V_s, and this process can be described by a simple chemical reaction

$$V_s + SiH_{4(g)} \xrightarrow{K_1} SiH_{4(a)} \tag{3.16a}$$

where the subscripts (g) and (a) refer to species in the gas phase and adsorbed onto the substrate respectively. The silane molecule will then decompose on the hot substrate:

$$SiH_{4(a)} \xrightarrow{K_2} Si_{(a)} + 4H_{(a)}. \tag{3.16b}$$

Finally, two molecules of hydrogen must desorb from the surface leaving a silicon atom behind:

$$4H_{(a)} \xrightarrow{K_3} 2H_{2(g)}. \tag{3.16c}$$

The equilibrium constants for the adsorption and desorption reactions, K_1 and K_3, are given by

$$K_1 = \frac{[SiH_{4(a)}]}{[SiH_{4(g)}][V_s]} \qquad K_3 = \frac{[H_{2(g)}]^{1/2}[V_s]}{[H_{2(a)}]} \tag{3.16d}$$

while the rate of growth of the silicon film, dx/dt, is given by

$$dx/dt = K_2[SiH_{4(a)}]. \tag{3.16e}$$

Rearranging these three equations, and assuming that molecular hydrogen is strongly absorbed on the substrate surface so that $[H_{2(a)}]$ is approximately 1, gives

$$\frac{dx}{dt} = K_1 K_2 K_3 \frac{[SiH_{4(g)}]}{[H_{2(g)}]^{1/2}}. \tag{3.16f}$$

The dependance of the silicon growth rate on the inverse root of the hydrogen partial pressure in the reactor has been experimentally observed by Duchenin *et al* (1978). It is clear from this analysis of a heterogeneous CVD reaction that the pyrolysis stage (equation (3.16b)) may not be the rate controlling step in the overall reaction. Adsorption and desorption thermodynamics and kinetics must be taken into account in any complete description of a reaction of this kind. For this reason, the measured activation energies for heterogeneous reactions contain significant contributions from the heats of adsorption of reactant species on the substrate. The overall heat of reaction is also an important parameter in a heterogeneous CVD reaction. If the reaction is exothermic (as it is in the case of the deposition of germanium by the decomposition of GeI$_2$ for example), then deposition will occur preferentially on colder surfaces away from the heated substrate. The whole CVD reactor must be heated if film growth is to be concentrated on the substrate surface, and these CVD systems are called 'hot-wall' reactors.

In this brief survey we have seen that the growth rate and film composition in APCVD processes are often controlled by gas transport kinetics in the reactor tube. In low-pressure CVD (and in many APCVD processes in which the precursor gases are highly diluted) the thermodynamics of the chemical reactions will often be more important than gas hydrodynamics and the composition of the gas mixture and the heats of mixing and reaction of the precursor gases can be used to predict the composition of the deposited layers (Beuchet 1985). In order to promote epitaxial growth of the deposit it is normally necessary to maintain the substrate at a high temperature. This has the additional effect of decreasing the sticking coefficient of gaseous molecules on the substrate, which decreases the rate of heterogeneous reactions. If the substrate temperature is sufficiently high, the reaction may shift from the substrate surface to the gas phase as a homogeneous reaction becomes more rapid than the heterogeneous one.

It has often been found that the electrical and optical properties of epitaxial layers deposited in CVD reactors depend strongly on all the parameters mentioned above: gas phase composition, flow conditions, substrate temperature and the character of the dominant chemical reaction. The first three of these can easily be controlled in most CVD equipment, and information on how to optimise the properties of the epitaxial layers is usually obtained empirically for each deposition facility. The geometry of the reactor is particularly important in determining gas flow conditions, and descriptions of the types of reactors which have been designed for CVD growth of semiconductor materials can be found in the reviews by Shaw (1975), Olsen (1982) and Pearce (1983).

The major advantage of CVD for the growth of thin epitaxial films of semiconductor crystals is that, once the optimum substrate temperature and precursor gas flow rates are determined, very many wafers can be processed in a single growth run to give layers with a high crystalline perfection. This ability to deposit layers on several substrates at a time, and the reasonably short cycle time for a growth run (as low as 1 h in some systems), make CVD a very efficient method for growing large numbers of epitaxial films, far outperforming LPE techniques in this respect.

3.7.1 CVD Growth of Epitaxial Silicon Layers

The use of CVD silicon layers in integrated circuits has been described in §2.3. Epitaxial layers are grown over the surface of relatively impure Czochralski silicon wafers and also over sapphire substrates for SOI devices. CVD silicon can be reproducibly grown with high resistivity and purity, as well as excellent crystalline perfection.

Silicon can be deposited from a range of compounds, including SiH_4, $SiCl_4$ and $SiHCl_3$. Silane is used mostly as the precursor gas from which to deposit polycrystalline silicon, as will be discussed in the next chapter. Silicon tetrachloride is the most popular compound for the CVD of epitaxial silicon films. The deposition proceeds by the overall reaction

$$SiCl_4 + 2H_2 \rightarrow Si + 4HCl \tag{3.17}$$

although there are many intermediate stages, as we would expect from the discussion above of the silane decomposition reaction. The decomposition of $SiCl_4$ has been described in greater detail by Sirtl et al (1974). Silicon tetrachloride is chosen as the precursor compound because it is available in a very pure form and is relatively cheap. However, a high substrate temperature is needed before reaction (3.17) begins, at least 900 °C. A high substrate temperature is also needed to stimulate epitaxial growth, because the activation energy for diffusion of silicon atoms on a silicon surface is very high, about 5 eV. We have already seen that extensive surface diffusion must occur if an epitaxial deposit is to be nucleated. Figure 3.25 shows the region of the substrate temperature–growth rate plot where epitaxial silicon layers can be grown on silicon wafers and also illustrates how strongly the temperature affects the rate of growth of the silicon film. Epitaxial silicon layers are normally grown on substrates at about 1200 °C and with fairly low concentrations of $SiCl_4$ in the gas phase to keep the growth rate around $1\ \mu m\ min^{-1}$. The gas mixture admitted into the reactor can be at atmospheric pressure, with a strong preponderance of H_2 over $SiCl_4$, or at a reduced pressure. 10^{-2} Torr is the lowest useful total gas pressure for reproducible film growth over the surface of many wafers in a reactor. Reducing the precursor gas concentration, and keeping the substrate temperature high, reduces the density of deposit nuclei on the substrate surface and so improves the crystalline quality of the epitaxial films.

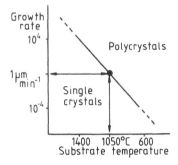

Figure 3.25 A schematic illustration of the way in which the substrate temperature and growth rate can control the structure of CVD silicon layers. Epitaxial layers are only grown when the substrate temperature is high or the growth rate low.

In situ doping of the silicon films during growth can be achieved by the inclusion of low concentrations of arsine, AsH_3, phosphine, PH_3, or diborane, B_2H_6, in the reactor gas mixture. The decomposition of these compounds on the hot substrate leads to the incorporation of the dopant elements in the growing silicon films. The kinetics of the processes by which the dopant species are included in the silicon lattice are complex and poorly understood. Under some circumstances the growth rate of the silicon epitaxial layer is markedly decreased by the addition of the precursor dopant species to the reactor. This may be because the precursor dopant molecules are strongly adsorbed onto the substrate surface, leaving few sites for the attachment of the $SiCl_4$ molecules. There also exists the possibility that dopant atoms like arsenic can react with the high concentration of chlorine gas released by the decomposition of the $SiCl_4$ and form volatile $AsCl_3$. It has been observed that the arsenic concentrations in silicon epitaxial layers decrease as the rate of deposition increases (Reif *et al* 1979) and the precise doping levels which are achieved in a CVD growth process can usually only be determined by direct measurement of the properties of epitaxial layers deposited in each particular reactor.

Impurity elements, from the materials of which the reactor is constructed and from the substrate itself, can be included in a epitaxial layer during CVD deposition. The reactor vessel is often made from quartz, while the susceptor on which the substrates are supported is conveniently made from graphite, which may be coated with silicon nitride in some cases. We expect that the passage of the reactant gas mixture over these components will result in the contamination of the deposit with oxygen and carbon. However, the relatively fast growth rates used in most silicon CVD processes limit the extent of this contamination to acceptably low levels. A much more serious impurity problem arises because of contamination from the substrate. I have already described in the section on the growth of silicon layers on sapphire substrates how the dissolution of the substrate by the silicon can result in the growth of degenerately doped material. A similar effect occurs when an epitaxial layer is grown on a heavily doped silicon wafer. Most epitaxial silicon layers are intended to have low doping levels, 10^{16} cm^{-3} or below, and yet solid state diffusion of the doping elements out of the substrate, and the evaporation and recondensation of dopant atoms during deposition, can result in the formation of unwanted heavily doped regions at the lower part of the deposited film. Figure 3.26 illustrates the result of this kind of autodoping process. Only at the top of the epilayer will the desired doping level be achieved. Autodoping can change the electrical properties of the whole of a thin epitaxial layer, and the higher the substrate temperature the worse this effect is likely to be. Device engineers have to take into account the possibility of autodoping when they are designing both the structures of the devices being grown and the CVD processes which are to be used to deposit the epitaxial silicon layers.

Finally in this section on the CVD growth of silicon layers, I should mention the kinds of defects which can be found even in epilayers of the highest crystalline quality. The occurrence of misfit dislocation networks at the interfaces between lightly doped silicon films and heavily doped

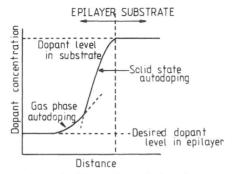

Figure 3.26 A schematic illustration of the phenomenon of 'auto-doping' in an epitaxial layer. Dopant species from the substrate can diffuse into the deposit layer and can also evaporate from the hot substrate and recondense on the growing surface.

wafer substrates has already been described in §3.5. In addition, dislocations in the substrates can propagate into the epilayers, i.e. threading dislocations. Impurity precipitates can also be formed in cases where the contamination levels are high, particularly SiC particles when excess carbon is incorporated from the susceptor. These precipitates will be surrounded by stacking faults or dislocation networks. The density of these kinds of defects must be very low in device-quality material and a carefully designed CVD process can be used to grow large numbers of these very perfect epitaxial silicon layers.

3.7.2 CVD Growth of Compound Semiconductor Layers

A very wide range of compound semiconductor materials can be grown in the form of epitaxial thin films by CVD processes. This range includes simple binary compounds like GaAs and InP, a huge number of ternary and quaternary III–V alloys, and some interesting II–VI materials including ZnO and ZnSe. In order to grow epitaxial layers of these materials, all that is required is that the mixture of gases introduced into the reaction chamber should include a precursor gas for each of the elements required in the deposit. Most compound layer growth is carried out at atmospheric pressure, which simplifies the design of the reactors. It is clear that the gas mixtures will be much more complex than the $H_2 + SiCl_4$ combination which is used for the growth of silicon layers. For instance, in the growth of a layer of a GaAsP alloy, a useful material in which to fabricate light-emitting diodes (see Chapter 8), the three constituent elements and at least one dopant element must be introduced into the reaction chamber in very carefully controlled pro-portions. The composition of the gas phase is important in determining the stoichiometry of the films which are deposited. It is frequently observed that the rate of growth of a compound film is controlled by the concentration of only the metal-carrying precursor gases. For example, in the growth of III–V materials like GaAs and GaAsP, the concentration of the gallium-containing precursor gas is the most important parameter in determining the growth rate. It is commonly assumed that all the metal atoms arriving on the substrate are incorporated into the growing film. This is very convenient since it means that the crystal grower does not need to pay too much attention to controlling the flow of the other precursors into the reactor, as long as the non-metal precursors are present in excess. In all CVD growth processes, a primary source of contamination is the reactor itself, and so silicon is a common impurity in the compound semiconductor films.

In figure 3.27 I have illustrated the kind of equipment which could be used for the deposition of a ternary III–V alloy, GaAsP in this case. Arsine and phosphine are common precursor gases when arsenic and phosphorus are to be included in a deposit, while the gallium is introduced to the gas flow by passing HCl gas over a molten gallium source to form volatile GaCl. The AsH_3 and PH_3 are pyrolytically decomposed in homogeneous reactions in a hot region of the reactor situated just before the substrate, so that GaCl, P_4 and As_4 molecules are all available in the gas flow over the substrate, where reaction occurs to form the ternary alloy. It is important to control the temperatures of source, decomposition and deposition zones carefully to ensure that the required gas phase chemical reactions proceed in the correct directions. These reactions are

source zone:$\qquad$ $2HCl + 2Ga \rightarrow 2GaCl + H_2$ $\qquad$ (3.18a)

decomposition zone: $4AsH_3 \rightarrow As_4 + 6H_2$ $\qquad$ (3.18b)

$\qquad\qquad\qquad$ $4PH_3 \rightarrow P_4 + 6H_2$ $\qquad$ (3.18c)

deposition zone:$\qquad$ $XAs_4 + YP_4 + 4GaCl + H_2$

$\qquad\qquad\qquad\qquad$ $\rightarrow 4GaAs_XP_Y + 4HCl.$ $\qquad$ (3.18d)

A knowledge of the equilibrium reaction constants for each of reactions (3.18a–d) is required if the correct mixture of gases is to be provided to the deposition zone. The ratio of arsenic to phosphorus in the epitaxial layer is controlled by the relative efficiencies of reactions (3.18b) and (3.18c). III–V alloys containing indium can be grown by producing $InCl_3$ in the same way as I have shown for GaCl. Dopant elements such as sulphur or selenium are conveniently added by the introduction of low concentration of H_2S or H_2Se into the reactor.

We can see that a most important requirement for preparing a compound semiconductor layer is that a volatile precursor should be available for each component element. These precursor gases are easy

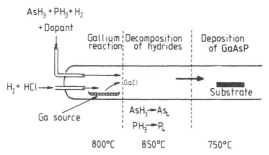

Figure 3.27 A sketch of the kind of CVD reactor in which epitaxial GaAsP layers can be grown. The gallium source and decomposition zones are maintained at different temperatures to ensure that the gas mixture reaching the deposition zone will grow a deposit layer of the desired composition.

to find for some of the elements. For example if arsenic is to be deposited, both AsH_3 and $AsCl_3$ are suitable compounds. Some elements have sufficiently high vapour pressures at reasonable temperatures for a solid source to function as an efficient source for CVD growth, e.g. mercury, tellurium and zinc. More unusual compounds must be used if elements like aluminium and cadmium are to be included in the deposited layers. Table 3.6 lists some volatile compounds from which it is possible to deposit these 'difficult' elements and also shows some of the alloys and compounds which they help to grow. These semiconductor materials include many of the most important II–VI compounds for electro-optical and surface acoustic wave devices. All the precursor compounds listed in table 3.6 are metallo-organic compounds and this has lead to the special CVD growth processes which use these materials being called metal organic CVD, or MOCVD. Most of these metallo-organic compounds are liquids at room temperature and special heating facilities may have to be installed on the CVD equipment to give a well controlled, and reasonably high, vapour pressure of the precursor for introduction into the reactor. A few of the common precursor compounds are extremely unstable and so tend to decompose or react with other gases before reaching the heated substrate. Indium-containing compounds are particularly prone to premature reaction. The purity of the metallo-organic compounds determines the quality of the epitaxial semiconductor layers and care has to be taken to exclude impurities which create deep states or which compensate for the dopants that are deliberately introduced. In the following chapter mention will be made of the use of some metallo-organic compounds in the deposition of metallic conductor films.

The growth of epitaxial layers of compound semiconductors by MOCVD techniques has become one of the most important methods for the

Table 3.6 A few of the metallo-organic compounds used as precursor gases in MOCVD growth processes, and some of the semiconductor compounds which they can be used to deposit.

Compound	Melting point (°C)	Semiconductor material
$(CH_3)_3Al$	15.4	GaAlAs, AlAs, AlN
$(C_2H_5)_3Al$	−58	
$(CH_3)_2Cd$	− 4.5	CdS, CdTe, CdSe
$(CH_3)_2Zn$	−42	ZnS, ZnSe, ZnO
$(C_2H_5)_2Te$	~10	CdTe, HgCdTe
$(CH_3)_3Sb$	−87	GaAlSb, GaAlAsSb
$(CH_3)_3Ga$	−16	GaAlAs, GaInAs, GaInAsP
$(C_2H_5)_3Ga$	−82	
$(CH_3)_3In$	88	InP, GaInAs, GaInAsP
$(C_2H_5)_3In$	−32	
$(CH_3)_4Sn$	−53	Dopant in III–V materials

preparation of microelectronic device material. This is because the electrical and optical properties of MOCVD films are often excellent. Once again, the gas hydrodynamics in the reactor and the substrate temperature are very important factors in controlling the composition and crystalline quality of the films, and we should not forget that the chemical reactions which occur during an MOCVD process are extremely complex and not fully understood. Considerable effort is currently being directed at developing new stable MOCVD precursor compounds which will be easier to handle than the poisonous, pyrophoric and unstable compounds listed in table 3.6. It may even prove possible to prepare volatile compounds which contain two or more of the elements required in the epitaxial films.

Another advantage of CVD growth techniques is that they have the potential to deposit films containing sharp changes in composition. This is because the composition of the film depends strongly on the relative proportions of the precursor gases immediately above the substrate. If we have independent control over the flow rate of each reactant gas into the CVD chamber, we can change abruptly the composition of the gas flow and the stoichiometry of the growing film. The deposit thickness over which these composition changes can be achieved can be as small as a few tenths of a nanometre because the composition of the gas above the substrate can be modified extremely rapidly. This is very useful if p–n junctions are to be grown *in situ* for instance, or in the production of some of the more complex compound semiconductor structures like the laser crystals described in §8.5. In addition, the composition, and so the lattice parameter, of an epitaxial layer can be graded from an initial composition which forms a perfect lattice match with the substrate at the bottom of the layer to material with the required optical or electronic properties at the top surface.

From this brief description of the use of CVD techniques for the growth of epitaxial layers of compound semiconductor layers, we can see that it is a most flexible process. The opportunity for growing epitaxial layers on many substrates in each run is a major advantage when high-volume production of epilayers is needed, but the successful design of multisubstrate epitaxial reactors has proved much harder for III–V materials than for silicon. Quite complex multiple epitaxial layer structures can be deposited, as long as the gas-handling facilities on the reactor are designed for rapid and accurate flow switching. However, there is a further use which can be made of the fact that CVD reactions in the vapour phase can lead to non-equilibrium growth, and this is very convenient for the preparation of an unusual semiconductor material. LPE techniques are commonly used for the growth of nitrogen-doped GaP epilayers for the production of light-emitting diodes. In a liquid phase growth process it is not possible to introduce a concentration of nitrogen which is in excess of its equilibrium solubility limit because of the essentially equilibrium nature of melt growth processes. In CVD, however, increasing the concentration of nitrogen in the reactant gases can result in the inclusion of a supersaturation of nitrogen in the growing GaP layer. Thus, while LEDs which operate only in the green part of the visible spectrum can be produced in LPE epitaxial GaP layers, the operating range of these devices in CVD material is extended into the yellow. The high melting temperature of GaP ensures that the excess nitrogen atoms remain electrically active at room temperature, although precipitation may occur if the epilayers are heated. This example illustrates the important point that a non-equilibrium growth process can produce non-equilibrium materials and that these materials may have unusual and useful properties which may be exploited in microelectronic devices. For some recent reviews of the use of CVD techniques for the growth of epitaxial layers of silicon and many III–V compounds and alloys the reader is directed to Pearce (1983), Mullin et al (1984) Ludowise (1985) and Stringfellow (1985).

3.8 MOLECULAR BEAM EPITAXY

Molecular beam epitaxy, MBE, is one of the newest of the epitaxial growth techniques and yet in some ways it is also the simplest. In essence, MBE is a sophisticated evaporation process, and since a section in the next chapter will be devoted to a description of the fundamentals of film deposition by evaporation, here I shall describe the features of MBE growth only very briefly. Figure 3.28 shows a sketch of an MBE reactor containing several elemental sources, each with its own independently operated shutter, and a heated substrate on which the film is deposited. Each source produces a molecular beam of deposit species directed at the substrate. At no point does this beam come into contact with the walls of the chamber and so there is no opportunity for contamination species from the deposition vessel to be included in the deposit. In addition, a high background vacuum is maintained in the chamber so that few collisions occur between residual gas species and the molecular beams. In this way, contamination from the gas phase can be avoided as well. MBE deposition processes are thus able to prepare films which are significantly cleaner than even the best CVD or LPE layers. Because of this, the growth rates in MBE deposition processes can be very low indeed, which gives the species absorbed on the substrate surface ample time to diffuse to a low-energy site. It is thus not surprising that films with very high crystalline quality can be prepared, even at relatively low substrate temperatures. The possibility of growing crystals at substrate temperatures as low as 300 °C is one of the most important features of MBE growth.

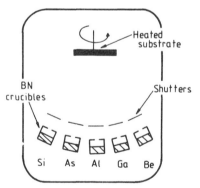

Figure 3.28 A sketch of the components of an MBE reactor which contains five elemental sources, each with an independently operated shutter, and a heated substrate.

The evaporation sources in most MBE systems are more properly described as effusion cells. In such a cell an element or compound is maintained at a carefully determined temperature within a crucible which contains only a small orifice directed at the substrate. These crucibles are commonly made from boron nitride. The vapour phase species subliming or evaporating from the charge in the cell will quickly reach equilibrium and then effuse in a highly reproducible manner into the MBE chamber to produce the molecular beam. Effusion sources of a wide range of elements are readily available, although two of the most widely used microelectronic elements, phosphorus and arsenic, have very high vapour pressures and are hard to evaporate as slowly as is required in MBE growth. Silicon has also proved difficult to sublime from an effusion cell because of its high melting temperature and so electron beam evaporation is often used to deposit this element in MBE systems. Some II–VI compound semiconductors evaporate to produce molecular species (e.g. PbTe evaporates as PbTe molecules), so the deposition of stoichiometric films of this material is very simple. Other II–VI compounds dissociate during evaporation, but the composition of the vapour is almost exactly the same as that of the solid phase, i.e. congruent sublimation. Figure 3.29 shows the equilibrium p–T diagram for the CdTe binary system which we have already seen in figure 1.3. Unlike the equivalent diagram for the hypothetical AB compound illustrated in figure 3.1, the vapour pressure loops of cadmium and tellurium overlap over a considerable temperature range. It is possible in a compound of this kind for congruent sublimation to occur where the equilibrium partial pressure of cadmium is exactly double that of the Te_2 species. The sloping lines in figure 3.29 show the variation in partial pressures of cadmium and tellurium with temperature during congruent sublimation, and it is possible to deposit stoichiometric layers of this compound from a single source. Most III–V compounds are much more conveniently deposited from elemental sources. This is because the vapour pressures of gallium and arsenic over solid GaAs for instance are hugely different, rather as illustrated for the AB system in figure 3.1. This means that

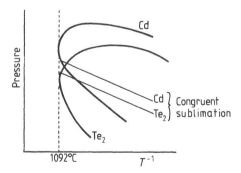

Figure 3.29 The p–T equilibrium phase diagram for CdTe, showing that congruent sublimation can occur over a wide temperature range.

congruent evaporation can only occur at very low source temperatures (637 °C for GaAs) where the equilibrium vapour pressures are too low for deposition at a useful rate. At high temperatures, the stoichiometry of the film will be difficult to control if a GaAs compound source is used. It is thus necessary to grow epitaxial films of GaAs from two effusion cells. An arsenic effusion cell will produce a beam of As_4 and As_2 molecules, while a gallium cell will produce only atomic species, and these species arriving on the substrate surface will react to form GaAs. The special problems which are associated with the MBE growth of III–V alloy semiconductor layers have been reviewed by Wood (1982).

Once appropriate effusion cells have been chosen for the deposition of semiconductor materials which do not evaporate congruently, it remains to determine the flux of material required from each source to grow a film with the desired stoichiometry. This is an exactly analogous problem to the choice of gas composition in CVD growth. In MBE processes, the choice of flux densities is complicated by the fact that the sticking coefficients of the vapour phase species on the substrate vary over a huge range. We should remember that at the very slow growth rates characteristic of MBE, the species arriving on the substrate surface have ample time to re-evaporate before they are covered by subsequent layers. Arsenic molecules have a very low sticking coefficient on most substrates, while gallium atoms have a sticking probability of almost unity. This means that the growth of stoichiometric GaAs can only be achieved if the flux of arsenic molecules is very much greater than of gallium atoms. The rate of film growth is controlled wholly by the gallium flux, the excess arsenic species rapidly re-evaporating from the substrate. In fact, Arthur (1968) has shown that arsenic molecules will only stick on a surface which is covered by gallium atoms, where they rapidly react to form GaAs. Similarly, the sticking coefficient of mercury atoms is very low indeed and large quantities of expensive pure mercury must be evaporated to grow even a thin layer of HgCdTe.

The final parameter which must be controlled to ensure the growth of high-quality epitaxial films is the substrate temperature. We have seen that the nucleation of epitaxial films will usually occur only when the substrate is hot enough to promote extensive surface diffusion. In general, the sticking coefficients of the deposit species decrease as the substrate temperature is increased, so the normal conditions for epitaxial growth are high deposit fluxes and a high substrate temperature. Under these conditions, a stoichiometric film of a compound semiconductor will be grown, even if the concentrations of the various deposit species on the substrate surface are not those in the semiconductor lattice. The excess species which are not able to react to form the compound semiconductor will simply re-evaporate from the hot substrate. There is thus some margin for error in setting the effusion rates for the component elements in a compound phase, but the flux of dopant elements must be controlled very exactly if films with the correct electrical properties are to be prepared. The choice of the dopant species to use in an MBE reactor is dictated by the equilibrium vapour pressures of suitable elements. Too high a vapour pressure will result in the dopant atoms re-evaporating from the heated substrate before being incorporated in the growing layer. Silicon, tin and tellurium can be used as donor impurities in III–V compounds, while beryllium, magnesium and manganese are effective acceptor dopant elements.

Figure 3.28 shows an MBE system with five elemental effusion sources suitable for the growth of epitaxial layers of both p- and n-type GaAlAs. It is also possible to use this system to deposit alternating thin epitaxial layers of GaAs and AlAs, simply by using the shutters to turn the various molecular beams on and off. This beam-switching action is extremely rapid and, when combined with the very slow growth rates characteristic of MBE processes, $0.1\text{--}0.5\,\mu\text{m}\,\text{h}^{-1}$, means that abrupt changes in composition or doping character can be grown into the films. Very accurately designed graded composition profiles can also be prepared if required. Figure 3.30 shows an example of the kind of multilayered epitaxial structure which can be deposited in CVD or MBE systems, using the ability to alter the composition very rapidly. These structures are examples of genuine 'nanometre engineering', which promises to produce material with some extraordinary and potentially very important new electronic properties. Some discussion of the use of 'quantum well' structures for the production of especially efficient laser diodes will be given in §8.5. The development of MBE and MOCVD techniques specifically for the deposition of these very perfect hetero-structures with extremely thin layers are described by Chang and Giessen (1985) and the special issue of the *Journal of Crystal Growth* **68** (1984).

MBE growth is, however, not a convenient method for the deposition of large numbers of epitaxial layers. The very stringent vacuum require-ments and the slow growth rates make each deposition process a lengthy operation. MBE equipment is also extremely expensive. Thus at present, this powerful growth technique is used primarily for the growth of complex layered structures of the kind shown in figure 3.30, which may be difficult to fabricate with sufficient accuracy by CVD or LPE techni-ques. The use of relatively low substrate temperatures also means that solid state diffusion problems like autodoping and the broadening of composition profiles are less damaging than in CVD and LPE processes. This may prove to be particularly important when IV–VI alloys, in which lattice diffusion is often extremely rapid, are being deposited. Further details on MBE growth techniques, and the quality of the material which can be grown, can be found in the reviews by Chang and Ludeke (1975), Chang (1980) and Tsang (1985).

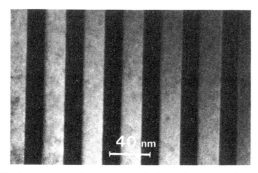

Figure 3.30 A cross-sectional transmission electron micrograph of a GaInAs/InP multiple quantum well structure. The contrast in the micrograph arises from the difference in the electron absorption efficiency in the two semiconductor materials. The individual layers are about 20 nm thick, although very much finer structures can be grown by MBE techniques if desired. (Courtesy of Mr P Augustus and Plessey Research (Caswell) Ltd.)

To conclude this discussion of techniques by which thin epitaxial semiconductor films may be grown, it seems useful to summarise the relative advantages of the three techniques which I have introduced: LPE, CVD and MBE. It is fair to assume that the crystalline quality of the epitaxial layers is comparable for all three growth processes. The three most important features of any commercial process for the deposition of large numbers of epitaxial films are: efficiency of throughput; initial expense of growth facilities; and the precision with which the growth can be controlled. We have seen that CVD provides the most efficient process for the growth of epitaxial layers on wafers, the best choice for a high-volume technique. LPE, by contrast, cannot be used to prepare very large numbers of epilayers, but is a cheap way of depositing III–V compounds for devices like light-emitting diodes. MBE is the most expensive technique, but gives the greatest control over the composition of the films and is a good choice for the deposition of the most complex structures.

REFERENCES

Abrahams M S and Buiocchi C J 1966 *J. Appl. Phys.* **37** 1973
Adachi S 1982 *J. Appl. Phys.* **53** 8775
Arthur J R 1968 *J. Appl. Phys.* **39** 4032
Ast D G, Cunningham B and Gleichman R 1984 in *Electron Microscopy of Materials*, *MRS Symp. Proc.* **31** (New York: North-Holland) p85

Bergh A A and Dean P J 1976 *Light Emitting Diodes* (Oxford: Clarendon)

Beuchet G 1985 in *Semiconductors and Semimetals* **22A** 261 ed. W T Tsang (New York: Academic)

Bisaro R, Mercenda P and Pearsall T P 1979 *Appl. Phts. Lett.* **34** 100

Bradshaw S E and Goorissen J 1980 *J. Cryst. Growth* **48** 514

Brice J C 1986 *Crystal Growth Processes* (Glasgow: Blackie)

Casey H C and Panish M B 1978 *Heterostructure Lasers* Part B (New York: Academic)

Chadwick G A and Smith D A 1976 *Grain Boundary Structure and Properties* (London: Academic)

Chalmers B 1964 *Principles of Solidification* (Chichester: Wiley)

Chang L L 1980 in *Handbook of Semiconductors* vol. 3 ed. S P Keller (Amsterdam: North-Holland) p563

Chang L L and Giessen B C 1985 *Synthetic Modulated Structures* (Orlando: Academic)

Chang L L and Ludeke R 1975 in *Epitaxial Growth* ed. J W Matthews (New York: Academic) p37

Cherns D and Pond R C 1984 in *Thin Films and Interfaces II* ed. J E E Baglin *et al* (New York: North-Holland) p423

Crossley I and Small M B 1973 *J. Cryst. Growth* **19** 160

Cullen G W 1978 in *Heteroepitaxial Semiconductors for Electronic Devices* ed. G W Cullen and C C Wang (New York: Springer) p6

Duchenin J-P, Bonnet M and Koelsch F 1978 *J. Electrochem. Soc.* **125** 637

Grovenor C R M 1987 *Materials for Semiconductor Devices* (London: Institute of Metals)

Hamilton D R and Seidensticker R G 1960 *J. Appl. Phys.* **31** 1165

Harmon T 1973 *J. Non-Metals* **1** 183

Hirth J P and Lothe J 1978 *Theory of Dislocations* (New York: McGraw-Hill)

Holmes D E, Chen R T, Elliott K R and Kirkpatrick C G 1982 *Appl. Phys. Lett.* **40** 1

Hobgood H M, McGuigan S, Spitznagel J A and Thomas R N 1986 *Appl. Phys. Lett.* **48** 1654

Horikoshi Y 1985 in *Semiconductors and Semimetals* **22C** ed. W T Tsang (New York: Academic) p 93

Hsieh J J 1974 *J. Cryst. Growth* **27** 49

Ishiwara H and Asano T 1984 in *Thin Films and Interfaces II* ed. J E E Baglin *et al* (New York: North-Holland) p393

Jordan A S 1980 *J. Cryst. Growth* **49** 631

Jordan A S and Ilegens M 1975 *J. Phys. Chem. Solids* **36** 329

Kern R, LeMay G and Metois J J 1979 *Currents Topics in Materials Science* ed. E Kaldis (Amsterdam: North-Holland) p131

Kressel H and Butler J K 1979 *Semiconductor Lasers and Heterojunction LED's* (New York: Academic)

Lawrence J E and Huff H R 1982 in *VLSI Electronics, Microstructure Science* **5** ed. N G Einspruch (New York: Academic) p 51

Leipold M H, O'Donnell T P and Hayen M A 1980 *J. Cryst. Growth* **50** 366

Lorenz M R 1967 in *Physics and Chemistry of II–VI Compounds* ed. M Aven and J S Prenner (Amsterdam: North-Holland) p73

Ludowise M J 1985 *J. Appl. Phys.* **58** R31
Matlock J H 1979 *Semicond. Int.* **2** October 33
Matthews J W 1975 in *Epitaxial Growth* (New York: Academic) p559
Metois J J, Gauch M, Masson A and Kern R 1972 *Surf. Sci* **30** 43
Metz E P A, Miller R C and Mazelsky R 1962 *J. Appl. Phys.* **33** 2016
Micklethwaite W F H 1981 in *Semiconductors and Semimetals* **18** ed. R K
    Willardson and A C Beer (New York: Academic) p 47
Mullin J B, Irvine S J C and Tunnicliffe J 1984 *J. Cryst. Growth* **68** 214
Mullin J B, Staughan B W and Bricknell W S 1965 *J. Phys. Chem. Soc.* **26** 752
Nakajima K 1982 in *GaInAsP Alloy Semiconductors* ed. T P Pearsall (New
    York: Wiley) p43
——— 1985 in *Semiconductors and Semimetals* **22A** ed. W T Tsang (New York:
    Academic) p 1
Nelson H 1973 *RCA Review* **24** 603
Nolder R and Cadoff I 1965 *Trans. Met. Soc. AIME* **233** 549
Olsen G H 1982 in *GaInAsP Alloy Semiconductors* ed. T P Pearsall (New York:
    Wiley) p11
Pamplin B R 1975 *Crystal Growth* (Oxford: Pergamon)
Patel J R 1977 in *Semiconductor Silicon 1977* (Pennington, NJ: Electrochemical
    Society) p521
Pearce C W 1983 in *VLSI Technology* ed. S M Sze (New York: McGraw-Hill)
    p11
Pfann W G 1966 *Zone Melting* (New York: Wiley)
Phillips J M and Gibson J M 1984 in *Thin Films and Interfaces II* ed. J E E
    Baglin *et al* (New York: North-Holland) p381
Pond R C, Gowers J B, Holt D B, Joyce B A, Neave J H and Larsen P K 1984
    in *Thin Films and Interfaces II* ed. J E E Baglin *et al* (New York:
    North-Holland) p423
Pond R C and Vlachavas D S 1983 *Proc. R. Soc.* A**386** 95
Porter D A and Easterling K E 1981 *Phase Transformations in Metals and
    Alloys* (London: Van Nostrand Reinhold)
Rae C M F, Grovenor C R M and Knowles K M 1981 *Z. Metall.* **72** 798
Ray B 1969 *II–VI Compounds* (Oxford: Pergamon)
Reif R, Kamins T I and Saraswat K C 1979 *J. Electrochem. Soc.* **126** 644
Shaw D W 1975 in *Epitaxial Growth* ed. J W Matthews (New York: Academic)
    p89
Sirtl E, Hunt L P and Sawyer D H 1974 *J. Electrochem. Soc.* **121** 919
Stephanov A V 1959 *Sov. Phys.–Tech. Phys.* **29** 339
Stolte C A 1984 in *Semiconductors and Semimetals* **20** ed. R K Willardson and
    A C Beer (New York: Academic) p 89
Stoyanov B 1979 in *Current Topics in Materials Science* vol. 3 ed. E Kaldis
    (Amsterdam: North-Holland) p 421
Stringfellow G B 1985 in *Semiconductors and Semimetals* **22A** ed. W T Tsang
    (New York: Academic) p 209
Sugita Y, Tamura M and Sugawara K 1969 *J. Appl. Phys.* **40** 3089
Sullivan P W, Farrow R F C and Jones G R 1982 *J. Cryst. Growth* **60** 403
Tsang W T 1985 *Semiconductors and Semimetals* **22A** (New York: Academic)
Van Gool W 1966 *Principles of Defect Chemistry of Crystalline Solids* (New
    York: Academic)

Van der Merwe J H 1963 *J. Appl. Phys.* **34** 117, 123

Van der Merwe J H and Ball C A B 1975 in *Epitaxial Growth* ed. J W Matthews (New York: Academic) p493

Venables J A and Price G L 1975 in *Epitaxial Growth* ed. J W Matthews (New York: Academic) p382

Wilkes J G 1984 *J. Cryst. Growth* **70** 271

Willardson R K and Beer A C 1984 *Semiconductors and Semimetals* **20** (New York: Academic)

Willardson R K and Goering H L 1962 *Compound Semiconductors* (New York: Reinhold)

Wood G E C 1982 in *GaInAsP Alloy Semiconductors* ed. T P Pearsall (New York: Wiley) p87

Zschauer K-H and Vogel A 1971 *Gallium Arsenide and Related Compounds (Aachen) 1970* (Inst. Phys. Conf. Ser. 9) p100

4

Polycrystalline Conducting Thin Films

4.1 INTRODUCTION

The semiconductor devices illustrated in Chapter 2 all contain thin conducting layers which provide electrical contacts between the devices and to the outside world via the packaging that supports the semiconductor chip. The structure of solar cells is discussed in Chapter 9 and in some of these devices thin polycrystalline films of semiconducting materials are the active components. It is clear that the deposition of thin polycrystalline conducting and semiconducting films is another important process in the fabrication of microelectronic devices. Conducting films which are used as interconnects must have resistivities which are as low as possible, while semiconductor thin films are normally needed with properties as similar as possible to those of the bulk materials. In most cases, the films are also required to have good adhesion to their metallic, semiconducting or insulating substrates. The details of the performance of practical metallisation layers, and thin polycrystalline semiconductor films in solar cells, will be considered in Chapters 5 and 9, respectively. Here the techniques which are used to deposit these films will be presented and their characteristic structures and electrical properties described.

There are a number of ways in which a thin polycrystalline layer of a conducting or semiconducting material can be deposited, and we will concentrate in this section on those techniques where the layer is grown from the vapour phase. These provide the most convenient and flexible ways of growing films with thicknesses up to a few tens of micrometres. All vapour phase deposition processes must be carried out in a vacuum system of one kind or another where the residual gas pressure will vary from about 10^{-2} Torr when the gas is an active part of the deposition process, as it is in sputtering, to the very good vacua needed for the evaporation of reactive metals, 10^{-8} Torr or better. It is out of place here to discuss the technology of vacuum equipment, but the interested reader is directed to the volumes by Holland (1960), Holland *et al* (1974) and Maissel and Glang (1983) for information on the techniques and practices that are important in the production of satisfactory thin films.

The techniques which are of particular interest for the deposition of polycrystalline thin films for microelectronic devices are evaporation, sputtering and chemical vapour deposition (CVD). A brief description of each of these techniques will now be given before the structure and properties of polycrystalline thin films are considered.

4.1.1 Evaporation

The basic principle of film deposition by evaporation is that increasing the temperature of a material will also increase its equilibrium vapour pressure. When this vapour pressure is significantly higher than the residual pressure in the deposition chamber, a flux of atoms or molecules of the material will be projected in all directions from the heated evaporation source. A cold substrate in the path of this vapour flux provides an efficient nucleation site for these species, which are now in a region of extreme supersaturation simply because the temperature difference between the evaporation source and the substrate is usually many hundreds of degrees. The equilibrium vapour pressure above the substrate will thus be very much lower than above the source, and condensation will readily occur. This process is exactly the principle of deposition from the effusion sources I have described in §3.8, except that we no longer need a heated substrate since no attempt is being made to induce epitaxial growth of the film on the substrate.

The most important property of any material which we wish to evaporate is thus its equilibrium vapour pressure, which determines whether it can be evaporated at reasonable source temperatures. The vapour pressure depends primarily on the enthalpy change on vaporisation, H_{vap}, as is clearly seen from the Clausius–Clapyron equation:

$$\frac{d(\log p)}{dT} = \frac{\Delta H_{vap}}{RT^2}. \tag{4.1}$$

Figure 4.1 shows the variation in equilibrium vapour pressure with temperature above the surfaces of a number of metals. This figure makes it clear that not all these metals will be at all easy to evaporate. The rate of film growth from a supersaturated vapour onto a cold substrate is roughly proportional to the vapour pressure of the evaporating material, and vapour pressures between 10^{-3} and 10^{-2} Torr are usually needed to achieve a sensible growth rate. We can see, therefore, that to evaporate aluminium at a reasonable rate will require source

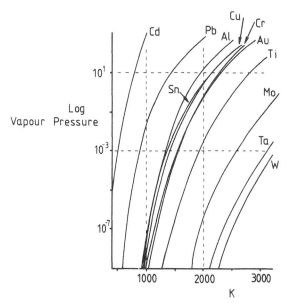

Figure 4.1 A plot of the equilibrium vapour pressures over pure metallic elements. The pressure range within which evaporation occurs at a reasonable rate is indicated by the horizontal broken lines, while the vertical broken lines show the source temperatures at which it is convenient to operate.

temperatures of the order of 1200 K, while to evaporate a refractory metal like tungsten the source will have to be heated to 3000 K or more. Most metals will melt before they evaporate at any very significant rate, but a few, like chromium for example, will sublime from the solid state. Some oxide compounds can be evaporated, usually forming vapour phase species which are fragments of the compound like AlO units from the decomposition of Al_2O_3. Other compounds evaporate without decomposing; BaF_2 is an example of such a material. A range of compound semiconductors can be successfully evaporated, for example most of the II–VI semiconductors (Cusano 1967, Chopra and Das 1983), but for III–V materials the control over the stoichiometry of the films is generally poor for the reasons discussed in §3.2.2. Evaporated CdS and ZnS films have been particularly comprehensively studied and their electrical and optical properties only approach those of bulk crystals if the substrate temperature is above 200 °C or the film is annealed after deposition (see §9.5.2). This is presumably because the electrically active point defects included in the film during evaporation are annealed out at elevated temperatures.

The precise stoichiometry of the evaporated film will depend on the vapour pressures of the various evaporant elements over the source. In the next chapter we will see that it is frequently necessary in the manufacture of microelectronic devices to deposit alloy materials, and the evaporation rates of the elements will depend on their relative concentrations in the alloy source as well as their equilibrium vapour pressures. The vapour pressure of component x in an alloy will be given by the product of c_x, the atomic fraction of x in the alloy, p_x^* the equilibrium vapour pressure above pure x, and f_x, the activity coefficient of x in the alloy:

$$p_x = f_x c_x p_x^*. \tag{4.2}$$

This means that for most binary alloys the deposited film will have a different composition to that of the source material. Only fortuitously would the vapour pressures of all the elements in the alloy be similar enough to ensure that the stoichiometry of the source material was reproduced in the deposit film. This problem can be overcome by overloading the source material with the alloying element with the lowest vapour pressure. Under these circumstances, however, the composition of the film will vary with thickness, the element with the highest vapour pressure being concentrated near the substrate. This is a particular problem with the deposition of Al–Cu alloy films, as will be described in §5.4.3. Another problem is that the evaporated film may not even have the expected equilibrium structure or mixture of phases. Metastable phases not included in the equilibrium phase diagram of the alloy can be formed, changing the properties of the films in an unpredictable manner. This kind of effect will be considered in §4.6.

The process of increasing the vapour pressure of the source material to a sufficiently high value to achieve a reasonably high rate of deposition requires only that the source be heated. This can be carried out in a number of simple ways. One of the most common of these is thermal evaporation from tungsten filaments or boats. These filaments are heated by the passage of a current and the source material hung over the tungsten will start to evaporate at a rapidly increasing rate as the current, and so the temperature, is increased. The vapour pressure of the tungsten at the evaporation temperature must be very much lower than that of the source material if contamination of the evaporated film with tungsten is to be avoided. This method is thus most successful for the evaporation of materials with relatively low melting temperatures like aluminium and the noble metals. However, since the volume of the charge of source material on a filament is limited, the coverage of large areas, or the growth of unusually thick film (more than 10 μm), is not easy. A number of microelectronic materials which can readily be resistance evaporated are listed in table 4.1.

Table 4.1 Techniques for the vapour phase deposition of conducting films of microelectronic materials.

Method	Materials most successfully deposited
Evaporation: thermal	Low-melting-point metals and alloys like Al alloys and gold Some II–VI semiconductors like CdS
electron beam	High- and low-melting-point elements, alloys and compounds including oxides and fluorides Si, W, Ta, SiO_2 and silicides
flash	Useful for thick layers and alloys with large differences in vapour pressure like Al–Cu alloys and III–V semiconductor compounds
Sputtering: DC, RF and magnetron	Wide range of metals, alloys and semiconductors Al alloys, refractory metals, silicon, oxides and silicides
reactive	Oxides and nitrides (TiN, SiO_2 and Si_3N_4)
Chemical vapour deposition:	Most used for polysilicon, refractory metals and silicides

Flash evaporation can be used to deposit alloy and compound materials. Here, small beads of the source material are dropped onto a ceramic substrate which is sufficiently hot to vaporise the alloy instantly. Thick deposits can be laid down by repeated evaporation of a large number of beads and the film composition will approximate quite closely to that of the source material, as any fractionation effects caused by differences in the vapour pressures of the alloyed elements are removed by the total evaporation of each small source bead. For instance, flash evaporation can be used reasonably successfully even for the deposition of III-V compound semiconductors where the difference in the vapour pressures of the two component elements above a molten source is often several orders of magnitude (see figure 3.1).

Induction heating of source materials in refractory boats, often made from boron nitride, is another simple method of inducing evaporation. In this case, the volume of the source material can be much larger than is comfortably accommodated on a tungsten filament and so thick films can easily be grown. Fractionation problems can be a problem in this process, causing quite dramatic changes in the stoichiometry of alloy films.

Electron beam evaporation involves the focusing of an intense beam of electrons onto the surface of a relatively large piece of source material. Not all the source is melted, so that contamination effects from the crucible are avoided since the melt is completely contained within the source material itself. (Contamination from BN crucibles can be significant in inductively heated evaporation systems.) A very wide range of materials can be successfully evaporated in an electron beam system, and these include low-melting-point and refractory metals, oxides and metallic alloys. Some of these materials are listed in table 4.1. Electron beam heating is the most flexible of these evaporation methods and is very commonly used for the deposition of metallic conduction layers onto microelectronic structures. For a discussion of the equipment used in these evaporation processes, the reader is directed to Maissel and Glang (1983).

4.1.2 Sputtering

The principle of sputter deposition is quite different to that of thermal evaporation. Here a plasma is struck above the source material in a low concentration of deliberately introduced gas. The high density of energetic gas ions formed in the plasma bombard the source surface, resulting in the abrasion of the surface material away into the vapour phase. This abrasion creates the high supersaturation of source atoms in the vapour necessary for the deposition of a film on a cold substrate elsewhere in the vacuum system. The plasma must be formed by an RF field when the source material is an electrical insulator, but DC plasmas can be used for metals or semiconductors. The particular features of a sputtering process which make it very attractive for the deposition of conducting layers are: that a very wide range of materials can easily be deposited; that the adhesion of these films on oxide, metallic and polymeric substrates is usually excellent; and that the composition of an alloy source is approximately reproduced in the films because the abrasion of the surface is less species' selective than an evaporation process. A comprehensive review of the sputtering process, specifically for the case of epitaxial film growth, has been given by Francombe (1975) but is relevant for the case of polycrystalline film growth as well.

The good adhesion of the sputtered films to their substrates may be a result of the cleaning of the substrate by the sputtering ions before the film starts to grow. Deposit atoms incident on the substrate have a much higher kinetic energy than those produced by high-temperature evaporation and will be able to transfer some kinetic energy to atoms on the substrate surface. The difference in the energy of incident species in evaporation and sputtering processes can result in the growth of films with quite different microstructures, even when the source material is the same in both cases. The most obvious problem with sputtering thin films is that a significant concentration of the sputtering ions is always incorporated into the growing layer because of the relatively high pressure of this gas in the chamber, often more than 10^{-2} Torr. For instance, a film sputtered in argon would normally contain at least 2 at.% of this element.

The noble gas sputtering process has been modified in two ways: to grow films more rapidly and to deposit compound materials like the oxides and nitrides. Imposing a strong magnetic field aligned perpendicular to the source surface gives a very much faster deposition rate since the deposit ions in the vapour phase are now constrained to diffuse directly to the substrate surface rather than randomly around the deposition chamber. This process is called 'magnetron sputtering'. Replacing the bombarding noble gas ions with reactive species like nitrogen or oxygen ions allows the direct formation of nitrides and oxides from pure metal sources. This process is called 'reactive sputtering' and can be used to grow compound films of many kinds with very well controlled stoichiometries. Table 4.1 gives some examples of the kind of materials which can be deposited in the various sputtering processes.

Both sputtering and evaporation have one major disadvantage when they are used for the deposition of continuous layers over rough substrate surfaces. Evaporation sources usually have areas smaller than 1 cm^2 which means that the layers are growing effectively from a point source. Sputtering targets can be as large as a few inches in diameter, and the sputtering gas can have a diffusing effect on the flux of sputtered particles, but the range of angles over which the deposit species are incident on the substrate surface is still relatively small. During deposition onto a rough substrate, shadowing effects can easily occur, giving films that are discontinuous or vary in thickness. Voids in a conducting film are obviously very unwelcome, and regions of unusual thinness offer greater resistance to the flow of the current, creating 'hot spots' in the film. These hot spots will be preferential sites for flux divergence in electromigration processes and can lead to crack formation and failure of the conductor (see §5.4). Typical shadowing effects are illustrated in figure 4.2 for a metallic film being deposited onto a silica layer containing an etched via leading down to the silicon wafer surface. It has been found that applying a voltage to the substrate in the sputtering system ('bias sputtering') can help to improve the conformal nature of metallic deposits to some extent. However, it is still difficult to ensure that all shadowing effects are removed in an evaporation or sputtering process. A deposition process which does not suffer so badly from this kind of non-uniform growth on a rough surface is chemical vapour deposition, and this technique is becoming increasingly more important in the growth of microelectronic thin films.

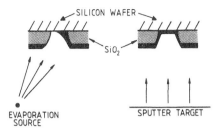

Figure 4.2 A schematic illustration of the phenomenon of shadowing during the sputtering or evaporation of a metallic film onto a via. In both cases the metal is of uneven thickness and may even be discontinuous.

4.1.3 Chemical Vapour Deposition

The CVD process has been introduced in the last chapter as a method for the growth of thin single-crystal semiconductor films, but it is equally useful for the growth of highly conformal polycrystalline conducting films. Chemical vapour deposition of metallic layers proceeds by the pyrolytic decomposition, or chemical reduction, of a gaseous species over a hot substrate. In either case, the non-volatile component of the decomposed gas will be deposited on the substrate. A few reactions which can be used for the deposition of thin conducting films are

$$WF_6 + 3H_2 \rightarrow W(s) + 6HF(g) \quad 250 - 500 \,^{\circ}C \qquad (4.3)$$

$$MoCl_5 + 2.5H_2 \rightarrow Mo(s) + 5HCl(g) \quad 800 \,^{\circ}C \qquad (4.4)$$

$$SiH_4 \rightarrow Si(s) + 2H_2 \quad 600 - 650 \,^{\circ}C \qquad (4.5)$$

$$6WF_6 + 3Si(s) \rightarrow 2W(s) + 3SiF_4 \quad \text{less than } 200 \,^{\circ}C. \qquad (4.6)$$

These examples show two reduction processes, a pyrolytic decomposition reaction and finally a very interesting low-temperature process for depositing thin tungsten films by the reduction of tungsten hexafluoride gas by the silicon surface itself. While most CVD processes will deposit layers indiscriminately all over the substrate surface, this last reaction will only occur where the WF_6 can come into contact with unreacted silicon. This allows the selective deposition of tungsten only at the bottom of vias through dielectric layers and not on the dielectric materials themselves. This selectivity arises because the enthalpy decrease for the reaction of WF_6 with silicon is some four times larger than that of WF_6 with SiO_2. The selective deposition process is also self-limiting in the sense that when a thin layer of tungsten is formed, the WF_6 can no longer come easily into contact with the silicon surface and the reaction rate drops dramatically. The selective thickness of tungsten is about 10–20 nm, and about twice this thickness of silicon is consumed in the reaction. Thicker layers of tungsten can be deposited by subsequent reduction of the WF_6 over the tungsten surfaces. This process is also selective because of the enhanced dissociation of hydrogen gas molecules on the tungsten surfaces. Further details on this fascinating and unusual way of depositing a thin conducting film are given by Broadbent and Stacy (1985), who also discuss the very stringent control over the cleanliness of the silicon surface necessary before satisfactory deposition is achieved. Residual oxide layers on the silicon surface result in very non-uniform tungsten layers with deep pits etched into the silicon surface.

Of the reactions (4.3)–(4.6) I have listed above, the deposition of polycrystalline silicon films by the pyrolysis of silane is currently the most important one in the fabrication of microelectronic devices. Polysilicon contacts and interconnects have found a variety of applications, of which the most common is probably as gate electrodes. The CVD reaction which is used to deposit these layers has therefore been studied in great detail, and much of this work has been reviewed by Adams (1983). The practical details of the CVD growth of polysilicon layers are quite similar to those described in Chapter 3 for the epitaxial growth of silicon, but the substrate temperatures are generally much lower (600–650 °C) as rapid surface diffusion is no longer required. The decomposition of silane to deposit polysilicon is a heterogeneous reaction (see §3.7) and usually carried out in a low-pressure CVD (LPCVD) process to give very conformal layers. It would be an advantage to be able to prepare polysilicon films with as low a sheet resistivity as possible, but it has been found that the inclusion of PH_3 or AsH_3 in the CVD reactor dramatically reduces the polysilicon growth rate. This will increase the processing time needed to grow doped layers of the required thickness to unacceptable levels. The use of alternative precursor gases may prove valuable in increasing the growth rate of doped polysilicon films, and compounds like disilane, Si_2H_6, have been studied for this application.

CVD methods can also be used for the deposition of aluminium films and compound materials like the metal silicides, which can be grown with very good control over the film stoichiometry. The purity of these layers can be excellent since metallo-organic gaseous precursor compounds are relatively easily purified by fractional distillation. The grain size in metallic films can be very small, 10 nm or less, and this can have unfortunate effects on stability as we will see in Chapter 5. For comparison, the grain size in evaporated films of the same materials can be as large as 5 μm. Some of the disadvantages of CVD techniques for the deposition of conducting layers include the toxicity and flammability of many of the precursor compounds and the relative complexity of the deposition process. The conditions for obtaining CVD films with good electrical and mechanical properties are usually more restrictive than in simple evaporation or sputtering processes and so the chance of producing unsatisfactory layers is higher. This point will be further emphasised when the CVD growth of dielectric films is discussed in Chapter 6. However, the conformal nature of the deposit, and the possibility of growing layers with positional selectivity, make it a very attractive process for the growth of complex metallisation systems, especially when devices with very small dimensions are being designed.

4.2 THE STRUCTURE OF VAPOUR-DEPOSITED THIN FILMS

At the end of the previous chapter we were concerned with methods for growing thin single crystals of semiconducting materials from the vapour phase onto substrates with lattice parameters and crystal structures well matched with those of the deposit. The situation is quite different when we are depositing thin conducting layers for microelectronic applications. Here it is usually not possible to grow a single-crystal film because the deposit structure will only rarely be matched with that of the substrate. In an extreme case metal layers are grown on amorphous substrates like dielectric glasses, a situation where of course no epitaxial relationship can exist. Under these circumstances, a polycrystalline film is always produced, although the films may be textured (the grains are aligned with a preferred direction normal to the plane of the film).

Even in the cases where an epitaxial relationship between substrate and deposit can be achieved, aluminium films on silicon surfaces for instance, the device engineer will not increase the temperature of the substrate to the level necessary to produce an epitaxial aluminium layer. This is because the electrical properties of the aluminium–silicon interface are only weakly dependent on whether the aluminium is a single crystal or polycrystalline, and so there is no advantage in growing an epitaxial interface. The aluminium will in any case be polycrystalline on the regions of the chip surface protected by oxide layers. The majority of the conducting layers deposited for microelectronic applications are thus polycrystalline, and we shall now consider in more detail the structure of these films. In order to do this it is necessary to consider the nucleation and agglomeration processes which occur during film growth from the vapour phase.

Thin film growth can be considered to be a process of impingement of deposit species on a substrate, the nucleation of clusters and the agglomeration of these clusters into a continuous film (see for instance Hirth and Pound (1963), or Rhodin and Walton (1964), and the description of film growth in §3.6). The atoms, or molecules, arriving from the vapour phase onto the substrate will not all stick to the surface; there will always be a fraction of the total incident flux that will immediately re-evaporate. For those atoms that do condense onto the surface, the next stage of film growth involves migration over the substrate prior to bonding of the atoms together to form a deposit nucleus. If the substrate temperature and the kinetic energy of the incident atoms are both low, then the amount of surface migration may be very limited, the atoms remaining roughly in the position on the substrate where they arrived from the vapour phase. Under these circumstances an amorphous deposit layer will normally be formed, as the individual atoms will not have sufficient energy even to arrange themselves into the equilibrium crystalline structure. Figure 4.3 is a schematic diagram indicating how the structure of a thin film grown from the vapour phase will depend on the growth rate and substrate temperature. High growth rates and low temperatures favour the formation of amorphous films. This is because the time available for each layer of deposit atoms to relax into a crystalline lattice before being covered by subsequent layers is small and the mobility of the atoms over the surface is also low. The mobility of deposit atoms depends partly on the incident kinetic energy of the deposit species, and a high incident kinetic energy may induce surface migration even when the substrate is cold. The strength of the interaction of the deposit atoms with the substrate may also influence the temperature of the amorphous/crystalline transition. In general, in the materials which we are interested in for microelectronic conducting layers, amorphous films can be formed easily only from silicon. Most metallic deposits have a fine-grained polycrystalline structure, even when deposited onto very cold substrates.

Semimetalloid elements like tellurium are easily deposited in an amorphous form, but the substrate temperatures have to be very low indeed (less than 10 K) to give any signs of an amorphous film for most transition metals. Amorphous layers of FCC metals like gold and aluminium cannot usually be deposited in any evaporation or sputtering process. An explanation of this difference in the structure of silicon and metal deposits is not very well developed. Part of the reason why silicon deposits are amorphous below the relatively high homologous temperature of half the melting point may lie in the crystal structure of silicon. The diamond cubic lattice is not close packed and so we might expect the transition between a random collection of silicon atoms on a substrate and a crystalline nucleus to involve more atomic diffusion and cooperative rearrangement than the equivalent process in a metallic deposit. It is also likely that the activation energy for surface diffusion of silicon atoms over substrate surfaces is much higher than the values of the same parameter for the metallic atoms. In fact, in a close-packed metal lattice, like that of gold or aluminium, the atomic movements

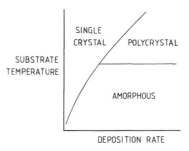

Figure 4.3 A schematic illustration of the relationship between deposition rate, substrate temperature and the structure of a film grown from the vapour phase. Figure 3.25 shows a plot of this kind for silicon films deposited by the CVD technique.

needed to create an FCC crystal from a random arrangement of deposited atoms may be so small that no diffusion at all is required, the transformation occurring by athermal shuffles. Whatever the reason for the ease with which silicon can be deposited in an amorphous form, the empirical observation is that the horizontal amorphous/polycrystalline phase boundary in figure 4.3 lies at a relatively much higher substrate temperature for silicon than for most metals. This has been exploited in the development of devices based on amorphous silicon (see §9.4).

4.2.1 The Development of Grain Structure

I have stated above that metallic thin films will in general grow as polycrystals, and in this section we will consider this structure more closely. During the discussion of the vapour phase growth of epitaxial films (§3.5) simple empirical rules concerned with reducing the lattice misfit between deposit and substrate and increasing the substrate temperature were introduced, by which the crystalline perfection of epitaxial layers could be improved. If these rules are not followed, or there is no low-energy epitaxial interfaces that can be formed between the deposit on the substrate, then very many nuclei are formed with a more or less random distribution of orientations. These nuclei grow and impinge to form a polycrystalline film as the deposit thickens. It is tempting to think that this is the end of the story, that this fine-grained polycrystalline film is the result of any non-epitaxial deposition process. Figure 4.4 shows a number of examples of the structure of metallic thin films deposited in very similar evaporation processes and the variety in the grain sizes is especially notable. So how do such very different structures arise in thin film growth?

(a)

(b)

(c)

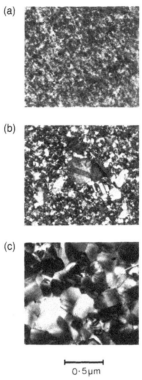

0·5μm

Figure 4.4 Transmission electron micrographs showing the grain structure in metal films 1 μm thick deposited onto rocksalt substrates: (a) tungsten; (b) gold; (c) nickel. The substrate temperature was around 300 K for (a) and (b), but was 500 K for (c).

Several theories have been developed to describe the manner in which the structure of a polycrystalline film depends on substrate temperature. These theories must explain the experimentally observed grain structures illustrated schematically in figure 4.5. This kind of plot of grain morphology against homologous substrate temperature was first presented by Movchan and Demchishin (1969) and modified more recently by Thornton (1977) and Grovenor *et al* (1984). It shows that four quite distinct kinds of grain morphologies are produced as the substrate temperature is increased: at low temperatures (zone 1) very many fine equiaxed grains are produced, often arranged into fibrous-looking bundles within which all the grains have roughly the same orientation. In zone T some very large grains are found amongst the equiaxed small grains characteristic of zone 1. In zone 2, the grains start to penetrate through the whole thickness of the film, a process Grovenor *et al* have called 'granular epitaxy', while the diameter of the grains does not increase very much. In zone 3, the substrate temperature is sufficiently high to allow extensive grain growth to occur even during the deposition of the film. The two sharp changes in film morphology occur at the zone 1/zone T and the zone T/zone 2 boundaries. The first change in structure is when a few grains grow very much larger than all the others, and this selective grain growth has been explained in terms of a few grain boundaries having especially high mobilities at this low substrate temperature, quickly giving a few grains a tremendous size advantage over their neighbours. This size advantage will encourage these grains to grow further, giving the bimodal grain size distribution illustrated in zone T. In zone 2, all the grain boundaries will have reasonably high mobilities and so all the grains will grow to about the same extent. The substrate temperature is now sufficiently high to allow considerable surface migration and to stimulate epitaxial growth of the deposit on the grains already deposited, i.e. granular epitaxy. These grains will then extend nearly all the way through the film, as illustrated in figure 4.5. The three thin film microstructures shown in figure 4.4 are characteristic of zones 1, T and 2 respectively.

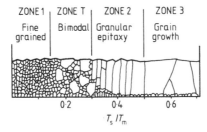

Figure 4.5 A schematic illustration of the variation in the grain size and morphology in thin metallic films with T_s/T_m, the ratio of the substrate temperature to the melting temperature of the metal.

The precise position of the zone boundaries in such a model of thin film structure depends on many deposition parameters other than the substrate temperature of course. Contamination and alloying elements usually decrease grain boundary mobility and so suppress the development of bimodal and zone 3 grain morphologies. The adsorption of gaseous contaminant species on the surface of the growing film can also suppress the phenomenon of granular epitaxy, encouraging the renucleation of small grains. In contaminated or alloyed films the fibrous zone 1 structure may thus be produced even at high substrate temperatures. The layer of small grains illustrated at the bottom of the films in figure 4.5, even at very high substrate temperatures, is a result of residual impurities on the substrate lowering the mobility of grain boundaries in the early stages of film deposition. The thicker the film, the longer it will experience the elevated temperature during growth and the larger the resulting grain size. It is very commonly assumed that the grain size in a vapour-deposited metal film will be roughly the same as the thickness and will increase linearly with the thickness. This relationship has been observed to be obeyed in a number of experiments on sputtered, evaporated and CVD films. Increasing the deposition rate will reverse this effect and shift the structure of the film towards the low-temperature end of the zone diagram. It is also quite generally observed that the grain structure of sputtered films is finer than in evaporated films, when the substrate temperatures are the same in both cases. This is probably a result of the higher kinetic energy of the incident sputtered species.

The growth of polyphase films increases the complexity of any grain growth process considerably. The interface mobilities will have quite different values to those expected in single-phase films and the heats of formation of the compound phases released during film growth can substantially alter the position of the zone boundaries. These effects have been discussed by Hentzell et al (1983). Even the angle of incidence of the deposit species onto the substrate surface can alter the grain morphology, exacerbating apparent preferential growth processes in grains with certain low index directions aligned along the incident material flux. I will not discuss here the multiplicity of factors that influence the morphology of deposited films any further, but more comprehensive treatments of the complex interactions between often rather poorly characterised processes can be found in the references above and in Maissel and Glang (1983).

In the conducting films deposited to form interconnection tracks and contacts in microelectronic devices there are only two common kinds of grain morphology, and examples of these are given in the cross-sectional micrographs shown in figure 4.6. CVD polysilicon has a fine-grained fibrous structure characteristic of zone 1, although some of the grains do penetrate all the way through the film. Evaporated aluminium films have a much larger grain size characteristic of zone 2, although a bimodal or zone T structure is sometimes found and can have important effects on electromigration processes as will be discussed in the next chapter. Annealing either grain structure after deposition will increase the grain size and this effect will be much more striking in polysilicon films than for aluminium layers where the grains are already quite large and the driving force for grain growth rather small. However, temperatures in excess of 900 °C are normally required to induce significant grain growth in polysilicon films, because the mobility of the grain boundaries is very low. This low mobility of grain boundaries in silicon, compared with those in aluminium, explains the very significant differences in their as-evaporated microstructures. I have explained that the grain structure of a film depends almost completely on grain boundary migration and when this is suppressed the lateral growth of the grains no longer occurs, but granular epitaxy caused by surface migration of the incoming species is still possible. This is why the structure of polysilicon films combines elements of the characteristic morphologies of zones 1 and 2. Evaporated films of III–V and II–VI compound semiconductors usually have grain structures more similar to that of polysilicon than of aluminium. In a situation where the grain boundaries are important in determining the electrical properties of a thin conducting film, the variation in microstructure which can be achieved by a careful choice of deposition conditions may have very significant technological advantages. We shall now consider what effect the lattice defects can have on the electrical properties of thin films.

(a)

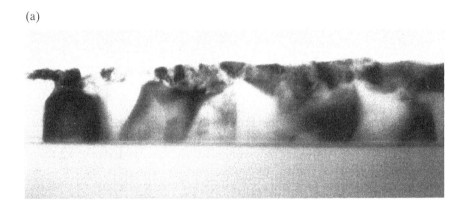

(b)

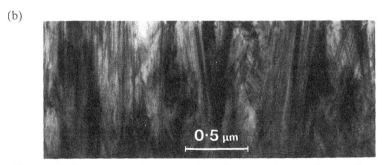

Figure 4.6 Cross-sectional transmission electron micrographs showing a comparison of the grain structure in as-evaporated (a) aluminium and (b) silicon films. (Courtesy of Mr D Gold.)

4.3 THE INFLUENCE OF LATTICE DEFECTS ON THE PROPERTIES OF THIN FILMS

Now that we have some idea of the characteristic structures found in most vapour-deposited semiconducting and metallic films, we can consider how the defects in these structures influence their electrical and optical properties. We shall begin with metallic films, for which rather few attempts have been made to investigate the properties of individual defects. We shall, therefore, have to content ourselves with a rather macroscopic treatment of the resistivity of thin metallic layers. A separate section on semiconducting materials will then show how the influence of grain boundaries in particular has been studied very extensively. We will go into some detail on the properties of semiconductor thin films as these are important not only in the understanding of the performance of polysilicon interconnection layers, but also for the polycrystalline semiconductors that are often active components in the solar cells discussed in Chapter 9. Consideration of the enhanced diffusion and chemical reaction effects associated with grain boundaries in polycrystalline layers will be deferred to Chapters 5 and 9, where particular problems associated with these processes can be discussed.

4.3.1 Metallic Films

The most important property of a metal film is its resistivity and so this is the parameter that we shall concentrate on in this section. (The properties of metal–semiconductor interfaces have already been discussed in Chapter 2.) However, before discussing the resistivity of vapour-deposited thin metal films, the basic mechanisms of resistivity, or electron scattering, in bulk metallic lattices will be very briefly reviewed. In pure metals at room temperature the electron scattering is dominated by interactions with phonons. At the operating temperatures of most microelectronic devices, we expect the resistivity of the metallic conductors to depend roughly linearly on temperature. To this intrinsic lattice resistivity the effects of vacancies, impurity atoms, dislocations, grain boundaries and surfaces can be added as a linear combination of contributions to a total resistivity. The effect of all these additional resistivity contributions can be measured for bulk metals since they persist even when the phonon scattering effects are frozen out at extremely low temperatures. These contributions are normally only a small fraction of the total room-temperature resistivity of a bulk metal. This is because the concentrations of vacancies and impurity atoms are rather low, the grain size relatively large (tens to hundreds of micrometres) and the surfaces only influence the resistivity of a bulk material in a layer of thickness given by the electron mean free path.

The mean free path for scattering of electrons in a metal is typically 10 nm at room temperature and so surface effects can be ignored in bulk metals.

The resistivity of bulk metals can be changed most effectively by alloying. This can be viewed as a very extreme impurity effect, the deliberate addition of atoms which distort the lattice. Solid solution alloying gives a roughly linear increase in resistivity with solute concentration (Matthiesson's rule), and the resistivity of a range of alloys in a binary system with complete mutual miscibility will be a maximum roughly half-way across the composition field. Even for elements as chemically similar as silver and gold, a 50/50 alloy has a resistivity approximately five times that of the pure elements. Two-phase alloys show a similar kind of resistivity variation with composition, the total resistivity being a weighted average of the resistivities of the two phases.

We have seen, however, that the structure of a vapour-deposited metal film is very different to that of a bulk material. The vacancy and dislocation concentration in an evaporated or sputtered film will usually be much higher than is normal in a bulk metal. Contamination from substrate, vacuum system or evaporation filament all contribute to give a high impurity content in deposited films. Even more strikingly, the grain size may be 1000 times smaller than in a bulk material. The residual stresses in thin films can also be very high, as will be discussed later in this chapter, and can influence the resistivity. In addition, the free surfaces of the film are now very much more important scattering centres than they were in the bulk. For films whose total thickness approaches that of the mean free path of electrons, diffuse scattering from the surfaces has been modelled by the Sondheimer theory. An approximate form of this theory states that the increase in resistivity due to surface scattering is given by the factor $3\lambda_x/8t$ where t is the film thickness and λ_x the mean free path for electrons in the bulk material. Only at very small film thicknesses will this effect increase the resistivity very greatly, and it is probably safe to assume that the extra electron scattering from the very high defect density in the thin films will be more effective at increasing the resistance than the surface scattering.

It is not easy to measure the relative magnitude of these additional scattering effects independently, but some authors have come to the conclusion that grain boundaries have the dominant influence on the resistivity of thin metallic films (see for instance Chauvineau et al (1969) and Mayadas and Shatzkes (1970)). The scattering of electrons by grain boundaries in metals has been modelled by Mayadas and Shatzkes, and an equation for the resistivity of polycrystalline films, ρ_g, derived of the form

$$\rho_g = \rho_{\text{lattice}} \left[1 - \frac{3}{2}\alpha + 3\alpha^2 - 3\alpha^3 \ln\left(1 + \frac{1}{\alpha}\right) \right]^{-1} \qquad (4.7)$$

where

$$\alpha = \frac{\lambda_\infty}{d}\left(\frac{R}{1-R}\right),$$

d is the grain size and R the reflection coefficient of the grain boundaries. R is defined as the fraction of the total incident electron flux that is reflected back by each boundary. In thin films of pure metals like aluminium or copper, the reflection coefficient of the grain boundaries has been estimated to be of the order of 0.2. The grain boundary resistivity effect is thus fairly small in these materials. However, in contaminated films, those grown in poor vacua or on dirty substrates, the reflection coefficients are estimated to be as high as 0.5 in tungsten films (Learn and Foster 1985). and a startling 0.9 in some tungsten silicide films (Campbell *et al* 1982). The resistivity of many polycrystalline metallic films will thus be dominated by the grain boundary scattering and the use of such materials in VLSI applications may be hampered by their poor conductivity. In refractory metal films and compound layers the grain boundary scattering parameters can be especially high, which may be due to the very high affinity for oxygen of metals like tungsten and titanium. Numerous experiments have demonstrated that considerable concentrations of oxygen can be found at the grain boundaries in refractory metal films deposited by all the vapour phase growth techniques considered in this chapter.

With the thin film structures shown in the previous section and the discussion above in mind, we should not be surprised that the measured resistivities of thin films, even those of nominally pure metals, are higher than those of the bulk materials from which they were evaporated. While this difference is usually relatively small, some metals can be evaporated with an enormous range of resistivities as a result of the growth of non-equilibrium film structures. The resistivity of evaporated Ta films with the low-temperature BCC structure approaches that of the bulk material, but if deliberately sputtered so as to grow films with the high-temperature tetragonal β phase, the resistivity is increased by more than a factor of ten. Some typical values of bulk and thin film resistivities for popular microelectronic conducting materials are given in table 4.2. It has to be remembered that the measured values of thin film resistivity are very dependent on the growth conditions (which control grain size and impurity levels), and so the values listed here are intended only as indications of the range of resistivities which can be achieved in vapour-deposited films.

There are two obvious points that can be seen in these data: the resistivities of films of some of the pure metals like aluminium and gold are very similar to the bulk values; and refractory metal films have to be annealed before their resistivities approach the bulk values. This is probably due to the rather high oxygen content at the grain boundaries, but how a simple annealing process reduces the electron reflection coefficient at the grain boundaries in these contaminated films is not fully understood. One possibility is that grain growth will occur, reducing the number of boundaries encountered in any given conduction path.

Table 4.2 Comparison of the resistivities of bulk and thin film microelectronic materials.

Material	Bulk resistivity ($\mu\Omega$ cm)	Thin film resistivity ($\mu\Omega$ cm)
Aluminium	2.67	2.8–3.3
Copper	1.7	1.7–1.9
Gold	2.2	2.4
Nickel	6.9	12.0
Tungsten	5.4	CVD 5.4–20.0
		Sputtered and annealed 13.0
Molybdenum	5.7	Sputtered and annealed 10.0
Tantalum	13.5	BCC Ta 15
		β phase 200
Tungsten silicide	12–55	Co-sputtered and annealed 35–100
Polysilicon		CVD (heavily doped) 500

4.3.2 Conduction in Very Thin Metal Films

So far we have been considering the properties of thin metal films which are continuous. However, very thin films are often observed to consist of disconnected nuclei if the film is too thin for the agglomeration stage of growth to be complete. The thickness at which this agglomeration occurs depends on the substrate temperature as well as all the other deposition conditions, but is in the range of a few tens of nanometres for most metals. Very thin metal films have a special application as the front contact in solar cells. Here they are required to act as an electrode, and so to have low resistivity, but must also allow as much of the incident light to pass through to the cell as possible. Gold is a popular choice for this application because of its low resistivity and high resistance to atmospheric corrosion. Studies of the optical properties of thin gold films have therefore been extensive. It has been found that there are three regimes of electrical and optical behaviour as the thickness of the films is increased. For layers grown on amorphous substrates (Gadenne 1979), films thinner than about 5 nm consist of islands of gold disconnected both mechanically and electrically, and so the resistivity is more than 10000 times that of the bulk element. At thicknesses of around 10 nm some agglomeration of the islands has begun and, although these larger islands are still separated physically from their neighbours, some electrical conduction is possible. This conduction is thought to be due to the tunnelling of the electrons between the islands. In these films the resistivity is reduced to 1000 times that of bulk gold, but the transmission of incident light is significantly reduced. Further thickening of the film to 30 nm results in a fully metallic conduction as the film becomes continuous, but the optical transmission is now very low. In the region between 10 and 30 nm the resistivity falls very sharply towards the bulk value and, curiously, there is a significant peak in the transmission. This may be a result of agglomeration of the gold nuclei into larger interconnected islands. It is in this range of thickness that the most advantageous properties of these gold films are obtained, where the resistivity is low enough to allow efficient operation as a contact while quite a substantial fraction of the total incident light can still pass through to the solar cell beneath. The variation of the transmission and resistivity of gold films with thickness is illustrated schematically in figure 4.7(a) and the variation in optical transmission with thickness in thin aluminium films is shown in figure 4.7(b). Here there is no anomalous transmission at a characteristic value of film thickness.

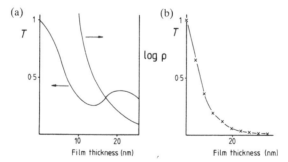

Figure 4.7 Experimental plots of optical transmission and film resistivity for (a) gold films on amorphous substrates (after Gadenne 1979) and (b) aluminium films.

4.3.3 Semiconducting Thin Films

The electrical properties of semiconductors depend very much more on point defects than do those of metals, as we have seen in Chapter 1. Thus while we have been able to conclude that grain boundaries may well have the dominant effect on the resistivity of thin metal films, point defects contributing a small additional contribution to this resistivity, it is not possible to make a similar assumption for semiconductor films. In addition, we must include the optical properties of the films in our consideration, as these are just as important as the electrical properties in many kinds of semiconductor devices. Here I will separate the effects of point defects from those of grain boundaries, but it should not be forgotten that they will both contribute to determining the properties of any polycrystalline semiconductor layer. The effect of dislocations on the optical and electronic properties is rather similar to that of grain boundaries, and some of the properties of dislocations have been considered in Chapter 1. Some specific problems associated with the presence of dislocations in semiconductor materials are described in Chapters 2, 8, 9 and 10. In most polycrystalline films, the grain boundary effects will overwhelm those of the dislocations, and so here we will only discuss the properties of boundaries.

4.3.4 Grain Boundaries in Semiconductors

The properties of grain boundaries in semiconductors that will concern us are those that influence the electrical and optical properties of the material, in particular the carrier mobility and the minority carrier lifetime. These two parameters are of direct interest when considering the performance of semiconductor thin films as interconnects in metallisation and when choosing active layers for solar cell applications (see Chapters 5 and 9). The presence of grain boundaries in a semiconductor can influence the properties in several ways, not all of them deleterious. For example, one effect of grain boundaries on the optical properties of semiconductors is to shift the absorption edge of materials like CdS to slightly lower energies (see for instance Clark (1980)). This can be a valuable modification of the properties of an absorbing semiconductor layer in a solar cell, allowing more of the solar spectrum to be absorbed and so increasing the efficiency of the device (see §9.2). By contrast, the luminescent efficiency of compound semiconductor materials is strongly degraded by rapid recombination at grain boundaries and so the performance of light-emitting diodes made from polycrystalline material is usually poor. We will see that in general the deleterious effects of grain boundaries on the electrical and optoelectronic properties of the material far outweigh any rather minor improvements in absorption characteristics.

In order to understand the effects associated with grain boundaries in semiconductors, we must first understand the changes in the local band structure of the material caused by the presence of the defect. The development of a model of the electrical properties of grain boundaries is based on the concept that distortion of the crystal lattice creates trapping states localised at the grain boundary, just as we have already assumed for dislocations in semiconductors (§1.6). Taylor *et al* (1952) suggested an analogy with Schottky barrier behaviour where charge flow from the grains will occur to fill these mid-gap states. This charge flow will equalise the position of the Fermi levels in grains and grain boundaries and create a symmetrical Schottky barrier around the boundary. The form of this symmetrical Schottky potential barrier is illustrated in figure 4.8(a). Seto (1975) has developed a simple model of how the height of this potential barrier varies with doping concentration, and this model predicts three regimes of behaviour:

(1) At low doping levels, there are only a few free carriers to be trapped in the grain boundary states, and the extent of band bending is small and the barrier height low.

(2) At intermediate doping levels, more carriers are available to be trapped and so the barrier height will increase.

(3) At high doping levels, there is an excess of carriers over and above those needed to saturate all the mid-gap grain boundary states. The Fermi level will therefore be freed from these states and will rapidly approach its equilibrium level in the grains, reducing the barrier height sharply.

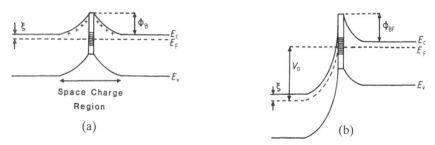

Figure 4.8 Schematic diagrams showing the proposed form of the band bending around (a) a grain boundary in n-type semiconductor material and (b) the same boundary with a potential V_a applied to the material.

The predicted variation of barrier height (Φ_B) with dopant concentration is shown schematically in figure 4.9. The effect of changing the grain size and the trap state density (N_t) can be added to this simple model if required (see for instance Baccarini *et al* (1978)). If the grain size is very small, all the free carriers will be depleted from the grain interiors before the boundary trap states are saturated and so the maximum value of Φ_B will be reduced. Increasing the grain size of the semiconducting material, increasing the doping level or increasing the trap density, N_t, will all increase the barrier height. Typical measured values of Φ_B have ranged from about 0.1 eV for CuInS$_2$ (Kazmerski *et al* 1975), to 0.4 eV in silicon (Seager 1982) and 0.73 eV in gallium arsenide (Seager and Pike 1982). These values do not represent the maximum values of the barrier heights in these materials; in fact the value quoted for CuInS$_2$ is an average of Φ_B taken from very many boundaries.

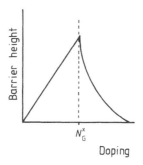

Figure 4.9 A sketch (after Seto 1975) of the variation in grain boundary barrier height with the doping level in the semiconductor.

In order to explain grain boundary barrier heights of the order of 0.5 eV an interfacial trap density of about 10^{12} cm^{-2} is required, or one trap state for every 1000 atom sites in the boundary. The structures of a few highly symmetric grain boundaries in silicon and germanium have been studied by high-resolution transmission electron microsopy (e.g. D'Anterroches and Bourret 1984, Bacmann *et al* 1982). These studies have shown that the cores of these boundaries consist of a small number of structural units repeated throughout the boundary plane, in agreement with the structural unit model for grain boundaries which will be described in §5.2. Most of these units apparently contain no dangling bonds; the tetrahedral coordination of all the atoms at the boundary plane is conserved. However, electron spin resonance experiments (Johnson and Biegelson 1982) have detected a concentration of 10^{12} cm^{-2} dangling bonds at grain boundaries in polycrystalline silicon, where most of the boundaries will not be of the highly symmetric kind studied in the microscopy experiments. Dangling bonds at grain boundaries may well have an associated mid-gap trapping state and there may also be electrically active structural units in the more highly symmetric boundaries, although they cannot be identified as such in an electron microscope image. Defect states produced by the presence of particular structural units in a grain boundary are called intrinsic states since they are characteristic of an uncontaminated boundary.

However, grain boundaries in semiconductors, and especially those in materials deposited in a rather poor vacuum onto a dirty substrate, are likely to be heavily contaminated. This contamination will consist of dopant atoms co-evaporated with the semiconductor and impurity species like hydrogen, carbon and oxygen. Polycrystalline CdS films often contain considerable quantities of oxygen at the grain boundaries, for example, and annealing the films after deposition to remove this oxygen can result in a dramatic decrease in resistivity. This implies that the presence of the oxygen atoms at the boundary increases the density of mid-gap states. Kazmerski (1982) has linked the presence of segregated titanium at silicon grain boundaries with an increased barrier height, and many other species segregated to semiconductor grain boundaries may have a similar effect. By contrast, hydrogen also segregates to silicon grain boundaries and has been observed to remove the electron spin resonance signal characterisitic of the dangling bond and lower the potential barriers at the boundaries. It is clearly the combined effect of intrinsic defects and extrinsic segregated species which determines the trap state density at any particular grain boundary. In compound semiconductors an additional influence on the barrier height is that grain boundaries can create local regions of non-stoichiometry. Very little direct evidence for this kind of stoichiometry variation has been found as yet, but the observation of asymmetric potential barriers around grain boundaries in gallium phosphide (Ziegler *et al* 1982) may be attributable to a non-uniform distribution of electrically active gallium or phosphorus vacancies.

We can see that although some knowledge of the structure and impurity concentration in a few semiconductor grain boundaries has been gained, we cannot determine precisely which defect produces which dominant mid-gap trapping state in any given interface. The dangling bond is one favoured candidate, but in common with many semiconductor studies the isolation of the point defect responsible for any particular electronic state is very difficult. Similarly, the distribution of the defect trapping states in the semiconductor band gap is poorly characterised since it depends on the boundary structure. It is important to remember that there is no such thing as the electrical property of silicon grain boundaries; all the boundaries have different properties that reflect their own individual structures. Some experiments have shown a broad distribution of trapping states across the middle of the band gap, while others analyse one or more narrow bands of states. Whether this is because semiconductor grain boundaries can indeed have all kinds of trap state distributions, each particular one being dependent on the details of the boundary structure and contamination, or whether the experimentally determined distribution of trapping states depends on the experimental technique used to make the measurement, as suggested by Broniatowski (1982), is not clear. There is a rough consensus, however, that the density of interface states is in the range 10^{11} to 10^{13} cm^{-2} in most semiconductor grain boundaries, in quite adequate agreement with the density of states needed to explain the observed grain boundary barrier heights.

Given this somewhat confused picture of the electronic structure of semiconductor grain boundaries, it may seem improbable that prediction of the conduction across, and recombination at, boundaries can be sensibly attempted. In fact both these parameters can be very successfully modelled. Figure 4.8(b) shows how a grain boundary with a double Schottky barrier might respond to an applied voltage. The Fermi level is pinned on the forward-biased side of the boundary by the unfilled boundary states which will trap all additional carriers. Most of the applied voltage is thus dropped across the reverse-biased side of the boundary. Eventual breakdown of the boundary barrier occurs either by the Fermi level breaking free of the saturated boundary states, or by tunnelling through the narrow reverse-biased barrier. A simple conduction model which includes contributions from thermionic emission over the barrier and tunneling has been constructed (e.g. Rai-Choudhury and Hower 1973, Fonash 1972). The thermionic component will dominate at high temperatures, or when the barrier is relatively low, and the current across the boundary will have the standard form

$$J = A^* T^2 \exp\left[\left(\frac{-\Phi_B + \zeta}{kT}\right)\left(\frac{eV_a}{kT} - 1\right)\right]. \qquad (4.8)$$

A^* is the effective Richardson coefficient and Φ_B, ζ and V_a are defined in figure 4.8(b). The form of a typical J/V curve characteristic of a grain boundary is shown in figure 4.10, where sections 1 and 2, the ohmic and resistive regions respectively, are very well predicted by an expression like equation (4.8). The exponential dependence of the transboundary current on Φ_B has been very frequently observed in experiments on a wide range of bicrystalline and polycrystalline semiconductor samples.

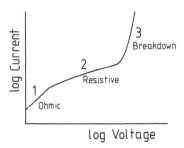

Figure 4.10 A plot of the variation in transboundary current with the applied voltage showing the ohmic, resistive and breakdown behaviour.

Conduction through polycrystalline films, where the current must pass across very many grain boundaries, can be modelled by a simple equation of the form

$$\sigma_p = \sigma_0 \mu_0 n \exp\left(\frac{-\Phi'_B}{kT}\right) \qquad (4.9)$$

and the carrier mobility in a polycrystalline material by

$$\mu_p = \mu_0 \exp\left(\frac{-\Phi''_B}{kT}\right) \qquad (4.10)$$

where σ_p is the conductivity of a polycrystalline material, σ_0 the conductivity of the single-crystal material, n the concentration of carriers in the grains, μ_0 the carrier mobility in the grain interiors and μ_p the mobility in the polycrystalline aggregate. In both these equations the parameter Φ'_B, or Φ''_B is not the average barrier height of the grain boundaries in the material, but an average value of $(\Phi_B + \zeta)$, as defined in figure 4.8(a).

Using Seto's model for the variation of grain boundary barrier height with doping (figure 4.9), and equations (4.9) and (4.10), remarkably good agreement is obtained with experimental data on the resistivity of, and carrier mobility in, polycrystalline semiconductors. Figure 4.11 shows that the values of resistivity and mobility predicted by the model fall very close to the experimental points obtained from polycrystalline silicon samples. This suggests that Seto's simple model is a very successful representation of the band bending at semiconductor grain boundaries. A striking feature of figure 4.11 is the deep minimum in carrier mobility at the doping concentration where the boundary barrier height reaches its maximum. At this same doping level the resistivities of the films fall rapidly towards the single-crystal values as the barrier height decreases.

Similarly, Seager (1983) has formulated equations for the recombination velocity of electron–hole pairs at grain boundaries in semiconductors. This is based on the assumption that minority carriers injected into a semiconductor close to a grain boundary will diffuse very rapidly down the potential gradients around the boundary and recombine with the majority carriers trapped there. This will result in the partial emptying of the trap states, and the empty states will then be filled by the majority carriers injected into the grains, completing the recombination cycle. The recombination velocity, S, can be calculated from:

$$S = q^{-1} u_{th}^{min} \sigma^{min} (2\varepsilon\varepsilon_0 N_G \Phi_B)^{1/2} \exp\left(\frac{\Phi_B}{kT}\right) \qquad (4.11)$$

where u_{th}^{min} is the thermal velocity of minority carriers, σ^{min} is the capture cross section of filled interface states for the minority carriers

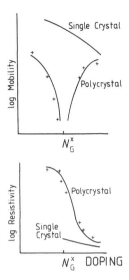

Figure 4.11 Experimental and theoretical plots of the variation in carrier mobility and the sample resistivity as a function of the dopant concentration in polysilicon and single crystal silicon (after Seto 1975).

and N_G is the dopant concentration. Once again the recombination behaviour of grain boundaries seems to be accurately modelled by this simple equation. These recombination processes will decrease the efficiency of solar cells, as I shall show in §9.5, and will also degrade the performance of luminescent devices because grain boundary recombination is generally a non-radiative transition (see Chapter 8).

We have shown, therefore, that grain boundaries in semiconductor materials have resistive and recombination properties that can be quite well modelled using the simple concept of the symmetrical Schottky potential barrier at the boundary. Average values of the barrier height can be measured reasonably easily in polycrystalline films and so the performance of the films as interconnect or solar cell materials can be predicted using equations (4.9) and (4.11). If we wish to optimise the performance of thin polycrystalline silicon films as interconnect materials in VLSI technology, figure 4.11 shows that the resistivity of polysilicon films falls to a value very close to that of the single-crystalline material if the doping level is sufficiently high. Under these circumstances, the barriers at the individual grain boundaries are reduced almost to zero. However, table 4.2 shows that the resistivity even of heavily doped polysilicon films is some 100 times higher than most metallic thin film materials and so long path lengths of polysilicon in a device structure must be avoided.

One additional effect of grain boundaries in semiconductors that has not yet been mentioned is that they can provide short-circuit conduction paths for carriers along the boundary plane. This is especially true if there is a high concentration of heavy metal impurities segregated to the boundary plane. A similar short-circuit conduction is also shown along some dislocations, as has been mentioned in Chapter 2. This effect in grain boundaries has been recently reviewed by Matare (1984) and can be very important in determining the performance of a device in which a grain boundary intersects a p–n junction.

This brief section has only introduced a very small fraction of the extensive literature and numerous models associated with grain boundaries in semiconductor materials. Some more detailed articles that give a wider range of views on the relative value of the multiplicity of models that have not been considered here are Matare (1984), Grovenor (1985), and chapters in Kazmerski (1980) and Fahrenbruch and Bube (1983).

4.3.5 Point Defects in Semiconductor Films

In Chapter 1 the fundamental properties of semiconductor materials were shown to depend heavily on point defect concentrations' in particular those of impurity atoms and ionised native point defects. It is thus to be expected that the relatively high concentrations of point defects formed during growth of thin films from the vapour phase can be very important in determining the optical and electronic properties of the films. The non-equilibrium point defect concentrations are a result of the extremely high supersaturation of deposit species in the vapour above the substrate and the fast quenching rate of the deposit onto the substrate. As well as describing some of the effects which can result from this high density of point defects, it is worth outlining which semiconductor properties we expect to be particularly sensitive to their presence.

We do not expect carrier densities to be strongly dependent on point defect concentrations unless the defect produces shallow states which compensate for the deliberately added dopant. However, in thin films of many compound semiconductors, the electrical properties are controlled almost completely by the high concentration of electrically active native point defects. A striking example of the effect of native point defects on the optical properties of a semiconductor is the shift in the absorption edge in vapour-deposited GaP films. When the substrate temperatures are low, the absorption edge is shifted from the bulk value of 2.34 eV to 0.8 eV (see figure 4.12). It has been assumed that this shift is caused by a very high density of both donor and acceptor point defect states introduced by the presence of vacancies, interstitial and antisite defects

on both sublattices, (§1.4). The effect of these defect states is to narrow
the band gap to 0.8 eV (Davey and Pankey 1969). Annealing these films
after deposition, or increasing the substrate temperature during deposition, allows the defects to annihilate one another and returns the
absorption edge to the position characteristic of the bulk material. It is
worth remembering that intrinsic self-compensating point defects of this
kind are rather easily removed by annealing and so any deleterious
effects that they might have on the properties of a thin film can be
avoided by a simple post-deposition heat treatment.

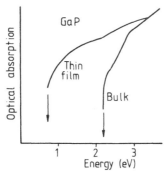

Figure 4.12 An illustration of the effect of native point defects on
the optical properties of GaP films (after Davey and Pankey 1969).

Native point defects are also present in deposited films of compound
semiconductors because of deviations from stoichiometry. Evaporated or
sputtered films of these materials often contain an excess of one of the
component elements, creating a high density of electrically active point
defects. This non-stoichiometry is related to the equilibrium between
semiconductor compound and gas phase, as described in §1.4, and also
to the sticking coefficients of the component elements on the substrate.
Thin CdS films, like the bulk material, usually contain an excess of
metal atoms and so are n-type, while $CuInSe_2$ films can show either n-
or p-type conduction characteristics depending on the substrate temperature during deposition and the exact details of the deposition process.
Here it is the relative concentrations of metal and selenium vacancies
which control whether the films are n- or p-type. The resistivities of
these films can be very low because of the high density of electrically
active point defects. Modifications in the electrical and optical properties
of semiconductor materials which arise as a result of non-stoichiometry
are more difficult to remove by a simple annealing process than the

excess concentrations of self-compensating point defects which I have described for GaP films. It is often impossible to produce thin films of compound semiconductors which exibit both n- and p-type character and this makes the production of certain types of thin film devices difficult, as I shall describe in Chapters 8 and 9.

There are two classes of impurity elements which may be found in vapour-deposited thin films: contaminants from the vacuum system itself (H, O, C and Ar in a sputtering facility); and impurities from the substrate or dirty source material. Of the vacuum contaminants, carbon is an electrically active dopant in some III–V materials and oxygen creates deep states in silicon. In addition, oxygen segregates strongly to grain boundaries in many semiconductor materials and increases the height of the potential barriers at the grain boundaries. These impurities can thus have a significant effect on the properties of polycrystalline semiconductor thin films.

Metallic impurities can have an even more serious influence on properties. For instance, Cu in GaAs produces deep states which act as particularly effective non-radiative recombination sites. The minority carrier lifetime, τ, is a parameter which is very sensitive to the presence of defects which create states deep in the band gap, especially if the defect carries a charge of the same sign as that on the majority carrier. It has been estimated that decreases in τ can be measured when the concentration of these kinds of defects exceeds the astonishingly low value of $10^4 \, cm^{-3}$. In Chapter 8 we will see that τ is an important parameter in the phenomenon of electroluminescence in semiconductor materials, and luminescent efficiency is often severely degraded by the presence of low concentrations of impurities which can be introduced from the source material, the vacuum chamber itself or the substrate. Diffusion of electrically active impurities from a substrate into a thin semiconductor film during growth can be an important effect and has already been described for the case of silicon growth on Al_2O_3 in §3.5.

It can be seen that we expect non-equilibrium concentrations of native point defects and impurities to be present in all vapour-deposited thin films and that some of these defects can modify the properties of these films significantly. Annealing treatments are routinely applied to many such films to increase their grain size, and so some of the effects seen in as-evaporated material will be removed before the film is included in a microelectronic device. Other kinds of point defects will not be removed by annealing, and these may seriously limit the practical use that can be made of thin films of some semiconductor materials. §9.5 includes a description of the properties of thin CdS films prepared by a variety of techniques, and it is shown that it is the combined effect of grain boundaries and point defects which determines the electronic properties of the films.

4.4 SILICIDE FILMS

Metal silicide compound formation is a rather special case of thin conducting film deposition since in many cases the pure metal is sputtered or evaporated onto a silicon surface and subsequently heated to form the silicide layer. A particular feature of this reaction which has made it useful for microelectronic applications is the flatness of the silicide/silicon interface. A typical interface is shown in figure 4.13, and the evenness of the polycrystalline silicide layer, with no spikes or protrusions at the bottom of the layer, is obvious. Silicon/silicide contacts can also be both chemically and electrically very stable, unlike the reactive aluminium/silicon interface which will be discussed in the next chapter. This means that Schottky contacts to silicon devices can be fabricated with very well defined barrier heights, a great advantage to the device engineer. Very generally, near noble metals like nickel and palladium react at low temperatures to produce M_2Si phases on silicon giving Schottky barrier heights of 0.65–0.95 eV, while refractory metals like tungsten, molybdenum and tantalum react at 600 °C to grow MSi_2 phases with slightly lower Schottky barriers, in the range 0.5–0.7 eV. Silicide formation on single-crystal silicon substrates is one of the most comprehensively studied reactions in all materials science and has been reviewed by Tu and Mayer (1978), Murarka (1983) and D'Heurle and Gas (1986) amongst many other authors. The silicide-forming reaction has been investigated extensively by Rutherford backscattering and x-ray diffraction techniques (see Chapter 10). In this section we shall try to understand the process by which a compound film is formed in a thin film reaction.

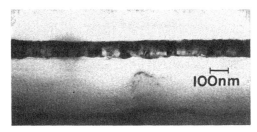

Figure 4.13 A cross-sectional transmission electron micrograph of a Pd_2Si layer formed by reaction of a thin palladium film with the surface of a silicon wafer.

The growth of a silicide layer in a via hole defined in a silica layer on a silicon wafer surface is an important technological operation and is illustrated in figure 4.14. The exposed silicon surface will be covered by a native oxide layer and the metal used to form the silicide compound must be able to penetrate through this oxide to react with the underlying silicon. Tu and Mayer (1978) have concluded that reactive metals form M–O bonds when deposited onto the native oxide layer and form a stable interface which acts as a barrier to low-temperature reaction between the silicon and the metal. This may explain why many reactive metals, titanium and tungsten for example, only form silicides at temperatures in excess of 400 °C, while aluminium forms no stable silicide at all. Noble metals like palladium and platinum are observed to form silicides at remarkably low temperatures (less than 300 °C) and this is assumed to be because there is no M–O bond formation. The noble metal atoms are thus free to penetrate through the oxide layer and to react with the silicon substrate. Taking the case of platinum, the result of evaporation of the metal and annealing at 550 °C is the formation of a thin layer of PtSi on the wafer surface, and often a thin layer of oxide is formed on the top of this silicide layer as well (figure 4.14(b)). The evaporated platinum layer on the silica surface remains unreacted. The oxide layer on the top surface of the PtSi can be rather useful, as it will protect the silicide during the aqua regia dip used to remove the unreacted platinum metal. This kind of process will result in the formation of a thin stable electrical contact on the silicon wafer.

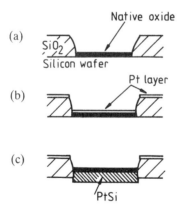

Figure 4.14 A schematic illustration of the process by which a PtSi contact is formed by the reaction of a platinum film with the surface of a silicon wafer.

A final point that must be mentioned when considering the process of forming silicide contacts is that some silicon is consumed in the reaction between evaporated metal and the wafer. The thickness of the silicon layer consumed obviously depends on the thickness of the metal layer and the composition of the silicide. The reaction of 100 nm of nickel to create Ni_2Si consumes a layer of the silicon surface about 91 nm thick for instance, while a similar reaction of 100 nm of titanium to form $TiSi_2$ will consume 227 nm of silicon. The device engineer must be very careful to take this into account when designing the depth at which shallow p–n junctions are placed in the wafer before silicide contact formation.

Some other silicide-forming metals, and particularly tantalum, do not react to form flat silicide/silicon interfaces and the electrical properties of a contact with a non-planar interface are poor. The reason for this is not properly understood. Co-sputtering tantalum and silicon can be used to grow silicide films which have very flat interfaces with the silicon wafer, and this is the second important way in which silicide films may be deposited for microelectronic applications. Co-deposited films with a very wide range of compositions can be made, but are usually amorphous after deposition and only crystallise on subsequent annealing. This range of compositions has made it possible to prepare polycrystalline films of many of the silicide phases in the equilibrium diagram, while avoiding the problem of eutectic melt formation and balling up of the deposit during the high-temperature reactions that would otherwise be needed to prepare these phases. This has made it possible to characterise the electrical properties of a large number of thin film silicide contacts on silicon. However, since the resistivities of co-sputtered films are usually higher than those of their evaporated and reacted counterparts, this silicide deposition process will not be further considered in this chapter. Murarka (1983) has given a comprehensive description of the fundamental properties of a wide range of co-sputtered silicide phases and some of these compounds have found important applications in MOS gate metallisations, as I shall mention in the next chapter.

4.4.1 The Thermodynamics of Silicide Formation

At this point we must attack the thorny problem of how one can predict which silicide phase will be formed first in a reaction between a thin metal film and silicon substrate, and which phases will form on subsequent heating. There are two practical reasons why we might wish to know the order of phase formation in these reactions. The most important one is that only with this information can the quantity of silicon consumed in the reaction be determined, and the second is to allow prediction of the electrical characteristics of the silicide/silicon contacts. In fact, to a first approximation the electrical properties, i.e. the Schottky barrier height, of these contacts do not vary strongly as a function of the stoichiometry of the silicide phase. Nevertheless, for the first reason it is still vital to be able to predict accurately which silicide phase will form first.

The equilibrium phase diagrams for the metal/silicon binary systems are mostly rather complicated, containing a large number of congruently and non-congruently melting compounds. We should bear in mind that these diagrams provide information only about an ideal system in equilibrium, that is to say after a reaction time which is infinitely long. Equilibrium phase diagrams may be unreliable guides to the kinetics of how a thin film system approaches equilibrium. In addition, much of the data required for a fundamental attack on the problem, using classical nucleation theory (the free energies of formation of all possible phases, the diffusion rates of both metal and silicon atoms through the various silicide phases as a function of temperature, and the energy of the interfaces between metal and silicon and the silicide phases), are simply not available. It seems likely, therefore, that we are not going to be able to use a first-principles approach to predict the stoichiometry of the first phase to form, although D'Heurle and Gas (1986) have pointed out that some progress can be made by considering the kinetics of compound nucleation. These authors have argued that nucleation phenomena are not likely to determine which phase forms first, since the enthalpies of formation of most silicide phases are large. This implies that the nucleation rate will be fast for every phase. The rate of diffusion of the metal and silicon atoms in the silicide phase is proposed as a parameter which will vary over a much wider range of values, and the phase with the highest diffusion coefficients should grow preferentially. This phase will then grow to a limiting thickness which depends on the value of the diffusion coefficients and the enthalpies of formation of the other equilibrium phases in the system. If the enthalpies of formation of the possible second phases are all rather similar, then the relative energies of the interfaces between the first and second nucleated phase become important and the phenomenon of epitaxial nucleation (to lower the interfacial energy) might be expected. However, these general conclusions rarely allow the prediction of first phases and an alternative strategy is to determine experimentally which silicide phase forms first. A huge body of literature has been devoted to just this kind of experimental observation. In its briefest form, the evidence which has been gathered can be summarised as follows:

(1) The first phase to form almost always grows to consume all of the metal thin film before the appearance of any second phase.

(2) The growth kinetics of this first phase are sometimes observed to be parabolic with time, which strongly suggests that a diffusion-controlled growth mechanism dominates the reactions. This is true of the formation of Ni_2Si and Co_2Si for example.

(3) Once the metal layer is consumed, further reaction between the silicide and the silicon occurs to form silicon-rich silicide phases. In some cases, these second phases also grow with diffusion-controlled kinetics, e.g. NiSi, CoSi and PtSi.

(4) In some other cases, the rate controlling step in the reaction to form the secondary phase has been shown to be the nucleation of the phase, not the rate of diffusion. Epitaxial relationships are often observed between grains of the first and second phases

(5) The formation of some silicide phases, e.g. $TiSi_2$ and WSi_2, still remain poorly characterised processes.

Observations (1)–(4) can be readily understood in terms of the diffusion-controlled growth selection of the first phase and the importance of interfacial energy in determining which phase grows second.

Let us now look more closely at the reaction of nickel with silicon. There are six equilibrium phases in the Ni/Si phase diagram: Ni_3Si, Ni_5Si_2, Ni_2Si, Ni_3Si_2, NiSi and $NiSi_2$. Now do we actually expect to detect all the phases in the equilibrium phase diagram in a thin film experiment? The answer to this should be yes, since each silicide phase will be in equilibrium only if in contact with those phases adjacent to it in the phase diagram. However, this assumes phases of infinite extent given an infinite amount of time to reach equilibrium, conditions which are not met in the course of a thin film reaction. In some novel observations on the lateral growth of silicide phases from a nickel source on a thin silicon film, the sequence of phases formed was $Ni/Ni_3Si/Ni_5Si_2/Ni_2Si/Ni_3Si_2/NiSi$ (Chen et al 1984). Figure 4.15 shows a micrograph from this work, and these experiments can be used to identify even an extremely thin layer of NiSi at the silicon side of the diffusion couple. These lateral growth experiments also demonstated that the first phase to form in the Ni/Si couple is Ni_2Si and that this phase grows to a thickness of 20 μm before any other silicide phase is detected. In a reaction between a thin Ni film and an Si substrate even fewer phases are detected, with the reaction sequence being

$$Ni + Si \rightarrow Ni_2Si \rightarrow NiSi \rightarrow NiSi_2. \qquad (4.12)$$

Annealing at 300 °C results in the formation of Ni_2Si and this phase grows with parabolic kinetics until all the nickel is consumed. Then, and only then, is NiSi observed to grow at the Ni_2Si/Si interface. These thin film experiments show no sign of the Ni_5Si_2 and Ni_3Si_2 phases which are clearly identified in both lateral and bulk diffusion couples. There are two possible reasons for the observed discrepancy between the thin film and bulk (or lateral diffusion) results. On the one hand, the reaction temperatures in the bulk experiments are higher than those used in the thin film cases and the nucleation of some phases may only be possible at these higher temperatures. On the other hand, NiSi seems to be unable to form between silicon and Ni$_2$Si until the first silicide phase has reached a thickness of at least 20 μm. This will automatically exclude the possibility of having coexisting NiSi and Ni$_2$Si when the intitial metal layer is only 10 nm thick. This second argument can be used to explain why relatively few of the equilibrium silicide phases are observed in thin film reactions. Interested readers will find that the experimental data on silicide growth collated by authors like Tu and Mayer (1978) and Murarka (1983), and the arguments of D'Heurle and Gas (1986), can be combined to give a cogent explanation of many of the unusual features of nucleation and growth phenomena in the metal/silicon systems and especially in understanding why so wide a disparity of behaviours is observed from system to system.

450 °C 12 hr.

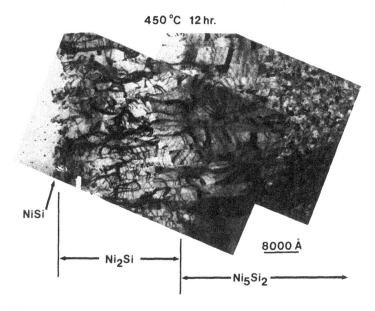

NiSi

Ni$_2$Si

8000 Å

Ni$_5$Si$_2$

Figure 4.15 A transmission electron micrograph of some of the silicide phases formed in a lateral diffusion couple of nickel on silicon. (Courtesy of Dr C Barbour and Professor J W Mayer.)

4.4.2 Contamination Effects

I have already mentioned that oxygen contamination on the silicon surface can have a strong effect on the silicide-forming reaction. The rate of reaction is often decreased, implying that the oxygen, or an oxide layer, blocks the diffusion of the reacting species. It is easy to imagine that a thin oxide layer will be an excellent barrier to diffusion of the reacting species. The observed first phase can also be changed, which can be understood if we assume that the oxygen contamination in the silicide phases alters the diffusion rates of reacting species more in some compounds than in others. If these diffusion rates select the first forming phase, as suggested above, then it is not surprising if some systems show a different first silicide phase when the reaction takes place under contaminated conditions. The V–Si system does indeed show a change in the first silicide phase under these conditions (Schutz and Testandi 1979).

Contamination effects have caused a certain amount of confusion in the microelectronic research community, since differing results on the phases formed and the reaction kinetics were found for experiments carried out under nominally the same conditions. It is now clear that unless great attention is paid to removing contamination from metal and silicon before reaction, the observed sequence of silicide-forming events will be dominated by this contamination, and it may be difficult to form an interface which is electronically and metallurgically stable.

4.4.3 Typical Silicide Parameters

The most important parameters of the silicon/silicide contacts used for microelectronic applications are the barrier height and the specific contact resistance. The resistivity of the silicide thin films is also important if they are to be used as conductor tracks. The Schottky barrier heights of all known silicides on n-type silicon fall roughly in the range 0.4–0.9 eV and the barrier heights on n-type silicon are relatively independent of the stoichiometry of the silicide phase formed—0.89 eV for PtSi and 0.85 eV for Pt_2Si for example. The barrier heights also depend slightly on the temperature of the annealing process used to form the silicide on the silicon surface. Specific contact resistances can be very low indeed, often less than $10^{-7}\,\Omega\,cm^2$, if the contact is prepared under clean conditions. It is always necessary to anneal the silicide films after reaction to achieve these excellent contact resistances. The silicide layers are usually fine-grained polycrystals, but remarkably their resistivities are often lower than those of the bulk silicide phases. This is quite different behaviour to that which has already been described for aluminium and gold thin films, where thin film resistivities are always higher than those of the bulk materials, and implies that the grain boundary reflection coefficients in uncontaminated silicide films are very low. It is possible that the measured values of the bulk resistivities are artificially high, since it is difficult in many cases to produce good-quality bulk material of these silicide phases.

Table 4.3 shows some typical values of bulk and thin film resistivities and barrier heights for a few silicide phases on n-type silicon. Especially noticeable is the low resistivity of $TiSi_2$ films. However, we can also see that the resistivities of the silicides are generally higher than those of the thin metal films in table 4.2. In the next chapter we shall see why silicides have become so important in integrated circuits.

Table 4.3 Electronic properties of some silicide compounds (after Murarka 1983).

Silicide	Thin film resistivity ($\mu\Omega$ cm)		Bulk resistivity ($\mu\Omega$ cm)	Schottky barrier height on n-type silicon (eV)
PtSi	30	(700 °C)	—	0.88
$TiSi_2$	13–16	(900 °C)	16–100	0.5
$TaSi_2$	30–70	(1000 °C)	8–46	0.59
WSi_2	35–100	(1000 °C)	12–55	0.65
$NiSi_2$	50	(900 °C)	118†	0.7

†This very high value may be a result of poor preparation of the bulk compound.

4.5 STRESS IN THIN FILMS

In this chapter on thin conducting films, it seems apt to include a brief description of one of the most universal properties of a vapour-deposited thin film—that it is under a remarkably high stress. Although this is found to be true for almost all thin films, relatively little attention has been paid to understanding the phenomena associated with the production and relief of these stresses. This is because in most circumstances the more immediately important properties of the film used for microelectronic applications—resistivity and chemical reactivity—are little affected by the state of stress in the film. The major problem associated with stress is that the film may crack or peel from the substrate surface, which is obviously undesirable when thin film conduction tracks are being deposited. The stress in thin films is also important in determining the integrity of insulating films, as will be mentioned in Chapter 6. In this section I will discuss some of the mechanisms by which stress can be induced in thin vapour-deposited films and give some examples of how high these stress levels can be.

There is a multiplicity of mechanisms which can cause a thin film to grow with a high internal stress, and of these intrinsic mechanisms two will be presented here. Intrinsic stress is taken to be that which results from properties or reactions of the film material itself, rather than from interactions with a substrate. Klockholm (1969) and Buckel (1969) have both suggested that part of the process of condensing a film from the vapour phase involves a transformation from a precursor amorphous phase to a crystalline deposit. This kind of transformation will generally involve an increase in density and shrinkage of the linear dimensions of the film. Such a process will result in a tensile stress being developed in the film. A similar effect occurs during thin film reactions like those used to form silicide layers by direct reaction between metal layers and silicon. Here the decrease in volume on transforming the metal/silicon couple into a silicide compound can be very large indeed (Murarka 1983) and, once again, tensile stresses will be developed in these films. Co-deposited silicide layers will tend to have lower intrinsic stress levels. Sputtered films are often under an intrinsic compression stress because of the energetic sputtered species being implanted under the surface of the growing film, causing the whole lattice to expand.

By contrast, there are three effects which can induce stress in thin films as a result of interaction with the substrate. Epitaxial growth of a deposit on a substrate with which it is nearly, but not quite, lattice matched will cause an elastic stress in the deposit film, as is described in §3.5. The sign of the stress will depend on the relative values of the lattice parameters in the two materials. A second extrinsic mechanism for inducing stress comes from the fact that most metallic materials have rather higher thermal expansion coefficients than typical microelectronic substrate materials like silicon or glassy films. This means that heating a metallic film on a silicon substrate will result in a compressive stress in the film, while cooling the same sample will reverse the sign of the mismatch removing the compressive stress or even causing a net tensile stress. Finally, Nakajima and Kinoshita (1974) have considered the grain growth that will occur in most films when heated after deposition. The grain boundaries are regions of lower density than the grain interiors and so removing the boundaries causes the film to shrink during annealing and sets up a tensile stress.

While the size of the stresses set up by the first two extrinsic effects can be estimated very simply from a knowledge of the difference in the lattice parameters and thermal expansion coefficients of film and substrate, a slightly more complex expression has been developed to model the increase in tensile stress in the film due to grain growth during annealing:

$$\Delta\sigma \propto \frac{E_f}{1 - v_f}\, a\left(\frac{1}{D_0} - \frac{1}{D}\right) \tag{4.13}$$

where E_f is Young's modulus of the film material, v_f is Poisson's ratio, a is the lattice parameter of the film and D_0 and D are the average grain sizes before and after heating respectively. It is possible to add to this model the effect of changes in crystal structure during heating, but the volume fractions of the phases present, their crystal structures and the texture in the film, if any, must be very accurately known. Estimation of the stresses set up by the intrinsic mechanisms is relatively easy in the case of thin film reactions like silicide formation, but rather hard in more general cases when so little is known about the precursor stage of thin film formation. However, it is clear that most thin films will be under stress as a result of one or more of these effects. It is not easy even to predict whether this stress will be tensile or compressive, because of the difficulty of estimating relative magnitudes of the stresses generated by the various mechanisms. Most metallic films are found to be under a net tensile stress, but on heating these samples a net compressive stress can be induced if the thermal expansion mismatch between metal and substrate is large. These stresses can easily exceed the yield stress of the film material, which will cause plastic flow of ductile metal films and possibly fracture of films of more brittle materials. Even after stress relief by these mechanisms, residual levels of tensile stress between 50 MPa and 200 MPa were measured in aluminium films by Klockholm (1969), while stresses in excess of 10^3 MPa are quite commonly observed in silicide layers (Ratajczyk and Sinha 1980).

The plastic relaxation of thin metal films under these very high stresses has been studied by Gangulee (1974). He measured stress in isothermal experiments and showed that grain boundary creep and plastic deformation occurred in copper and lead films. The plastic deformation should lead to very high densities of dislocations in all metal films which have been cooled to room temperature from an elevated deposition temperature. Dislocation densities in excess of 10^9 cm^{-2} would not be surprising. More recently, Hieber (1984) studied dynamic stress relaxation in thin films and came to the same conclusion: that both grain boundary creep and plastic flow occur during cooling. It is extremely fortunate that metallic and silicide films can plastically deform to reduce the extremely high intrinsic and extrinsic stresses developed during film growth. Without these stress relief mechanisms the films would almost certainly crack or decohere from the substrates. In fact, some poorly bonded silicide layers do crack off the substrate during heating or cooling. Apart from this, fortunately rare, effect the mechanical and electrical properties of thin conducting films are not affected very strongly by the stress-inducing reactions introduced above. Only in the cases where the conducting film has a very high yield stress are the unrelieved stress levels likely to damage the integrity of the conducting paths. When designing long conduction tracks it is wise to use highly ductile materials like aluminium, where the plastic flow will reduce the stresses to levels where cracking of the films is very unlikely. A problem which arises as a result of stress relief in aluminium conductor films will be described at the beginning of the next chapter. Plastic deformation cannot occur in the dielectric films discussed in Chapter 6, leading to catastrophic fracture of these brittle layers when the intrinsic stresses generated during deposition, or the extrinsic stresses formed in the cooling of the films from the elevated temperatures, are high.

4.6 NON-EQUILIBRIUM STRUCTURES IN THIN FILMS

We have already seen in this chapter that defect concentrations in thin films can be very much higher than in bulk materials. This section will consider the additional possibility that phases exist in thin films which are not found in the equilibrium phase diagram. The most common form of non-equilibrium structure has already been mentioned in §4.2, i.e. the formation of amorphous elemental and compound films by vapour deposition when a crystalline deposit is the equilibrium state of the film. These amorphous phases are the result of the enormous undercooling experienced by deposit atoms on arrival on the substrate and in some cases the influence of contamination species reducing the atom mobility on the surface and inhibiting the rearrangement of the atoms into the equilibrium crystal structure. Thin film 'phase diagrams' have been constructed which show the regimes of thickness and deposition rate under which these non-equilibrium amorphous phases can be formed. Figure 4.3 shows a schematic illustration of this kind of diagram.

Similar studies of the structures formed in co-deposited alloy films have resulted in the generation of thin film phase diagrams for these systems. Figure 4.16 shows a comparison of the bulk and thin film phase diagrams for the nickel–aluminium alloy system. There are quite significant differences in the positioning of the phase lines below about 500 °C in the two diagrams. In addition, the grain size in the thin films will be much smaller than in bulk materials and the point defect and dislocation densities will be high. The rapid quenching of the deposited alloy film can be used to explain why these discrepancies are observed, but a prediction of the nature of the differences would require a much more complete understanding of the deposition process than is currently available.

A further example of non-equilibrium phase formation during thin film deposition is found in the evaporation of CdS or CdTe layers. The stoichiometry of these II–VI films is usually fairly close to that of the source materials, but the films often consist of a mixture of crystal phases. CdS and CdTe films are found to consist mostly of grains of the hexagonal wurtzite structure, but can also contain some grains with the sphalerite structure. The equilibrium low-temperature crystal structure of both these semiconductors is the sphalerite lattice and so the thin films are in non-equilibrium conditions in the sense that they contain grains with structures characteristic of the high-temperature forms of the compound. It is often rather hard, even with very extensive annealing, to persuade these films to transform fully to the equilibrium sphalerite structure.

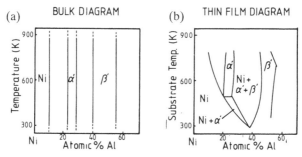

Figure 4.16 A comparison of (a) bulk and (b) thin film phase diagrams for the alloy system Ni–Al. The thin film diagram has been generated for sputtered films and will probably be slightly different for evaporated films because the conditions of deposition will alter (after Hentzell 1981.)

These three examples clearly indicate that vapour-deposited thin films do not necessarily have the structure or relative phase abundance which would be predicted from the relevant composition/temperature point in the equilibrium phase diagram. I shall now describe a case where a thin film reaction produces a metastable phase which is not present at all in the relevant phase diagram. Gold–germanium alloys are a popular ohmic contact alloy on gallium arsenide and so considerable pains have been taken to identify both the structure of vapour-deposited gold and germanium layers and the reaction products of these elements with the semiconductor surface. Some of these experiments have been described in §2.2. Sequential evaporation of germanium and gold, and alloying at 400 °C, results in the formation of two distinct alloy crystal phases, although the gold/germanium phase diagram shows no compound phases. A detailed study of one of these phases shows it to have a hexagonal structure, and 4.17 shows a transmission electron micrograph of a polycrystalline region of this phase in a thin film sample. Similar observations of metastable crystalline phases have been made in splat-cooled samples in the same alloy system.

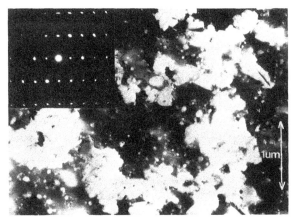

Figure 4.17 A transmission electron micrograph of a metastable phase formed by sequential evaporation of aluminium and germanium and annealing at 400 °C.

Thus we cannot rely on the equilibrium phase diagram to give an accurate prediction of the structure of vapour-deposited thin films, even when the stoichiometry of the films is very well controlled. It has proved necessary to determine the structure and electrical properties of most of the conducting films which are used in microelectronic devices entirely empirically. Once a method is found by which a useful film can be prepared, great trouble is taken to ensure that the deposition process always takes place under these conditions. In addition, the high density of lattice defects which are always present in these polycrystalline films means that the electrical and mechanical properties may be significantly different to those of the bulk material. For the same reason, the reaction rates in thin films can be much faster than in bulk materials, and this will have important consequences for the chemical stability of the metallisation films which I shall discuss in the following chapter.

REFERENCES

Adams A C 1983 in *VLSI Technology* ed. S M Sze (New York: McGraw-Hill) p 93

Baccarini G, Ricco B and Spadini G 1978 *J. Appl. Phys.* **49** 5565

Bacmann J J, Papon A M and Petit M 1982 *J. Physique Coll.* **43** Cl 15

Broadbent E K and Stacey W T 1985 *Solid State Technol.* December 51

Broniatowski 1982 *J. Physique Coll.* **43** Cl 63

Buckel W 1969 *J. Vac. Sci. Technol.* **6** 606

Campbell D R, Mader S and Chu W K 1982 in *Thin Films and Interfaces* ed. P S Ho and K N Tu (New York: North-Holland) p 351

Chaviuneau J P, Croce P, Devant G and Verhaeghe M F 1969 *J. Vac. Sci. Technol.* **6** 776

Chen S H, Zheng L R, Barbour J C, Zingu E C, Hung L S, Carter C B and Mayer J W 1984 *Mater. Lett.* **2** 469

Chopra K L and Das S R 1983 *Thin Film Solar Cells* (New York: Plenum)

Clark A H 1980 in *Polycrystalline and Amorphous Thin Films and Devices* ed. L L Kazmerski (New York: Academic) p 135

Cusano D A 1967 in *Physics and Chemistry of II–VI Compounds* ed. M Aven and J S Prener (Amsterdam: North-Holland) p 709

D'Anterroches C and Bourret A 1984 *Phil. Mag.* A **49** 783

Davey J E and Pankey T 1969 *J. Appl. Phys.* **40** 21

D'Heurle F and Gas P 1986 *J. Mater Res.* **1** 205

Fahrenbruch A L and Bube R M 1983 *Fundamentals of Solar Cells* (New York: Academic)

Fonash S J 1972 *Electron* **15** 783

Francombe M H 1975 in *Epitaxial Growth* ed. J W Matthews (New York: Academic) p 109

Gadene P 1979 *Thin Solid Films* **57** 77

Gangulee A 1974 *Acta Metall.* **22** 177

Grovenor C R M 1985 *J. Phys. C: Solid State Phys.* **18** 4079

Grovenor C R M, Hentzell H T G and Smith D A 1984 *Acta Metall.* **32** 773

Hentzell H T G 1981 *Doctoral Dissertation* (No. 64) Linkoping Institute of Technology

Hentzell H T G, Andersson B and Carlsson S E 1983 *Acta Metall.* **31** 2103

Hieber H 1984 in *Electronic Packaging Materials Science* ed. E A Geiss, K N Tu and D R Uhlmann (Pittsburgh PA: MRS) p 191

Hirth J P and Pound C M 1963 *Condensation and Evaporation* (Oxford: Pergamon)

Holland L 1960 *Vacuum Deposition of Thin Films* (New York: McGraw-Hill)

Holland L, Steikelmacher W and Yarwood J 1974 *Vacuum Manual* (London: Spon)

Johnson N M and Biegelsen D K 1982 *Appl. Phys. Lett.* **40** 882

Kazmerski L L 1980 *Polycrystalline and Amorphous Thin Films and Devices* (New York: Academic)

Kazmerski L L, Ayyagari M S and Sandborn G A 1975 *J. Appl. Phys.* **46** 4865

—— 1982 *J. Vac. Sci. Technol.* **20** 423

Klockholm E 1969 *J. Vac. Sci. Technol.* **6** 138

Learn A J and Foster D W 1985 *J. Appl. Phys.* **58** 2001

Maissel L I and Glang R 1983 *Handbook of Thin Film Technology* (New York: McGraw-Hill)

Matare H F 1984 *J. Appl. Phys.* **56** 2606

Mayadas A F and Shatzkes M 1970 *Phys. Rev.* B **1** 1382

Movchan B A and Demchishin A M 1969 *Fiz. Met. Metalloved.* **28** 83

Murarka S P 1983 *Silicides for VLSI Applications* (New York: Academic)

Nakajima Y and Kinoshita 1974 *Japan. J. Phys.* suppl. **2** 575

Rai-Choudhury P and Hower P L 1973 *J. Electrochem. Soc.* **120** 1761

Ratajczyk J F and Sinha A K 1980 *Thin Solid. Films* **70** 241

Rhodin T N and Walton D 1964 in *Single Crystal Films* ed. M H Francombe and H Sato (Oxford: Pergamon) p 31

Schutz R J and Testandi L R 1979 *J. Appl. Phys.* **50** 5773

Seager C H 1982 in *Grain Boundaries in Semiconductors* ed. H J Leamy, G E Pike and C H Seager (New York: North-Holland) p 85

—— 1983 in *Defects in Semiconductors* ed. S Mahajan and J W Corbett (New York: North-Holland) p 343

Seager C H and Pike G E 1982 *Appl. Phys. Lett.* **40** 471

Seto J W Y 1975 *J. Appl. Phys.* **46** 5247

Taylor W E, Odell N H and Fan H Y 1952 *Phys. Rev.* **88** 867

Thornton J A 1977 *Ann. Rev. Mater. Sci.* **7** 239

Tu K N and Mayer J W 1978 in *Thin Films, Interdiffusion and Reactions* ed. J M Poate, K N Tu and J W Mayer (New York: Wiley) p 359

Ziegler E, Siegel W, Blumtritt H and Breitenstein O 1982 *Phys. Status. Solidi* a **72** 593

5

The Metallurgy of Aluminium- and Gold-based Metallisation Schemes

5.1 INTRODUCTION

In this chapter I will discuss the metallic materials commonly used to form the conducting layers in integrated circuitry. The necessity for providing a dense array of conducting tracks on the surface of a chip has been discussed in Chapter 2 and a schematic diagram of the fabrication of a simple metallisation scheme given in figure 2.7. We have also seen how a lithographic process can be used to pattern the conduction tracks on the chip surface (§2.4). The techniques for the deposition of thin metal films, and the microstructures commonly found in these films, have been presented in Chapter 4. It only remains to discuss how the material is chosen to make these thin film conduction tracks. Most of this chapter will concentrate on the metallisation systems used on silicon integrated circuits, but in §5.8 some attention will be paid to the metals used in circuits on GaAs wafers and high-temperature devices.

There are two obvious requirements for a conductor material: that the resistivity of the thin films be as low as possible; and that the films should adhere strongly to both oxide and silicon surfaces. The electrical properties of the contact between the metal and the silicon must also be carefully controlled. There are a number of additional technological requirements for a successful metallisation material, including easy patternability in a lithographic process and very low cost. Table 4.2 shows that metals like aluminium, gold and copper can be deposited in thin film form with resistivities very little higher than those characteristic of the bulk material. Refractory metals like tungsten and molybdenum have rather higher bulk resistivities and thin films often have to be annealed at high temperatures before their resistivities approach those of the bulk materials. Thin film interconnect materials are normally required to have resistivities below about $50 \, \mu\Omega \, cm$ to be useful in integrated circuit applications, and we can see from table 4.2 that a wide range of metallic thin films can be prepared with resistivities much lower than this value. The thin film resistivity is thus not a very stringent restriction on the choice of an interconnect material.

This material should also adhere strongly to the surfaces exposed by wet or dry lithographic etching processes. These surfaces may be the silicon wafer exposed through vias, the oxide layer over the remainder of the wafer, or other metallisation layers. The silicon surface will quickly be covered by a thin native oxide layer when exposed to air. Bond formation with oxide materials is thus necessary before satisfactory adhesion between the deposited metal and the substrate is obtained. Noble metals like gold, and to a lesser extent copper, form only weak bonds with most oxides and so adhesion of thin films of these metals to the surface of a microelectronic circuit can be poor. Intermediate 'glue' layers of titanium or chromium are often used to improve the adhesion of gold conduction tracks. Aluminium is a much more reactive metal, and readily forms M–O bonds with silicon oxides. Likewise, refractory metals like tungsten react strongly with oxide surfaces. Aluminium can decompose the silicon oxide because the heat of formation of alumina (per mole of oxygen) is larger than that of silica. The aluminium will thus reduce the silicon oxides:

$$3SiO_2 + 4Al \rightarrow 3Si + 2Al_2O_3. \tag{5.1}$$

The complete reduction of the native oxide layer on the silicon wafer usually requires an annealing cycle at about $450 \, ^\circ C$ for at least a few minutes; which results in the formation of a contact with excellent electrical properties. However, annealing can also have some unfortunate side effects as will be discussed in the next section.

The final criterion which must be taken into account when choosing a material for metallisation interconnections is the electrical properties of the contact formed between the thin film and the silicon. Both ohmic and Schottky contacts are required for different applications in microelectronic devices (see §2.3). The device engineer will usually specify a stable barrier height for Schottky contacts and as low a contact resistance as possible for ohmic contacts. The contact resistance depends on the barrier height and the doping concentration in the semiconductor, as we have seen in §2.2, so that reducing the barrier height will lower the contact resistance as well. Aluminium forms a Schottky or rectifying contact to lightly doped n-type silicon, with a barrier height of 0.7 eV. Aluminium contacts on p-type silicon have a much lower barrier height, 0.4 eV, and usually behave in a 'pseudo-ohmic' manner at room temperature. Contact resistivities in the range 10^{-4} to $10^{-7} \, \Omega \, cm^2$ can be achieved for aluminium films on p-type silicon, but similar figures can be achieved for many other metals and compounds as well (see table 2.2).

It is clear that the properties of contacting and interconnecting materials needed for integrated circuitry can be found in a range of metallic materials. However, we should note that aluminium films adhere strongly to silicon and silica, have very low resistivities and reduce native oxide layers on silicon to give low contact resistances. This combination of properties explains why the majority of integrated circuit metallisation schemes are based on aluminium alloys. Historically, the ease with which thin films of these alloys can be deposited was also important in deciding which material to choose for this application. For most of the remainder of this chapter I shall only consider the development of aluminium alloy metallisation schemes. I will mention a few of the advantages and problems associated with the use of gold conductors in later sections. The corrosion resistance of gold films makes them particularly attractive in devices which are not to be hermetically packaged and in which aluminium thin films would be very prone to atmospheric attack (see §5.5).

At this point, it is as well to mention that there is a problem that can be encountered with aluminium thin film metallisation which is nothing to do with its electrical or chemical properties. Aluminium is a relatively soft material and so will deform easily under any applied stress. This kind of stress can arise during the heating of a metallisation structure where aluminium tracks run across a silicon wafer surface, because of the widely different thermal expansion coefficients of the two materials — $2.6 \times 10^{-6}\,K^{-1}$ for silicon and $24 \times 10^{-6}\,K^{-1}$ for aluminium. A substantial compressive stress is thus developed in the aluminium films during heating and these stresses are easily relieved by grain boundary sliding. This process will cause grains to be forced out of the film surface as sketched in figure 5.1. If this sliding is too severe, it can result in the formation of an open circuit in the metallisation, and, at an earlier stage in the process, roughening of the film surface makes it difficult to deposit subsequent metal or insulator layers with even thicknesses. Reduction of the grain boundary sliding rate by addition of a few atomic per cent of solute elements which segregate to the grain boundaries, or constraining the film under a glass layer, are both quite effective ways of limiting this problem. Electromigration damage in conduction tracks is also reduced by alloying additions, as will be discussed in §5.4.3. Alternative choices for interconnection metals like gold or tungsten suffer less from this kind of problem because the grain boundary diffusion and sliding rates are much lower that they are in aluminium at the maximum temperature reached during the processing of integrated circuit metallisation.

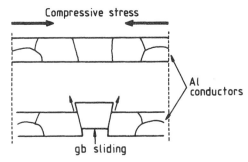

Figure 5.1 A schematic illustration of the effect of compressive stress on thin film conductor tracks. Grain boundary sliding can even lead to the formation of open circuits by grains 'popping' out of the surface of the film.

5.2 DIFFUSION IN POLYCRYSTALLINE THIN FILMS

In the previous section I described the reasoning that led to the choice of aluminium as an attractive thin film interconnection material. However, it has also been found that there are a number of mechanisms by which the metallurgical and electrical properties of thin aluminium interconnections can be degraded. Some of these will be presented in subsequent sections in this chapter, but we should note here that in all these reactions the diffusive transport of material through the films is important in controlling the rate at which the reaction can proceed. We must, therefore, spend some time considering the process of diffusion in polycrystalline thin films. In Chapter 4 I showed that the density of grain boundaries in a polycrystalline thin film is usually much higher than is found in the bulk material. It is rare to find grains that have diameters much larger than 1 μm in films which have approximately this thickness. It is thus important to consider diffusion in both grain boundaries and in the grain interiors when estimating the total flux of material transported through a thin film, or the rate at which two thin films will react together to form compound phases. The extent to which grain boundary diffusion will contribute to a total diffusion flux will depend on the value of the average grain boundary diffusion coefficient and on how it compares with that of lattice diffusion. It is often stated that the activation energy for grain boundary diffusion in metals is roughly one-half that for lattice diffusion and so the grain boundary diffusion coefficients will be correspondingly much higher. We must look rather more closely at the structure of grain boundaries in metals in order to try and understand this observation.

5.2.1 Diffusion in Grain Boundaries in Metals

The first established model of diffusion in grain boundaries was proposed by Turnbull and Hoffman (1954). It was well known at that time that low-angle grain boundaries consisted of arrays of discrete dislocations separated by regions of perfect crystal. Turnbull and Hoffman considered that all grain boundaries were made up of arrays of similar dislocations and were then able to model the grain boundary diffusivity as the sum of contributions from pipe diffusion down all the dislocations. The core of each dislocation was taken to be a highly disordered region where the activation energy for diffusive jumps was lower than in the perfect crystal. The grain boundary diffusion coefficient is then of the form

$$D_{gb}\delta = \frac{D_p^0 A_p}{b} \sin\left(\frac{\theta}{2}\right) \exp\left(-\frac{Q_p}{kT}\right) \tag{5.2}$$

where $D_{gb}\delta$ is the product of grain boundary diffusion coefficient and boundary thickness and is the quantity that is usually measured experimentally. D_p^0 is the pre-exponential factor for dislocation pipe diffusion, A_p the cross-sectional area of the dislocation core, $b/\sin(\frac{1}{2}\theta)$ the dislocation spacing in a simple low-angle boundary, b the Burger's vector of the dislocations and Q_p the activation energy of the diffusive jumps along the dislocation cores. θ is the angle of misorientation of the two grains across the grain boundary.

This model for grain boundary diffusion predicts that the diffusion coefficient, D_{gb}, should be extremely anisotropic, since the rapid pipe diffusion will occur only along the line of the dislocations in the boundary. In addition, the activation energy for grain boundary diffusion should depend only on the character of the dislocations and not their separation, and so should be independent of the angle of misorientation across the boundary, θ. Measurements of diffusion in low-angle grain boundaries, where the misorientation angle is less than about 16°, have shown both strong anisotropy and a roughly constant value of Q_p in agreement with the model. We can also predict that the diffusion coefficient should increase with the misorientation angle since the density of dislocations is proportional to $\sin(\frac{1}{2}\theta)$. Once again this dependence has frequently been observed in the low-angle regime. Finally, it has been shown that the activation volume for silver self-diffusion in grain boundaries is that same as for silver self-diffusion in the bulk lattice (Martin et al 1967). This is extremely strong evidence for diffusion being controlled by a simple vacancy mechanism in metallic grain boundaries, just as it is in the grain interiors. All this indicates that our understanding of diffusion in low-angle grain boundaries is rather comprehensive. Unfortunately, in the high-angle regime, the experimental data are less easy to interpret.

Figure 5.2 shows some data from measurements of the penetration of zinc along grain boundaries in aluminium. The grain boundaries have a wide range of misorientation angles around a [001] axis, and this range extends far out of the low-angle regime (Herbuval *et al* 1973). Large variations in penetration depth can be seen and indications of misorientations at which there is particularly slow grain boundary diffusion. These same experiments also demonstrated that zinc diffusion along aluminium grain boundaries is extremely anisotropic. In order to explain this more complex variation of diffusivity as a function of the misorientation across the grain boundaries, we must consider whether the structure of high-angle grain boundaries can be satisfactorily described by the dislocation model used above. As the misorientation of the grain boundaries is increased, the dislocation cores will have to move closer and closer together, and eventually they will overlap. At this point it is difficult to describe the boundary as consisting of an array of discrete dislocations, as is required if the pipe diffusion model is to be applied. We need another model for the structure of grain boundaries, but can immediately discard the notion that boundaries can be considered to be homogeneous slabs of amorphous material. Neither the variations of D_{gb} with θ, nor the observed anisotropy of grain boundary diffusion, could be explained by such a model.

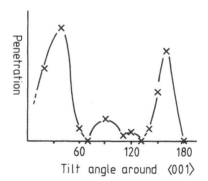

Figure 5.2 An experimental plot of the distance of penetration of zinc atoms along grain boundaries in aluminium. It is clear that diffusion is very much faster along some boundaries than others (after Herbuval *et al* 1973).

The most recent theory of high-angle grain boundary structure is based on the so-called structural unit model, SUM. In this model, high-angle grain boundary cores are built up of a relatively small number of close-packed units. These units are very similar to the structures proposed by Bernal (1964) as the building blocks of a liquid.

The identification of these units at grain boundaries has been made by two methods: the first involving the direct observation of grain boundary structures in high-resolution electron microscopy; and the second as a result of very extensive work on computer-generated relaxed grain boundary structures. A few of the simple Bernal units which have been isolated in grain boundary structures are sketched in figure 5.3. The SUM concept is that the cores of some grain boundaries can be constructed wholly from only one of these favoured structural units. These 'favoured' boundaries will be found at characteristic misorientations and it is possible to consider each structural unit as accommodating a particular angle of misorientation across the boundary. In the range of misorientations between two such favoured structures, the grain boundary cores will consist of a mixture of structural units from the two nearest favoured boundaries. By mixing the two units together in the correct proportions, all the grain boundaries with misorientations between the two favoured structures can be created. In this way, the structure of any given boundary can be predicted if the structural units of the adjacent favoured boundaries are known. A much more comprehensive description of this elegant model can be found in the review by Sutton (1984).

It is possible to make a qualitative link between the structure of a grain boundary and the rate of diffusion along the boundary plane. Each structural unit can be thought of as having an intrinsic diffusivity, and this will not be the same for all the possible structural units. Some of the Bernal units have rather open structures, like the Archimedean square antiprism, while others offer no easy path for diffusion. From the shape of the units illustrated in figure 5.3, diffusion through grain boundaries would also be expected to be anisotropic. A favoured grain boundary core, which consists of only one structural unit, must have a diffusivity characteristic of that unit. A non-favoured boundary, in which the core is made of a mixture of structural units, will have a diffusivity that is the sum of all the contributions from transport through the individual units. We can envisage the diffusion process as a gradual migration of atoms through the easy paths along the grain boundary plane, exactly as was originally proposed for diffusion along dislocations, but with the structural units replacing the dislocation cores. With this kind of model we can explain the variation of diffusivity with misorientation angle, θ, because the character and density of the structural units will change with θ. Boundaries which consist wholly of structural units with large open volumes, like the Archimedean square antiprism, are expected to have activation energies for diffusion which are very much lower than the value for diffusion in the perfect lattice. Other grain boundaries with more closely packed structural units will be expected to have lower diffusivities and thus we are able, at least conceptually, to explain the variations in diffusion rate with θ illustrated in figure 5.2. The anisotropy of diffusion in high-angle grain boundaries is also explained by the intrinsic anisotropy of diffusion through the structural units.

(a)   (b)   (c)

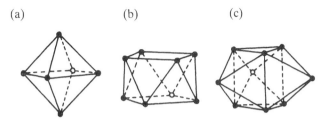

Figure 5.3 Some of the Bernal units which have been identified at the cores of 'favoured' grain boundaries in metals: (a) the octahedron; (b) the Archimedean square antiprism; (c) the capped trigonal prism.

Unfortunately, we are not able at present to predict the diffusivities in particular grain boundaries simply from a knowledge of the boundary misorientation. This is because values for diffusion parameters like the pre-exponential factors and activation energies of individual diffusive jumps in arrays of structural units are not available. Computer calculation techniques may in the future provide some of these values, as has been discussed by Martin and Peraillon (1980).

It is clear from the discussion above that there is no such thing as a single value of the grain boundary diffusion coefficient. Each boundary will have a diffusivity that depends on its structure, and even the measured activation energy of diffusion in a grain boundary is usually an average value because of the multiplicity of structural units making up the boundary core. The grain boundary diffusion process will, in general, involve diffusive jumps through a number of different structural units, each characteristic jump having its own activation energy. In the next section we will consider diffusion through very large arrays of grain boundaries in a polycrystalline thin film, but we should not forget that the total diffusion flux is made up of unequal contributions from each of the boundaries.

5.2.2 Diffusion in Polycrystalline Films

In the previous section I argued that diffusion in grain boundaries is much faster than in the lattice if the boundary core consists of structural units in which diffusive jumps have a low activation energy. In this section we must determine the conditions under which grain boundary diffusion is the dominant material transport mechanism in a polycrystal-line thin film. The two alternative available diffusion paths are through the bulk lattice and along dislocations in the grains. I have mentioned in the previous chapter that dislocation densities in vapour-deposited thin films are often high and so the contribution to a total diffusion flux from dislocation pipe diffusion may be significant, although it is normally ignored in polycrystalline samples. Harrison (1961) has suggested that there are three regimes of diffusion kinetics in polycrystalline materials, distinguished by the relative values of grain boundary and lattice diffusion coefficients. These three regimes are illustrated in figure 5.4 and show the cases for solute diffusion from a surface source into a polycrystalline material where:

(a) Grain boundary diffusion is very little faster than diffusion in the lattice, so there is almost no difference in the penetration distance at boundaries and in the bulk lattice.

(b) Grain boundary diffusion is much faster than in the bulk and so the solute penetrates further along the boundaries, but the diffusion fields around adjacent boundaries do not interact significantly.

(c) Lattice diffusion is negligible and penetration only occurs along the grain boundaries.

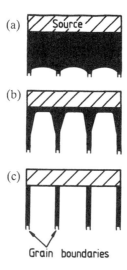

Figure 5.4 A schematic illustration of Harrison's three regimes of diffusion in polycrystalline materials. (a), in which grain boundary diffusion is hardly faster than in the bulk; (b), in which grain boundary diffusion is significantly faster than in the bulk; (c), in which grain boundary diffusion dominates the transport processes.

Thin film diffusion couples offer a very convenient way of investigating grain boundary diffusion kinetics because it is possible to find conditions under which all three of these regimes are observed. Bulk diffusion couples usually exibit type (b) kinetics because the grain sizes and the diffusion distances are both large.

Mathematical models for the planar concentration of solute carried into a polycrystalline material by the combination of grain boundary and lattice diffusion have been constructed for all three of the Harrison regimes. These models have analytic solutions which offer the experimentalist the opportunity to use measured solute diffusion profiles to calculate values of the average grain boundary diffusion coefficients. These solutions often have quite complex mathematical forms which will not be given here. Interested readers are directed to the articles by Whipple (1954), LeClair (1963), Gilmer and Farrell (1976) and Hwang and Balluffi (1979), who have obtained solutions for the diffusion of solutes in polycrystalline thin films for type (a), (b) and (c) kinetics.

Before we discuss some of the data that have been obtained from experiments on grain boundary diffusion in thin films, I should briefly mention the two most popular techniques by which diffusion kinetics are measured in these specimens. Analytical solutions have been obtained relevant to two specific kinds of experiments: accumulation of solute on the back side of a thin film usually in Harrison's regime of type (c) kinetics (Hwang and Balluffi 1979); and depth profiling to measure planar solute concentrations through a polycrystalline specimen after diffusion of the solute from a surface (Gilmer and Farrell 1976). Surface accumulation experiments on thin films are carried out with samples like that shown in figure 5.5(a), where a polycrystalline thin film has a solute source layer deposited on one side. Observation of the back side of the film with a sensitive and surface-specific analytical technique like Auger spectroscopy (see §10.3) allows measurement of the coverage of the surface by solute atoms which have diffused along the grain boundaries in the film. These measurements are usually carried out at low temperatures to ensure that lattice diffusion is frozen out and type (c) kinetics are obeyed. The variation of back-surface coverage with time has the form shown in figure 5.5(b), and matching this kind of data with the relevant analytical solution allows calculation of $\bar{D}_{gb}$, the average diffusion coefficient in the array of grain boundaries in the film.

Depth-profiling measurements are normally carried out on thicker polycrystalline samples into which a solute or a radiotracer isotope has been diffused. Sectioning the sample is carried out by controlled sputtering or chemical dissolution processes and the average concentration of the indiffused species measured from radioactivity counting when a radiotracer has been diffused, or Auger analysis for ordinary solute diffusion. The form of the data obtained from such an experiment is

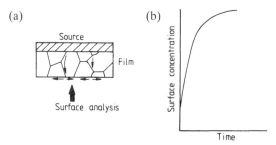

Figure 5.5 (a) A sketch of the kind of sample used in surface accumulation experiments; (b) an illustration of how the surface solute concentration changes with time during the experiment.

sketched in figure 5.6, and the contributions from lattice and grain boundaries to the overall diffusion profile can be separated. The analytical solution to this diffusion process is of the form

$$\bar{D}_{gb} \propto \left(\frac{\delta(\ln\bar{C}(x))}{\delta x^{6/5}}\right)^{-5/3} \tag{5.3}$$

where $\bar{C}(x)$ is the average solute concentration at depth x. Plotting the natural logarithm of the average concentration against the 6/5 power of depth gives straight-line regions from which the value of $\bar{D}_{gb}$ can easily be obtained. Once again, the measured value of the diffusion coefficient is the average of the contribution from a large number of grain boundaries.

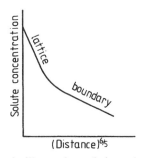

Figure 5.6 A schematic illustration of the solute concentration profile measured in a depth profiling experiment in a polycrystalline sample.

At this point a note of caution should be sounded about the interpretation of these kinds of experiments on grain boundary diffusion. Figure 5.7 shows a compilation of data on the activation energy of grain boundary diffusion, Q_{gb}, of silver in copper specimens. These results were obtained at a range of temperatures, mostly on polycrystalline samples, but including one measurement on silver diffusion in a copper bicrystal. There is a surprising scatter in the measured values and some indication of a systematic increase in activation energy with the temperature at which the measurement was made. This could possibly be due to differing contamination levels in the copper samples, but can be more convincingly explained by there being a range of activation energies for diffusion in copper grain boundaries of different structure. At low temperatures, only the diffusion through those boundaries where the activation energy is low will be measured, while at higher temperatures diffusion can occur through most of the boundaries and the measured average activation energy will be higher. This indicates that extrapolating diffusion data on a polycrystalline material outside the temperature range within which they were measured can give a very inaccurate prediction of the rate of material transport through the film. In particular, the relatively high activation energies for grain boundary diffusion measured at high temperatures are likely to lead to a severe underestimate of the diffusion rates at low temperatures.

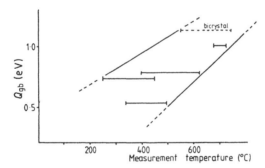

Figure 5.7 Values of the activation energy of grain boundary diffusion of silver in copper samples measured over a number of temperature ranges. The value of Q_{gb} seems to increase with the measurement temperature.

A further complication can arise when concentration-profiling experiments are used to measure D_{gb}, and grain growth occurs during the experiment. Grain boundary migration will result in the formation of alloyed volumes, since solute atoms are left in the grains behind the moving grain boundaries. This process can increase the apparent diffusion flux which is measured in depth-profiling or surface concentration experiments. Glaeser and Evans (1986) have calculated that simple grain growth during diffusion will affect the concentration profile to such an extent that quite significant inaccuracies in the calculated values of D_{gb}

and Q_{gb} will result. A high-temperature anneal before the diffusion source is deposited onto the sample has been used by many experimenters to equilibrate the grain structure before carrying out the diffusion experiment. This kind of precaution will limit the possibility of grain growth during diffusion, but a relatively recently discovered phenomenon called diffusion-induced grain boundary migration (DIGM) will still occur.

DIGM was first identified by Cahn *et al* (1979) and the basic feature of this phenomenon is that the diffusion of a solute along a grain boundary will cause the boundary to migrate even if there is a negligible driving force for grain growth. Thus any experiment where solute diffusion along grain boundaries is deliberately encouraged can result in quite extensive boundary migration and the production of the alloyed regions behind the moving boundaries that so significantly modify the measured diffusion kinetics. An example of how a DIGM process influences the concentration of a diffusing species behind a migrating grain boundary is illustrated in figure 5.8(a). Here diffusion of copper along a grain boundary in a thin gold film has caused migration of the boundary. The copper concentration in the volume swept through by the grain boundary is very much higher than elsewhere in the film where little lattice diffusion has occurred (figure 5.8(b)). This figure illustrates how effectively a DIGM process can increase the amount of solute which penetrates into a material during a diffusion experiment. The mechanism of DIGM is still the subject of some debate, although observations of the phenomenon have now been made in a considerable number of metals, semiconductors and even minerals. The effect of DIGM on the performance of thin film barrier layers will be discussed in §5.7.

Subject to the inaccuracies mentioned in the last few paragraphs, depth-profiling and surface accumulation experiments have been used to compile an extensive database on the grain boundary diffusion coefficients for self- and solute diffusion in a number of materials (see for instance Peterson (1980)). Table 5.1 lists a small number of examples taken from these data. It can be seen that the activation energies for grain boundary diffusion in low-melting-point metals like gold and aluminium lie roughly between 0.5 and 1.0 eV/atom. For reference, self-diffusion in the gold lattice has an activation energy of about 1.7 eV/atom. In general, the presence of any solute elements will reduce the rate of grain boundary diffusion and the addition of small concentrations of tantalum to gold increases the activation energy for grain boundary diffusion substantially. Grain boundary diffusion in refractory alloys also has a high activation energy, as shown by the data for silicon diffusion in tungsten grain boundaries. The variation in the activation energy for grain boundary self-diffusion with melting point for some thin film materials is shown in figure 5.9. These data show that grain

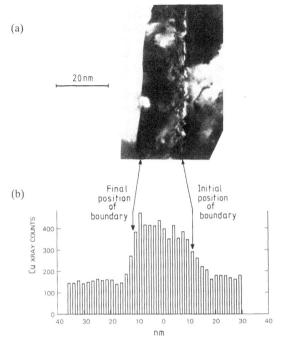

Figure 5.8 (a) A transmission electron micrograph showing a region of a thin gold film where a grain boundary has migrated as a result of the indiffusion of copper. (b) A copper concentration profile across the region in (a); the enhanced copper concentration in the volume swept by the boundary is clearly visible.

Table 5.1 Some selected data on grain boundary diffusion kinetics in thin film materials.

Matrix	Diffusant	Temperature range (°C)	D_0 (cm² s⁻¹)	Q_{gb} (eV/atom)	Reference
Au	Cr	210–293	3×10^{-3}	1.09	Holloway et al (1976)
Al	Cu	150–300	30	0.81	Howard et al (1976)
Al	Si	200–600	0.02	0.85	van Gurp (1973)
Au	Au	117–177	0.02	1.0	Gupta and Asai (1974)
Au 1.2%Ta	Au	200–400	10.0	1.26	Gupta and Rosenberg (1975)
Au	Ag	30–270	10	0.62	Hwang et al (1979)
W	Si	670–850	—	2.6–3.2	Chang and Quintana (1976)

boundary diffusion will be a rapid transport mechanism at all but extremely low temperatures, since the activation energies for this process are indeed roughly half those for lattice diffusion. It is also

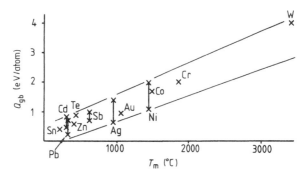

Figure 5.9 A plot of the measured activation energies for grain boundary diffusion in a number of metals plotted against the melting temperature, T_m. These values are roughly half the activation energies of bulk diffusion in the same elements.

apparent that the use of an alloyed film, or a refractory metal film, will reduce the rate of grain boundary diffusion. I will give examples later in this chapter where limiting the rate of diffusion in thin films can be very useful in controlling the stability of thin film metallisation structures.

We can now return to our original question: which path for diffusion in a thin film contributes the most to a diffusive flux — grain boundaries or the lattice? Balluffi and Blakely (1975) have used measured values of grain boundary and lattice diffusion parameters to estimate the dominant diffusion mechanism in films of FCC metals with a variety of grain sizes. Figure 5.10 shows two regimes in a plot of grain size against temperature, one where lattice diffusion is the most important diffusion mechanism and the other where grain boundaries contribute more to the total flux. As we would expect, low temperatures favour grain boundary diffusion even in films with a fairly large grain size, because lattice diffusion is effectively frozen out. A point has been added to this diagram showing the approximate position of an aluminium film with an average grain size of 1 μm held at 100° C. Grain boundary diffusion is predicted by the Balluffi and Blakely plot to be the only significant diffusion process, and we can take this as being a general feature of diffusive transport in metallisation structures. We shall see in the following sections that grain boundary diffusion is a very important phenomenon in the degradation mechanisms which operate in these structures.

5.3 REACTIONS IN ALUMINIUM THIN FILMS

In this section I shall describe reactions of aluminium films with other materials which are used in integrated circuit metallisation structures.

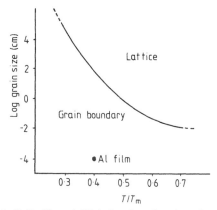

Figure 5.10 A Balluffi and Blakely plot showing the regimes where grain boundary and lattice diffusion dominate the transport process. The point for a thin film of aluminium at 100 °C has been added, indicating that only grain boundaries are effective diffusion paths in this material.

These reactions will include the dissolution of silicon in aluminium at metal–semiconductor contacts and intermetallic phase formation between aluminium and other metal thin films. Explanations for why the additional thin metal films are included in practical metallisation systems will be given later in this chapter.

5.3.1 Interdiffusion and Spiking at the Al–Si Interface

In Chapter 4 it was shown that metal films deposited from the vapour phase are usually polycrystalline with a grain size roughly equal to their thickness. Let us now consider the effect of evaporating an aluminium film with this structure onto a silicon wafer surface exposed through a via etched in an oxide layer (figure 5.11(a)). The silicon surface will be covered by a thin native oxide layer and this must be reduced by the metal before intimate contact between the silicon and the aluminium can be made. A short anneal above 300 °C will start this reduction process and it is reasonable to assume that some patches of the silicon surface will be cleared of oxide before the others. There will thus be localised regions of the contact where the silicon and the aluminium are able to interact. Figure 5.12 shows the aluminium-rich corner of the Al/Si phase diagram and it can clearly be seen that the solubility of silicon in aluminium increases quite sharply at about 300 °C. In the regions of the contact where the native oxide layer has been dissolved, the silicon will rapidly go into solution in the aluminium film. The solubility of aluminium in silicon is very low at these temperatures and so no significant reverse solution of aluminium in the wafer will occur. Figure

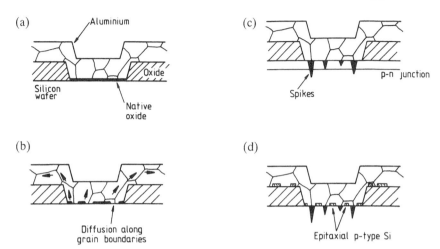

Figure 5.11 A schematic illustration of the reactions which can occur at an aluminium/silicon interface. (a) The as-deposited contact structure with the aluminium film separated from the silicon by a native oxide layer. (b) The results of an annealing treatment which partially reduces the native oxide and allows some dissolution of the silicon surface and transport of silicon atoms along grain boundaries in the aluminium film. (c) The spikes formed in the silicon wafer surface by this dissolution process. (d) The precipitation of epitaxial silicon on the wafer during cooling.

5.11(b) illustrates the process by which the native oxide layer is partially reduced and the dissolution of the silicon wafer begins. Silicon atoms are shown to be diffusing along the grain boundaries in the aluminium film, originating from the regions where the oxide film has been perforated. The diffusion rate of the silicon along the aluminium grain boundaries will be very rapid at temperatures above 300 °C and transport of silicon away from the wafer surface will be especially effective around the edge of the contact where the whole of the aluminium conduction track is available to act as a sink. Saturation of the aluminium film with silicon will only occur when very significant amounts of the wafer surface have gone into solution. The result of this localised dissolution process is the formation of voids in the silicon wafer under the contact, and these voids will fill with aluminium to create metal 'spikes' in the wafer surface. At the tip of these spikes the electric field across the contact will be greatly enhanced, which can alter its electrical properties. In more severe cases the spikes can even short out shallow p–n junctions, as illustrated in figure 5.11(c). Figure 5.13(a) shows an example of the morphology of the silicon wafer surface under an overaged aluminium–silicon contact after removal of the metal film.

The deep spikes chewed into the silicon surface are very obvious. If the contact area is small then this spiking process will be especially severe, since a large volume of aluminium exists to act as a sink for the diffusing silicon.

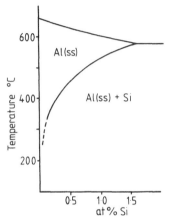

Figure 5.12 A detail of the aluminium/silicon binary phase diagram, showing that the solubility of silicon in aluminium increases sharply at about 300 °C.

If the annealing time is sufficiently long, then we must expect the aluminium film around the via to become saturated with silicon. On cooling, the equilibrium solubility of silicon in aluminium falls sharply (see figure 5.12) and so silicon will precipitate onto the wafer, insulating oxide surfaces and the grain boundaries in the aluminium film. Figure 5.13(b) shows an example of reprecipitated epitaxial silicon islands on the silicon wafer surface. These small islands delineate the shape of the grain boundaries in the aluminium layer, even after the aluminium has been removed by chemical dissolution. This reprecipitated silicon will be saturated with aluminium at a level given by the equilibrium solubility of aluminium in silicon at the temperature of precipitation and will be very heavily p-doped. Card (1974) has considered the effect of this p-type material on the properties of the aluminium–silicon contact, concluding that the effective barrier height of the contact will be increased. This is shown schematically in figure 5.14.

The effect of the spiking process, followed by epitaxial regrowth of p-type silicon during cooling, is to decrease the barrier height (spiking) and then gradually increase it again as the p-type layer is formed. This kind of variation of contact barrier height with annealing time has been

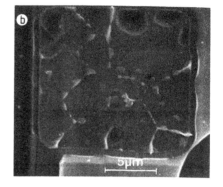
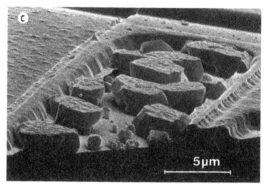

Figure 5.13 Scanning electron micrographs showing: (a) an example of the spikes formed under an overaged Al/Si contact; (b) epitaxial islands of p-type silicon precipitated on the wafer surface in a pattern which delineates the grain boundary morphology in the aluminium film; (c) very large p-type silicon precipitates formed as a result of overaging Al 1 at.% Si contacts. (Courtesy of Dr B Richards. © GEC plc. Reproduced with permission of *GEC Journal of Research*.)

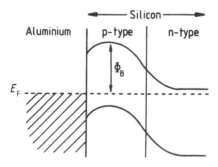

Figure 5.14 A sketch of the band structure at a contact where p-type silicon has precipitated over the surface of an n-type wafer.

shown to occur for aluminium/n-type silicon contacts at 400 °C, and the range of variation of the barrier height is of the order of 0.1 eV. This may sound a trivial change in electrical properties, but it is quite sufficient to alter the performance of Schottky diodes and disrupt the operation of an entire integrated circuit. Aluminium ohmic contacts to p-type silicon do not suffer from epitaxial regrowth effects, but spiking can still short out shallow junctions. The shorting across shallow junctions by spikes is, of course, even more damaging to device performance than small changes in the contact barrier height.

The obvious answer to the problem of spiking is to include some silicon in the evaporated aluminium conduction tracks; a film composition of aluminium 1 at.% silicon is commonly used. Reference to Figure 5.12 will show that at this composition the aluminium is supersaturated with silicon at all temperatures below about 520 °C and so we must expect the film to contain some silicon precipitates at lower temperatures. During the annealing cycle used to allow the aluminium to reduce the native oxide layer, silicon will no longer dissolve in the aluminium and so spiking cannot occur however long the annealing may continue. The supersaturation of silicon in the film, and the elevated temperature which encourages rapid diffusion, combine to allow Ostwald ripening of the undissolved silicon particles. The few large silicon particles that will result are then very favourable nucleation sites for silicon forced out of solution in the aluminium film as the contact is cooled. Annealing aluminium 1 at.% silicon/silicon contacts will create large isolated aluminium-doped precipitates on the wafer surface, and an example of the morphology and size of these islands is shown in figure 5.13(c). Once again, these islands are of p-type material. The addition of the silicon to the aluminium film has stopped the formation of spikes, but the increase in contact barrier height will still occur. The unsatisfactory state of affairs has forced the development of a new concept in interconnection technology, that of separating the contacting and interconnecting functions and giving them to two materials. A new set of materials are chosen to form stable contacts to the silicon devices and aluminium alloys are retained as the interconnection material. This kind of metallisation structure will be discussed in §5.6.

5.3.2 Thin Film Reactions Between Aluminium and Other Metals

In this section, the reaction of thin aluminium films with gold, hafnium and chromium films will be described. The gold–aluminium interface is formed when gold wires are bonded to aluminium interconnection tracks (see §7.4) and when gold films are deposited over aluminium to afford protection from atmospheric corrosion. Hafnium and chromium films can be added to metallisation structures to improve the electromigration resistance of the interconnection tracks, as will be described in §5.4. A comprehensive list of thin film reactions of relevance to microelectronic device fabrication engineers has been compiled by Baglin and Poate (1978).

5.3.2(a) Al–Au reactions

The gold/aluminium equilibrium phase diagram contains five intermetallic phases: Au_4Al, Au_5Al_2, Au_2Al, $AuAl$ and $AuAl_2$. Philofsky (1970) studied the phases formed in Au/Al bulk diffusion couples at temperatures below 500 °C and identified all these phases except AuAl. He suggested that there is a large nucleation barrier for the formation of this phase, but it is also possible when excess aluminium is present that $AuAl_2$ forms very freely and simply consumes all the AuAl grains. The dominant phase in terms of the volume formed is Au_5Al_2. The mechanical strength of the joint between the two metals is excellent providing that no voids are formed during the interdiffusion process. Philofsky found, however, that fatigue cracks are easily formed in these rather brittle intermetallic phases, especially if Kirkendall voids are formed in the Au_5Al_2 and $AuAl_2$ layers. The voids are created by an imbalance in the diffusion fluxes of aluminium and gold across these phases. The mechanical integrity of the intermetallic joint falls off very sharply once these voids have formed.

The most detailed study of the interaction of thin films of gold and aluminium has been carried out by Magni et al (1981). They showed that even at room temperature Au_5Al_2 forms between films of the two metals and that annealing at around 80 °C usually produced Au_2Al, either replacing or as well as the Au_5Al_2. Higher-temperature annealing results in the formation of the other intermetallic phases, the dominant phase in each case being dictated by the relative starting thickness of the aluminium and the gold layers. For instance, if the gold is present in excess, the initial Au_5Al_2 intermetallic layer reacts with the gold to form the most gold-rich phase, Au_4Al. If the gold and aluminium layers are of roughly equal thickness, Au_2Al and $AuAl_2$ are formed together and then react to give a single layer of AuAl — the end phase dictated by the overall stoichiometry of the bilayer sample. This complex behaviour makes it difficult to predict exactly what will happen when an aluminium film and a gold wire are brought into intimate contact. We certainly expect the formation of Kirkendall voids, which will threaten the mechanical and electrical integrity of the contact. On a more fundamental level, it is difficult to predict accurately the stoichiometry of the first phases, Au_5Al_2 and Au_2Al. In §4.4 I showed that it is hard to use bulk thermodynamic data to predict the first phases formed in a thin film reaction and concluded that the phase in which the reacting elements have the largest diffusion coefficients will grow in preference to all the others. It is interesting to note that Magni *et al* showed that Au₅Al₂ forms at room temperature, which implies a low activation energy for the rate controlling diffusion process, and quote a growth rate for the Al_2Al phase at low temperatures more than two orders of magnitude greater than for $AuAl_2$ or AuAl. These observations are in good agreement with a model which supposes that the phase with the highest diffusion coefficients grows to an observable thickness before all the others. At higher annealing temperatures it is not surprising that the dominant phase tends to be that dictated by the overall stoichiometry of the thin film couple, as all the diffusion processes will be rapid.

What then is the importance of understanding the microstructure of the thin film gold–aluminium interface? One of the most common problems in metallisation and packaging used to be the notorious 'purple plague', where the interface between gold wires (or films) and aluminium conductors was very prone to brittle failure. Such joints could fail after almost no time in service, exposing a purple fracture surface. The phase $AuAl_2$ has a bright purple colour and it was at first thought that just the formation of this phase was sufficient to cause joint failure. The bulk $AuAl_2$ phase was known to be rather brittle. However, Philofsky's bulk experiments indicated that it is the formation of Kirkendall voids during interdiffusion and compound formation that results in the poor mechanical strength of some thin film and wire-bond joints. The purple colour of the fracture surfaces could simply be due to the formation of the voids preferentially in the $AuAl_2$ phase, which we have seen to be formed freely in gold/aluminium diffusion couples.

Figure 5.15(a) shows an optical micrograph of a metallographic cross section through a gold wire/aluminium film joint. The sample has been etched to highlight the intermetallic phases formed between the bulk gold wire and the aluminium film, and in this case the reaction products form a very uniform layer with no obvious Kirkendall voids visible. Figure 5.15(b), however, shows another very similar bond in which large Kirkendall voids have been formed, extending almost all the way across the bond area. If the reaction between the gold and aluminium occurs at low temperatures, where grain boundary diffusion dominates the diffusive transport process, then there will be a much higher density of diffusion paths in the aluminium film than in the relatively large-grained gold wire. Thus the interface between a gold wire and a fine-grained aluminium film is especially likely to suffer from Kirkendall void formation. The diffusive flux of gold into the aluminium film will be much greater than the flux of aluminium in the reverse direction. The reaction between a bulk gold wire and an aluminium thin film can also result in the consumption of the aluminium around the wire bond as well as directly underneath the gold wire. Figure 5.15(c) shows a gold/aluminium joint where much of the aluminium bonding pad has reacted to form intermetallic phases. The electrical and mechanical properties of this contact will probably be very poor. In cases where a thin gold film is used to passivate the surface of an aluminium conductor, we would expect the intermetallic phases to be formed in the order determined by Magni et al (1981). Kirkendall voids may still be formed in this thin film system and we shall see that barrier layers are often used to separate gold and aluminium films in metallisation structures (see figure 5.30).

The problem of wire-bond failure as a result of Kirkendall void formation is considerably reduced if the processing temperatures which the bond will experience during device packaging are kept as low as possible. Low temperatures will slow up the rate of diffusion and chemical reaction, but will of course have no effect on the thermodynamics of the reaction between the gold and the aluminium. For this reason, the gold–aluminium interface will always be a potential site for bond failure in microelectronic systems. For many years, the 'purple plague' was almost eliminated from integrated circuits by the use of aluminium contact wires and extremely low processing temperatures. However, the introduction of plastic potted devices has encouraged the use of gold wires because of their excellent resistance to corrosion. (We will see in Chapter 7 that plastic potting materials are not very effective at stopping the indiffusion of water and other corrosive species.) Problems associated with the purple plague have thus reappeared in integrated devices. Since the consumption of the aluminium conductor and the formation of Kirkendall voids are intrinsic features of the reaction between gold wires and aluminium thin films, the only way of avoiding this kind of failure altogether is by selecting a completely different combination of materials to form the wire/pad bond.

5.3.2(b) Reactions of hafnium and chromium with aluminium
Reactions between aluminium and transition metals like chromium and hafnium have been studied as part of the drive to produce composite conductor structures where the lower resistivity of the aluminium films can be combined with the strength and stability of intermetallic compounds formed between the two elements. It is important to understand both the reaction sequence and the reaction kinetics in these structures to enable the optimum combination of aluminium conductor and compound support to be chosen. In later sections in this chapter we will see that this kind of conductor design can give excellent resistance to deleterious processes like electromigration. Howard et al (1976) have carried out an extensive study of the reactions between thin films of aluminium and various transition metals. They found that although an aluminium-rich intermetallic phase was formed rapidly when hafnium and chromium films were reacted with aluminium films, there were

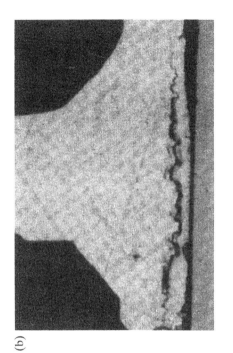

(b)

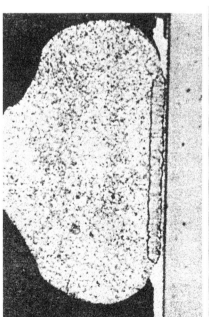

(a)

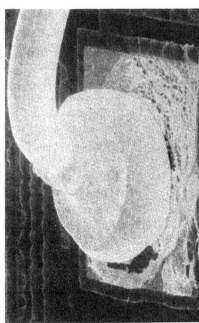

(c)

Figure 5.15 Scanning electron micrographs showing the reactions at interfaces between gold wires and aluminium thin films. In each micrograph the ball at the end of the wire is about 50 μm across. (a) A bond at which intermetallic phases have formed, but no Kirkendall voids are visible. (b) A bond in which a Kirkendall void runs almost the whole way across the field of view. (c) An example of excessive reaction between the wire and the contact pad. (Courtesy of Dr B Richards. ©GEC plc. Reproduced with permission of *GEC Journal of Research*.)

indications that the morphology of the phase was different for the two metals. $CrAl_7$ is formed between aluminium and chromium and grows with parabolic kinetics indicating a simple diffusion-controlled reaction. This reaction seems to form a smooth and discrete layer of the compound phase. The transport of aluminium and chromium atoms through the grain boundaries in the $CrAl_7$ may be the rate-determining step in the reaction, although some workers believe that lattice diffusion dominates the transport process. $HfAl_3$, however, seems to grow along the grain boundaries in the aluminium and to form with Harrison type (c) growth kinetics when the grain size of the aluminium is reasonably large. This intermetallic layer was thought to be much less smooth than the $CrAl_7$ layers, and the difference in the suggested growth morphologies of the two phases is illustrated in figure 5.16(a). The observed difference in reaction morphology seems to suggest that grain boundary diffusion in $HfAl_3$ is less easy than in $CrAl_7$, which may be related to the large difference in the melting points of the two compounds, about 1400 °C and 790 °C respectively. We have already seen that the activation energy for grain boundary diffusion varies monotonically with the melting point. If we assume that diffusion across the $HfAl_3$ layer is slow, then it seems plausible that the hafnium diffuses preferentially into the aluminium grain boundaries and reacts there to form elongated particles of the intermetallic phase.

More recent investigations have forced us to modify the idea that a rough or needle-like intermetallic growth morphology will always result when the intermetallic phase has a high melting point and shows low diffusivity. In fact, it is possible that the grain structure of the aluminium and transition metal films plays an important part in controlling both the morphology and the grain structure of the intermetallic phase. Reaction couples prepared by sputtering aluminium and zirconium films can have very small grain sizes indeed and can react to form very smooth intermetallic layers (Mingard 1987). This is in direct disagreement with the results of Howard *et al*, who predict that this interface should have a rough morphology. Part of the reason for the disagreement must lie in the variety of deposition techniques, sample purities and analytical methods which have been used in the study of these reactions. In particular, rather few direct observations have been made on the morphology of the intermetallic layers, presumably because of the difficulty of producing cross-sectional samples for observation in a transmission electron microscope (see §10.3). Mingard (1987) has shown that an ultramicrotome can be used to prepare this kind of specimen relatively rapidly, and an example of the detail which can be observed in a reaction between an aluminium and a transition metal thin film is shown in figure 5.16(b). This work has suggested the intriguing possibility that the Al/Zr reaction proceeds in two distinct stages, with the

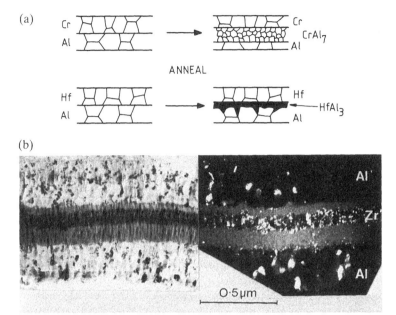

Figure 5.16 (a) A sketch of the proposed morphologies of the reaction products formed between aluminium and chromium or hafnium films (after Howard *et al* 1976). (b) A bright and dark field pair of transmission electron micrographs showing one stage of the reaction between two Al 4 at.% Cu films and a central zirconium film. This sample has been annealed at 325 °C, and a very fine-grained reaction product can be seen between the aluminium alloy and the zirconium. (Micrographs courtesy of Mr K Mingard.)

formation of a metastable phase before the equilibrium $ZrAl_3$ compound.

The kinetics of these intermetallic phase-forming reactions have been measured and 500 nm thick films of most refractory metals are completely consumed after a few hours at 400 °C. At lower temperatures, the reactions will be very much slower and long anneals are required to convert all of a transition metal layer into the aluminide phase. In these three-layer structures, it is the slow diffusion rates in the aluminide reaction products which give the composite film considerable chemical stability. The morphology of the reaction product plays an important role in determining the stability as well, since the further the reacting atoms have to diffuse in the aluminide phase the slower the reaction. This explains why hafnium is sometimes observed to be consumed rather faster than a similar thickness of chromium when sandwiched between aluminium layers; the irregular shape of the hafnium aluminide illustrated in figure 5.16(a) will only present a short diffusion barrier to the reacting atoms. The thick $CrAl_7$ layer will be a much more severe obstacle to diffusing atoms, even though the rate of diffusion may be higher than in the $HfAl_3$. We will see in the following section that the addition of a small amount of copper to the aluminium thin film results in a considerable change in the kinetics of the reaction which forms the intermetallic phases, possibly because the copper acts to block the grain boundary diffusion paths.

5.4 ELECTROMIGRATION IN CONDUCTOR FILMS

The phenomenon of electromigration involves the transport of material in response to the passage of an electric current. While this process was first studied in bulk materials, a great deal of work has been directed at understanding electromigration in polycrystalline thin films. Before discussing the mechanism of electromigration, let us be clear as to why it is so important in thin films. We have seen in §5.2 that diffusion in polycrystalline films is much faster than in a bulk sample of the same material because grain boundary diffusion has a lower activation energy than diffusion in the bulk. A transport process which is dominated by diffusion will thus be very rapid in a thin film, even at rather low temperatures. The large surface area in thin films means that surface diffusion can be another effective way of transporting material. The activation energy for surface diffusion is often even smaller than that for grain boundary diffusion. The second point about a thin film conductor that distinguishes it from a bulk material is that the current density carried in a conductor track can be very high indeed, $10^6 \, A \, cm^{-2}$ or more. It is impossible to pass such a high current density through most bulk conductors because Joule heating would melt the material. A thin film conductor is so closely connected to a heat sink along the whole of its length (the substrate and any covering layers) that Joule heating is severe but not catastrophic, even at these very high current densities. Any process which is driven by the passage of the current must be expected to be extremely rapid in a conduction track passing these high electron fluxes. Thus in a thin film conductor we have both a high driving force for electromigration and the possibility of transporting material effectively at low temperatures by diffusion along the grain boundaries. It is thus not surprising that electromigration is often an important phenomenon in metallisation systems. In §5.4.1 I will consider the basic process of electromigration, while the failure modes in metallisation layers and methods for increasing the stability of conductors are discussed in §§5.4.2 and 5.4.3 respectively.

5.4.1 The Physics of Electromigration

In any transport process, the flux of material, J_e, can be described by the Nernst–Einstein diffusion relationship

$$J = \frac{NDF}{kT} \qquad (5.4)$$

where N is the density of moving species, D their mobility and F the driving force for migration on each of these species. In electromigration, F is the force exerted on a metallic atom by the passage of an electron flux and this force is made up of two contributions. The ionic core of the metal atom experiences a force due to the potential gradient across the conductor. This force is proportional to the valence of the metal and is directed in the opposite direction to the electron flux. The second contribution to F comes from the rather mysterious 'electron wind' force, which may be thought of as being due to collisions between the electrons and polarised vacancy–metal ion complexes. The momentum transfer between electron and ion usually results in a force directed in the same direction as the electron flux. In gold and aluminium the electron wind force is measured to be much greater than the field-ion force and so dominates the electromigration process.

F can be taken to be equal to EeZ^*, where E is the electric field across the conductor, e the electronic charge and Z^* the effective charge on the metal ions taking both electron wind and valency effects into account. For instance, in gold the valence is 1 but the measured value of Z^* is about -7, so the electron wind force must contribute an effective valence of -8 (Herzig and Cardis 1975). The total electro-migration force can be estimated from an equation of the form

$$F = EeZ^* = EeZ\left[1 - \left(\frac{\rho_v}{N_v}\right)\left(\frac{N_l}{\rho_l}\right)\gamma\right] \qquad (5.5)$$

where Z is the normal valence of the metal, ρ_v/N_v the specific resistance of a moving vacancy, ρ_l/N_l the specific resistance of the lattice and γ a parameter related to the polarisability of the vacancy–ion complexes (Huntington and Grone 1961). This equation makes clear the close relationship between the fundamental process of electron scattering which contributes to electrical resistivity and the electronic scattering event that is the cause of the electron wind force. It is also clear that the vacancy–ion complex is important in electromigration just as it is in ordinary diffusional transport. Electromigration requires both a force on the metallic atom to encourage it to migrate and a mechanism for migration, in this case vacancy diffusion.

Equation 5.5 was developed to describe electromigration in bulk materials where the transport process is lattice diffusion by a vacancy mechanism. However, diffusion in polycrystalline films, especially at low homologous temperatures, will be primarily through the grain boundaries. Some of the parameters in the electromigration equations must be modified if they are to describe this new transport process:

$$J_b = \frac{N_b \delta D_{gb} Z_b^* eE}{dkT} \tag{5.6}$$

where the electromigration flux along grain boundaries, J_b, is now a function of the grain boundary diffusivity, D_{gb}, the grain diameter in the conductor, d, the width of the boundaries, δ, the effective charge on the moving ions in the boundary, Z_b^*, which may not be the same as this parameter in the lattice, and the density of these moving ions in the grain boundaries, N_b. Once again we are forced to assume that all the grain boundaries in the polycrystalline film have the same diffusion coefficient. While some knowledge of these average values of D_{gb} can be obtained as described in §5.2.2, the parameters N_b and Z_b^* are not easy to measure or to estimate. The first requires a very detailed knowledge of the structure of the grain boundaries and the second can only be obtained by measuring J_b and measuring, or guessing, values for all the other parameters in equation (5.6). Another problem which arises when measuring data on grain boundary electromigration is that migrating impurity atoms will often segregate to the boundaries as well. This will certainly modify the value of N_b and possibly of Z_b^* as well. D'Heurle and Ho (1978) have assumed that $N_b = \beta N$, where β is an estimated enrichment factor for solute in the boundaries, but no confirmation of the accuracy of this assumption has been obtained.

It is important to emphasise that if we are to model the electromigration process in thin films, it is the inconvenient equation (5.6) that we must use, not the much simpler equations for bulk electromigration. D'Heurle and Ho state that the contributions made by lattice diffusion to the electromigration transport of copper and aluminium in aluminium thin films at temperatures below 250 °C fall into the range of 10^{-6}–10^{-4} of the total flux, a negligible fraction. A glance at figure 5.10 confirms that grain boundary diffusion is the only significant transport mechanism at this kind of temperature in an aluminium film. The thin oxide layer which is present on all aluminium surfaces prevents surface diffusion being an important phenomenon in aluminium conductor films. We shall see later in this chapter that surface electromigration can be a severe problem on gold conductors.

So far we have seen that the passage of a current through a thin film conductor will encourage migration of metallic ions along the grain boundaries. Both aluminium and gold have values of Z_b^* which have been measured to lie in the range -1 to -40, although these values depend somewhat on the measurement temperature, decreasing roughly linearly as the temperature increases. This implies that the electromigration flux is directed towards the anode. In fact there is still some disagreement as to the direction of this flux in gold films, experiments which only measure Z^* indirectly suggesting that the flux is directed towards the cathode. This discrepancy is discussed further by D'Heurle and Ho (1978). Whichever the direction of the electromigration flux, we still expect a flux of metallic species along the grain boundaries when a current flows through the conductor. To complicate further our understanding of the phenomenon of electromigration, Luby *et al* (1980) have shown that a randomly fluctuating current causes more electromigration than a simple DC current. This appears to be a result of variations in local current densities and an enhancement in grain boundary diffusivities under these conditions. This work raised the interesting possibility that the extent of electromigration will depend on the character of the electrical signals propagating in a thin conductor track.

Electromigration thus causes a displacement of the metallic species along a thin film conductor, but such a displacement cannot create a discontinuity in a film. In order for a discontinuity to be formed there must be an imbalance in the electromigration flux at some point along the conduction path, which is usually referred to as 'flux divergence'. Flux divergence will occur whenever there are changes in F, the driving force for electromigration, or in D_{gb}, the mobility of the diffusing species in the grain boundaries. F depends on Z_b^* and this parameter can vary both from grain boundary to grain boundary in a polycrystalline film and at contacts between two dissimilar metals. The rate of grain boundary diffusion depends on the temperature of the conductor and so variations in temperature along the conductor will cause flux divergence. In addition, the grain boundary diffusivity also varies from boundary to boundary and across bimetallic joints.

Figure 5.17 summarises some of the more important sites for flux divergence in polycrystalline metallisation structures. The first of these, figure 5.17(a), is a grain boundary triple point where the incident electromigration flux along boundaries 1 and 2 is greater than can be removed along boundary 3. This can arise from low values of either D_{gb} or Z_b^* in boundary 3. Under these circumstances, accumulation of material must occur at the triple point, causing hillock formation. A similar situation where $J_b^1 + J_b^2 < J_b^3$ will cause material depletion and eventual void formation at the triple point. Figure 5.17(b) shows a long conduction track connected to large bonding pads at both ends. Here Joule heating will increase the temperature along the narrow part of the track and so increase the grain boundary diffusion rate in this region. The result is flux divergence at both ends of the track, depletion at the cathode and accumulation at the anode in the case of an aluminium track where Z_b^* is negative. A third possible place in a metallisation system where flux divergence can occur is at a bond between an aluminium conduction track and a gold wire (figure 5.17(c)). Here the activation energy for grain boundary diffusion in the aluminium will be about 0.5 to 0.7 eV, while in the gold wire this figure is more than 1 eV. Grain boundary diffusion will thus be very much faster in the aluminium than in the gold. In addition, there is a much lower density of grain boundaries in the bulk wire than in the thin aluminium film. Under these circumstances, electromigrating species will accumulate under the joint and result in distortion and crack formation. A similar kind of effect can be found at aluminium–silicon contacts, where the diffusional transport of aluminium in the single-crystal silicon will be very limited indeed and depletion of aluminium at the cathode and accumulation at the anode is likely to occur.

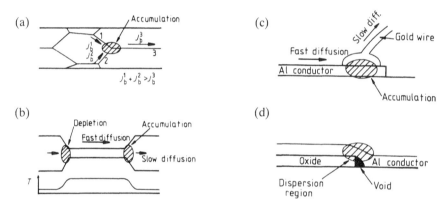

Figure 5.17 Schematic illustrations of several sites at which flux divergence is expected in metallisation systems: (a) triple points in conductor layers; (b) regions where the conductor changes in width; (c) a gold wire/aluminium film contact; (d) at defects in a conductor track.

A final place where electromigration divergence is expected to be unusually severe is where a thin aluminium track is deposited over a sharp feature. A simple example of this is at the edge of a via hole opened through an insulating layer down to a silicon wafer surface. Here there will be a thinning of the conductor track directly over the step, and a void may even be formed under the aluminium film because of shadowing during the deposition process (see figures 4.2 and 5.17(d)). The current density through this thinner region of the film will be higher than in the surrounding areas and so the temperature and the rate of electromigration will be increased locally.

5.4.2 Electromigration Failure

In this section we shall consider the two most important modes of failure in metallisation systems, concentrating on failures at the sites illustrated in figure 5.17. The character and density of grain boundaries in the conductors often dictate the manner in which the electromigration failure occurs. We have already seen that local depletion of material during electromigration can result in the formation of voids in conductor tracks. When grain boundary diffusion is the only significant transport mechanism, these voids will form along the boundaries themselves. Agglomeration of several of these voids can create a crack that extends all the way across the conductor, i.e. an open circuit. While all grain boundary triple points are possible sites for flux divergence in the manner mentioned above, the ones which actually suffer local material depletion depend in a complex manner on the grain boundary geometry and the values of D_{gb} and Z_b^* in the intersecting boundaries. We can certainly say that void formation is statistically likely to occur somewhere along a conduction track carrying a high current density, but predicting the triple points at which the first crack will appear is difficult. D'Heurle (1971) has shown some elegant micrographs of crack formation at grain boundaries in current-stressed aluminium tracks, which emphasise the essentially random nature of the crack-forming process. An example of the kind of voids which can nucleate in aluminium thin film conductors is shown in figure 5.18(a).

Any morphological inhomogeneity in the conductor film will act as a site for flux divergence and will be an exceptionally favoured site for crack formation. The thinner regions of conductor tracks at steps, illustrated in figure 5.17(d), are obvious sites for void nucleation and crack formation. Local variations in grain size will also be likely to provide sites for material depletion. The bimodal grain structure characteristic of zone T (see §4.2 and figure 4.4(b)) is a perfect example of an imhomogeneous grain structure, and films with this grain morphology might be expected to be particularly prone to rapid electromigration failure at the triple points around the large grains. Observations on the formation of cracks at the grain boundaries in thin film conductors suggest that equiaxed large-grained polycrystalline films — zone III structure in §4.2 — should be less susceptible to open-circuit failure than fine-grained ones. This method for improving the electromigration resistance of conduction paths will be considered further in the next section.

It is worth emphasising again the importance of the role played by grain boundaries in the propagation of voids in polycrystalline conductors. Figure 5.18(b) shows a transmission electron micrograph of part of a narrow conductor track in which voids produced as a result of locally enhanced electromigration rates have coalesced to form a continuous crack along some grain boundaries. In similar specimens the transport of material from one end of the conductor track to the other during current stressing can be very clearly shown. Figure 5.18(c) shows an example where voids have been formed at grain boundaries on the right-hand side of the image, and isolated thicker regions of the film on the left-hand side indicate where the excess material has accumulated.

(a)   (b)

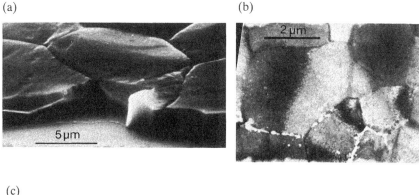

(c)

Figure 5.18 (a) A scanning electron micrograph showing the kind of void which can be produced in a conductor track during electromigration. (Courtesy of Dr F M D'Heurle.) (b) A transmission electron micrograph of part of a conductor track showing that the voids formed during electromigration occur preferentially on the grain boundaries. (c) A lower magnification image of the same kind of track as seen in (b) showing the depletion of aluminium from one end of the track and accumulation at the other end. (Figures (b) and (c) courtesy of Dr S Luby.)

Much of the work on electromigration failure has concentrated on crack formation at regions of material depletion. However, there are also regions of material accumulation in metallisation structures, leading to the growth of spikes or hillocks. Figure 5.19 shows an example of a very large aluminium spike produced by electromigration. Protuberances of this kind can easily grow to form a short-circuit path between adjacent conductor tracks, and this is the second important mode of electromigration failure. These adjacent tracks can be vertically above one another in the metallisation structure (separated by an insulating layer of course), or lying in the same plane. Hillock and spike growth can also crack passivation layers, exposing the sensitive aluminium conductors to corrosive attack by the free atmosphere. An example of a passivation layer cracked by hillock growth in the underlying conductor track is shown in figure 5.20.

Figure 5.19 A scanning electron micrograph of a large metal protrusion created in an accumulation region of a conductor track during electromigration. (Courtesy of Dr B Richards. © GEC plc. Reproduced with permission of *GEC Journal of Research*.)

Figure 5.20 A scanning electron micrograph of a passivation layer cracked by hillock growth in an underlying conductor track. (Courtesy of Dr B Richards. © GEC plc. Reproduced with permission of *GEG Journal of Research*.)

These micrographs illustrate that electromigration is a very real practical problem in microelectronic metallisation, causing both short-circuit and open-circuit failures. We should remember that Joule heating will increase the temperature of the thin conductor tracks well above room temperature in many systems, and pure aluminium conduction tracks typically fail after only a few hours at 250 °C when passing a current of 10^6 A cm^{-2}. It has thus proved very important to discover methods by which the resistance of the conductors to electromigration failure can be increased.

5.4.3 Methods for Increasing Electromigration Resistance

The simplest method for increasing the resistance of thin film conductors is to remove the principal transport paths, the grain boundaries. Increasing the grain size of thin films by a heat treatment after deposition can reduce the density of grain boundaries and has been found to increase the electromigration lifetimes of conduction tracks. Single-crystal aluminium films have been shown to have lifetimes as long as 36 000 hours at 175 °C while passing 2×10^6 A cm^{-2}, while tests in unusually large-grained aluminium alloy films have also shown very long lifetimes before electromigration failure (Gangulee and D'Heurle 1973). This result is not surprising, as once the rapid diffusion paths are removed the electromigration process can only proceed by lattice diffusion, which is very slow at the low temperatures at which these tests are performed. However, in Chapter 4 I showed that it is difficult to deposit single-crystal metallic films on amorphous substrates like silica. It is also hard to induce grain growth to proceed sufficiently far in a metal thin film to give grains larger than about 10 times the film thickness, even with extended annealing. Some workers have found methods of increasing the grain size of thin films, usually involving the addition of a solute element to the conductor material which increases the grain boundary mobility in the polycrystalline film. These methods are interesting from the point of view of understanding exactly what the interaction is between solute and grain boundaries, but are generally rather hard to use effectively when a complete metallisation scheme is being fabricated.

The most elegant way of producing a thin film conductor with a grain structure that will resist electromigration failure is to grow tracks with the 'bamboo' structure illustrated in figure 5.21. Here narrow conductors are deposited and annealed such that almost all the grain boundaries run perpendicular to the long direction of the track. The boundaries will tend to migrate into this configuration to minimise their surface area. Boundaries running across the conductor track cannot contribute to electromigration along the conductor, and by this very simple change in the film morphology the electromigration resistance can be improved hugely. The preparation of conductors with the bamboo structure is usually only possible if the tracks are rather narrow, but since this is the trend in modern integrated circuit metallisation design, this is not an inconvenience.

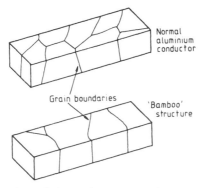

Figure 5.21 Sketches of the grain structure in a normal polycrystalline metallic conductor track and in a track with the 'bamboo' structure.

A second method for reducing the rate of electromigration depends on the reduction of grain boundary diffusivity by the addition of a solute element to the conductor material. I have shown in table 5.1 that a small concentration of tantalum in a gold film increases the average activation energy for grain boundary diffusion quite markedly. Additions of copper, magnesium and nickel have a very similar effect in aluminium, reducing the grain boundary diffusion coefficient of aluminium in the boundaries by as much as two orders of magnitude. Rosenberg *et al* (1972) have suggested that any solute species which segregates strongly to the grain boundaries in a conductor film will reduce boundary diffusion rates by filling most of the sites in the boundary along which diffusive transport occurs. This is equivalent to saying that the solute segregates to 'vacancy' sites in the structural units at the boundary core where diffusive jumps will be relatively easy. Whether this is correct or not, it is certainly observed that the electromigration lifetime of aluminium conduction tracks can be increased from about 30 h for pure aluminium to tens of thousands of hours at 175 °C and 2×10^6 A cm^{-2} by additions of 4 at.% of copper or 2 at.% of magnesium (D'Heurle and Ho 1978). Hillock formation in aluminium films under compressive stress (§5.1) is also substantially reduced by the same solute additions.

We have already seen in §5.3.2 that silicon is added to aluminium metallisation to stop spiking under the contact. It is now expedient to add another alloying elements to increase the electromigration resistance. A particularly popular alloy for integrated circuit conductors is aluminium with about 1.5 at.% silicon and 4 at.% copper. The silicon can have a small beneficial effect in reducing the grain boundary diffusion rate as well. This composition is very similar to a range of well known age-hardening alloys, and some of the properties of the thin film conductor films can be understood by consideration of phase transformations in alloys of this kind. From the phase diagram and an understanding of the phenomenon of heterogeneous nucleation in polycrystalline materials, we can predict that the equilibrium structure of aluminium 4 at.% copper thin films should, after annealing at about $150°$ C, consist of large $CuAl_2$ precipitates distributed primarily along the grain boundaries in a matrix which is locally depleted in copper. The regions from which the copper has been depleted will be much more prone to electromigration failure than those which preserve the full 4% copper level. This is because once the copper atoms are no longer segregated to the grain boundaries, the rate of electromigration will immediately increase and cracks will start to nucleate at any flux divergence site. Thus the equilibrium structure of Al–Cu alloy films does not have the optimum distribution of copper for resistance to electromigration failure.

However, evaporation from Al–Cu alloy sources does not result in homogeneous films because the copper has a higher vapour pressure than the aluminium (see §4.1). The thin film will have most of the copper concentrated close to the substrate and at the top will be severely depleted in this protective element. In addition, the as-evaporated films will certainly not have the equilibrium structure which we have predicted above. It has been shown that Al–Cu films contain some metastable precipitates immediately after evaporation, but a considerable fraction of the copper remains in solid solution in the aluminium matrix. The nucleation of the $CuAl_2$ phase in evaporated alloy films has been studied by Vavra and Luby (1980) and Thomas *et al* (1986) and shown to occur in a very inhomogeneous fashion because of the variations in composition already present in the film and the tendency of this phase to nucleate on the surface of the film and significantly roughen the conductor. These reactions will be greatly accelerated if the films are heated at any temperature below the solvus in the phase diagram. The copper will itself electromigrate away along the grain boundaries, resulting in eventual depletion of this protective element. The diffusion of copper in grain boundaries in aluminium is quite rapid (see table 5.1). Thus, while these alloy conduction tracks do indeed have improved electromigration resistance by virtue of decreased grain boundary self-diffusion rates, this resistance is eventually degraded by phase transformations, inhomogeneous film structures and electromigration away of the protective element.

We can see that these alloy conductor films are thermodynamically and kinetically extremely unstable, but they do have very good electro-migration resistance for the period of time that the copper remains in solid solution, or as a distribution of fine-scale precipitates on the grain boundaries. Al–Si–Cu alloys are very widely used for conducting tracks in integrated circuit metallisation, and the improvement in electromigration resistance which they offer, compared with that of pure aluminium films, is shown in figure 5.22. Sputtering techniques (§4.2) are normally used to deposit these alloys because of the improved homogeneity of the films compared with those that are evaporated. We should remember that the alloying element concentration is very high in these alloys and so we might expect their bulk resistivities to be significantly greater than that of pure aluminium. Fortunately, the increase in resistivity is only about 10%, from $2.86 \, \mu\Omega \, \text{cm}$ for pure aluminium to $3 \, \mu\Omega \, \text{cm}$ for Al–4%Cu–1.5%Si for instance. This level of increased resistivity does not cause any problems in conventional circuit design. However, the dry etching of Al–Cu alloys is significantly more difficult than of pure aluminium (see §2.4) and the increased hardness of the films makes wire-bond joints less reliable as well.

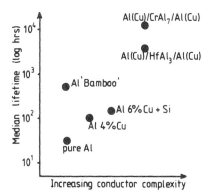

Figure 5.22 A comparison of the measured electromigration lifetime of a variety of aluminium-based conductor alloys tested under identical conditions (after Pramanik and Saxena 1983).

Two other methods for increasing the electromigration resistance of aluminium conductors deserve a brief mention: glassing and refractory layer additions. The second of these has already been introduced in §5.3.2 where the reactions between aluminium and hafnium or chromium layers were described. Howard *et al* (1977) have shown that if three-layer conductor tracks are deposited, Al–4%Cu/Hf/Al–4%Cu for instance, then the electromigration lifetime of the tracks is longer than for single-layer Al–4%Cu tracks. These three-layer structures react to form intermetallic phases like $HfAl_3$ or $CrAl_7$ at the centre of the conductors, so that there is a refractory alloy layer separating two aluminium alloy films. The improvement in the electromigration lifetime of these tracks is quite simply because with two parallel conductors the chance of producing an open circuit is much reduced. The refractory layer acts to block void propagation from one conductor to the other and a crack in one of the aluminium layers does not damage the electrical integrity of the conductor as a whole. Electromigration in the intermetallic phases at the centre of the track is very slow and so the possibility of crack formation at the grain boundaries in these layers during electromigration is small. However, Grabe and Schreiber (1983), amongst others, have shown that $Al/TiAl_3/Al$ composite conductor tracks are particularly prone to the formation of large spikes like those illustrated in figure 5.19. These extrusions can cause short circuits between adjacent conductors even when the conductors themselves shown no sign of failing by any electromigration mechanism. In the case of Al–Cu/refractory metal/Al–Cu conductor structures, it has been shown that the presence of the copper in the aluminium films increases the activation energy for formation of the intermetallic compounds at the centre of the track. This effect will increase the stability of the multilayered conductors in service, and it is suggested that the copper decreases the rate of grain boundary diffusion in the aluminium because of the second-phase precipitates blocking these diffusion paths.

Simply adding an adherent, chemically stable and refractory layer to the conventional thin film conductors increases the average electro-migration lifetime by at least a factor of 10. Some typical data on the lifetimes of a range of aluminium-based conductor tracks are shown in figure 5.22, illustrating how both the structure and chemistry of the films have been modified to develop the very stable interconnection systems that are being tested for use in the next generation of integrated circuits.

Finally, let us consider the effect of passivating glass layers on conduction tracks in which electromigration is occurring. Blech and Herring (1976) have shown that a significant stress gradient exists along an aluminium film in which electromigration is occurring. Aluminium films on a silicon substrate will always be under compression when at elevated temperatures because the thermal expansion coefficient of the metal is much larger than that of silicon. An electromigration flux of metal ions towards the anode will gradually relieve the stress at the cathode by the depletion of material from this end of the conductor. The stress will not be relieved at the anode, which results in the development of the stress gradient along the conductor. Aluminium will tend to backdiffuse along the track under the driving force of this pressure gradient, and this flux will eventually balance the electromigra-tion flux. When this occurs, damage to the track due to electromigration flux divergence will be very much reduced. A glassy passivation layer over the top of the conductor track will also stop the anode end of the conductor plastically deforming by grain boundary sliding and hillock formation under the high local stress. This will increase the stress gradient which is set us along the conductor and the electromigration flux will cease at an earlier stage of the process of material transport. Typical stress levels generated in the soft aluminium are about 10^3 MPa, far in excess of the yield stress. However, when the metal is encapsulated in a relatively 'stiff' glass layer, deformation of the aluminium cannot occur. Eventually sufficient stress is generated to crack the passivation layer, and short-circuit failure and atmospheric corrosion of the track can then occur. Lloyd and Smith (1983) have demonstrated that the electromigration resistance of passivated Al–4%Cu conductors depends on the thickness of the glassy layers, and that increases in lifetime of between 5 and 10 times are readily achieved by using sufficiently thick passivation to contain even the highest stresses generated in the metal films. Thus the glass layer which is frequently used to protect aluminium alloys from corrosion has the fortuitous additional advantage of increasing the electromigration lifetime of integrated circuit conductors.

The use of one or more of these methods for improving electromigration lifetimes has resulted in the production of very reliable metallisation conductor layers. We have seen that it is by a combination of tailoring the alloy compositions very precisely, and by carefully controlling the structure of the conductor tracks, that these improvements have been achieved. While this section has concentrated on aluminium alloy conductors, we should not forget that simple gold conductors have intrinsically a higher electromigration resistance than aluminium. This is in part due to the higher activation energy for grain boundary diffusion in gold — about 1 eV as opposed to 0.5 eV in aluminium. The electromigration resistance is one reason why gold is still preferred to the aluminium alloys in some metallisation schemes, and one or two of these will be described in §5.8.

5.5 CORROSION OF ALUMINIUM ALLOY CONDUCTORS

Aluminium is generally considered to be very corrosion resistant when it is used in metallurgical applications and yet it is rather susceptible to corrosion attack by species which can dissolve its protective oxide layer. A very small amount of corrosion will completely destroy a conductor track in a microelectronic system because the track may only be 1 μm wide and often only a few hundreds of nanometres thick. Because of this, corrosive attack of aluminium alloy conductors in metallisation layers must be avoided at all costs.

Nguyen and Foley (1980) have suggested that chloride ions are the species which most effectively dissolve the protective oxide on aluminium films by the formation of ions like $Al(OH)_2Cl^-$. This results in the gradual exposure of the reactive aluminium metal and rapid dissolution of the conductor by the following overall reactions:

$$Al + 4Cl^- \rightarrow AlCl_4^- + 3e^- \tag{5.7}$$

$$AlCl_4^- + 3H_2O + 3e^- \rightarrow Al(OH)_3 + 4Cl^- + \tfrac{3}{2}H_2. \tag{5.8}$$

Here the chloride ions are not bound into stable compounds which protect the surface of the metal and so the aluminium surface can be rapidly dissolved. The final corrosion product is aluminium hydroxide, $Al(OH)_3$, usually seen as a whitish gel on the surface of the conductor. The chloride ions necessary for the initial reaction can be left on the metallised wafer during degreasing with solvents like trichloroethylene and also by the plasma etching of aluminium lines with gases which contain chlorine (table 2.6). During the operation of the device in service, the residual chloride ions will migrate over the surface to the positively charged aluminium tracks. These conductors act as anodes in the corrosion reaction and are dissolved much faster than the relatively cathodic negatively charged conductors.

However, chloride ions are not the only species that the device manufacturer has to control. An excessive concentration of phosphorus in passivating glass layers deposited over the top of metallisation films can result in the formation of phosphoric acid by reaction with atmospheric water vapour. The useful properties of phosphosilicate glasses for this passivating application are described in §6.2 and table 6.2. The phosphorus content is usually kept as low as possible to limit the amount of the acid that is formed. HPO_3 seems to assist in the breakdown of the passivating oxide layer on aluminium cathodes and then the conductor track is consumed by reaction with OH^- ions which are produced by the electrolysis of water molecules at the cathode. Grain boundaries in the aluminium conductors are particularly strongly attacked in this reaction.

Straightforward chemical dissolution of the aluminium alloy conductors is only one of the common corrosion problems in microelectronic metallisation structures. These films also make contact with numerous other metals, for instance gold wires and copper and chromium films at solder bonds (see figure 5.30). We have also seen that Al–Cu films are thermodynamically unstable, decomposing to form a distribution of copper-rich particles in a depleted aluminium matrix. A metallisation structure thus consists of a very large number of bimetallic couples, and at any such couple the dissolution rate of a region which is anodic relative to its surroundings will be very fast indeed. This enhanced corrosion will be most severe when small areas of anodic material are exposed to an electrolyte when surrounded by a large area of relatively cathodic material. The electrolyte can be formed by solutions of contaminants like Cl^-, Na^+ and P^- ions in water on the metallisation surface. Table 5.2 shows the standard electropotentials of some metals and alloys in a solution of chloride ions with a calomel reference electrode. The value for gold is not included here, but this noble metal will be cathodic with respect to all the materials listed, giving it a large positive electropotential. From this table it is clear that an aluminium alloy conductor film will be anodic with respect to copper, chromium and gold. This means that it will be preferentially dissolved when it is in electrical contact with these metals in an environment which contains chloride ions. In addition, precipitation of $CuAl_2$ at the grain boundaries in an Al-Cu alloy film will reduce the copper content in the surrounding regions. These depleted regions will be anodic relative to both phase precipitates and the undepleted grains in the film. Table 5.2 shows that an increasing copper concentration in solid solution in aluminium makes the alloy more cathodic.

Table 5.2 Electrode potentials of some metallisation materials, after Dix *et al* (1961) (measured against a calomel electrode at 25 °C).

Al–5% Mg	−0.88 V
Al	−0.85 V
Al–1% Si	−0.81 V
Al–2% Cu	−0.75 V
$CuAl_2$ θ phase	−0.73 V
Al–4% Cu	−0.69 V
Cr	−0.4 V
Cu	−0.2 V

There are thus two preferential electrocorrosion processes which can occur in exposed aluminium alloy conduction tracks: dissolution around the grain boundaries where second-phase precipitates have formed; and sacrificial anodic corrosion of the whole of the conductor when in contact with more noble elements like gold and chromium. These processes are illustrated schematically in figure 5.23. The conductor alloy can be exposed to a corroding solution when there are shadowing effects so that the thin films of the protective metals are not of even thickness, as shown in figure 4.2, by solid state diffusion of chloride ions through a Cr–Al interface or through pinholes and by cracking of a passivation layer. Once the corrosion reaction has begun, the formation of $Al(OH)_3$ will soon distort the protective metal or glassy films and expose new areas of conductor for corrosive attack. Voids in the conductors will be rapidly formed even in the very first stages of corrosion.

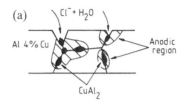

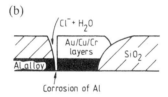

Figure 5.23 A schematic illustration of two places in an aluminium-based metallisation scheme where corrosion problems are expected to be particularly severe: (a) around $CuAl_2$ particles in Al 4 at.% Cu conductors; (b) at cracks or pinholes in protective metal or oxide layers which expose the aluminium for anodic corrosion.

It is obviously important to reduce the chance of corrosive attack as much as possible if long-term stability of the metallisation system is to be ensured. Improving the cleanliness of the fabrication procedures, and especially removing chloride ions very carefully, is only the first stage in this process. The deposition of adherent and mechanically strong glass passivation layers can reduce the chance of chloride ions having access to the conducting films, but these layers are quite brittle and can be fractured by strains generated in the metallisation systems, as has been mentioned earlier in this chapter. An example of a cracked passivation layer is shown in Chapter 10. The most effective method of protecting the metallisation system from corrosion is to place the whole device or chip in a hermetic package containing an inert atmosphere. The structures of some of the most common hermetic packages are described in Chapter 7. However, the need to package devices in this way is an admission that it is not easy to fabricate a metallisation scheme based on aluminium alloys which is intrinsically corrosion resistant. This is because aluminium conductor films are extremely susceptible to corrosive attack, as a result of a combination of the electrochemistry and structure of these alloys. One of the great advantages of gold as a material for conductor tracks is its resistance to corrosion, and the use of gold-based metallisation schemes can remove the need for hermetic packaging. This will often result in a significant reduction in the total cost of the component, even though the gold is extremely expensive.

5.6 SILICIDE CONTACTS

In the previous chapter I described the reactions that occur between metal films and silicon substrates to form stable silicide phases, and in §5.3 the detrimental effects that reactions between aluminium interconnection films and silicon devices can have on the contact properties were presented. We can now combine these two sets of observations to explain why silicide layers are chosen to replace the aluminium–silicon contacts in many modern integrated circuits. The properties in which we will be especially interested are the electrical character of the silicide contact and its long-term metallurgical stability. The electrical properties of some silicide films, and the Schottky barrier heights of the contacts they make to n-type silicon, are listed in table 4.3. The specific contact resistance depends on this barrier height and the doping concentration in the silicon (see §2.2) and so reported values of this parameter vary widely. However, silicide contacts with specific contact resistances at least as good as those of aluminium contacts have been prepared. Part of the reason for these excellent contact resistances is that the silicide-forming reaction creates the contact under the original silicon surface, as illustrated in figure 4.14. The contamination species which are present on all silicon surfaces exposed to air can degrade the electrical properties of simple unreacted metal–silicon interfaces by inhibiting the formation of an intimate contact between the two materials. Chemical reaction between a metal and silicon at higher temperatures can result in the rejection of the contamination species to the top of the silicide layer (figure 4.14(c)) and the formation of a very clean silicide/silicon interface. We should not forget that the presence of the contamination can alter both the kinetics of the silicide-forming reaction and the particular silicide phases formed (see §4.4.2). In addition, if care is taken to ensure that the end phase in the chain of chemical reactions between the metal and the silicon is reached, the contacts are also very stable in a metallurgical sense. Popular silicide compounds for applications in metallisation schemes are $PtSi$, $TiSi_2$, WSi_2 and $NiSi_2$, which are all end phases in the reactions of a thin metal film with an excess of silicon. These phases are thus in thermodynamic equilibrium, at least as far as reactions between the silicide and the silicon substrate are concerned.

The contacts formed between end-phase silicide compounds and silicon devices can thus have excellent electrical properties and can be chemically very stable. This combination of properties has led to the development of a metallisation structure with many of the contacts to the silicon wafer made by a silicide phase, keeping the aluminium alloy film for use as conductor tracks. This involves the formation of numerous silicide/aluminium interfaces, and it is the stability of these that we must now consider. Hosack (1973) studied the metallurgical reaction of aluminium with PtSi and identified that the phase $PtAl_2$ is formed after annealing above 300 °C. This phase is thermodynamically more stable than PtSi and so will be preferentially formed at the interface. The silicon released by the decomposition of the silicide remains as large aluminium-doped precipitates in the $PtAl_2$. At the same time, the barrier height of the contact between the PtSi and the underlying silicon decreases from the starting value of 0.85 eV characteristic of the clean PtSi/Si contact to 0.6 eV. As the annealing process is continued, the barrier height increases to a stable level of 0.72 eV (figure 5.24). An explanation for this behaviour has been suggested as follows.

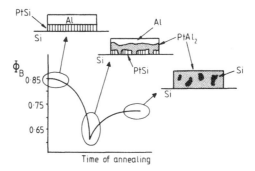

Figure 5.24 An illustration of the effect of annealing a PtSi contact which is covered by an aluminium film. The barrier height of the contact decreases from 0.85 eV to about 0.6 eV and increases back up to about 0.7 eV as the reaction to form $PtAl_2$ proceeds (after Rosenberg *et al* 1978).

The $PtAl_2$ compound forms unevenly, so that at an intermediate stage of the annealing process both $PtAl_2$ and PtSi grains are in contact with the silicon. As the reaction between the aluminium and the silicide phase proceeds, all the PtSi is converted to the aluminide and a $PtAl_2/Si$ contact is formed. The Schottky barrier height at this interface is assumed to be 0.72 eV, corresponding to the final state of the annealed contact.

The intermediate region, where the lowest barrier height of 0.6 eV is recorded, is due to the formation of small areas of low barrier height contact, $PtAl_2$ (0.72 eV), in a larger contact area of higher barrier height, PtSi (0.85 eV). In a conventional $C-V$ measurement to determine the barrier height averaged over the whole contact area, a mixed contact of this kind will give a much lower effective barrier than would be measured if either of the compounds was present over the whole contact area, a minimum value of 0.6 eV in this case (Rosenberg *et al* 1978). The stages in this reaction, and the resulting electrical characteristics of the contact, are illustrated in figure 5.24. This kind of variation in contact barrier height is just as damaging to the operation of the devices in the silicon wafer as the reactions between aluminium and the silicon surface described in §5.3. A similar kind of reaction has been observed between Pd_2Si contacts and aluminium films. Here a ternary phase, Al_3Pd_4Si, forms at the silicon/silicide interface, gradually consuming the whole of the silicide layer (Koster *et al* 1982). While this reaction delays the diffusion and reaction of aluminium with the silicon surface, it will not stop it completely, and the attendant modifications to the barrier height of the Pd_2Si/Si contact will still occur eventually.

These two examples show that while many silicides give excellent contacts to silicon, these contacts may not retain their stability when the silicide is covered by an aluminium conduction track. Murarka (1983) has compiled a list of the lowest temperatures at which there has been observed to be significant reaction between silicide phases and aluminium. These temperatures for some of the most popular silicide phases are: PtSi, 250 °C; Pd_2Si, 300 °C; $TiSi_2$, 450 °C; and $TaSi_2$, 500 °C. These data show that it would be possible to use the more refractory silicide phases like $TiSi_2$ or $TaSi_2$ as contacts to silicon devices and directly covered by aluminium conductors as long as no processing stage required temperatures in excess of about 400 °C. However, diffusion of both aluminium and silicon through the grain boundaries in these more refractory silicides can still cause gradual contact degradation at lower temperatures. The outdiffusion of silicon to go into solution in the aluminium conductors will cause spiking of the silicon wafer surface unless the diffusion paths in the silicide layer are blocked (Ting and Wittmer 1982).

Thus we have seen that even the silicide phases which form such excellent contacts to silicon devices cannot remain stable when in contact with reactive aluminium interconnection tracks. Once again we are at the stage where we must separate these two films, the contact and the conductor, making use of their individually valuable electrical properties, but not allowing them to interact. At this point I shall introduce the concept of the barrier layer which is used to separate thin films which react in undesirable ways. The structure and properties of these layers will be discussed in the following section.

5.7 BARRIER LAYERS

The concept of the use of barrier layers in metallisation systems is extremely simple: two materials between which there is an unwanted chemical reaction or interdiffusion process are kept separate by an intervening layer. Nowicki and Wang (1978), Nowicki et al (1978) and Ting and Wittmer (1982), amongst others, have described the features that should be possessed by the ideal barrier layer:

(1) If the barrier layer separates materials A and B, the barrier should be thermodynamically stable when in contact with both A and B. As an unfortunate side effect, this lack of chemical reactivity between barrier and metallisation films can lead to poor adhesion at the interfaces on either side of the barrier.

(2) Most barrier layers will be polycrystalline, since this is the only structure with which twin films can easily be deposited (see Chapter 4). However, we generally require that there should be no rapid diffusion of materials A or B along the grain boundaries in the barrier film. It is obviously important that there are no pinholes in the barrier, which implies that scrupulous cleanliness needs to be maintained during the thin film deposition process.

(3) The barrier layer should form low-resistance contacts with both A and B. The resistivity of the layer itself is usually insignificant as the barriers are generally less than 200 nm thick.

(4) The barrier layer must adhere well to all the materials with which it is in contact. These include glassy insulating layers as well as the contact and conducting films, and this may be incompatible with (1).

(5) The material of the barrier layer should not have an electrochemical potential very different from that A and B, or galvanic corrosion cells can be set up in the metallisation layers.

There are many places in an integrated metallisation system where barrier layers serve an important function and a few practical examples will be described below. Before considering these we should first understand the basic principles of how a barrier layer works in a thin film structure.

5.7.1 Fundamentals of Barrier Layer Properties

Sections 5.3 and 5.6 have shown that both diffusion and chemical reaction processes can degrade the performance of contacts to silicon devices. These processes occur because there is a strong thermodynamic driving force in each case, a concentration gradient in the case of grain boundary diffusion, and the lowering of free energy on the formation of $PtAl_2$ in the reaction between aluminium and PtSi for instance. There is nothing we can do to change the thermodynamics of these processes, but use of a barrier layer can alter the rate at which they occur in two ways: (i) by slowing down diffusion by using a material for the barrier layer in which all diffusion processes have a high activation energy; and (ii) by introducing an alternative chemical reaction path. Nicolet (1978) has termed these diffusion and sacrificial barriers respectively. Figure 5.25 illustrates the two ways in which a barrier layer can impede the reaction or interdiffusion of layers A and B, and we will now consider these two kinds of barrier layers separately.

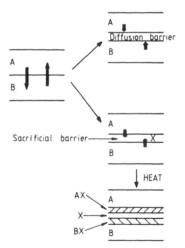

Figure 5.25 A schematic illustration of the two ways in which a barrier layer can impede any reaction or inderdiffusion between two materials A and B; by providing only slow diffusion paths, or by reacting with the diffusing A and B atoms.

5.7.2 Diffusion Barriers

The use of simple diffusion barriers appears to be a convenient method of slowing down unwanted thin film reactions, and here the principle is that the introduction of an additional diffusion step in the overall chemical reaction process will delay the onset of any deleterious changes in electrical or mechanical properties. However, just putting a layer of any material between A and B does not necessarily result in any significant reduction in the rate at which they react. There are several reasons for this. Firstly, almost all vapour-phase-deposited thin conducting films will be polycrystalline (see Chapter 4) and the high density of grain boundaries in these films will provide pathways for rapid diffusion of A to B, and vice versa. It has also been argued that using a barrier film in which neither A nor B is soluble (according to the equilibrium bulk phase diagram) would stop transport of these species through grain boundaries in the barrier film as well. Nicolet (1978) has shown that it is not possible to form an efficient diffusion barrier simply by selecting a barrier layer material in which the diffusant is insoluble. This is because the diffusant atoms may be soluble in the grain boundaries in the barrier layer, even though they are not soluble in the grain interiors to any significant extent. The relatively open structure of the grain boundaries allows them to accommodate atoms which are insoluble in the bulk by virtue of different atomic size or valence. Molybdenum layers may appear at first sight an attractive material for use as a barrier to gold diffusion; gold is insoluble in molybdenum grains. However, Nicolet observed the rapid failure of polycrystalline molybdenum barrier layers by grain boundary diffusion of gold through the molybdenum thin films. Nicolet and Bartur (1981) have given several other examples of ineffective barrier layers, where a failure to understand the nature of the grain boundary diffusion process in thin films contributed largely to a poor choice being made of a barrier layer material.

A second process which can be important in determining the rate of transport of material across a polycrystalline thin film is DIGM (§5.2). In this phenomenon, the diffusion of solute atoms along the grain boundaries in a thin film can cause grain boundary migration and accelerated solute transport across the film. The effect of this kind of DIGM process is shown in figure 5.26(a), where the formation of alloyed volumes by migration of the grain boundaries is illustrated. We normally expect grain boundary diffusion only to contribute a diffusive flux along a stationary boundary plane, a much less effective mechanism of solute transport than DIGM.

What we require in an effective diffusion barrier is a material in which the diffusive transport mechanism is intrinsically slow. Nothing else will stop atoms diffusing down the concentration gradients which always exist in thin film structures. The ideal way of slowing down diffusive transport is to grow single-crystal barrier layers. Unfortunately it is impossible to grow single-crystal films in the desired places in practical metallisation schemes. We are forced, therefore, to look for thin film materials in which grain boundary diffusion is slow. The basic approach which has been taken is to choose materials that are refractory, on the principle that the higher the melting point the higher the activation energy for grain boundary diffusion. This follows the conclusion presented in §5.2 that refractory metals like tungsten have much higher measured values of Q_{gb} than a metal like gold.

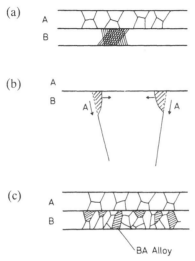

Figure 5.26 A schematic illustration of how the phenomenon of DIGM can increase the rate of material transport across a thin film diffusion barrier by the formation of alloyed volumes. (a) Two thin films of A and B in contact; (b) the start of the process of diffusion of A into the grain boundaries in B causing grain boundary migration; (c) the end result of a DIGM process with alloyed volumes formed in the B layer.

Many authors have discussed the performance of refractory elements as barrier layers and tungsten and molybdenum in particular have been tested for this application. A very popular refractory barrier material is an alloy of titanium and tungsten with a composition in the range 10–30 at.% titanium. Thin films of this material sputter deposited from alloy targets are polycrystalline with a BCC lattice structure. This is a metastable structure because the equilibrium solid solution limit of titanium in tungsten is only about 10 at.% at 600 °C. These metastable alloy films have been found to be relatively stable and do not decompose at the temperatures that are usually experienced during the processing of the metallisation layers. However, even these refractory diffusion barrier layers have been shown to be relatively ineffective if they are deposited under clean conditions in the sputtering apparatus, and the failure of molybdenum to stop the diffusion of gold has already been mentioned. The breakdown of the refractory barriers by grain boundary diffusion is not surprising once the structure of the films is investigated. The grain diameter is normally around 30 nm in a typical sputtered film and so the density of grain boundaries is exceptionally high. We should remember that the grain size developed in a vapour-deposited film depends on the grain boundary mobility (see §4.2) and this decreases with increasing melting temperature. Thus we have chosen barrier layer materials in which the grain boundary diffusion rates are low but the grain size will also be small; these two effects cancel each other out to some extent.

Nowicki *et al* (1978) and Nowicki and Wang (1978) have shown that deliberately depositing Ti–W films contaminated with oxygen or nitrogen results in very much improved barrier layer properties. This is thought to be due to the blocking of the grain boundary diffusion paths with contaminant atoms, as discussed in §5.2, or even by the formation of oxide or nitride precipitates along the grain boundaries. This process is known as 'stuffing' the grain boundaries to decrease the diffusion rate across the barrier. Diffusion barriers have also been prepared of thin films of TiN, MoN and TiB_2. These films generally have a fine-grained structure, but the activation energy of grain boundary diffusion in these refractory compounds is expected to be very high indeed. Sinke *et al* (1985) have shown that even for refractory nitride films like TiN, exposure to oxygen will increase their effectiveness as barrier layers, presumably by further blocking of the grain boundary diffusion paths. Thus a gradual improvement in barrier layer properties is obtained by first choosing refractory materials, then stuffing the grain boundaries with impurity species and finally by growing very stable refractory compound films in which the activation energy for grain boundary diffusion is even higher. In many cases refractory barrier layers are grown in the grey area between the stuffed refractory metal film and the true compound layer. It is then very difficult to say whether a film has good barrier layer properties because the grain boundaries in the metal are blocked with contamination, or because the film is a refractory compound in which grain boundary diffusion is intrinsically slow. The microstructures of these successful barrier layers have not been very carefully investigated and may be a field of interesting research for the materials scientist in the future.

An additional advantage of Ti–W and TiN barrier layers is their relative stability when in contact with aluminium or silicide phases. Titanium, and to a lesser extent tungsten, will react with aluminium, as is discussed below, but TiN is extremely stable and unreactive and will function as a true diffusion barrier reacting only very slowly with aluminium at elevated temperatures to form AlN and $TiAl_3$ (Wittmer

1984). The structure and properties of refractory metal nitrides for microelectronic applications have been reviewed by Wittmer (1985) and these materials are amongst the most promising for the preparation of barrier layers with exceptional long-term stability. The deposition of these compounds is normally carried out by one of the reactive sputtering techniques mentioned in §4.1.2.

A very recent development in diffusion barrier research has been the use of thin films of metallic glasses for reducing diffusional transport. At first sight it seems improbable that diffusion in these disordered alloys should be slow, because similar kinds of structural units are thought to make up both grain boundary cores and amorphous glasses. In fact, recent measurements have indicated that diffusion coefficients in metallic glasses can be orders of magnitude lower than in the grain boundaries of crystalline alloys of similar composition. The diffusion coefficients of gold in a variety of amorphous alloys are shown in figure 5.27. For comparison, the diffusion coefficients of gold in some polycrystalline and single-crystalline materials are also included. It is clear that the gold diffuses at about the same rate in the glasses and the single-crystal samples. Amorphous alloy layers are relatively easy to deposit by sputtering but may be difficult to pattern by lithographic techniques because they tend to be very unreactive. Relatively little testing of amorphous alloys for barrier layer applications has been completed, although some indication of their properties has been given by Nicolet *et al* (1983) and Todd *et al* (1984), who show that they can form exceptionally stable barriers in metallisation schemes on high-temperature devices.

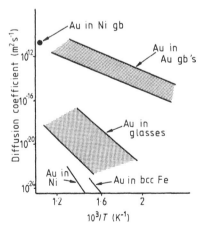

Figure 5.27 A comparison of the experimental data on the diffusion coefficients of gold in a variety of amorphous and crystalline materials (after Cantor *et al* 1985).

5.7.3 Sacrificial Barriers

The second kind of barrier layer commonly used in metallisation structures is a sacrificial barrier where a flux of diffusing species from layer A, that would react with layer B in the absence of a barrier layer, is intercepted to form an intermetallic compound AX (see figure 5.25). This phase formation may not be as thermodynamically favourable for the system as a whole as the reaction between A and B, but the geometry of the thin films forces the A/X and B/X reactions to occur to completion before free A and B atoms can come into contact and react. Once the whole of the sacrificial barrier is consumed to form AX and BX phases, the only impediment to the A/B reaction is diffusion of the reacting species through these new phases. At this stage, the new intermetallic phases function only as diffusion barriers and will probably not delay the onset of the deleterious A/B reaction for very long. We can see from this description that a sacrificial barrier fails to satisfy point (1) in the list at the beginning of this section, because it reacts with the A and B films it is intended to separate. However, Nicolet (1978) has pointed out that a knowledge of the kinetics of the A/X and B/X reactions that consume the barrier and the operating temperature of the component allows a design engineer to add a sufficiently thick sacrificial barrier to the metallisation system to give it stability for the desired component lifetime.

The most common materials for sacrificial barriers have been transition or refractory metals interposed between silicide contacts and aluminium alloy conductors. Here the metal will react with the aluminium to form aluminide phases and will also decompose the silicides to form new silicide phases in some cases. Metals like tungsten, chromium and vanadium have been tested, and the result of annealing such a barrier layer structure is shown in Figure 5.28. Other metallic sacrificial barriers will react in slightly different ways: only $TiAl_3$ is formed when titanium is interposed between aluminium and PtSi for instance, and failure of the barrier layer occurs by the eventual diffusion of the aluminium through the grain boundaries in the $TiAl_3$ to decompose the silicide (Salomonson *et al* 1981). These sacrificial barriers usually fail after a few minutes at temperatures above 400 °C, placing an upper limit on the temperature at which any subsequent processing of the metallisation can be carried out. The relevance of the maximum temperature that can be sustained in the metallisation without breakdown of the barrier layers will become clear in Chapter 6 where the conditions for the effective deposition of insulating films are reviewed. The rather poor performance of these metallic sacrificial barriers is due to the fact that the aluminide phases form very rapidly at quite low temperatures and consume the barrier layers. This occurs in exactly the fashion described for chromium and hafnium in §5.3. The aluminium is then free to diffuse through the aluminide phase and react with and degrade the performance of the silicide contact.

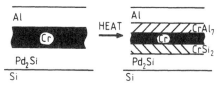

Figure 5.28 A schematic illustration of the sacrificial reactions which occur at a chromium barrier layer between an aluminium conductor track and a Pd_2Si contact (after Olowolafe *et al* 1977).

However, the transition metal itself can be used as the electrical contact to the silicon devices. Titanium, for instance, can form ohmic and Schottky contacts to silicon, reducing the native oxide and giving very low specific contact resistivities. Aluminium indiffusion through the barrier layer towards the silicon surface is relatively harmless; the aluminium will pile up at the metal–silicon interface but does not react with the silicon in any way. The outdiffusion of silicon through the barrier is a far more damaging process since it results in spike forma-tion, but many refractory metals are excellent barriers to silicon diffu-sion. Thus it is possible to design a metallisation scheme that discards the silicide layer altogether. An $Al/Ti/Si$ contact metallurgy will even-tually fail when all the titanium is converted to $TiAl_3$, and this point can be delayed by the addition of about 20 at.% of tungsten to the titanium layer (Babcock and Tu 1986).

It is possible to do away with the silicide contacts described in the last section only if care is taken over the choice of a material which can function both as a barrier layer to silicon outdiffusion and as an electrically stable contact. The deposition of tungsten films by the CVD process has been described in §4.1.3, and this simple low-temperature deposition method seems to be ideal for the preparation of contact/barrier layers with the required properties. This discovery may lead to the use of refractory contacts and conductors in the future, with all the attendant advantages of not having to design fabrication proces-ses which keep the temperature of the metallisation below rather low maximum values.

5.7.4 Barrier Layers Between Aluminium and Silicon

In the previous two sections I have mentioned several kinds of barrier layers which can be used to separate aluminium conductor tracks from silicon device surfaces. In this section I shall follow the path taken by the device engineers in deciding what the best barrier is for this interface. Silicide layers have been shown to be very poorly resistant to decomposition by reaction with aluminium, and many refractory metals used as barrier layers both react with the aluminium and fail to impede indiffusion of aluminium to the silicide contact very effectively. Ti–W alloys have much better diffusion barrier properties, especially when deposited under 'dirty' conditions so that the grain boundaries are stuffed with contaminants. Refractory nitride or carbide layers are even better diffusion barriers, preventing both silicon outdiffusion and aluminium indiffusion. Table 5.3 shows a selection of data comparing the approximate maximum temperature at which diffusion barriers will survive for at least a few minutes. The usual criterion for efficient operation is that the Schottky barrier height and resistance of the underlying contact to the silicon should be invariant during the heating process. The excellent performance of the refractory alloy and compound diffusion barriers can be clearly seen in this comparison. The maximum operating temperature of the TiN barriers is less than 100 K below the melting point of the aluminium conductor film. In all the cases shown in table 5.3, a silicide layer is used to make electrical contact to the devices. The performance of Al/W/Si contact structures may rival that of the compound barriers in the future.

Table 5.3 A comparison of the properties of a selection of barrier layers used at aluminium/silicon contacts (after Wittmer 1984).

Metallisation structure	Approximate maximum operating temperature (°C)	Observed mode of failure
Al/PtSi/Si	350	Compound formation
Al/TiSi$_2$/Si	400	Diffusion
Al/Cr/PtSi/Si	450	Compound formation
Al/Ti/PtSi/Si	450	Compound formation
Al/Ti$_{30}$W$_{70}$/PtSi/Si	500	Diffusion
Al/TiN/PtSi/Si	600	Compound formation
Al/TiC/PtSi/Si	600	Compound formation

5.7.5 Further Examples of Practical Barrier Layers

The semiconductor–metal contact is a particularly sensitive region of an integrated circuit or individual device because very small changes in chemistry can alter the electrical properties of the contact, as we have seen in Chapter 2. So far I have only presented problems that arise due to the metallurgical instability of the contact/silicon interface, but contacts to devices in III–V materials have to be just as carefully designed to give a reproducible and stable performance. Barrier layers are a vital part of any metallisation system to III–V devices, and I will illustrate this with the specific example of the Pt–GaAs Schottky contact.

5.7.5(a) Pt/GaAs reactions

As-deposited platinum thin films give an excellent Schottky contact to GaAs, but if the contact is annealed, reactions occur to form Pt/As and Pt/Ga phases. As a result of these reactions the interface becomes roughened, and if the degree of roughness becomes too severe the performance of the Schottky barrier is degraded and the surface of the GaAs wafer, containing the implanted or diffused regions, is consumed in the chemical reactions. Mahoney (1975) has showed that if a Pt/W/Pt/GaAs multilayer contact is deposited, where the lower platinum layer is very thin and is completely reacted with the GaAs substrate, then the tungsten barrier layer stops further indiffusion of the platinum. The initial reaction between the metal and the GaAs will roughen the interface a little, but the barrier layer limits the extent of the reaction so that the contact properties are reasonably stable. The important properties of the barrier layer material are that it must impede the diffusion of platinum and should not react with GaAs. Tungsten films have both of these properties and, in addition, limit the indiffusion of gold which is often used as an interconnection material over the top of the platinum. Gold has a very deleterious effect on the Schottky contact properties. The structure of this successful metallisation scheme, which incorporates a simple diffusion barrier layer, is shown schematically in figure 5.29. Breakdown of this kind of structure seems to occur in service by the outdiffusion of gallium to react with the gold conductors.

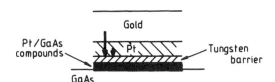

Figure 5.29 A schematic illustration of the use of a tungsten barrier layer to separate a gold conductor track from a platinum contact on a GaAs device.

5.7.5(b) Reactions between solder and aluminium

As another example of a place in a metallisation scheme where barrier layers are important, let us consider joints between solder and aluminium conductor tracks. Devices interconnected by thin film conduction tracks in an integrated circuit are finally connected to the outside world by some form of relatively macroscopic contact, either a wire-bond or solder joint. (The design and properties of these two kinds of contacts are considered in Chapter 7.) Tin–lead solder alloys are frequently used for this application, and it would be convenient if solder balls could be joined directly to the top layer of aluminium metallisation. However, solders wet aluminium very poorly, and the tin–aluminium phase diagram has a eutectic at 232 °C and at 99.5 at.% tin, well below the temperature at which tin–lead solder alloys become sufficiently fluid for effective solder joining. A eutectic melt will form very rapidly if a molten tin–lead solder wets an aluminium alloy thin film. This melt will dissolve the aluminium conductor tracks almost instantaneously, irreparably damaging the metallisation structure. Howard (1975) has shown that a eutectic melt will preferentially dissolve the grain boundaries in the aluminium alloy films, rapidly creating open circuits in the conductors.

A barrier layer between the solder and the aluminium conductors is obviously needed, which must both protect the aluminium from the molten solder and stop any subsequent solid state outdiffusion of aluminium which could create a eutectic melt at a local hot spot on the metallisation surface. Thin chromium films have been found to be quite effective at separating these materials, as long as the chromium can be deposited with no pinholes or cracks. These layers act to sweep up all the diffusing aluminium atoms into aluminide compounds, reducing the chance of forming the Sn–Al eutectic melt by solid state diffusion, and also stop the molten solder coming into contact with the aluminium conductors. Chromium is only very slightly soluble in molten tin–lead solders. The chromium is thus functioning as a combined sacrificial and diffusion barrier layer and has the further useful property of aiding the adhesion of copper and gold films which are added to the contact structure. The copper layer is added to improve the adhesion of the solder to the contact area and in some cases a thin gold film is used to stop the oxidation of the copper surface which would reduce the wetting of the contact window by the solder. The gold layer must be separated from the aluminium conductors or there is a risk of the purple plague reaction, and the chromium barrier layer serves this function at the relatively low temperatures of the soldering operation. The structure of this kind of multilayered contact between a solder ball and an aluminium conductor track is sketched in figure 5.30.

These few applications illustrate some of the ways in which barrier layers can be used to improve the stability of metallisation structures. The drive to fabricate ever smaller devices to increase the speed at which electronic systems can operate has exacerbated the problems faced by the device engineer, since the amount of interdiffusion or chemical reaction required to destroy completely an expensive device or chip is reduced with the conductor linewidth and the contact areas. This has necessitated the development of extraordinarily stable barriers which can separate reactive films at high temperatures for extended periods without breakdown. Although these barrier layers play no active part in determining the electronic properties of the devices, or in the propagation of the electronic signals through the metallisation layers, they are vital for the long-term operation of almost all microelectronic circuits. We should not forget amongst all the emphasis placed on the 'silicon revolution' that the humble stuffed Ti–W alloy barrier layer is one of the most successful of all microelectronic materials in terms of doing the job for which it was designed. The production of efficient integrated circuitry would not have been possible without the development of metallurgically stable barrier layers.

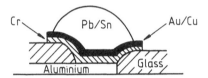

Figure 5.30 A possible structure for a solder/metallisation joint. The roles of the various films in this illustration are described in the text.

5.8 INTEGRATED METALLISATION SCHEMES

So far in this chapter I have concentrated on the metallurgy of the aluminium conductor on silicon wafer technology introduced in Chapter 2 and outlined the ways in which this sample structure must be modified to give suitable electrical characteristics and long-term stability. This has required the introduction of silicide and barrier layers as well as the patterned conducting tracks and insulating films which were originally envisaged. We can now sketch out a composite three-layer metallisation structure for the interconnection of devices on a silicon chip (figure 5.31). This kind of complex multicomponent structure, or something very similar, is at the heart of many current integrated circuits, although of course not all circuits require three, or even two, levels of conduction tracks. We should note that Ti–W barrier layers are often included not only between the aluminium alloy conductors and the silicide contacts but at every via as well. This is because it has been found that the contact resistance and adhesion at the vias can be improved by avoiding aluminium-to-aluminium joints, where oxide contamination can make intimate metal-to-metal contacts hard to achieve.

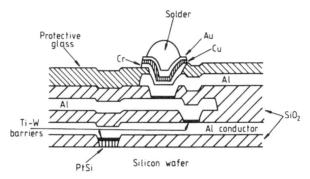

Figure 5.31 A sketch of a three-layer aluminium-based metallisation scheme, illustrating how the various thin film components described earlier in this chapter can be brought together to form a set of stable interconnects for an integrated circuit.

We have so far considered the use of ohmic or Schottky contacts in metallisation and should now describe the modifications needed when oxide–semiconductor interfaces are included as the gate contacts in MOS devices. These devices have been introduced in Chapter 2 and form a very important class of components in large-scale integrated circuitry. The deposition of the thin gate oxide layers is described in the next chapter, but here we can consider how these contacts can be integrated into a complete metallisation scheme. Figure 5.32(a) illustrates the structure of a single MOS field effect transistor showing the two PtSi ohmic contacts and the gate electrode metallurgy. Gate contacts are relatively simple in a metallurgical sense since the conducting layers do not come into contact with the semiconductor surface. Thus the choice of the first contacting layer on the gate oxide depends more on the ease of deposition and patterning of the material than its electrical properties. However, the speed at which an array of MOS devices can operate depends very strongly on the sheet resistivity of the interconnecting material. As the linewidths of the conductor tracks are decreased in the drive to pack ever more devices onto the wafer surface, it becomes even more important to reduce the resistivity of the conductors.

Polycrystalline silicon films are a very common choice for the gate contact material because they are simple and cheap to deposit, give good adhesion and conformality, and are stable during high-temperature processing in subsequent stages of the fabrication of the metallisation system. Their major disadvantage is their high resistivity (see table 4.2 for a comparison of the sheet resistivity of polysilicon and metal films). This high resistivity of even the most heavily doped polysilicon means that it is not possible to use this material for interconnection paths longer than about 100 μm. Alternative gate contact materials, refractory metals and refractory silicide phases like $TiSi_2$, $TaSi_2$ and WSi_2, have been considered and used with reasonable success. However, a favourite gate metallisation scheme is the composite polysilicon/silicide structure illustrated in figure 5.32(a). Here the polysilicon forms the contact to the gate oxide, an interface that has electrical properties which are both very stable and well understood. A refractory silicide layer is then deposited above the polysilicon to give low-resistivity conduction paths. The three silicide compounds listed above have all been used for the so-called 'polycide' gate contacts and after annealing can have sheet resistivities as low as 20–50 $\mu\Omega$ cm. The deposition and properties of these refractory silicide compounds have been described in §4.4. The overall resistance of a gate contact metallisation can be reduced very substantially by the switch from a simple polysilicon conductor to a polycide structure. In some cases, the silicide layers are also found to be compatible with conventional MOS processing techniques, meaning that no new fabrication steps need to be added in the switch to polycide MOS technology. This is especially true of $TaSi_2$ which is almost completely resistant to dissolution in HF (one of the most severe stages the polycide layers must be able to withstand in manufacture). The silicide layer in polycide contacts can also help protect the polysilicon contact from reaction with aluminium.

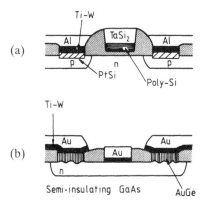

Figure 5.32 A schematic illustration of: (a) the structure of a silicon field effect transistor with PtSi ohmic contacts and a $TaSi_2$ polycide conductor over the gate oxide; (b) a GaAs field effect transistor with Au–Ge ohmic contacts, a Ti–W barrier layer and gold conductors.

Finally, we should remember that gold-based conductors are used in the metallisation structures to some kinds of discrete silicon devices, especially those which have an operating temperature above about

100–150 °C. In these structures, a glue layer is normally used to assist the adhesion of the gold films to the insulator layers, and so the conductor metallurgy is of the form Cr/Au, Ti–W/Au or Ti/Pt/Au. Many reactive metals make very effective glue layers, but in some cases they are sensitive to atmospheric attack. The oxides of molybdenum are soluble in water, for instance, and so a metallisation system containing this metal must be carefully hermetically packaged. Gold also reacts with some reactive metals, for example titanium, to form intermetallic compounds. Reactions of this kind can increase the resistivity of the conduction tracks, and platinum layers may be added between the gold and titanium to act as a diffusion barrier. In fact, the Ti/Pt/Au trilayer conductor is not very stable because of grain boundary diffusion across the platinum layer. Nowicki and Nicolet (1982) have shown that a TiN barrier is much more effective at separating the gold conductor and titanium glue layer.

It is worth reiterating that gold conductors are more resistant than aluminium alloys to electromigration failure by a grain boundary mechanism and considerably more stable with respect to corrosive attack. However, a common failure mechanism in metallisation systems containing gold conductors is the surface diffusion of the gold and the formation of dendrites. In the presence of moisture, and CN^-, Br^- or Cl^- ions, gold can be dissolved at conduction tracks which are at a positive potential to form ions of the type $(AuCl_4)^-$ (English and Mellior-Smith 1978). These ions migrate over the device surface to cathodic conductor tracks where the gold is plated onto the track, often in the form of dendrites which can easily create short circuits. An example of the result of this kind of corrosion and electromigration process is shown in Chapter 10. It is thus very important to ensure that ionic contaminants, like the chloride ions used in cleaning and dry etching treatments, are not left on the device surface when gold conductors are used. Another mode of corrosion failure involves the formation of $Au(OH)_3$ in the presence of excess moisture. This compound can grow to form short-circuit paths between adjacent conductor tracks (Frankenthal and Becker 1979). In addition, the gold layers can act as efficient large-area cathodes when in contact with electropositive metals like Ti, Mo and Ta, and this can lead to severe local anodic corrosion of the underlying glue layers (see §5.5). A more detailed discussion of some of the factors that have dictated the choice of materials for metallisation conductors on silicon devices has been given by Ghate (1982).

5.8.1 Integrated Metallisation on GaAs Devices

Gallium arsenide is the only III–V compound on which large-scale integrated circuitry has been successfully produced to date, and both microwave and digital circuits have been prepared on GaAs wafers. Indium phosphide is another semiconductor on which large-scale device integration may be possible in the future. The reason why these two semiconductors can be considered for integration lies in the successful development of crystal growth techniques which can produce semi-insulating material of very high quality (see Chapter 3). It might be expected with the enormous amount of experience gained on the design and deposition of aluminium-based metallisation schemes for silicon integrated circuits that this same technology would be used for GaAs circuits. In fact, gold-based metallisation schemes are almost universally used in this developing technology.

Figure 5.32(b) illustrates a typical structure of a GaAs field effect transistor, a device which requires two high-quality ohmic contacts to source and drain regions and a stable gate contact if the high operation speeds possible in GaAs devices are to be realised. Many devices of this kind would be included in a microwave circuit. In §2.2 we saw that Au–Ge alloys can be used to form low-resistance ohmic contacts to GaAs, but the interface between the contact and the wafer will be roughened by the chemical reaction needed to produce these excellent properties. Gate contacts to silicon devices are made with thin SiO_2 layers, but the thermal oxides grown on GaAs have none of the electrical or thermodynamic stability of silica. Vapour-deposited dielectric layers rarely give GaAs/insulator contacts with properties as good as those of the Si–SiO_2 interface. For this reason, Schottky gate contacts are normally used in GaAs FETS, and I have shown a titanium layer being used as a Schottky contact in figure 5.32(b). In §2.2 we saw that the high density of electronic states localised on GaAs surfaces means that almost any metal can be used to form a stable Schottky barrier. Titanium and chromium are particularly suitable for this application as they adhere very well to the underlying GaAs. It is important to exclude oxygen during the deposition of these reactive elements, or an oxide layer will form at the metal–GaAs interface and the electrical properties of the contact will be degraded. Ti–GaAs and Cr–GaAs contacts have been shown to have stable Schottky barrier properties at temperatures below about 400 °C (Sinha and Poate 1978). Gold contacts are to be avoided because of the rapid reaction with the GaAs to form intermetallic phases, which results in unstable Schottky barrier performance.

Gold is often used as the material for the conductor tracks in integrated circuits on GaAs, but gold and titanium react vigorously at relatively low temperatures, which may allow the undesirable penetration of the gold down to the GaAs wafer. Tungsten or platinum barrier layers are used to stop this reaction, and it has proved convenient to deposit Ti/W/Au multilayers both to form the Schottky gate contact and to make interconnections with the ohmic contacts. This is the structure illustrated in figure 5.32(b). The GaAs/Ti–W/Au metallisation system can survive several hours at 700 °C before the contact properties are degraded. An alternative metallisation structure uses an aluminium gate contact which is then protected from reaction with the gold conductors by Ti–Pt or Ti–W barriers. Some manufacturers are investigating the use of amorphous alloy contacts to GaAs, and Ta–Ir and Ni–Nb alloys have been shown to produce very stable Schottky barriers. To complete the metallisation structures, insulating layers of silica, silicon nitride, or even polyimide can be deposited just as for silicon integrated metallisations, as described in Chapter 6. A more complete list of metallisation schemes used on GaAs devices has been given by Palmstrom and Morgan (1985).

Let us now briefly consider some of the mechanisms by which these metallisation schemes on GaAs devices can degrade in service. Most of these mechanisms will be the same as those encountered in metallisation schemes on silicon devices. The surface electromigration of ionic species produced by corrosive attack of the gold conductors seems to be assisted by the outdiffusion of gallium from the wafer and is a particularly common mode of failure. Short-circuiting gold dendrites are formed, as described above. The ohmic and Schottky contact properties also degrade if overheated, because of interdiffusion and chemical reaction processes in these complex systems. In many cases the exact nature of the degradation mechanism is poorly characterised. In addition, electrochemical cells are easily formed in metallisation structures where electronegative elements (Al, Mo and Ta for instance) are found in contact with noble elements like gold and platinum. In the presence of moisture and Cl^- or P^- ions, any exposed region of a barrier layer under a gold conductor will corrode very rapidly. The difference in the thermal expansion coefficients of the contact metals and the GaAs can also be very high, leading to decohesion at the metal–semiconductor interface. These kinds or failure mechanisms have been reviewed by Davey and Christou (1981) and Morgan (1981) and some of the techniques which have been developed for identifying which failure mode has occurred in a particular faulty component will be described in Chapter 10.

Gallium arsenide integrated circuits have not yet been developed to the complexity of silicon circuits, but it has been shown that large processing arrays can be produced using the above interconnecting and contacting metallurgies. These circuits make quite effective use of the high carrier mobility, and hence the fast device operation, in GaAs. A photograph of a large GaAs circuit is shown in figure 5.33. More information on the materials and processing techniques used in the fabrication of monolithic GaAs circuits can be found in Willardson and Beer (1984) and Howes and Morgan (1985).

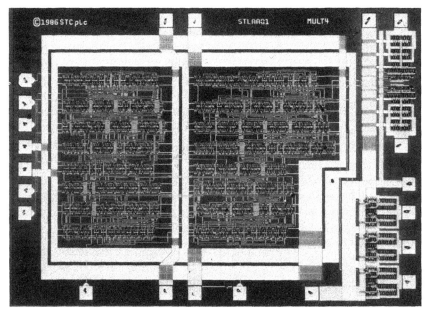

Figure 5.33 An optical micrograph of the top surface of an integrated circuit fabricated on a GaAs wafer. This circuit is a 4 × 4 multiplier. (Courtesy of Dr S Kitching and STC Technology Ltd.)

5.8.2 Metallisation to High-temperature Devices

To conclude this chapter, we should briefly review the materials chosen to make contacts and interconnection tracks on devices which operate at temperatures above 200 °C. It is often convenient to be able to operate semiconductor devices at high temperatures—inside engines for example, or in space where solar irradiation increases the temperature of a component significantly. Silicon devices operate poorly at these temperatures because of the large number of free carriers which are thermally excited across the narrow band gap. Minority carrier devices are particularly severely affected by high operating temperatures. It is thus necessary to use semiconductors with wider band gaps for these applications to avoid the production of thermal carriers. Gallium arsenide solar cells are commonly used in space, while power transistors and high-temperature devices can be fabricated in GaP or SiC. It is thus necessary to make reliable contacts to these materials.

Table 2.2 shows some of the materials which can be used to form ohmic contacts on GaAs devices, and the GaAs/Ge/Ti–W ohmic contact scheme in particular has excellent stability at temperatures up to 350 °C (Anderson *et al* 1982). Here the germanium ensures that a heavily doped n^+ layer is formed on the top surface of the GaAs, resulting in a low contact resistance, and the Ti–W forms refractory conductor tracks. Gold–germanium or gold–beryllium alloys can be used to make ohmic contacts to GaP devices, but the contact resistances are often rather high. This is because the surface states on GaP pin the Fermi level at the centre of the band gap and it is difficult to form a contact with a small potential barrier even after alloying. More details on the problems associated with contact formation to high-temperature GaP devices have been given by Zipperian *et al* (1982). Amorphous metal alloys may prove useful for stable metallisations on GaAs and GaP devices, but can also have high sheet resistances.

REFERENCES

Anderson W T, Christou A, Giuliani J F and Dietrich H B 1982 *IEEE Trans. Ind. Electron.* **IE29** 149
Babcock S E and Tu K N 1986 *J. Appl. Phys.* **59** 1599
Baglin J E E and Poate J M 1978 in *Thin Films, Interdiffusion and Reactions* ed. J M Poate, K N Tu and J W Mayer (New York: Wiley) p305
Balluffi R W and Blakely J W 1975 *Thin Solid Films* **25** 363
Bernal J D 1964 *Proc. R. Soc.* **280A** 299
Blech I A and Herring C 1976 *Appl. Phys. Lett.* **29** 131
Cahn J W, Pan J D and Balluffi R W 1979 *Scr. Metall.* **13** 503
Cantor B 1985 in *Rapidly Quenched Metals* ed. S Steels and H Worlimont (Amsterdam: Elsevier) p595
Card H C 1974 in *Metal-Semiconductor Contacts* (Inst. Phys. Conf. Ser. 22) p129
Chang C C and Quintana G 1976 *Thin Solid Films* **31** 265
Davey J E and Christou A 1981 in *Reliability and Degradation* ed. M J Howes and D V Morgan (Chichester: Wiley) p237
D'Heurle F M 1971 *Proc. IEEE* **59** 1409
D'Heurle F M and Ho P S 1978 in *Thin Films, Interdiffusion and Reactions* ed. J M Poate, K N Tu and J W Mayer (New York: Wiley) p243
Dix E H, Brown R H and Binger W W 1961 in *Metals Handbook* (Metals Park, OH: ASM) p916
English A T and Mellior-Smith C M 1978 *Ann. Rev. Mater.* **8** 459
Frankenthal R P and Becker W H 1979 *J. Electrochem. Soc.* **129** 1718
Gangulee A and D'Heurle F M 1973 *Thin Solid Films* **16** 227
Ghate P B 1982 in *Thin Films and Interfaces* ed. P S Ho and K N Tu (New York: North-Holland) p371
Gilmer G H and Farrell H 1976 *J. Appl. Phys.* **47** 3792, 4373
Glaeser A M and Evans J W 1986 *Acta. Metall.* **34** 1545
Grabe B and Schreiber H-H 1983 *Solid State Electron.* **26** 1023
Gupta D and Asai K W 1974 *Thin Solid Films* **22** 121
Gupta D and Rosenberg R 1975 *Thin Solid Films* **25** 171

van Gurp D J 1973 *J. Appl. Phys.* **44** 2040

Harrison L G 1961 *Trans. Faraday Soc.* **57** 1191

Herbuval I, Biscondi M and Goux C 1973 *Mem. Sci. Rev. Met.* **70** 39

Herzig Ch and Cardis D 1975 *Appl. Phys.* **5** 317

Holloway P H, Amos D E and Nelson G C 1976 *J. Appl. Phys.* **47** 3769

Hosack H H 1973 *J. Appl. Phys.* **44** 3476

Howard J K 1975 *J. Appl. Phys.* **46** 1910

Howard J K, Lever R F, Smith P J and Ho P S 1976 *J. Vac. Sci. Technol.* **13** 68

Howard J K, White J F and Ho P S 1977 *J. Appl. Phys.* **49** 4083

Howes M J and Morgan D V 1985 *Gallium Arsenide, Materials, Devices and Circuits* (Chichester: Wiley)

Huntington H B and Grone A R 1961 *J. Phys. Chem. Solids* **20** 76

Hwang J C M and Balluffi R W 1979 *J. Appl. Phys.* **50** 1339

Hwang J C M, Pan J D and Balluffi R W 1979 *J. Appl. Phys.* **50** 1349

Koster U, Ho P S and Lewis J E 1982 *J. Appl. Phys.* **53** 7436, 7445

LeClaire A D 1963 *Br. J. Appl. Phys.* **14** 351

Lloyd J R and Smith P M 1983 *J. Vac. Sci. Technol.* **A1** 455

Luby S, Lobotka P and Besak V 1980 *Phys. Status Solidi* a **60** 539

Magni G, Nobile C, Ottaviani G, Costato M and Galli E 1981 *J. Appl. Phys.* **52** 4047

Mahoney G E 1975 *Appl. Phys. Lett.* **27** 613

Martin G, Blackburn D A and Adda Y 1967 *Phys. Status Solidi* **23** 223

Martin G and Peraillon B 1980 in *Grain Boundary Structure and Kinetics* (Metals Park, OH: ASM) p239

Mingard K 1987 *Part II Thesis* University of Oxford

Morgan D V 1981 in *Reliability and Degradation* ed. M J Howes and D V Morgan (Chichester: Wiley) p151

Murarka S P 1983 *Silicides for VLSI Applications* (New York: Academic) p157

Nguyen T H and Foley R T 1980 *J. Electrochem. Soc.* **127** 2563

Nicolet M-A 1978 *Thin Solid Films* **52** 415

Nicolet M-A and Bartur M 1981 *J. Vac. Sci. Technol.* **19** 786

Nicolet M-A, Suni I and Finetti M 1983 *Solid State Technol.* 129

Nowicki R S, Harris J M, Nicolet M-A and Mitchell I V 1978 *Thin Solid Films* **53** 195

Nowicki R S and Nicolet M-A 1982 *Thin Solid Films* **96** 317

Nowicki R S and Wang I 1978 *J. Vac. Sci. Technol.* **15** 235

Olowolafe J O, Nicolet M-A and Mayer J W 1977 *Solid-State Electron.* **20** 413

Palmstrom C J and Morgan D V 1985 in *GaAs, Materials, Devices and Circuits* ed. M J Howes and D V Morgan (Chichester: Wiley) p195

Peterson N L 1980 in *Grain Boundary Structure and Kinetics* (Metals Park, OH: ASM) p209

Philofsky E 1970 *Solid-State Electron.* **13** 1391

Pramanik D and Saxena A N 1983 *Solid State Technol.* 131

Rosenberg R, Sullivan M J and Howard J K 1978 in *Thin Films, Interdiffusion and Reactions* ed. J M Poate, K N Tu and J W Mayer (New York: Wiley) p13

Rosenberg R, Mayadas A F and Gupta D 1972 *Surf. Sci.* **31** 566

Salomonson G, Holm K E and Finstad T G 1981 *Phys. Scr.* **24** 401

Sinha A K and Poate J M 1978 in *Thin Films, Interdiffusion and Reactions* ed. J M Poate, K N Tu and J W Mayer (New York: Wiley) p407

Sinke W, Frijlink G P A and Saris F W 1985 *Appl. Phys. Lett.* **47** 471

Sutton A P 1984 *Int. Met. Rev.* **29** 377

Thomas M E, Keyser T K and Goo E K W 1986 *J. Appl. Phys.* **59** 3768

Ting C Y and Wittmer M 1982 *Thin Solid Films* **96** 327

Todd A G, Harris P G, Scoby I H and Kelly M J 1984 *Solid-State Electron.* **27** 507

Turnbull D and Hoffman R E 1954 *Acta Metall.* **2** 419

Vavra I and Luby S 1980 *Czech. J. Phys.* B **30** 175

Whipple R T P 1954 *Phil. Mag.* **45** 1225

Willardson R K and Beer A C 1984 *Semiconductors and Semimetals* **20** (Semi-insulating GaAs) (New York: Academic)

Wittmer M 1984 *J. Vac. Sci. Technol.* **A2** 273

—— 1985 *J. Vac. Sci. Technol.* **A3** 1797

Zipperian T E, Chaffin R J and Dawson L R 1982 *IEEE Trans. Ind. Electron.* **IE29** 129

6

Deposition and Properties of Insulating Films

6.1 INTRODUCTION

The growth, structure and properties of thin films of metals and semiconductors have been discussed in earlier chapters. Here I shall consider the last important thin film component in an integrated circuit metallisation system—dielectric, or insulating, films. In Chapters 1 and 2 three dielectric materials were suggested for applications such as electrical isolation, surface passivation and atmospheric protection: silicon oxide, SiO_2, silicon nitride, Si_3N_4, and polymer layers. These materials can play a vital role in a metallisation structure, even though they have no active electronic properties, and it has proved necessary to develop well characterised, reproducible and reasonably inexpensive techniques by which they can be deposited as thin films. The parameters that must be controlled include layer thickness, stoichiometry, impurity content, pinhole density, adhesion, conformability, intrinsic stress and, of course, dielectric properties. The details of the deposition conditions often influence many or all of these properties very strongly and an enormous amount of research has gone into understanding how to control these processes to achieve the desired film properties. A major difference between the crystalline thin films discussed in earlier chapters and amorphous inorganic or polymeric films is that there is the possibility of predicting, at least crudely, the electronic properties of crystalline materials from observation of the defect concentration in the crystal lattice. Defects such as dislocations, grain boundaries and second-phase precipitates in metals and semiconductors influence the electrical properties in ways which are reasonably well understood (see Chapters 1, 2, 4 and 8). Determining the concentration of these lattice defects with techniques like electron microscopy or defect etching thus allows us to predict the electrical properties of a film or epitaxial layer. A high dislocation content is almost certainly an indication that a semiconductor material is not suitable for the fabrication of integrated devices, for example. Similarly, the resistivity of a metal layer will be increased if a second phase is present. It is not easy to make the same kind of prediction for amorphous layers, since modifications to the electronic properties are associated with changes in bonding configurations, which are themselves the result of changes in the local atomic chemistry. For instance, the introduction of alkali metal ions into glassy silica rapidly degrades the dielectric properties of the materials by breaking up the network structure. This kind of atomic defect is rather difficult to detect (compared with a dislocation or a grain boundary) and isolating the reason why a dielectric layer has been grown with unsatisfactory properties is often very hard. It would be fair to say that the well characterised techniques for growing dielectric films which are in use today have been developed as a result of many years of patient trial and error, rather than a complete understanding of the often very complex relationship between the deposition conditions and the electrical properties of these layers.

The insulating layer on which by far the most attention has been lavished in the field of microelectronic device fabrication is silicon dioxide, SiO_2. Similarly, the growth of this oxide by a thermal oxidation process is the best characterised of the popular methods of preparing an insulating layer. This process will thus be discussed first and in the greatest detail. In subsequent sections the deposition of other dielectric layers for applications in silicon- and gallium arsenide-based devices will be described, including the vapour phase deposition of silicon oxides and nitrides and the deposition of thin polymer layers.

6.2 THERMAL OXIDATION OF SILICON

It is fortunate for the microelectronics industry that the amorphous oxide of silicon is both easily produced, by straightforward furnace heating in an oxidising atmosphere, and exceptionally thermodynamically stable. An oxide layer prepared by thermal oxidation forms an excellent dielectric even when very thin, as is shown by the successful operation of MOS devices in which oxide gates are only about 30 nm thick. The schematic diagrams of simple devices in Chapters 2 and 5 show that SiO_2 is almost always the bottom layer of a metallisation structure on a silicon wafer. This thin layer is most conveniently grown by thermal oxidation. Thermally grown SiO_2 layers can also act as efficient diffusion barriers during dopant diffusion processes (§2.4.4).

Oxidation of a silicon wafer can be carried out either in a dry oxygen atmosphere or in one containing water vapour. The overall reaction during oxidation is different in these two ambients:

$$\text{dry oxidation:} \quad Si_{(s)} + O_{2(g)} \rightarrow SiO_{2(s)} \tag{6.1}$$

$$\text{wet oxidation:} \quad Si_{(s)} + 2H_2O_{(g)} \rightarrow SiO_{2(s)} + 2H_{2(g)}. \tag{6.2}$$

Because of the relatively open network structure of amorphous SiO_2, the volume change during these reactions is large, a 1 μm thick layer of oxide being formed for every 0.44 μm of silicon consumed. This is illustrated in figure 6.1. Careful radiotracer measurements have shown clearly that the rate-determining step in the oxidation process is the indiffusion of O_2 or H_2O molecules from the oxide surface to react with the silicon at the Si–SiO$_2$ interface. The volume increase is thus localised at this interface and can generate considerable interfacial stresses. When thick oxide layers are being grown, these stresses can even distort the shape of small device features, and we will see an example of this effect below.

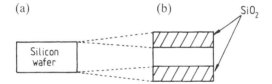

Figure 6.1 A schematic illustration of the fact that there is a large volume increase when a silica layer is formed by thermal oxidation of a silicon wafer surface. This has important implications for the structure of devices where the silicon wafer has to be selectively oxidised (see figure 2.8).

The normal temperature range used in the thermal oxidation of silicon lies between 850 and 1100 °C if a reasonable growth rate is to be achieved. This illustrates the point that the oxidation rate is very slow, about 60 nm h^{-1} at 1000 °C in dry oxygen for example, and so very accurate control of the thickness of the oxide layers is possible.

6.2.1 The Kinetics of Silicon Oxidation

A model of the oxidation rate of silicon has been constructed by Deal and Grove (1965) by considering the balance of fluxes across the gas–oxide and oxide–silicon interfaces, and diffusion of the oxidising species through the growing oxide layer. Let us be clear what these fluxes are in physical terms. At the gas–oxide interface it is the rate of dissolution of O_2 or H_2O in the oxide layer, while at the oxide–silicon interface it is the reaction of O_2 or H_2O with the silicon substrate that determines the 'flux'. The Deal–Grove solution to the mass balance equations gives the thickness of the oxide layer, x, at any time t:

$$x = \frac{A}{2}\left[\left(1 + \frac{t + \tau}{A^2/4B}\right)^{1/2} - 1\right] \tag{6.3}$$

where $A = 2D(1/k_s - 1/h)$ and $B = 2DC^*/N$. D is the diffusion coefficient of the oxidising species through the oxide, k_s the rate constant of the reaction to form the oxide at the interface, h the gas–oxide mass transfer coefficient, C^* the equilibrium bulk concentration of the oxidising species in the oxide, N the number of O_2 or H_2O molecules in unit volume of the oxide, and τ a time constant added to the equation to take into account the oxide that pre-exists on the silicon surface before oxidation commences, i.e. the native oxide.

The Deal–Grove equation predicts that there are two regimes of oxidation kinetics. For short oxidation times, $t + \tau < A^2/4B$,

$$x = \frac{B}{A}(t + \tau) \tag{6.4}$$

which is a linear grown law and normally considered to be associated with the reaction at the oxide–silicon interface. Assuming that the temperature dependence of the rate constant B/A can be described in terms of an Arrhenius equation, an activation energy for the interface reaction E_a, can be defined by the equation

$$\frac{B}{A} \propto \exp\left(\frac{-E_a}{kT}\right). \tag{6.5}$$

Deal and Grove measured E_a to be about 2 eV and linked this activation energy with the temperature dependence of the interface rate constant, k_s. It is interesting to remember that the energy required to break a Si–Si bond is 1.83 eV, as pointed out by Katz (1983). This bond breaking must be one of the fundamental processes in the oxidation reaction and may be the rate-determining step for oxidation.

For long oxidation times, $t > \tau$, equation (6.3) reduces to

$$x = (Bt)^{1/2} \tag{6.6}$$

a parabolic, or diffusion-controlled, rate law. The rate constant B contains the diffusion coefficient, D, and measurement of the temperature dependence of B yields an Arrhenius activation energy which it is reasonable to call E_d, the activation energy for diffusion of O_2 or H_2O through the oxide. This activation energy is about 0.7 eV for wet oxidation and 1.24 eV for dry oxidation. These figures are in comfortingly close agreement with those measured for diffusion of H_2O and O_2 through crystalline silica. The transition between the linear and parabolic rate laws for oxidation occurs at a thickness of about 100 nm. A schematic plot of the oxide thickness against time is shown in figure 6.2.

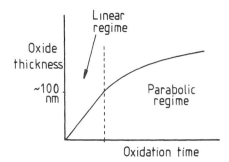

Figure 6.2 A schematic illustration of the variation in oxide thickness with time during the thermal oxidation of a silicon wafer, showing the linear growth kinetics when the oxide is thin, and the parabolic, or diffusion-controlled, growth of the thicker oxides.

Comparison of the predictions of the Deal–Grove model with direct measurements on oxidation kinetics have shown that there is, in general, remarkably good agreement. The linear and parabolic stages of oxidation are clearly observed—the reaction and diffusion-controlled stages of the oxide growth process respectively. This has allowed empirical values of the rate constants A and B to be catalogued for a whole range of oxidation temperatures, ambients and partial pressures of the oxidising species. This catalogue is a most valuable technological tool, permitting the accurate selection of particular values of the process parameters to give any required oxide thickness. Only in the value of τ is the Deal–Grove model inaccurate, predicting rather too low a value in dry oxidation conditions. It appears that the initial oxidation rate in dry oxygen is very much faster than is predicted by the model, some 25 nm of oxide being produced over and above that allowed for in the term τ. Such an effect could be a result of a small concentration of O_2^- ions increasing the initial rate of oxidation. It is also possible that the stress-free growth of an initial oxide layer is much faster than that which is possible once the interfacial stress is fully developed in a thicker oxide. This initial growth transient is still unexplained and a considerable amount of research is being directed at understanding the very early stages of silicon oxidation. Some of the most recent ideas on the mechanism of the initial oxidation process can be found in Stoneham *et al* (1987). It seems that the SiO_2 at the oxide–silicon interface is both chemically and structurally different from bulk SiO_2, and this material has been termed 'reactive SiO_2'. It is this reactive layer which controls the initial rate of oxidation. At present the effect of this initial rapid oxidation has to be added empirically to the Deal–Grove equation before accurate estimates can be made of the thickness of oxide layers produced by short anneals under dry oxidation conditions.

Finally in this section we must try and account for the well documented fact that wet oxidation is always faster than dry oxidation at the same temperature and in the same partial pressure of oxidising species. The rate constant B appears in both linear and parabolic rate laws (equations (6.4) and (6.6)) and this constant contains the parameter C^*. The equilibrium solubility of H_2O in SiO_2 is some three orders of magnitude higher than that of O_2 and so the value of B will be greater by that factor for wet oxidation when compared with dry oxidation. Deal and Grove have made a strong argument for this fact, explaining the observed rapid oxidation of silicon under wet conditions.

6.2.2 Further Effects on Oxidation Kinetics

The Deal–Grove equation has been shown to give a reasonably accurate framework from which to predict the oxidation rates of silicon, but the precise conditions of oxidation alter the observed values of the rate constants A and B very significantly. As a result of this, A and B have been measured as a function of changes in parameters like the plane of the silicon lattice exposed to the oxidising ambient, the impurity or dopant concentration in the silicon, and the composition and pressure of the oxidising gas. Table 6.1 shows how changing these parameters can affect the oxidation process. Increasing the pressure of the oxidising gas can result in some significant improvements in the oxidation process, and this effect is currently being exploited in device fabrication. High-pressure oxidation allows the oxidising temperature to be reduced, with attendant advantages in preserving carefully tailored dopant profiles, without undesirable increases in the time needed to produce an oxide of the required thickness.

6.2.3 Deleterious Effects of Thermal Oxidation

Although the thermal oxidation of silicon is a swift and efficient way of growing a dielectric layer, it does have some attendant problems which the device engineer must control. In particular, a high level of interfacial stress can be developed, which results in the creation of point defects and dislocations in the silicon substrate and possibly encourages dopant redistribution as well. As mentioned above, the large volume change on

Table 6.1 Effects of some further parameters on the oxidation rate of silicon.

Oxidation parameter	Effect on oxidation kinetics	Reference
Wafer orientation	The oxidation rate of {111} planes is greater than that of {100}; this is considered to be due to the lower concentration of Si bonds exposed on the {100} surface for attack by the oxidising species, i.e.'steric hindrance'	Ligenza (1961)
Dopants or impurities in the silicon (only effective at very high dopant concentrations)	Boron preferentially accumulates in the growing oxide and weakens the glass network encouraging further diffusion and an enhanced oxidation rate; phosphorus segregates to the oxide/silicon interface and appears to increase the rate of the interface reaction	Deal and Sklar (1965) Ho *et al* (1978)
Impurities in the oxidising gas	Very small concentrations of water vapour increase the 'dry' oxidation rate considerably; HCl additions to O_2 slightly increase the dry oxidation rate and improve the gettering of metallic impurities by removing them from the surface as volatile chlorides	Katz (1983)
Pressure of the oxidising gas	High-pressure oxidation can be carried out at much lower temperatures than atmospheric-pressure processes	Razouk *et al* (1981)

the oxidation of silicon inevitably means that stress is developed at the oxide–silicon interface, unless the temperature of oxidation is suffiently high to allow viscous flow of the oxide layer to relieve the stress. EerNisse (1979) has measured the intrinsic interfacial stress at temperatures between 850 and 1000 °C, showing that above 950 °C only very low levels of stress are developed because of this viscous flow process. Below this temperature, however, enormous compressive stresses were measured in the oxide layer, about 200 MPa at 900 °C. The critical shear stress of silicon at 800 °C is some two orders of magnitude below this figure, and so we should not be surprised that simple thermal oxidation of silicon can create considerable microstructural damage in the top layers of a silicon wafer. In addition, thermal expansion mismatches between the oxide and the silicon wafer can produce even higher compressive stress levels during cooling.

Two kinds of lattice defects are produced during the thermal oxidation of silicon: stacking faults and dislocation arrays. The stacking faults are thought to arise from the diffusion of silicon self-interstitials away from the oxide–silicon interface, where an excess of silicon atoms is inevitably produced by the interfacial reaction of silicon to form SiO_2. These interstitials agglomerate into long stacking faults at the high temperature of oxidation. Lower oxidation temperatures limit the diffusion range of the interstitials and can remove the stacking fault defects altogether. However, dislocation arrays are created near the top surface of the silicon to relieve the interfacial stresses generated during oxidation but are not observed in samples oxidised at temperatures above 950 °C. We can see that the conditions for the removal of these two different kinds of oxidation defects are not compatible and in practice some microstructural defects are always to be expected near the oxide–silicon interface. Some of the deleterious electrical effects observed when these defects are present in the active regions of silicon devices have been described in Chapter 2. The oxidation-induced defects are produced exactly where they are least wanted—in the top layer of the wafer where the devices are fabricated.

When oxidising doped silicon we must also consider what happens to the dopant atoms in the silicon as they come into contact with the incoming oxide–silicon interface. The macroscopic dopant distribution should be controlled by the equilibrium solubility of the dopant in the oxide and silicon, in a manner exactly analogous to solute redistribution during crystal growth from a liquid. The solute distribution around the moving oxide–silicon interface depends on the distribution coefficient, k_i (the ratio of the equilibrium dopant concentration in silicon and in the oxide) and the diffusion coefficients of the dopant in both oxide and silicon. Use of a modified version of equation (3.4) allows calculation of the dopant distribution, if k_i and these two diffusion coefficients are known.

Boron segregates strongly to the oxide; k_i is much less than 1 and this leads to boron distributions after oxidation like that shown schematically in figure 6.3(a). By contrast, the distribution coefficient for phosphorus is greater than 1 and so the phosphorus is rejected ahead of the growing oxide layer. This leads to a phosphorus distribution around the oxide–silicon interface like that shown in figure 6.3(b). Arsenic behaves in a manner similar to phosphorus, being rejected ahead of the growing oxide layer, i.e. the 'bow-wave effect'. In both these diagrams the dopant concentrations are not uniform in either the oxide or the silicon because the bulk diffusion rates of the dopants are relatively slow even at the temperature of oxidation. The phenomenon of dopant redistribution is thus an example of incomplete mixing, as defined in §3.2, and this explains why equation (3.4) is chosen above for k_1 and not equation

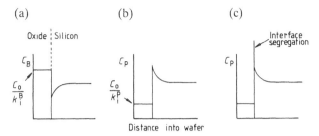

Figure 6.3 Schematic diagrams showing a hypothetical distribution of dopant elements at the oxide/silicon interface during thermal oxidation when: (a) the dopant segregates preferentially to the oxide; (b) the dopant segregates to the silicon; (c) both segregation to the oxide/silicon interface and to the silicon occur.

(3.3). The oxide–silicon interface is also a possible site for the equilibrium segregation of dopant atoms. This kind of segregation will produce a dopant profile at the interface somewhat similar to the bow-wave effect. However, the bow wave is a result of the dopant species having a lower chemical potential in the silicon, while interfacial segregation is due to the chemical potential of the dopant atoms being minimised at the oxide–silicon interface itself. Figure 6.4 shows an example of the direct analysis of the interfacial segregation of arsenic. The very narrow arsenic peak, and its roughly symmetrical shape, are characteristic of interface segregation rather than a dopant redistribution phenomenon during oxidation. It is quite possible for both solute redistribution around the oxide–silicon interface and interfacial segregation to occur during the same oxidation process, and a composition profile of the kind shown schematically in figure 6.3(c) would result. The practical significance of this kind of impurity redistribution during oxidation is that it

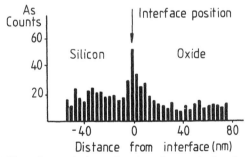

Figure 6.4 Experimental data showing the analysis of a significant level of arsenic segregation at an oxide/silicon interface (after Wong *et al* 1985). The very narrow interface peak indicates that this is a case of equilibrium segregation and not the bow-wave effect (see figure 6.3(c)).

can disturb carefully designed dopant profiles which form the active regions of the devices in the silicon wafer.

In this section we have seen that the thermal oxidation of silicon is a convenient way of producing a passivating or insulating layer on the surface of a silicon wafer. The growth process is well characterised, but some damage to the crystal structure of the silicon and to the dopant profiles within the wafer can be caused during oxidation. However, if the oxidation process is carefully controlled, the electrical properties of the oxide–silicon interfaces can be excellent, and this kind of technique must be regarded as an industry standard for the preparation of a dielectric layer directly on a silicon surface.

6.2.4 Electrical Properties of the Thermal Oxide/Silicon Interfaces

It is worth at this point considering in a little more detail what we mean when we define an oxide/silicon interface as having excellent properties. Here we are concerned specifically with the properties of the oxide interfaces which are used as gate contacts in devices like the field effect transistor structure illustrated in figure 5.32. It is particularly important that the Fermi level at the oxide/silicon interface should be free to move when a small potential is applied to the oxide film. In order for this to be achieved the total amount of charge trapped in the oxide layer and at the oxide–silicon interface must be as low as possible. This implies that the density of carrier trapping states at the interface must be minimised and that care must be taken to avoid the introduction of charged impurities into the oxide. These charged interface states and oxide defects have been classified by Deal (1980) and are identified in Figure 6.5.

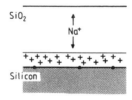

• Interface states
+ Fixed oxide charges

Figure 6.5 A schematic illustration of the nature of charged defects near the oxide/silicon interface. Interface states may be associated with silicon dangling bonds, and the fixed oxide charges with broken bonds in the oxide near the interface. The mobile ions are usually alkali metals like sodium and potassium.

Electronic states lying in the silicon band gap are created at the oxide–silicon interface either by the segregation of heavy metal impurities to the interface or by structural defects like dangling bonds on silicon atoms. Oxidation procedures have been designed which result in the preparation of oxide layers in which the interface state densities are as low as $10^9 \, cm^{-2} \, eV^{-1}$ in the silicon band gap and with a minimum in the density of states lying roughly at the centre of the band gap. Annealing oxidised wafers in hydrogen at low temperatures, 400–500 °C, can also lower the interfacial state densities, presumably by saturating all the dangling bonds with hydrogen.

A significant density of charged defects is also found in the oxide layer close to the SiO_2/Si interface. These charges are usually positive and immobile and are called fixed oxide charges. It is known that the stoichiometry of the oxide layer very close to the oxide–silicon interface differs significantly from SiO_2 (and in part this is thought to explain the 'reactive layer' behaviour described above) and the fixed oxide charges may be created by broken bonds on silicon or oxygen atoms in the silica network. The density of these charges is usually in the range 10^{10}–$10^{12} \, cm^{-2}$.

Mobile charges exist in the oxide layer as well and are normally attributed to the presence of sodium, potassium or lithium ions as contaminants in the oxide network, although some heavy metal impurities may also contribute to the mobile oxide charges. Sodium is introduced as a result of many of the washing and etching processes described in Chapter 2. Alkali metal ions are able to diffuse extremely rapidly through the oxide when a potential is applied and can cause serious instability in MOS devices. There are two ways in which these ionic species can be kept away from the oxide/silicon interface: by including a thin layer of silicon nitride in the gate insulator structure; or by adding a phosphosilicate glass layer over the top of the gate oxide. The diffusion of alkali metals in Si_3N_4 is extremely slow so that the ions cannot penetrate to the oxide/silicon interface, while the glass layer acts as a preferential segregation site for the alkali metal ions.

A discussion of the effect of these oxide charges and interfacial trapping states on the performance of MOS devices has been given by Wood (1981) and in Chapter 7 of Sze (1981). Sze also includes a description of some of the electrical techniques which have been used to measure the densities of these various defects. It has been suggested that the involuntary inclusion of sodium ions in the oxide is the most common cause of failure in some MOS devices (Edwards et al 1979). An excellent review on all aspects of the structure and properties of SiO_2/silicon interfaces has recently been prepared by Schwarz and Schulz (1985).

6.3 DIELECTRIC LAYERS DEPOSITED BY VAPOUR PHASE TECHNIQUES

Figure 5.31 clearly shows that dielectric layers are required at several places in a multilayer metallisation system on a silicon wafer other than directly on the surface of the silicon wafer. These insulating and protective layers cannot, of course, be produced by thermal oxidation, but are generally deposited from the vapour phase. In addition, the oxides formed by annealing most compound semiconductors in oxidising environments have little mechanical or chemical stability, as we will see in §6.5. Effective passivation or dielectric masking of these materials cannot be accomplished by a simple thermal oxidation process, and vapour phase deposition of dielectric layers can sometimes be used as a substitute, although the electrical properties of the interfaces are often poor. A variety of vapour and electrolytic deposition techniques have been used to grow insulating films on gallium arsenide devices.

Sputtering, reactive sputtering, evaporation and anodising are relatively little used in silicon integrated circuit technology because film composition, thickness uniformity and conformability are often hard to control. (Anodising can be a convenient way of growing oxide layers on III–V semiconductors like GaAs; see §6.5) The most popular, and well characterised, technique for the vapour phase deposition of thin dielectric films is chemical vapour deposition, CVD. This technique has been introduced in Chapters 3 and 4, and dielectric films with a wide range of compositions and properties can be deposited in relatively simple CVD reactions.

Two quite distinct kinds of dielectric layers are required in a multilayer metallisation system. A thick oxide layer is often deposited over the whole top surface of the metallisation to protect the aluminium alloy conduction tracks from mechanical damage and atmospheric corrosion. This kind of protective layer is called a passivation oxide and should impede the indiffusion of water, metallic ions and alkali metals. At lower levels in the composite structure, thin insulating layers are needed that conform well with the rather corrugated surface of the exposed conductor layer. The device engineer must thus be able to deposit thin even layers of dielectric material with high conformability. Another important characteristic of the 'perfect' interlayer dielectric is the ability to smooth out rough etched surfaces to make the deposition of continuous conductor tracks easier. It is sometimes necessary to use a third kind of dielectric layer as an oxidation mask in the early stages of device processing. Silicon nitride is an especially suitable material for this kind of application because of its very slow oxidation rate. All these dielectric layers are required to adhere to oxide and metallic surfaces, which can be a problem when gold conductors are used in the metallisation.

While dielectric layers for oxidation masking can be deposited at high temperatures, those for VLSI insulation and passivation applications can only be deposited at temperatures well below that at which serious breakdown of the barrier layers occurs in the underlying metallisation structure (see §5.7 and table 5.5). This puts a very severe restriction on the temperature at which the substrate can be held during vapour phase deposition. This is particularly important for CVD processes in which the chemical reaction is stimulated by the elevated substrate temperature. Considerable effort has therefore been put into discovering effective low-temperature processes for the deposition of silicon dioxide and silicon nitride layers. Should the use of refractory metal conduction tracks become common (see §5.9), the necessity for keeping the substrate temperature low during dielectric deposition would be removed, which would be a very significant processing advantage in integrated circuit manufacture.

The level of intrinsic stress developed in the dielectric films must also be limited as much as possible, since excess stress can damage metal conductors and lead to cracking of the insulating layers. High stress levels in dielectric layers may also alter the performance of some GaAs devices, because GaAs is a piezoelectric material. If high stress levels are unavoidable in the dielectric films, a compressive stress will probably cause less damage to underlying metallisation than a tensile stress.

A comparison of some of the relevant properties of a number of dielectric materials which can be deposited as thin insulating films for use in microelectronic devices is given in table 6.3.

6.3.1 CVD of Dielectric Layers

The basic principles of CVD growth were introduced in §3.7. There we saw that it was possible to divide CVD processes very roughly into two groups: (i) those in which the total gas pressure was high (around atmospheric pressure, APCVD) and the rate of transport of reactant species to the substrate controlled the structure and composition of the films; and (ii) low-pressure reactions in which the thermodynamics of the chemical reactions become more important, LPCVD. APCVD processes can suffer from problems of uneven coverage in different parts of the reactor, because the growth rate is so dependent on the local gas flow conditions. LPCVD processes give rather slower growth rates (1–100 nm min^{-1}) but a more even coverage in general. Kern and Rosler (1977) have described how the variables in a CVD reaction—substrate temperature, gas pressure and concentration of precursor species—can influence the rate of growth of a dielectric film.

6.3.2 Deposition of SiO$_2$

A few of the overall reactions which can be used to deposit silicon oxide layers are given below:

$$SiH_4 + O_2 \rightarrow SiO_2 + 2H_2 \quad 400\,°C \qquad (6.7a)$$

$$SiH_4 + 4NO \rightarrow SiO_2 + 2H_2O + 2N_2 \quad 700\,°C \qquad (6.7b)$$

$$Si(OC_2H_5)_4 \rightarrow SiO_2 + \text{complex organic species} \quad 700\,°C \qquad (6.7c)$$

$$SiCl_2H_2 + 2N_2O \rightarrow SiO_2 + 2N_2 + 2HCl \quad 900\,°C. \qquad (6.7d)$$

In most cases the reactant gases are extremely toxic and are often flammable as well. The care that must be taken to contain and control these chemicals when operating a CVD facility is one of the major disadvantages of the technique. It is apparent from these examples that most of the available CVD reactions require a much higher substrate temperature than can be withstood by the aluminium alloy metallisation systems. Only reaction (6.7a) is suitable for the deposition of the dielectric layers in the conventional aluminium-based integrated circuit metallisation structures discussed in the previous chapter. However, MOS devices with polysilicon (or polycide) metallisation (figure 5.29), or metallisation structures with refractory metal conduction tracks, can stand much higher processing temperatures and in these circumstances the other reactions may be suitable for depositing the dielectric layers.

It is important to understand why the reaction (6.7a) is not used for the deposition of all SiO$_2$ interlayer dielectrics. The conformal coverage of stepped surfaces is a vital feature of successful dielectric growth and this CVD reaction does not give very good coverage. This is not surprising when we recall from §3.7 that it is a homogeneous reaction and so takes place primarily in the vapour phase above the substrate. Heterogeneous reactions will, in general, give much better conformal coverage than homogeneous reactions. For instance, Wilkes (1984) has demonstrated that the silane/nitric oxide reaction (6.7b) is heterogeneous and that it produces a deposit with much better coverage of surface steps than the homogeneous reaction (6.7a). Reactions (6.7b, c and d) all give especially good conformal coverage and so are often preferred when high-temperature processing can be tolerated.

Some of the properties and areas of application of SiO$_2$ films grown by CVD reactions are given in table 6.2. This table gives some typical values for important properties of these films, but we should remember that these values can be dramatically changed by alterations in the precise conditions in the CVD reactor. The major difference between CVD oxide layers and those grown by thermal oxidation is in the concentration of impurities like hydrogen and chlorine which are incorporated into the oxide films during the CVD growth processes. This change in the composition of the CVD films does not seem to degrade their dielectric properties significantly. CVD oxides can be very satisfactory dielectric layers for a whole variety of integrated circuit applications. LPCVD growth is often used for the deposition of interlayer dielectrics where conformal growth is more important, while APCVD is more commonly applied to growing thicker passivation layers. It is clearly important to select the most appropriate chemical reaction and gas pressure for the particular application required of the dielectric layer.

Table 6.2 Process parameters, properties and applications of silicon oxide layers.

Reaction	$SiH_4 + O_2$ CVD	Plasma CVD	$SiCl_2H_2 + N_2O$ CVD	Thermal oxidation
Temperature (°C)	450	100–400	900	800–1100
Reaction product	SiO_2 (+H)	SiO_x (+H) $(x < 2)$	SiO_2 (+Cl)	SiO_2
Conformal quality	Poor	Good	Good	Good
Resistivity (Ω cm)	10^{16}	10^{16}	10^{16}	10^{16}
Intrinsic stress	Tensile	Compressive	Compressive	Compressive at low temperature
	Stress levels are controlled by the details of the deposition process, but are usually above 100 MPa			
Dielectric strength (V cm⁻¹)	8×10^6	5×10^6	10^7	10^7
Dielectric constant	3.8–4.2	4.6	4.2	3.9
Density (gm cm⁻³)	2.1	2.3	2.2	2.2
Applications	Passivation layers on top surface of metallisation, often containing P or B; deposited at atmospheric pressure		Highly conformal interlayer dielectrics; deposited at low pressure	Gate oxides and field oxide layers on silicon wafers

One of the most attractive features of CVD layer growth is that the composition of the films can be altered simply by changing the composition of the reacting gas mixture. Phosphorus-doped silica glass layers can be deposited by adding PH_3 to the SiH_4 and O_2 in reaction (6.7a), and silicon oxynitride layers by reacting SiH_4, N_2O and NH_3 for instance. Phosphosilicate glass layers are useful passivation layer materials, as indicated in table 6.3, and oxynitride layers with low intrinsic stress levels can be used to replace simple nitride layers.

Phosphorus-doped glass layers, with phosphorus concentrations in the range 2–8 at.%, can be deposited at low temperatures. The value of such a material is that it will flow easily at temperatures around 1100 °C if the phosphorus content is greater than 6 at.%. An initial deposit that is only poorly conformal can thus be induced to flow over the substrate to give a smoother and much more conformal layer (Adams and Capio 1981). These doped films have also been shown by Kern and Rosler (1977) to have a lower intrinsic stress than pure SiO_2 layers and so can be grown as thick protective films with less risk of cracking. These materials are often referred to as PSG (phosphosilicate glasses). Sodium ions, which can cause problems by migrating under field stress in oxide layers, are effectively trapped in PSG and it is a distinct advantage if this damaging element can be kept away from interlayer dielectrics and gate oxides at the bottom of a metallisation structure. The problem of the production of corrosive HPO_3 by reaction of these glasses with water vapour has been described in §5.5. Glass layers containing both phosphorus and boron are also sometimes used, as surface passivation and protective layers. The reaction of SiH_4, B_2H_6 and PH_3 with oxygen can be used to deposit a borophosphosilicate glass, often called BPSG.

6.3.3 Deposition of Silicon Nitride Layers

Silicon nitride is a very efficient diffusion barrier for H_2O and sodium ions, much better than silica for example, and also oxidises extremely slowly. It is thus a versatile material for atmospheric protection and oxide masking applications. Nitride layers are also used as protective 'caps' to stop the decomposition of III–V surfaces during the heat treatment required to remove implantation damage (see §2.4.4). A typical overall LPCVD reaction for the deposition of Si_3N_4 is

$$3SiH_4 + 4NH_3 \rightarrow Si_3N_4 + 12H_2 \quad 700 \,°C. \qquad (6.8)$$

A reaction of this kind can be used to deposit a stoichiometric Si_3N_4 layer containing about 8 at.% of hydrogen. These layers are very resistant to dissolution in hydrofluoric acid solutions and have excellent dielectric properties. The only problem is that the films often contain enormous intrinsic tensile stresses of the order of 10^3 MPa. This extremely high stress can result in cracking of the film or damage to the semiconductor substrate by plastic deformation. Dislocation generation at the edge of Si_3N_4 windows is commonly observed during oxidation. It is also clear that the relatively high temperature of deposition, 700 °C, makes this process unsuitable for placing protective layers over aluminium alloy metallisation schemes. However, the reaction is very effective for the production of oxidation masking layers, especially if a thin layer of SiO_2 is deposited between the substrate and the nitride layer to absorb some of the excessive tensile stress. Silicon nitride layers can also be deposited by sputtering, but are usually found to be non-stoichiometric SiN_x with rather low breakdown voltages ($10^7 \, V \, cm^{-1}$). However, the stress levels in sputtered films can be extremely low.

6.3.4 Plasma-enhanced CVD (PECVD)

We have seen that high substrate temperatures are often required for the efficient deposition of oxide and nitride layers and have explained why this makes these processes unsuitable for the deposition of inter-layer dielectrics in an aluminium-based metallisation scheme. However, if a glow discharge is struck in the gas phase above the substrate, then these same chemical reactions will proceed at very much lower substrate temperatures. In effect, the plasma excites the electrons in the reactant gas species to very high temperatures (about $10^5 \, K$), which encourages the chemical reactions to proceed without any additional thermal energy being required from the substrate. The use of this kind of CVD process would seem to be a way of overcoming the problems associated with excessive substrate temperatures. Both silicon oxide and nitride films can be deposited by this method with substrate temperatures as low as $200-300 \, °C$. This is well below the melting point of aluminium alloys and in the temperature range where many barrier layers retain their integrity for long periods (see table 5.5). Typical PECVD reactions are of the kind

$$SiH_4 + 4N_2O \rightarrow SiO_x + N_2 + H_2O \qquad (6.9a)$$

$$SiH_4 + NH_3 \rightarrow Si_xN_yH_z + H_2. \qquad (6.9b)$$

The dielectric layers deposited in these processes have been shown to be non-stoichiometric mixtures of silicon and oxygen, or of silicon, hydrogen and nitrogen. Just as in conventional thermally assisted CVD processes, the rate of deposition and the stoichiometry of PECVD films depends heavily on all the process parameters. The frequency of the plasma field, the plasma power and the choice of electrode materials are especially important, as has been discussed by Rand (1979). The most striking novel feature of PECVD is the huge range of stoichiometries with which the films can be prepared by changing these deposition parameters. Silicon nitride films can be readily deposited with silicon-to-nitrogen ratios between 0.7 and 1.7 and with as much as 25 at.% hydrogen incorporated. Silicon oxide layers are usually of lower oxygen content than LPCVD films and can contain as much as 9 at.% of hydrogen. Stoichiometry variations as wide as this might be expected to alter the dielectric properties of the films quite significantly. It is certainly true that both the dielectric strength and the resistivity of $Si_xN_yH_z$ films are lower than for stoichiometric Si_3N_4 films deposited by conventional CVD processes. However, the intrinsic stress in PECVD $Si_xN_yH_z$ films is two or three times lower than that in the more conventionally deposited films. In fact, with very careful choice of deposition conditions it is possible to produce PECVD films of both nitrides and oxides with dielectric properties that closely resemble those of the best LPCVD films. The low temperature of deposition, and the possibility of achieving layers with good conformability with the substrate, make PECVD a valuable growth technique. Because of these advantages, plasma-assisted techniques may in the future become the dominant way of producing dielectric thin films.

6.4 THE GROWTH OF VERY THIN OXIDE LAYERS

In the fabrication of some MOS devices it is necessary to grow uniform oxide layers as thin as 30 nm on the wafer surface. In order to prepare these thin layers reproducibly, very slow oxidation processes have been developed which allow precise control to be achieved over the oxide thickness. We should remember that this kind of oxide is only about 100 atoms thick. In such very thin layers the native oxide layer that is always present on a silicon surface will provide a significant fraction of the required total thickness. In fact, the layer of native oxide on silicon surfaces, which can be as thick as 2.5 nm, may be sufficiently stable and have satisfactory dielectric properties without any further oxidation.

Irene (1978) has shown that low-temperature dry thermal oxidation can produce the required thin oxides on silicon with great reproducibility. Growth rates in thermal oxidation can be lowered by reducing the partial pressure of the oxidising species. A similar effect is observed in LPCVD growth, and can be used to grow thin oxide layers on wafers of silicon and III–V materials. The dielectric properties of thin films grown by these two techniques are often excellent. A third method for preparing even thinner oxide layers is chemical oxidation at temperatures below 100 °C. Films with thicknesses of less than 2 nm, and with remarkable uniformity, have been grown in these chemical solutions. An electron micrograph of such a layer is shown in figure 6.6 and gives a very clear indication of the uniformity of these extremely thin films.

The composition of these films may not be very close to the 'perfect' SiO_2 we would expect in a thicker thermal oxide. In fact, it is very difficult to determine accurately the composition of such a layer because it contains so few atoms. There are indications that SiO may be a more accurate stoichiometry for at least part of the layer, and an example of the variation in oxygen-to-silicon ratio in a very thin chemical oxide is

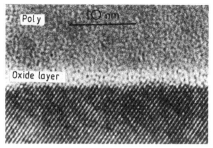

Figure 6.6 A cross-sectional transmission electron micrograph showing the extreme uniformity that can be achieved in very thin oxide layers. In this case an oxide layer about 1.4 nm thick has been chemically grown on a silicon wafer and then capped by a polysilicon layer. The stability of the oxide layer is demonstrated by the fact that it has survived firing at 900 °C. (Courtesy of Dr N Jorgenson.)

given in figure 6.7. However, the most important feature of the layers as far as their use in microelectronic devices is concerned is that they should have the correct electrical properties. It has been shown that layers deposited by the growth techniques outlined above can be successfully used as gate oxides in highly efficient MOS devices. When the stability of these very thin oxide layers is not sufficiently high for device fabrication, multiple-layered structures can be grown by very accurately controlled CVD techniques. Figure 6.8 shows an electron micrograph of a three-layer dielectric structure, $SiO_2/Si_3N_4/SiO_2$, where all the layers are less than 10 nm thick.

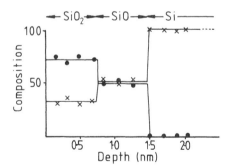

Figure 6.7 A composition profile through a 1.5 nm thick anodic oxide grown on a (100) silicon surface. This profile has been obtained with the pulsed laser atom probe technique (see Cerezo *et al* 1986) and indicates that the oxide layer consists of an outer region of stoichiometry SiO_2, with an inner layer of SiO.

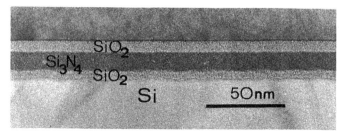

Figure 6.8 A transmission electron micrograph showing a trilayered $SiO_2/Si_3N_4/SiO_2$ structure where the individual layers are only a few nanometers thick. (Courtesy of Drs H Cerva and H Oppolzer, and Siemens Ltd.)

6.5 DIELECTRIC LAYERS ON GALLIUM ARSENIDE

We have seen that oxide layers can be grown directly on silicon wafers by simple thermal oxidation, and it would be convenient if oxides with similarly good dielectric properties (and low interface state densities) could be grown on gallium arsenide as well. However, the thermal oxidation of compound semiconductors is a much more complex phenomenon than the equivalent process on silicon and even the stoichiometry of the bulk thermal oxide is not completely determined. The equilibrium stoichiometry of a thermal oxide grown on a GaAs crystal might be expected to be a mixture of equal amounts of Ga_2O_3 and As_2O_3, but it is usually found that the bulk of the oxide is gallium rich with an excess of elemental arsenic at the oxide–semiconductor interface. The remainder of the arsenic is lost from the surface of the oxide during growth. As the oxidation temperature increases the gallium oxide crystallises to form β-Ga_2O_3. The thermochemistry of the Ga–As–O ternary system at typical oxidation temperatures between 300 and 600 °C has been studied in some detail by Thurmond *et al* (1980).

The electrical properties of these non-stoichiometric thermal oxides are generally rather poor, with a resistivity of about 10^{14} Ω cm, a breakdown field of only 5×10^5 V cm^{-1} and, even more seriously, a very high density of surface states, often in excess of 10^{12} cm^{-2}. This is not surprising when we recall that the GaAs (100) and (111) surfaces have a high density of intrinsic surface states and that the exposure of these surfaces to oxygen results in even more effective pinning of the Fermi level in the centre of the band gap (see for instance Spicer *et al* (1980)). Some improvement in these properties can be achieved by using As_2O_3 as the oxidising species, which results in the inclusion of a higher concentration of arsenic in the thermal oxide films. Breakdown fields in excess of 5×10^6 V cm^{-1} have been reported for these oxides and interface state densities as low as 5×10^{11} cm^{-2} (see Takagi *et al* 1979). Oxides of this quality are almost suitable for use as gate contacts on GaAs field effect transistors.

Oxide layers can also be grown on GaAs by anodic treatments and water is often used as the oxidising agent. The rate of anodic oxidation can be very slow and well controlled, allowing the deposition of adherent and continuous oxide layers. The anodising solutions are often complex mixtures, and some of the most commonly used compositions can be found listed in Croydon and Parker (1981). Anodic oxide layers have compositions rather similar to those of thermal oxides and so are usually deficient in arsenic. Once again, an excess of elemental arsenic is frequently detected close to the oxide–GaAs interface. Annealing treatments can improve the electrical properties of the oxide–GaAs interface, but will also lead to the further loss of arsenic from the bulk of the oxide and the crystallisation of β-Ga$_2$O$_3$. Oxide layers with very similar compositions can be grown by exposing the surface of a GaAs wafer to an oxygen plasma. The electrical properties of these anodic and plasma-grown oxides are generally slightly better than those characteristic of thermal oxides. Breakdown fields of 5×10^6 V cm^{-1} and interface state densities around 10^{12} cm^{-2} are quite commonly achieved. It is the high interface state densities which are the most serious obstacle in using these oxides as active components in MOS devices.

An alternative way of preparing oxide–GaAs interfaces is to evaporate thin metal layers onto the GaAs surface and to oxidise these layers to create a stable oxide layer. Aluminium is often chosen as the metal overlayer because of the ease with which continuous alumina films can be formed by oxidation. Vapour phase deposition techniques, and in particular CVD reactions, can be used to lay down layers of insulating materials like SiO$_2$, Si$_3$N$_4$, SiO$_x$N$_y$, AlN or Al$_2$O$_3$. The dielectric properties of these materials are usually better than those of the GaAs native oxides and the interface state densities can be reduced well below 10^{12} cm^{-2} as well. It is thus not surprising that many GaAs devices are fabricated with these 'foreign' dielectric materials. It has even proved possible to use polyimide, a polymer which is described in the next section, as an effective dielectric material. For a detailed discussion of the applications and relative importance of these various dielectric materials in GaAs devices, the reader is directed to Croydon and Parker (1981) and Howes and Morgan (1985).

6.6 THE DEPOSITION AND PROPERTIES OF POLYMER LAYERS

One of the major advantages of polymeric films for use as insulating, passivating or lithographic layers on microelectronic circuits is that they can be deposited by exceedingly simple techniques. Thermal oxidation and CVD technology is well proven for the deposition of oxide and nitride layers, but high-temperature furnaces and vacuum equipment which can handle corrosive and flammable gases are rather expensive pieces of equipment. By contrast, polymers can be 'spun' onto surfaces to give very good conformal films using almost no equipment at all. This very simple way of depositing a thin film can be described as follows:

(1) A filtered and carefully purified drop of the polymer precursor is dropped onto the substrate, which is then spun at high speed to give a thin even liquid layer. The thickness of the layer depends on the rate of spin and the solids content of the particular precursor mix.

(2) A low-temperature bake is then usually sufficient to remove the solvent from the film. Polymer films for lithography are heated to about 90 °C, a process often referred to as the 'softbake'. This softbake treatment is a very important stage in the lithography process, since the chemical reactions stimulated in the polymer by the exposing radiation will only proceed correctly once most of the solvent has been removed. Polymer dielectric films require higher curing temperatures to achieve the best properties, usually above 250 °C.

Let us now look at the composition of the polymers which have found most application as dielectric layers in metallisation systems. Lithographic polymers will be described in §6.6.2.

6.6.1 Polymers for Insulating and Protective Applications

The polymers which have been most widely investigated for applications as interlayer dielectrics and passivating layers are the class of materials called polyimides. These are formed from precursor solutions of polyamic acids whose general formula is

Heating these precursor acids causes removal of two water molecules per formula unit to give a polymer of the type

The bake temperature required for full imidisation is about 400 °C for most polyimides, very much lower than is required for either thermal oxidation or most CVD growth processes and also well below the melting temperature of aluminium alloys. Polyimides with a wide variety of R groups have been prepared to try and lower the imidisation temperature as much as possible and to improve their resistance to dissolution in solvents and their adhesion to wafer surfaces. This has led to the development of a number of commercially available polyimide precursor acids from which layers with well-defined properties can be deposited.

Once the polyimide has been baked, the films have dielectric properties very similar to those of SiO_2, particularly in their low dielectric constant, 3.5, and high breakdown voltage. In addition, there is almost no intrinsic stress in a polymer dielectric film, in contrast with the rather high stresses generated in both thermally oxidised and CVD oxide films. Thermal expansion mismatch stresses are generated during cooling from the curing temperatures, but these are never more than about 70 MPa (Goldsmith *et al* 1983), lower than the stress levels in thermal or CVD SiO_2 films given in table 6.2. Polyimides are usually rather resistant to chemical dissolution as well. A final pair of advantages of polyimide films is that they can easily be lithographically patterned using plasma etching techniques like those outlined in §2.4 and, in common with all polymer films, are highly conformal with the substrate.

It is thus very easy to grow polyimide layers with properties similar to those that a device engineer requires of the silicon oxide or nitride layers currently used in metallisation structures. In some ways polymers may even have definite advantages, including low-temperature processing, cheapness and low stress levels. It seems, however, that there must always be disadvantages associated with any material, and polyimide is no exception. The purity of most precursor materials is not as good as is required in direct proximity to semiconductor integrated circuits and the diffusion of contaminants through the fully cured layers is considered unacceptably fast. The diffusion rate of water molecules through cured polyimide is rather rapid, for instance, and this can lead to moisture reaching the metallic conductors and encouraging blistering and corrosion. (We should remember that the diffusion of most contaminant species through silicon oxide and nitride layers is very slow indeed.) In addition, the adhesion of polyimide to some metal surfaces can be poor.

Even with these problems, the increasing use of polyimides, or similar polymeric materials, as interlayer dielectrics and as protective capping layers in integrated circuit technology seems very likely because of the major economic and technological advantages we have mentioned above. Polyimide is particularly widely used as an interlayer dielectric in metallisation to GaAs devices. Table 6.3 gives a general comparison of the strengths and weaknesses of the materials we have discussed above for applications as dielectric layers in microelectronic devices.

Table 6.3 Dielectric thin film materials: properties and applications.

	Advantages	Disadvantages	Applications
SiO$_2$	Low dielectric constant LPCVD layers are highly conformal	High intrinsic stress levels	Interlayer dielectric Thermal gate oxide
Si$_3$N$_4$	Low diffusion rates for contaminants Low oxidation rates	High dielectric constant (6–7) High intrinsic stress	Capacitor dielectric Annealing cap Diffusion barrier Oxidation mask
AlN	Low dielectric constant Thermal expansion coefficient same as GaAs	Deposition process not well developed	Annealing cap
Polyimide	Cheap Conformal Low dielectric constant	Stable only below 400 °C Rapid indiffusion of water	Interlayer dielectric (especially on GaAs devices) Protective coatings on device surfaces
PSG	Flows at high temperatures Traps sodium ions	Corrosion problems due to HPO$_3$	Passivation layers

6.6.2 Lithographic Polymer Materials

Although the polymer materials used in lithography are not components in completed semiconductor devices, they play a vitally important role in the fabrication of these devices. The ease with which conformal resist layers can be deposited onto uneven substrates is just as important in lithography as it is when a material like polyimide is being considered as an interlayer dielectric. However, for lithographic applications the optical and chemical properties of the polymers are more important than the dielectric properties which were of central concern in the previous section. In Chapter 2 the fundamental process of transferring device features onto semiconductor wafers by lithography was described. In this section some of the materials which are used as resists will be introduced.

The principle that underlies all current lithographic processes is that radiation incident on polymeric materials can cause dramatic changes in the molecular structure. The particular importance of polymer materials for this application is that they are very sensitive to many kinds of radiation, more so than most inorganic materials for instance. The incident radiation can be light, electrons or x-rays, and the lithographic processes based on each of these incident radiations have been described in Chapter 2. Here we shall consider the effect that the radiation has on the polymers.

The negative resist shown in figure 2.13 clearly becomes less soluble in developing solutions after exposure (or more resistant to dry etching). The basic process underlying this increase in chemical stability is radiation-induced cross linking of the polymer chains. This cross linking increases the molecular weight (MW) of the polymer in the exposed regions. Similarly, positive resist materials undergo chain scission when irradiated. This scission normally occurs at particular molecular sites on the chains and the detailed structure of the macromolecules determines the rate of decrease in the molecular weight. The dissolution rate of a polymer in a developing solution is a complex function of the particular chemical reactions that occur, but in many cases is roughly proportional to $MW^{-0.7}$. The ratio of the dissolution rates of exposed and unexposed areas of polymer can be as high as 100, which is quite sufficient for effective lithographic patterning. Dissolution of negative resists is necessarily rather slower than for positive ones because of their higher molecular weight, and this can be a disadvantage in processing. Many other factors also influence the dissolution rates strongly, such as the tactic form of the polymer chains (Ouano 1984). In some cases scission can occur to such an extent in positive resists that the material completely depolymerises and spontaneously evaporates from the exposed regions, thus removing the need for a developing process.

The radiation dose needed to stimulate these cross-linking or bond scission reactions varies widely with the composition and structure of the resist and the type and wavelength of the incident radiation. A resist which is sufficiently sensitive to be practically useful when exposed to electrons may be very insensitive to x-rays. (Sensitivity is here defined as the rate at which the MW changes with radiation dose.) It is obviously convenient if the resist is as sensitive as possible to the exposing radiation, or the time taken for each exposing step will be unsatisfactorily long. In addition, the resists are often required to protect the substrates from the chemical or dry etching techniques used to remove materials in the patterned windows (see §2.4). It is no good producing a resist, however sensitive, for this kind of lithographic process if it is rapidly etched away. This dual requirement on a resist is often difficult to satisfy. For instance a positive resist has a high sensitivity only when it rapidly undergoes scission when exposed, but this frequently means that it can be rapidly removed in a dry etching process as well. (Resists used for lift-off metal patterning processes do not of course need to be etch resistant.)

In order to improve performance, resists are often produced by mixing a number of polymer materials which have different properties. A polymer with a good etch resistance will be used as the bulk of the resist layer, with the addition of a sensitiser which absorbs the incident radiation efficiently and stimulates cross linking or scission in the matrix. Synthetic rubber compounds are commonly used as the etch-resistant matrix in negative optical resists. The sensitiser polymer can act in a number of different ways, including: (i) absorption of the energy of the incident radiation and the transferral of this energy to the breaking of bonds in the matrix polymer; and (ii) decomposition when irradiated to produce chemicals which stimulate cross linking or depolymerisation. The first of these sensitiser effects is nicely illustrated by the following example taken from the work of Taylor (1980). The sensitivity of the common positive electron resist polymethylmethacrylate, PMMA, when used as an x-ray resist is very low. (PMMA is a thermally degrading polymer and so its use in lithographic processes relies on the heating effect produced by the exposing radiation.) The sensitivity can be improved by incorporation of chlorine atoms into the PMMA, since these atoms have an absorption edge just below the wavelength of Pd Kα x-rays. The chain scission reaction rate is hugely increased by the additional thermal energy absorbed by these atoms. Other sensitisers for positive resists are simply insoluble in the developing solutions until irradiated and so protect the soluble matrix polymer from dissolution in the unirradiated areas.

A huge number of polymer resists have been formulated for optical, x-ray and electron lithography, and a few of these are given in table 6.4. Elliott (1982) has given a more complete list of the properties of commercial lithographic polymers. Most efficient polymer resist recipes are closely guarded commercial secrets and so little detailed information on their compositions is available.

6.6.2(a) Processing resist polymers

The sequence of steps which is needed to define a pattern in a resist layer is deposition, softbake, exposure, development and, after the etching, doping or coating of the underlying material, resist removal (see figure 2.14). The deposition and softbaking of resist polymer layers has already been described, as have some aspects of the exposure process. In this section we will look in a little more detail at the development and removal processes. Once again, Elliott (1982) is recommended to all readers who wish to investigate this area of device fabrication more thoroughly.

The development stage is basically the removal of the exposed regions of a positive resist, or unexposed regions of a negative resist, and is usually a wet chemical process. The time for full development depends

Table 6.4 A few of the many kinds of polymer materials developed for resist applications.

Polymer	Type	Reference
Epoxy resins plus sensitiser	UV − ve	Schlessinger (1974)
Polyisoprene plus sensitiser	UV − ve	Long and Walker (1979)
Novolak resin plus diazonaphthoquinone	mid UV + ve	Babie et al (1984)
Polymethylmethacrylate (PMMA)	electron + ve	McGillis (1983)
Polybutenel sulphone	electron + ve	McGillis (1983)
Polytetrafluorochloropropylmethacrylate	electron + ve and − ve	Bednar et al (1984)
Polyvinylnaphthalene	electron − ve	Ohnishi et al (1984)
Polyglycidylmethacrylate-co-ethylacrylate	electron − ve	Elliott (1982)
PMMA plus Cl atoms	x-ray	Taylor (1980)
Polyglycidalmethacrylate-co-ethylacrylate	x-ray − ve	Elliott (1982)

on the exposure time, developer concentration and temperature, and thickness and softbake temperature of the resist. It is particularly important to avoid overdevelopment, or the resist may be thinned in the unsensitised regions and fail during an etching process. Negative resists are generally developed in solvents, while positive resists require developing in basic solutions which may contain metal ions—sodium ions are especially common. It is important to remove these ions from the wafer surface after the development process is complete because of the serious effect that these contaminant elements can have on the performance of MOS devices. There is an interesting difference between the development process for positive and negative resists. The solvent developing solutions penetrate the exposed (and so heavily cross-linked) regions of a negative resist, causing swelling of the resist and consequent loss of resolution in the device features being patterned. By contrast, positive resists are developed by a surface dissolution process which results in no swelling. As a result, the resolution which can be achieved with positive resists is far better than that obtained with negative resists.

An alternative method of developing resists is by use of an oxygen plasma, where the plasma power is carefully chosen to ensure that only the exposed regions of a positive resist are etched away at any significant rate. The development and substrate etching processes can then be carried out in the same equipment, with only the gas composition being changed to modify the etching process from resist etching in pure oxygen to aluminium etching in CCl_4 for instance (see table 2.6).

Resist removal, or 'stripping', is the final stage in a lithographic process and must be designed to leave all the sample surfaces as clean as possible, with no residual polymer films or flakes of resist material. Several methods can be used to strip resist, including simple dissolution in organic solvents or sulphuric-acid-based solutions, and plasma etching in oxygen. It is important to be sure that there are no metallic ions in the chemical strippers, or that these damaging species are thoroughly washed from the sample surface. An oxygen plasma can be used to oxidise the resist polymers, with the formation of CO, CO_2 and H_2O. However, exposed silicon surfaces can also be oxidised during the stripping process and, unless care is taken to remove the oxide layer, this can cause problems of high contact resistances when the metallisation layers are added.

6.6.3 Comparison of Positive and Negative Resist Polymers

It may seem odd that two kinds of lithographic resists have been developed in parallel, but this can be explained by looking at figure 6.9. It is quite easy to produce a very sensitive negative resist material such that the thickness of the resist remaining in the irradiated regions, when all the unexposed regions are completely developed, exceeds half the original thickness. It is much harder, even with the use of sensitisers, to achieve similar sensitivities in positive resists where all the resist is to be removed in the exposed regions. The radiation dose required to ensure complete development of positive resists is thus rather high. This would seem to indicate that negative resists have an enormous advantage over their positive counterparts when speed of throughput is an important process parameter. The difference in exposure time between the two kinds of resists can be as much as a factor of 1000 (Greeneich 1980). However, we have seen that significant swelling occurs during the development of negative resists, with consequent loss of resolution. Thus there is a balance that needs to be struck between the requirement for very-high-resolution features, which can only be obtained with positive resists, and the high sensitivity needed for rapid processing, which means using negative resists. (The resolution achieved with negative resists can be improved if the development stage is a plasma etching process, but this is a significantly more complicated and expensive operation than wet development.)

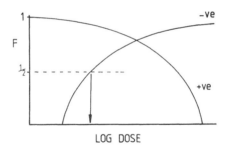

Figure 6.9 A schematic illustration of the relationship between the fraction of a resist layer remaining after a development process, F, and the exposure dose. A relatively low dose will allow sufficient cross linking to occur in a negative resist so that one-half the original thickness remains when the unexposed regions are completely dissolved. By contrast, positive resists can only be fully developed in the exposed regions if the exposure dose is very high.

The preparation of both positive and negative resists has thus been necessary to offer a wide range of materials that can cover all lithographic requirements as completely as possible. Where high resolution of the features is the prime requirement, positive resists can be used; where the feature size is coarser, negative resists are adequate. The resolution limit in electron beam lithography is about 0.1 μm in positive resists and is limited by beam spreading in the polymer layer. It is difficult to achieve resolutions as good as 1 μm in negative resist layers. Significant improvements in resolution can be achieved by using multi-layer resist structures, in which the top layer of resist acts as a high-resolution mask for a lower layer. More details on these multi-layered resists can be found in McGillis (1983).

REFERENCES

Adams A C and Capio C D 1981 *J. Electrochem. Soc.* **128** 423
Babie W T, Chow M-F and Moreau W M 1984 in *Polymers in Electronics* ed. T Davidson (Washington, DC: American Ceramic Society) p 41
Bednar B, Devaty J, Kralicek J and Zachovel J 1984 in *Polymers in Electronics* ed. T Davidson (Washington, DC: American Ceramic Society) p 130
Cerezo A, Grovenor C R M and Smith G D W 1986 *J. Microsc.* **141** 155
Croydon W F and Parker E H C 1981 *Dielectric Films on Gallium Arsenide* (New York: Gordon and Breach)
Deal B E 1980 *IEEE Trans. Electron Dev.* **ED27** 606
Deal B E and Grove A S 1965 *J. Appl. Phys.* **36** 3770
Deal B E and Sklar M 1965 *J. Electrochem. Soc.* **112** 430
Edwards J R, Bhatti I S and Fuller E 1979 *IEEE Trans. Electron Dev.* **ED26** 43
EerNisse E P 1979 *Appl. Phys. Lett.* **35** 8
Elliott D J 1982 *Integrated Circuit Fabrication Technology* (New York: McGraw-Hill)
Goldsmith C, Geldermans P, Bedetti F and Walker G A 1983 *J. Vac. Sci. Technol.* **A1** 407
Greeneich J S 1980 *Electron Beam Technology in Microelectronic Fabrication* ed. G R Brewer (New York: Academic) p 60
Ho C P, Plummer J D, Meindl J D, Deal B E 1978 *J. Electrochem. Soc.* **125** 665
Howes M J and Morgan D V 1985 *Gallium Arsenide, Materials, Devices and Circuits* (Chichester: Wiley)
Irene E A 1978 *J. Electrochem. Soc.* **125** 1708
Katz L E 1983 *VLSI Technology* ed. S M Sze (Singapore: McGraw-Hill) p 131
Kern W and Rosler R S 1977 *J. Vac. Sci. Technol.* **14** 1082
Ligenza J R 1961 *Phys. Chem.* **65** 2011
Long M and Walker C 1979 *Proc. Kodak Interface* 125
McGillis D A 1983 in *VLSI Technology* ed. S M Sze (Singapore: McGraw-Hill) p 283
Ohnishi Y, Tanigaki K and Furata A 1984 in *Polymers in Electronics* ed. T Davidson (Washington, DC: American Ceramic Society) p 191
Ouano A C 1984 in *Polymers in Electronics* ed. T Davidson (Washington, DC: American Ceramic Society) p 79
Rand M J 1979 *J. Vac. Sci. Technol.* **16** 420
Razouk R R, Lie L N, Deal B E 1981 *J. Electrochem. Soc.* **128** 2214
Schlessinger S I 1974 *Polymer Eng. Sci.* **14** 513

Schwarz S A and Schulz M J 1985 in *VLSI Electronics Microstructure Science* **10** ed. N G Einspruch and R S Bauer (New York: Academic) p 30

Spicer W E, Lindau I, Skeath P and Su C Y 1980 *J. Vac. Sci. Technol.* **17** 1019

Stoneham A M, Grovenor C R M and Cerezo A 1987 *Phil. Mag.* **B55** 201

Sze S M 1981 *Physics of Semiconductor Devices* (New York: Wiley)

Takagi H, Kano G and Teramoto I 1979 *Surf. Sci.* **86** 264

Taylor G N 1980 *Solid State Technol.* 73

Thurmond C D, Schwartz G P, Kammlott G W and Schwartz B 1980 *J. Electrochem. Soc.* **127** 1366

Wilkes J G 1984 *J. Cryst. Growth* **70** 271

Wong C Y, Grovenor C R M, Batson P E and Isaac R D 1985 *J. Appl. Phys.* **58** 1259

Wood J 1981 in *Reliability and Degradation* ed. M J Howes and D V Morgan (Chichester: Wiley) p 191

7

The Packaging of Semiconductor Devices

7.1 INTRODUCTION

The structures of semiconductor devices and metallisation schemes are described in Chapters 2, 4 and 5, and it is tempting to think that once the problems associated with fabrication of these complex structures have been overcome we have succeeded in creating devices which are suitable for packaging into electronic systems of all kinds, both large and small. However, it is in the word 'packaging' that a multitude of new problems are hidden, covering the processes of mounting and interconnecting individual devices or large integrated circuits. A considerable part of the materials science involved in the manufacture of microelectronic components is concentrated in these processes. First we should define why it is that a device needs be packaged at all. Some reasons can be listed as follows:

(1) To provide electrical contact from the devices on the chip to the outside world. In practice, this means that a multiplicity of contacts must be made from individual devices to a relatively more massive array of contact-forming components on the outside of a package.

(2) To provide a pathway for excess heat to escape from the active region of the devices.

(3) To seal the chip away from chemical species which can attack the delicate connection paths on the chip surface.

(4) To render the chip capable of being handled. A normal device chip is very small (millimetres across at most) and also very fragile.

We can see that a package is required to provide a dense array of contacting paths from chip to a larger contact array, good thermal conduction paths from the chip surface, high mechanical strength to support the chip, and these components must often be hermetically sealed as well. There is a huge number of ways of designing such packages, each device manufacturer having developed their own preferred methods. The performance required of a package is highly dependent on the kind of devices which it contains and so many packages are now custom designed for particular chips. The range of commercial package types is thus dauntingly large and I shall not attempt to describe all these technologies. Instead, I shall present some of the most popular types of package to illustrate this diversity and then concentrate on the materials which are needed to construct these complex multi-component assemblies.

One of the most severe problems associated with the packaging of modern circuits is that the signal transmission times through the very complex array of wires, pins and metallisation tracks are the limiting factors on the performance of the packaged components. Similarly, very highly integrated devices often generate a great deal of heat and it can be very difficult to dissipate this through the package. Most of the mechanisms of failure in metallisation systems which have been discussed in Chapter 5 are thermally activated, so that it is very important to make the package from materials with the highest possible thermal conductivities. It has been common in the past to assume that packaging was the 'low-tech' part of the process of designing and fabricating microelectronic devices. However, it has become clear that the performance of some devices is now limited not by the properties of the semiconductor material or metallisation scheme, but by the package itself. This has resulted in an enormous amount of basic research being directed at understanding the properties of package materials and how they may be joined into high-performance components.

7.2 MICROELECTRONIC PACKAGE TYPES

The simplest form of package is one that takes a single semiconductor chip and encloses it in a sturdy and easily handled container suitable for plugging into an array of other electronic components, i.e. the single-chip can. The basic structure of such a package is illustrated in figure 7.1. The chip die is separated from the array of very many identical devices on a processed wafer (see §2.3) by a careful sawing operation and is bonded to a header plate. Contact pins are mounted through, but electrically isolated from, this header, the isolation being provided by a glass sealing material. Electrical contact between the chip and the pins is made with a set of very fine wires. The final stage of manufacture is the hermetic sealing of the can by bonding a metal top to the header in an inert atmosphere, nitrogen or helium for example. Figure 7.1(b) shows the components inside an opened single-chip can.

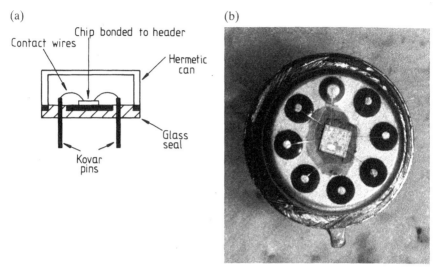

Figure 7.1 (a) A sketch of the structure of a single-chip can package identifying the basic components: header, pins, wire connections and hermetic top. (b) An optical micrograph showing the structure of an opened single-chip can.

A second packaging structure which has become very popular with manufacturers, because it is easy to produce in highly automated processes, is the leadframe package. The components of such a package are shown in figure 7.2. A leadframe consists of an array of contact leads and a central chip-holding plate, or paddle. The chip die is bonded to this paddle, and wire connections are made between chip contacts and the pins on the outside of the package. (In some more recent packages of this type, the chip-to-leadframe connections are made as an integral part of the process of bonding the chip to the substrate. This method will be described in the section on tape bonding.) Finally, the chip and leadframe assembly can be 'potted' into a solid block of polymer or glass to protect the delicate contacts and corrodible surfaces, especially those of the aluminium alloy contacts at the top of the chip metallisation structure, from atmospheric attack. The particular lead-frame assembly shown in figure 7.2 is of the kind called dual-in-line packages, DIPS (or DILS), because of the two parallel lines of contact leads left protruding from the potting material. For many kinds of low-cost components, this is the most common packaging method. Leadless polymer potted chip packages for surface mounting on complex printed circuit boards are a relatively new development in packaging design, and both the packing density of chips on substrates, and the reliability of automated assembly operations, can be improved by the use of these surface-mounted components.

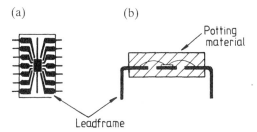

Figure 7.2 A sketch of the structure of a dual-in-line plastic potted leadframe package in (a) plan and (b) cross-sectional projections.

When the thermal conductivity and strength of the package are required to be high, or where the complexity of interconnection wiring cannot be accommodated in a leadframe assembly, chips are packaged on ceramic substrates. The chip die is bonded to the ceramic substrate and contacts are made between the chip surface and metallic conduction tracks on the ceramic surface. Wire bonding, or a variety of more specialised connection processes, are used to form these contacts. The deposition of conducting tracks onto ceramic substrates is considered in §7.7. These tracks make contact with an array of pins, leads or other external contacts for electrical communication with the outside world. Ceramic chip carriers can be hermetically sealed with ceramic or metal tops, or potted in glass or polymer blocks, as for leadframe packages. The contact arrays can take many forms and some simple ones are illustrated in figure 7.3. The pin grid array shown in figure 7.3(b) has the potential for providing a very dense array of contacting paths and can offer excellent thermal dissipation properties, and so is used to package many of the most complex integrated circuits. Leadless ceramic chip carriers (figure 7.3(a)) are also produced for the surface-mounted technologies mentioned above.

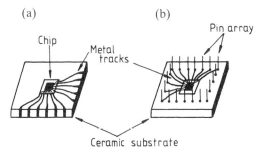

Figure 7.3 A sketch of the kind of contact arrays which can be used on a ceramic substrate. A leadless ceramic substrate is illustrated in (a) and a pin grid array in (b).

One of the great advantages of ceramic packaging is that the conducting tracks on the ceramic surface can form a complex pattern, and many electronic components, including resistors and capacitors as well as a multiplicity of individual chips, can be added to the surface to complete a complex circuit on a single substrate. This is the principle of hybrid circuitry which is described in §7.7. In general, ceramic packaging is reserved for the more sophisticated and expensive chips, where a high density of conduction tracks and improved hermeticity are needed. These cannot easily be provided in polymer potted leadframe packages. A recent development is to replace ceramic substrates with polymer substrates covered with the same metallic conduction tracks. These assemblies have excellent properties and are very cheap, but may not satisfy the most stringent hermeticity standards. The final stage of assembly of a microelectronic system requires the bonding of individually packaged components to a printed circuit board, PCB. The structure of PCBS, and the properties of the solder materials which are used in these joining processes, are described in §§7.10 and 7.6 respectively.

7.3 SUBSTRATE MATERIALS

In the previous section, three different kinds of substrates which can be used to support chips in packages were described: headers in cans, leadframes and ceramic sheets. The materials that are used for the first two applications are broadly similar and will be considered together.

In many device structures, the chip-carrying substrate is required to provide electrical contact from the back surface of the chip to one of the contact leads and so must be a good electrical conductor. It is often an advantage if its thermal conductivity is high as well, to help dissipate the excess heat generated in the devices. From figures 7.1 and 7.2 we can see that the substrate may be bonded to an insulating glass material and so adhesion between glasses and the substrate material must be good. The first two requirements are fulfilled by very many metals and alloys, and it is the condition of good adhesion between substrate and glass insulating materials that has, in practice, dictated the choice of header and leadframe materials. It has been observed that good integrity in a glass/metal seal is achieved only if there is a difference of less than 1 part in 10 000 in their thermal expansion coefficients over the whole temperature range the component will encounter in service. The kinds of glasses used to seal packages usually have thermal expansion coefficients of about $4 \times 10^{-6} \, \text{K}^{-1}$. An alloy of iron, nickel and cobalt has been designed with a thermal expansion coefficient which is well matched with these glasses between 100 and 500 °C. The alloy contains roughly 54% Fe, 29% Ni and 17% Co, and a whole range of Fe/Ni-based alloys are used as contacts and headers in microelectronic packages under the general name of Kovar. The disadvantages of this alloy are that it has a rather high resistivity (49 $\mu\Omega$ cm) and a thermal conductivity which is much lower than either gold or aluminium. In relatively low-performance packages these are not serious disadvantages, and the excellent seal between Kovar and glass gives a package good integrity in terms of long-term hermeticity and electrical insulation properties. When bonding wires to the Kovar pins, it has been found that plating a thin layer of gold or silver over the Kovar improves the wire adhesion enormously. The explanation of this improvement in adhesion is given in the section on bond forming below. Headers and leadframes are easily stamped out of thin Kovar sheets and so these components can be mass produced in huge quantities. Figure 7.4 shows some examples of typical commercial leadframes. Electroplating the components with noble metals or solder is a simple process that gives fine-grained, soft and adherent surface layers. More recently, selective plating of only the contact areas has been practised to reduce the consumption of expensive noble metals.

Figure 7.4 Some examples of commercial leadframe components. These are Kovar leadframes for plastic packages. (Courtesy of Heraeus GmbH.)

In leadframe packages where polymer potting compounds are used, and no glass materials are included, the thermal expansion coefficient does not have to be controlled so carefully and other materials can be used to fabricate the leadframes. A range of copper alloys like the bronzes and phosphor-bronzes are commonly used, because their high thermal and electrical conductivities give significant improvements in the performance of the package. Leadframe components of these alloys can be stamped from sheet and plated in the same way as for Kovar.

7.3.1 Ceramic Substrates

In the early stages of designing a packaging technology based on insulating substrates, the requirement for chip-carrying materials that were mechanically strong and chemically stable pointed clearly to ceramics. A well developed technology already existed dedicated to the production of ceramic components for a wide variety of applications. The necessary features of the ideal substrate material, however, go much further than those few mentioned above. A more complete list, and explanations of their importance for an electronic substrate, is given in table 7.1. Also included in this table is a comparison of the properties of two mass-produced oxide materials—alumina and common glass. It is

Table 7.1 Some properties of an ideal electronic substrate material.

Property	Importance	Glass	Alumina
High strength High elastic modulus High thermal shock resistance	To support chip and metallisation	Poor	Good
High resistivity	For complete electrical isolation of chip	Prone to surface conductivity	Good
Chemical stability	To ensure no reactions with metal tracks or processing chemicals	Can be attacked by HF	Good
High thermal conductivity	To remove waste heat from chip	Very poor	Poor
Thermal expansion coefficient	Should be matched to silicon to reduce stresses on chip	Can be matched	Poor match
Low dielectric constant	To lower microwave losses	Glass lower than alumina	
Surface smoothness	For ease of thin film deposition	Very good	Rough
Low porosity	Reduces outgassing after packaging	Good	Good
Low cost	To reduce the cost of complete device packages	Cheap	Fairly cheap

evident that alumina has very much more attractive strength and stability properties than glass, but some of the other requirements are better fulfilled by glass. This is especially true when the excellent smoothness and controllable thermal expansion coefficients available in the glasses are compared with the equivalent properties of alumina. The excessive brittleness of most glasses, and their reaction with moisture to give ionic species which are mobile on the surface leading to high surface conductivities, are the reasons why they are usually neglected when choosing materials for high-performance electronic substrates.

Alumina substrates are very commonly used in ceramic chip packages because they offer a range of properties giving a reasonable approximation to the ideal outlined in table 7.1. The properties which are least satisfactory are: the thermal expansion coefficient which should be lower to match that of silicon more closely; the low thermal conductivity; and the high dielectric constant which will lead to capacitative loss of high-frequency signals. Table 7.2 gives some selected properties of a wider range of ceramic and glassy materials and shows that there are some ceramics with properties more closely matched to the ideal than alumina. Beryllia, for instance, has a lower dielectric constant and excellent thermal conductivity, but can be rather weak. Aluminium nitride is a relative newcomer to the commercial ceramics field, but is very strong and has a high thermal conductivity. Silicon nitride has an attractive range of properties as well. The major problem with these materials is that they are very much more expensive than alumina. Beryllia is the only one of these materials used at all frequently for microelectronic substrates, usually when good heat dissipation from the chip is an overriding consideration. This is not to say that some of these

Table 7.2 Selected properties of ceramic and glassy substrate materials.

Material	Dielectric constant	Thermal expansion coefficient $(10^{-6}\ K^{-1})$	Flexural strength (MPa)	Thermal conductivity $(W\ cm^{-1}\ K^{-1})$
Al_2O_3	9.3	5.5	250	0.2
BeO	6.3	6.5	90	2.0
AlN	8.8	2.6	300	1.5
Si_3N_4	6.0	3.1	650	0.25
Glasses	3.8–7.0	0.6–4.0 ·	~70	~0.001
Glass ceramic†	7.5	4.2	230	0.003–0.03

†Glass ceramics are partly or completely crystallised glasses and usually consist of a number of ceramic phases in a retained glassy matrix. The data given here are for one of an infinite range of such materials. The particular value of glass ceramics when considering materials for electronic substrates is that their thermal expansion coefficients can be tailored to any required value by selecting the mixture of crystalline phases correctly.

other ceramics may not find applications in the future. As the complexity of chip and packaging design increases, the demands on substrate performance become more severe, and the current compromise choice of alumina may not always provide the most satisfactory substrate material.

7.3.2 Production of Alumina Substrates

Alumina sheets for microelectronic substrates are commonly produced by a continuous casting process and some of the other important ceramic materials can be prepared by the same techniques. The basic features of the process are shown in Figure 7.5. A liquid precursor mixture of fine ceramic particles, Al_2O_3 plus MgO and SiO_2 which act as sintering aids, in an organic matrix is placed in contact with a flexible support tape. The liquid mixture is typically about 65–75% powdered ceramic in a solution of binder and plasticiser polymers in a solvent like toluene or ethanol. This mixture is called 'slip'. As the tape is passed under the liquid reservoir it draws a liquid layer past a 'doctor blade' positioned just above the tape surface. In this way, a thin even sheet of slip is retained on the tape. This liquid is dried very carefully to avoid forming bubbles of solvent vapour, leaving the ceramic particles bonded together by the partly cured plasticiser. The binding polymer gives the 'green sheet' enough structural integrity to be removed from the supporting tape as a free-standing material. An alternative method of producing the green sheet is by dry pressing a slip mixture, but this cannot be made into a continuous process. The green-sheet material can be fired at temperatures of about 1550°C to sinter the ceramic particles into a solid sheet and to drive off the plasticiser polymer.

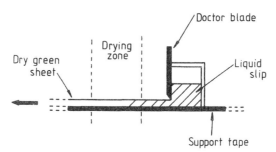

Figure 7.5 A schematic diagram showing the process of tape casting used in the production of ceramic substrates. The green sheet is fired once it has been separated from the supporting tape.

During this firing process, the sintering agents, MgO and SiO_2, form a liquid phase when solid state diffusion has produced an

$MgO/Al_2O_3/SiO_2$ alloy. There is a considerable concentration field in the SiO_2-rich corner of the ternary phase diagram which is liquid at 1550 °C. This means that the alumina powder can be effectively sintered into a solid material well below its melting point. The microstructure of such a sintered alloy consists of alumina grains a few micrometres in diameter held together by MgO- and SiO_2-rich phases at the grain boundaries and triple points. During the sintering process, the dimensions of the sheet decrease significantly because of the loss of the plasticiser material and the densification of the powder into a solid sheet. The ceramic sheets produced by such a process are called alumina but do of course contain a small fraction of the sintering aids. Commercial alumina materials are produced with a range of carefully controlled compositions, such as $Al_2O_3(96)$ in which 4% of the volume is taken up by the sintering aids.

The mechanical properties of the fired ceramic sheet are very dependent on the retained porosity, since small pores are nucleation sites for brittle failure. Great care must therefore be taken with the drying and sintering processes to minimise porosity, and to a large extent this has been satisfactorily achieved. Handling of the sheets during the chip mounting and wiring processes can easily cause cracking and chipping of the substrates, as any point impact will be likely to promote brittle failure. However, the advantage of alumina substrates over their glass counterparts is most easily seen in their resistance to brittle fracture. The work to fracture in a brittle material is dependent on both its elastic modulus and surface energy; the surface energy of alumina is about twice that of a typical glass, while its elastic modulus is about six times as large. The problem of brittle fracture of substrates has been partly cured by producing ceramic sheets with smoothly curving edges so that the probability of point impacts is reduced. (Some other methods of reducing the brittleness of ceramic substrates have been reviewed by Flock (1985)).

7.3.3 Impurities in Ceramic Substrates

We have seen that in the production of alumina substrates some impurities are deliberately added to the slip to act as sintering agents. There are, however, other impurities which can have a significant effect on the operation of devices on a chip mounted on the substrate. Most commercial ceramic powders contain very small concentrations of the naturally radioactive elements uranium and thorium. These concentrations do not normally exceed 0.1 ppm, but this is sufficient to give an α particle flux of 0.01 α particles $cm^{-2} h^{-1}$ passing through any chip mounted on a substrate fabricated from this powder. This may sound like a very low flux of α particles, but it can have a significant effect on the performance of some kinds of semiconductor devices.

Ziegler and Lanford (1979) have considered the effect of an α particle passing through a silicon chip. The fission decay products in uranium have very high kinetic energies, and it has been estimated that more than one million electron–hole pairs can be produced by a single α particle. Most of these carriers immediately recombine, but many memory chips are especially sensitive to the presence of low concentrations of electron–hole pairs. The information content of an individual memory 'bit' can be reversed (giving a 1 where originally a 0 was recorded) by a local increase in the concentration of electron–hole pairs. This is called a soft error. Simple estimates of the maximum α particle flux which can be tolerated around these sensitive devices gives a value of about 0.01 α particles $cm^{-2}\ h^{-1}$. We can see, therefore, that the radioactive impurities in the ceramic powders give α particle fluxes dangerously close to this acceptable level. This problem was first pointed out by May and Woods in 1979 and since then a considerable amount of research has been directed both at measuring these very low particle fluxes and at screening incoming ceramic powders to reject those with excessive uranium and thorium concentrations. Morris *et al* (1985) have shown that the radioactive elements are present in widely varying concentrations in commercial ceramic powders, and also in polymer precursors, and that casting and firing processes do not seem to change the concentration of these species. Careful choice of the source of ceramic materials can result in substrates with an average α particle flux well below the critical level. However, measurements of average α activity can obscure the fact that the radioactive elements are often concentrated into localised 'hot spots' where the ceramic substrate will have a very high activity (Mapper *et al* 1981).

Ideally, feedstock Al_2O_3, MgO and SiO_2 powders with zero α particle activity are required, but are very difficult to obtain. Novel methods for preparing ceramic sheets, such as the sol-gel process, are under active study since the precursor ceramic materials can be very highly purified in this kind of technique.

7.4 JOINING PROCESSES IN PACKAGES

In the fabrication of any kind of microelectronic package, a wide range of joining operations have to be performed. These may be the soldering of a wire to a bulk contact pin, the bonding of a chip to a substrate, or the reaction of a thick conductive film with a substrate material. Although these processes are very different in terms of the materials that they join, and often the mechanisms by which the joint is created as well, there are some fundamental aspects of all joining processes which should be considered before the individual bond-forming operations are described. The first property of a joint that it is important to specify is the kind of bonds formed at the interface. There are two kinds of bonds which can be formed in the joining operations in which we will be interested: van der Waals' and chemical bonds. Van der Waals' bonds are always formed when two materials are brought into contact, but they are relatively weak. Joints with reasonable mechanical strength cannot be created solely by these electrostatic bonds, but unfortunately the adherence of dust or contamination particles to mask or wafer surfaces by van der Waals' forces is sufficiently strong to make absolute cleanliness very difficult to achieve. Polymer layers deposited on non-metallic surfaces (like ceramic substrates) can only adhere by these weak bonds, and so this interface is not expected to be able to support any large stress.

Chemical bonds of many kinds are formed in package fabrication processes: for instance, metallic bonds between wires and contact pads and ionic bonds between glasses and metal oxides or ceramics. These chemical bonds are much stronger than the van der Waals' forces. There a several kinds of joining processes which are chosen for the fabrication of mechanically stable microelectronic packages because they result in the formation of these strong chemical bonds. Soldering, compression joining and diffusion bonding techniques are particularly important in the package types described in this chapter.

In order for a sufficient density of these chemical bonds to be formed, intimate contact between the two materials must be achieved. One simple way in which this can be effected is by having a liquid phase at the joint during the bond-forming process, either by having one of the materials to be joined in the liquid phase (as in glass/ceramic joining where the glass may be molten), or by placing a molten layer between the two materials. This second process is the principle of solder joining of course, and the solder wets both materials to form an intimate contact even when the surfaces are rather rough or the components irregular in shape. Metallic solders are used to join metal components and molten glass acts as a very efficient solder between many metals and ceramics. Intimate surface-to-surface contact can also be achieved by pressing two soft-metal components firmly together. Plastic flow under the applied stress forces the surface regions of both samples to deform, ensuring that the materials are in good atomic contact over a large fraction of the total bond area, i.e. a compression joint.

Once intimate contact has been achieved, it is still necessary to form chemical bonds between the two surfaces. In very many joining processes this bond formation is spontaneous once the surfaces are in atomic contact. This implies that the free energy change on creating a bonded interface to replace the two free surfaces must be negative. Before considering interface thermodynamics, we should think a little more about the bond-forming process. It is intuitively obvious that materials with like bonding character can bond together more easily than those with completely dissimilar character. Metal/metal bonds and glass/ceramic bonds are likely to be rather easy to form because the types of bonding in the two materials are already the same—metallic and ionic respectively. A more complex interface which is very frequently encountered in microelectronic packages is the metal/ceramic, or metal/glass, joint where the bonding characters of the materials are quite different. I shall assume for the moment that these glassy or ceramic materials are all oxides, although nitride compounds are also found in some packages. We might expect that we could divide these interfaces into two classes: those where the metals have some affinity for reaction with oxygen and those between noble metals and ionically bonded materials. Base metals are readily oxidised and so there can exist a thermodynamically stable oxide phase between the metal and the ceramic. Under these circumstances, bonds can easily be formed across the ceramic–metal interface. The metal oxide and the ceramic may react together to form an intermediate layer, and an example of this kind of interface is nickel on alumina, where the NiO present on the metal reacts to form the spinel phase $NiAl_2O_4$ which is very firmly bonded to both nickel and alumina and provides a very strong joint. The important interface in this case is that between nickel and the spinel phase, where even though one is metallic and the other ionic in character, the strong Ni–O bonds that stretch across the interface give the joint good mechanical integrity.

On the basis of this argument, we would expect that noble metals would not bond easily to ceramics, because of their low affinity for oxygen. In other words M–O bonds would not be formed under normal processing conditions. However, both platinum and palladium bond strongly to many oxide ceramics, forming intermetallic compounds with the metal species in the ceramic. This reaction is particularly rapid in reducing atmospheres. Strong bonds also appear to be formed at gold–alumina interfaces (Bailey and Black 1978), which may imply that some Au–O bonds are stable however unlikely this may seem. The mechanism by which these bonds are formed is still unclear, although the partial pressure of oxygen seems to be an important factor.

A very similar bonding mechanism seems to operate in polymer/metal joints. Reactive metals like chromium and nickel form M–O bonds readily with oxygen atoms on the polymer chains, while copper forms rather less stable bonds that can still give considerable interface strength (Burkstrand 1981). No such bonds are formed across polymer–ceramic interfaces, as there is no analogue of the M–O bond in this kind of joint. These interfaces are only held together by the van der Waals' forces.

The thermodynamics of interface formation is very complex and includes contributions from interfacial phase formation, changes in free energy caused by diffusional mixing of the materials, and surface curvature (the Gibbs–Thompson effect). In general, it is hard to predict the strength of an interface because of the subtle interaction of these effects and the lack of detailed information on the values of thermo-dynamic parameters, such as free energies of mixing, and the changes in chemical potential on diffusional mixing. The joints which are used in microelectronic packaging usually show strong chemical bonding in the cases where intimate surface contact can be ensured. Particular cases where adhesion problems are found will be presented in the relevant sections below.

From this discussion of bond formation across interfaces, some rather simple criteria for predicting whether a joint will be strongly adherent can be proposed. Some of these are given in Table 7.3 and only the last of them deserves further comment. Many joints between dissimilar materials are made by the so-called 'diffusion bonding' process. The condition for formation of a strongly bonded interface is taken to be the

Table 7.3 Factors which may influence interfacial strength.

Surface roughness	Reduces the contact area and encourages void formation in the interface for solid/solid bonds
	Has little effect when a liquid phase bonding material like a solder is used and can even give lock-in mechanical strengthening of the interface
Contamination—oxides or carbonaceous layers	Impedes interface bond formation in metal/metal joints
	Intermediate oxide layers are often necessary for the integrity of metal joints to glasses or ceramics, i.e. M–O bonds
High chemical reactivity of the materials to be joined	Enhances interface bond formation often by partial decomposition of one of the materials, e.g. $Ti + SiO_2 \rightarrow TiO_2 + Si$
	However, the formation of brittle phases at an interface will lower the adhesion strength
Mutual solid solubility of materials, leading to diffusion across the interface	Often an indication of a strong bond having been formed, but the observed diffusion is not always part of the bonding process

mutual interdiffusion of the two materials across the interface. Solution of one material in another is an excellent indication of bond formation between them, as a solute atom in a host solvent must be bonded to its neighbours in a manner similar to the solvent atom that it replaces. Thus the diffusion is an indication of the potential for forming a mechanically sound interface, but it is not the diffusion process itself which creates these strong interfacial bonds (unless a diffusion-induced change in interfacial free energy is the dominant thermodynamic factor in the bond-forming process, as it may be for Au–alumina interfaces). More generally, the observed mutual solubility would lead us to expect satisfactory adhesion simply because of the ease of interface bond formation that it signals. Some of the mechanisms of diffusion bonding have been described by Derby and Wallach (1982).

Finally, the possibility of corrosion effects must always be considered when dissimilar metals are in contact. Here the important criteria are the local concentration of moisture, or corrosive ions, and the relative positions of the two metals in the electrochemical series. In Chapter 5 we considered the corrosion of aluminium alloy conduction tracks in some detail, and it is these components that are the most sensitive to corrosive attack in a packaged chip. Metals used to improve adhesion like titanium and chromium form protective oxides which are not susceptible to dissolution in the presence of chloride ions (unlike aluminium oxide) and the noble metals do not corrode to any extent under normal conditions. Soft solders like the Pb/Sn alloys can be corroded when in contact with more noble metals like gold, silver and even copper, because these metals have rather low positions in the electrochemical series. However, anodic corrosion processes are most aggressive when the area of exposed anode is very much smaller than that of the cathode (see figure 5.23). Corrosion processes are thus much more likely to promote failure in metallisation conductors than in solder components, because of the large difference in the size of a solder ball and an individual conductor track. A very small amount of corrosion attack of an aluminium conductor will cause an open circuit, while very considerable corrosion would have to occur before a solder joint was significantly damaged. Materials which will promote galvanic corrosion when in contact with aluminium include copper, nickel, silver and gold, and so it may not be wise to place these metals in contact with a thin aluminium conductor pad.

7.5 CHIP BONDING TO SUBSTRATES

In the package types illustrated in figures 7.1–7.3, the first fabrication operation once the device chips have been separated from the wafer is to attach them to the package. The substrates have been shown to be made from either a metal alloy or a ceramic material. There are three common ways of attaching the chip die to these substrates: back bonding, flip chip and beam lead technologies. Since these methods use very different materials, and are quite different in concept, they will now be considered separately.

7.5.1 Back Bonding

The properties which are required of the materials used to make this kind of bond are: a reasonable thermal conductivity to assist in dissipating waste heat from the chip surface; a thermal expansion coefficient which is reasonably well matched with that of silicon to avoid excessive stresses being generated on the chip during thermal cycling; and in some components, a low electrical resistivity as well. The effect of thermal expansion coefficient mismatch can also be reduced by using a very soft bonding material that will flow to accommodate the stresses rather than transfer them to the chip. There are three kinds of materials which are commonly used to make these bonds: simple solders; solders based on the Au/Si eutectic system; and glass or polymer adhesives which can be made conducting or insulating as required.

The operation of simple solder back bonding is shown in Figure 7.6. Solder preforms about 50 μm thick are laid on metal contact pads on the substrate surface and the back surface of the chip placed over the preform. A simple heating cycle above the melting point of the solder will then bond the chip securely to the substrate. Electroplated metal layers are often deposited onto both chip and substrate surfaces to encourage efficient wetting by the molten solder alloy (solders will not wet ceramic surfaces at all). The integrity of the chip adhesion depends on the strength of the bonds formed between the metal layers and the chip or substrate. Reactive metals like chromium or nickel are often used as bonding layers to ceramic substrates and these layers are

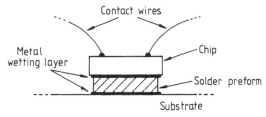

Figure 7.6 A schematic diagram of the use of a solder preform for the back bonding of a chip to a substrate material. The back surface of the chip and the ceramic are shown covered with a thin metal layer to improve the solder wetting.

sometimes gold plated as well to limit oxidation of the reactive metal surface which will reduce solder wetting and adhesion. Pb/Sn or Au/Sn solders are frequently used for chip back bonding.

The second method of chip bonding is very similar to this simple solder process, but here one of the components of a eutectic solder is silicon dissolved from the back surface of the chip die. A thin layer of gold, or gold/silicon alloy, electroplated or in the form of a preform, is placed between the chip and the substrate surface. The gold/silicon equilibrium phase diagram has a eutectic at about 3.6% silicon and 370 °C. Annealing above 370 °C causes the silicon from the back of the chip to diffuse rapidly into the solid solder alloy. The silicon concentration in the alloy increases until a melt is formed at the interface, and at this point the whole of the solder is rapidly consumed by the melt, the back surface of the chip acting as the source of additional silicon. A eutectic gold/silicon solder is formed on cooling and securely bonds the chip to the substrate. This is an extremely popular method of die bonding as it is very simple and reliable, although the use of gold makes it rather expensive. Similar back-bonding operations are not yet in wide use for the attachment of gallium arsenide chips, partly because of the greater complexity of the chemical reactions of a metal with a compound semiconductor.

The cheapest die-bonding method is to use a polymer or glass adhesive, which can be made electrically conducting by filling with silver particles. Epoxy-resin-based adhesives may degrade at temperatures above 300 °C and so can only be used when subsequent fabrication operations are carried out below this temperature. Polymer adhesives were initially treated with some suspicion for this kind of application, because the chloride ion concentration was often rather high and they shrank during curing, thus putting unacceptably high stresses on the chip. New formulations of these adhesives have reduced the chloride ion concentrations to around 10 ppm and shrinkage to about 4%, making them very attractive materials for low-cost die bonding. In situations where a high-temperature process follows die bonding, silver-filled glass adhesives provide satisfactory alternatives to the temperature-sensitive polymers.

Table 7.4 illustrates some typical properties of some of these die-bonding materials. Gold/silicon eutectic solders are rather stiff, but are reasonably well matched in thermal expansion coefficient with silicon and gallium arsenide, alumina and most metal substrates. Both gold/tin and lead/tin solder alloys are soft, but poorly matched in thermal expansion coefficient with silicon and gallium arsenide. Polymer adhesives are extremely soft, very poorly matched in thermal expansion coefficient with silicon or gallium arsenide, and their thermal conductivity is by far the lowest of these materials. Using this selection of properties, a die-bonding material can be chosen that is most suitable for any particular fabrication process. Once a die-bonding operation has been performed, an upper limit is placed on the temperature of any subsequent fabrication process. A temperature some 20–30 °C below the liquidus of eutectic solders is usually chosen as this upper limit, to ensure the solder bonds remain partly solid. The relatively high eutectic temperature of gold/silicon (370 °C) is a distinct advantage here as it allows further, lower-temperature, soldering operations, with lead/tin solders for instance, to be used during package fabrication.

Table 7.4 Properties of materials used for back bonding (after Steidel 1983).

Material	Melting point (°C)	Thermal expansion coefficient (10^{-6} K^{-1})	Elastic modulus (GPa)	Thermal conductivity (W cm^{-1} K^{-1})
Au 3%Si	370	12.3	83	0.27
Au 20%Sn	280	16.0	59	0.57
Pb 5%Sn	314	29	7.4	0.63
Silver-loaded epoxy	Cure at 150 Degrades above 300	53	3.5	0.008
Silicon	—	2.6	130	1.5
Gallium arsenide	—	5.5	110	0.46
Alumina substrates	—	6.5–8.1	260–350	0.2
Kovar substrates	—	4.1	147	0.15
Copper alloy substrates	—	17	120	2.6

7.5.2 Wire Bonding

The die-bonding process described in the previous section is used to attach the chip to the substrate and in some cases to provide an earth contact for the devices on the chip surface. All further electrical connection of these devices to the package must be made from the top surface of the chip. We will now consider the most common way of making these connections, i.e. wire bonding. Figure 7.6 shows the end result of a wire-bonding process. Very fine wires are bonded to the top metallisation layers on the chip surface and then to the metallic pins which provide contact with the outside world. Similar wires connecting devices on the chip with a leadframe are sketched in figure 7.2. The wire-bonding process requires the formation of a mechanically and electrically stable interface, which can be achieved in the following way.

Very thin metallic wires, often of a diameter of 32 μm, are brought into mechanical contact with the area to which the bond is to be made. The end of the wire may have been melted to form a ball if the surface tension in the molten metal is sufficiently high. This balling is possible in gold wires, but not in aluminium. A special tool now puts energy in the form of heat, pressure or ultrasonic vibrations into the joint. Heat and pressure are used to cause plastic flow of the wire and contact pad, which ensures intimate contact is made between the two materials. Ultrasonic energy can be used to break up any oxide or contamination layers on the surfaces. Simple heating or pressure processes are defined as thermocompression bonding, while a combination of ultrasonic and thermal energy as a thermosonic process. Following the earlier discussion of the fundamentals of joining, we expect that bonding operations which ensure intimate contact between metal surfaces will allow strong metallic bonds to form across the interface.

It is possible to suggest some simple criteria for the 'joinability' of two metals in a wire-bonding process. The higher their ductility, the more plastic flow will occur for any given applied pressure and so the more intimate the interface contact. The more noble the metals, the correspondingly less important will contamination and oxidation effects be in determining bond stability. Metals of dissimilar crystal structure are generally rather harder to compression bond than those of the same structure, perhaps because of a limited mutual solubility which may indicate a low tendency to form stable metallic bonds. Finally, systems that form brittle intermetallic phases may not give joints with good long-term stability.

In order to choose a material to make a reliable wire bond, we must first consider the metals to which the bonds are to be made. Depending on the design of the chip, the bonding areas may be aluminium alloy metallisation tracks, hybrid circuit thick or thin film noble metal conductors (see §7.8), or a whole variety of components which are gold plated to improve bondability, Kovar pins for example. The wires must therefore bond readily with aluminium and a range of noble metal alloys. The most common wire materials are gold and aluminium, although both copper and silver have been considered as alternatives. Gold wires are almost universally used in plastic potted devices where the hermetic sealing may be poor and can be thermocompression bonded to most materials since they are very ductile and free from surface oxide. Aluminium wires are more often bonded by ultrasonic techniques to break up the oxide layer on the wire surface. The poor corrosion resistance of aluminium wires can be improved by the addition of a few per cent of a more noble metal like palladium. The bond strength of gold-wire joints is usually much higher than that of aluminium wires on aluminium conductors, especially to shearing deformation. This is an important point when polymer packaging of devices is being considered (§7.7).

A particularly common bonding configuration is a gold wire onto an aluminium alloy pad and, as we have seen in Chapter 5, there is a rapid and potentially disastrous chemical reaction at this kind of interface to form intermetallic phases—the purple plague. Although the rate of intermetallic growth can be limited by addition of a few per cent of palladium to the gold wire for instance, these phases will always grow under these bonds, with the concurrent possibility of void formation at the wire–contact interfaces. Examples of the degradation of Au–Al interfaces have been given in figure 5.15 and show how serious the reaction between the gold and the aluminium can be for the mechanical integrity of these joints.

Another interesting effect at the Al–Au wire bond interface occurs when copper is added to the aluminium alloy to improve the electro-migration resistance (see §5.4). The inclusion of a few atomic per cent of copper causes the hardness of the alloy pad to increase dramatically, since what we have now is a material with composition very close to the optimum age-hardening alloy. The thermocompression bonding of gold wires to this hard alloy is much less successful than to pure aluminium, presumably because plastic flow to create an interface in intimate contact is restricted. There are some indications that overageing the Al–Cu alloys, by giving them a longer time at elevated temperatures, gives a softer alloy and consequently better bonds with gold wires, although this can also decrease the electromigration resistance.

7.5.3 Tape Bonding

We have seen that back-bonding and wire-bonding processes offer one way of solving the problem of how to mount and form connections to a chip. An alternative method is called tape bonding, and here the principle is to attach all the contacts to a chip surface in a single operation, instead of the many individual operations that are needed to wire-bond these same connections. To replace a leadframe, thin contact arrays very similar in shape to leadframes are patterned on polymer sheets. The conduction paths are often formed in lithographically patterned copper sheet supported on a polyimide tape. The copper layers are very much thinner than the contacts on a leadframe assembly and most of the mechanical strength of the assembly is now provided by the polyimide. The copper contacts can be gold plated to improve bondability. The separation of individual copper 'fingers' can be as little as 50 μm, very much closer together than can be reliably achieved in stamped leadframes. In order to allow this tape assembly to be directly bonded to the chip surface, the bonding sites on the chip must be raised above the level of the final passivation layer covering the metallisation. This is achieved by putting 'bumps' on the bond sites, as shown in figure 7.7. These bumps are rather thick (about 25 μm) plated layers of gold (or gold-plated copper) and the gold is isolated from the aluminium alloy conduction tracks in the metallisation by thin Cr/Cu barrier and adhesion layers, as described in §5.7. The tape contacts can then be bonded to these raised bumps either by direct gold-to-gold thermocompression joining, or by a simple soldering process. All the contact fingers are attached to the chip bumps in a single operation.

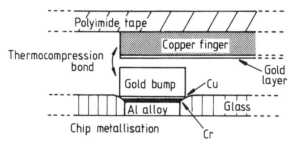

Figure 7.7 A schematic illustration of the process of tape bonding, where copper conductors on a polyimide sheet are brought into contact with a gold bump at the top of a metallisation structure.

This attachment process can be fully automated and results in chips securely bonded to arrays of copper contact fingers on the polymer supporting sheets. The whole assembly can be directly potted into polymer compounds to give a complete package, with the ends of the copper fingers protruding from the polymer block. Dissipation of waste heat can only occur along these copper fingers, as there is now no substrate to act as a conduction path. A wide variety of device chips can be simply packaged in this way, and it has the great advantage that the chip/tape assembly can be fully electrically tested before the final packaging operations are completed.

7.5.4 Flip Chip Mounting

The two methods for mounting and forming contacts to a chip which have been considered so far, back bonding plus wire contacts, and tape bonding, are used to produce individually packaged chips. These devices are then independent components which can be further packaged together to make a larger electrical system, as is described in later sections. The concept of a flip chip is to combine the potential for forming all the electrical connections to the chip surface in one operation, which is such an advantage of the tape–bonding process, with packaging many chips on a single substrate. This kind of packaging reduces the length and number of the conduction paths between individual chips, with consequent improvements in operating speed and the reliability of components in service.

Flip chip technology was pioneered by IBM in the 1970s and replaces the copper-covered tape of tape bonding by more rugged ceramic substrates supporting metallic conductor tracks. The concept introduced in the last section of raising the contact pads on the chip surface above the level of the passivation layers to allow bonding to all the pads at once is also used for flip chip technology. Here, however, the gold-bonding bump is replaced by a different kind of contact pad. The most common flip chip contacting method is to use solder pads to form the electrical and mechanical joints between chip and substrate. In order to do this, the fact that solder will not wet glass or ceramic surfaces is important. The chip contact areas are prepared in a manner similar to that already described for tape-bonding operations; the aluminium alloy conduction tracks are exposed through vias in the passivation glass layers and coated with $Cr/Cu/Au$ or $Cr/Ni/Au$ adhesion and barrier layers. The gold layer is very thin and only added to improve the corrosion resistance of the exposed surface prior to solder bonding. Thick solder layers are then evaporated through masks so that the solder overlaps the glass passivation layer on either side of the exposed contact area. On heating, the solder melts and because of its high surface tension forms a roughly hemispherical ball on the contact, avoiding the glass surface. The area of the exposed contact thus controls the height of the solder ball, which is typically around 100 μm. During the heating process, the thin gold layer is dissolved in the molten solder, and the solder bonds to the copper layer underneath. The solder materials which are commonly used to attach flip chips to substrates are alloys in the systems Pb/Sn, Au/Sn or Pb/In. The chip, with solder layers in place, is then inverted and placed above the contact fingers on the substrate to form the required bonds, as illustrated in figure 7.8.

A very important advantage of this kind of bonding is that even if the chip solder balls and substrate contact pads are somewhat misaligned, the surface tension of the solder will draw the chip contact pads and substrate connectors into registry above one another. This is called self-alignment and is part of the reason why flip chip mounting gives an extraordinarily high yield of correctly bonded chips. The size of the molten solder ball must be carefully controlled, as this ensures that all the bonds can be made at the same time without fear of some balls not being large enough to touch the substrate contact pads.

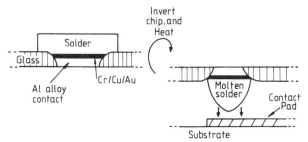

Figure 7.8 A schematic illustration of 'flip chip' joining. The metallisation scheme on the chip has thick solder layers which can be melted to form solder balls which form the connections to the substrate conductor pads.

7.5.5 Beam Lead Joining

One further method for attaching a chip to a substrate deserves a brief mention. The beam lead process involves the plating of thick contact 'beams' directly onto contact pads on the chip surface. These beams extend from the active central region of the chip out to the edge of the die where there are no semiconductor devices. The die edges are then chemically etched away, leaving the beams extending away from the chip surface. The reasonable mechanical strength of these beams allows the chip to be turned over and the beams soldered or thermocompression joined to substrate contact pads. Here it is the beams which are acting in the same contacting and interconnection role as the solder balls in the flip chip technology. The beams are very commonly made of gold, because of the ease with which this very ductile noble metal can be joined in a thermocompression process.

7.6 SOLDER MATERIALS

The use of solders to fabricate electrical and mechanical joints in microelectronic packaging has now been described in operations like die bonding and flip chip joining and is also very widely used for the bonding of discrete devices onto printed circuit boards. Mechanical failure of solder joints is a common cause of breakdown in complex electrical systems, both because there are such an enormous number of such joints and because much of the strain in complex packages is taken up by the solder material. A mismatch in the thermal expansion coefficients of semiconductor chip and ceramic or metallic substrates is an important source of strain. Considerable effort has been put into understanding the failure resistance of solder joints and some of the deformation models which have been developed will be presented below. In this section I shall discuss the methods by which soldering processes are carried out and then consider the thermomechanical properties of the solder materials themselves. The most common solder alloys are in the tin/lead system, although the precise composition chosen depends on the mechanical properties that are required. Metals like antimony, cadmium or indium are sometimes added to give improved physical properties and corrosion resistance.

First we should understand how soldering operations are carried out in a microelectronic fabrication process. Manual soldering is of little use when assembling chips on substrates because of the very small size of all the components. Automated soldering processes have been developed which are both efficient and suited for mass production. There are two such processes which are in common use—solder dip and solder reflow. The first of these uses the fact mentioned above that most molten metallic solders will not wet ceramic materials. Thus an assembled array of plastic potted dual-in-line packages with their pins pushed into metal-coated sockets on a non-metallic substrate can be dipped into molten solder and only the pins and substrate socket surfaces will be coated by the solder. This means that the solder joints will be made exactly where they are wanted, all at the same time. This principle can be used to attach all kinds of packaged devices to a whole range of substrates quickly and efficiently. However, the molten solder passes all the way round the packed devices, and the heating of the encapsulated components can cause damage as a result of enhanced chemical reaction rates or the generation of thermal expansion mismatch strains.

Solder reflow is often used for the attachment of chips to substrates and has in fact already been illustrated in the section on back bonding and flip chip mounting. Here a solid layer is placed over the areas where the joints are to be effected, the surfaces to be joined brought together, and a heating cycle used to melt the solder and form the bond. The solder layer can be in the form of a mechanically positioned preform, or an evaporated, plated or screen-printed layer. The method chosen for deposition of the solder will depend on the geometry of the components to be joined. Screen-printed solder layers often give very reliable joints between individually packaged chips and a printed circuit board. Once again, the fact that the solder will not wet non-metallic surfaces is useful in that it stops the solder running over the substrate or chip surface and shorting out contact pads or wire bonds. Horne (1986) has given a description of some of the more common processes for solder reflow joining. In both of these mass-production soldering operations, the cleanliness of the metal pins, pads and contact holes is a critical factor in determining the yield of successful joints.

7.6.1 Chemical Reactions with Solder Materials

The reactions which are important in determining the integrity of a solder joint are of two kinds: chemical reaction of the solder to form intermetallic phases with the materials being joined; and simple dissolution of these materials in the molten solder. We should remember that both the reaction and dissolution processes will be rather rapid during the soldering operation simply because the solder is liquid. This is not to say that the relatively slow solid state diffusion processes which will occur after the solder joint is cooled and solidified cannot influence the properties of the joint. The reaction between the common Pb/Sn solder alloys and copper surfaces is a particularly well documented one and nicely illustrates some of the features of this kind of chemical reaction.

Molten Pb/Sn solders react very rapidly with copper to form the intermetallic phase Cu_3Sn. After longer times, the phase Cu_6Sn_5 also forms, but the rate of growth of this phase is now controlled by solid state diffusion either of copper through the Cu_3Sn layer or of tin through the Cu_6Sn_5 layer. Parabolic growth kinetics are observed for both these phases, as we would expect for diffusion-controlled processes. Even at room temperature, Cu_3Sn and Cu_6Sn_5 will continue to grow, although the rates of growth are obviously slower than at the soldering temperatures. It is important to know whether this intermetallic phase formation has any deleterious effect on the mechanical integrity of the solder joints. Both of the Cu_3Sn and Cu_6Sn_5 phases are rather brittle, so that we might expect that the interface would become more susceptible to brittle failure as the thickness of the layers increases. In fact, the fracture strength of Pb/Sn–Cu interfaces has been shown by Marinis and Reinert (1985) to be only slightly dependent on the thickness of the Cu/Sn alloy phases. The solder joints usually fail by the propagation of cracks close to the solder–copper interface, but not actually through the intermetallic layers. The formation of Au–Sn alloys at the Pb/Sn–gold interface, or $PbPd_2$ at the Pb/Sn–palladium interface, are further examples of chemical reactions which are not as damaging to solder joint integrity as we might expect at first glance. The observation that cracks propagate through the bulk solder material means that it is the mechanical properties of the solder itself which we must understand if we are to be able to design more reliable solder joints.

The second solder/metal reaction which is important is the dissolution of the contact pad material in the molten solder. Many noble and near-noble metals are attractive materials for contact pads on substrates but dissolve freely in lead/tin solders. Silver and gold are particularly prone to this dissolution for instance. When the contact layer is rather thin, as it is in packaging technologies like the thick and thin film processes discussed in §7.8, all the contact material can rapidly be dissolved in the molten solder. This will result in a complete loss of bond integrity as the solder will not wet or adhere to the non-metallic substrate. It is obvious that this dissolution process must be controlled in a reliable soldering operation. Conductor alloys which have a reduced tendency to dissolution can be produced, and some of these will be discussed below. The solder composition can also be modified to limit noble metal dissolution. The addition of a few per cent of silver to conventional Pb/Sn solder alloys can reduce the rate of dissolution of silver from a contact pad. Indium-based solders, or Ag–Pb alloys, also lead to reduced dissolution rates and so are sometimes used where the problem is particularly severe. An alternative way of limiting this effect is to minimise the length of time that the molten solder is in contact with the soluble material.

7.6.2 The Mechanical Properties of Solder

Before considering the mechanical properties of solder materials, we should understand the nature of the deformation processes which occur in microelectronic packages. The geometry of a flip chip solder contact is shown in figure 7.9 and will be used as an example to illustrate the deformation forces on solder joints in all kinds of package geometries. When the devices on the chip are operating they dissipate some heat, quite a significant amount if they are logic or power devices and much less if the chip is a memory array. This heat will be lost primarily through the solder contacts into the substrate. The whole assembly will thus heat up when the chip is 'switched on' and a typical operating temperature could be as high as 150 °C. The thermal expansion coefficient of silicon, α_c, is about 2.6×10^{-6} K^{-1}, while a substrate material like alumina has a thermal expansion coefficient, α_s, of 6×10^{-6} K^{-1}. There is thus a significant expansion mismatch between the chip and substrate during the heating and cooling cycles experienced in operation. The shear strain rate, $\dot{\gamma}$, on the solder ball which connects these the components can be estimated from

$$\dot{\gamma} = (\alpha_s - \alpha_c)\frac{\Delta T\, l_c}{\Delta t\, h_s} \qquad (7.1)$$

while the rate of change of shear stress across the joint is given roughly by

$$\dot{\tau} = [GS(\dot{\gamma} - \dot{\gamma}_p)] \qquad (7.2)$$

where $\dot{\gamma}$ is the total shear strain rate and $\dot{\gamma}_p$ is the plastic strain rate in the solder. In these equations, ΔT is the temperature excursion during heating or cooling the assembly, Δt the time taken to reach thermal equilibrium, l_c the distance of the solder ball from the centre of the chip (the neutral point where chip and substrate do not move relative to one another) and h_s the height of the solder ball. G is the shear modulus of the solder and S is the fraction of the total strain excursion accommodated by the solder rather than the other components in the assembly. A simple estimate of a typical shear strain rate in a solder ball can be made using equation (7.1). If the parameters in the equation are chosen as $\Delta T = 100\,^{\circ}\mathrm{C}$, $\Delta t = 5$ s, $l_c = 100\,\mu\mathrm{m}$ and $h_s = 100\,\mu\mathrm{m}$, a strain rate of $\dot{\gamma} \sim 7 \times 10^{-5}\,\mathrm{s}^{-1}$ can be calculated, which is rather high.

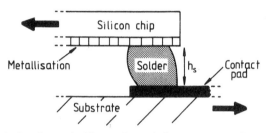

Figure 7.9 A schematic illustration of the structure of a solder ball assembly in a flip chip joint, showing the kind of stresses which can be generated by thermal expansion mismatches in substrate and chip.

It is tempting to assume, therefore, that the principal deformation mode of a solder joint is in response to the cyclic shear stress experienced as the chip is switched on and off. Failure of the solder material by a fatigue mechanism would seem the most likely result of this kind of deformation. However, the solder may also experience relatively long periods of static stress when the chip is in either the on or off condition, unless all the stress generated during the heating and cooling cycles is relaxed by dynamic fatigue deformation. Stone *et al* (1985a) have modelled the stress levels in a solder joint as a function of time using an equation like (7.2) and conclude that the joint will experience extended periods of static stress. Under static stress, creep deformation will occur, and to predict the failure mechanism of any particular solder joint now requires an understanding of the relative importance of the fatigue and creep deformation mechanisms. We can simplify the rather complicated deforming assembly shown in figure 7.9 by assuming that all the plastic deformation and creep processes are localised in the solder ball itself. This is a reasonable assumption because the yield stress of the solder will be very much lower than the yield stress of the other components of the assembly around the joint, i.e. the silicon chip, ceramic or metal substrate, copper contact pad and glass passivation layers in the chip metallisation. The solder also has a very low melting point when compared to these materials and thus creep processes in the solder are likely to be rapid. However, not all the applied stress is taken up in the solder ball because the rest of the assembly will strain elastically; in other words S in equation (7.2) will in fact be slightly less than 1.

We are now at the stage where we can start to build up a model to describe the failure mechanism of solder joints when subjected to a high-temperature high-strain-rate deformation process. It is simplest to begin by considering pure fatigue deformation. There is a very well known empirical law which predicts the number of cycles to failure of a material undergoing low-cycle high-strain fatigue, namely the Coffin–Manson equation (Coffin 1959):

$$N_f = C\left(\frac{1}{\Delta\gamma_p}\right)^{\beta} \tag{7.3}$$

where N_f is the number of temperature cycles before fatigue failure, $\Delta\gamma_p$ the plastic strain excursion, β an empirical factor for each material, and C the fatigue ductility coefficient of the material. This relation has been shown to be valid for a number of eutectic alloys (Norris and Landzberg 1969, Lawley 1976) and so might reasonably be used to predict the time to failure in solder alloys undergoing pure fatigue. The condition under which pure fatigue deformation might be expected to occur is when the package is cycled relatively rapidly between room temperature and the elevated operating temperature, leaving little time for creep to take place. This process should be dominant in soft solders where the yield stresses are low. However, very many electrical systems operate under rather infrequent 'off–on' conditions, sometimes with hours between turning the chips on and off again. Under these circumstances, it is very unlikely that the Coffin–Manson equation alone will provide a reliable prediction of solder joint lifetime. We must therefore extend our model to include creep deformation and the influence of microstructural changes in the solder material at the operating temperature.

The first decision we must make is which of the many mechanisms of creep we are going to consider. The choice is made much simpler if we recall that the microstructure of a eutectic solder consists of very many fine lamellae about 1–10 μm across. The precise size of the microstructure will depend on parameters of the soldering process like the rate of cooling of the solder ball from the molten state. What is important is that there are very many interphase boundaries in the solder down which diffusion will be rapid and in which the vacancy concentration is likely to be higher than within the lamellae. It seems reasonable to assume that creep deformation by interphase boundary diffusion, Coble creep, or even interfacial superplastic creep, will dominate over dislocation creep. In a creep equation relating strain rate to stress applied, the exponent of the stress, n, is a sensitive indication of the creep mechanism:

$$\dot{\varepsilon} \propto \sigma^n \tag{7.4}$$

where n is greater than 2 when dislocation creep is operating (Wu and Sherby 1984), n = 1 for pure interface diffusion creep (Coble 1963), and for superplastic creep where interface sliding, Coble and dislocation creep are all operating at once, n is about 2 (Ashby and Verrall 1973). Grivas *et al* (1979) have shown very convincingly that Pb/Sn eutectic alloys will deform by superplastic interfacial creep at high strain rates, as long as the stress is not sufficiently high to initiate dislocation slip. In near-eutectic Pb/Sn solder, the stress level at which dislocation creep is initiated is about 10 MPa in the temperature range 0–125 °C. We might, therefore, expect superplastic creep deformation to be dominant in a solder joint if the heating cycle frequency is low so that fatigue effects will be minimised and if this stress level is not exceeded. The lower the deformation stress, the more chance there is of avoiding dislocation creep. If either superplastic creep or ordinary Coble creep does dominate the deformation process, cavities will be formed at the interlamellar boundaries by interface sliding. This cavitation may well be an important mechanism in nucleating solder failure. Diffusion creep and the resulting boundary cavitation seems a reasonable choice for the creep failure mode in solder balls when the temperature cycling rate is slow. However, we cannot rule out the possibility of dislocation creep operating, especially if high stresses are generated in the solder by large thermal mismatches, or at the periphery of large chips or packages. Dislocation creep can reduce the likelihood of interface cavitation and so actually assist in protecting the solder from failure caused by a Coble creep mechanism.

A modified equation for the number of cycles a solder joint will survive before failure has been suggested by Clatterbaugh and Charles (1985):

$$N_f \sim C\left(\frac{1}{\Delta\gamma_p}\right)^\beta \nu^\alpha \exp\left(\frac{-E_i}{kT}\right). \tag{7.5}$$

Here ν is the 'off–on' cycle frequency and α an empirical parameter which contains the effect of all frequency-dependent deformation processes like interfacial creep and also recovery during the static stress periods. The exponential term is used to introduce the effect of grain growth or recrystallisation during the extended heating cycles. The parameter E_i is not the activation energy of the grain growth process, but the activation energy of the composite process of grain growth and joint weakening and so will include any temperature-dependent contribution from the failure mechanism. Here I have assumed that intermetallic phase formation does not contribute significantly to joint failure.

This equation provides a framework on which to base predictions of solder joint lifetime in service. However, values of the empirical parameters α, β and E_i must first be determined for the solder material under consideration under strain conditions similar to those which will be experienced by the solder joint in service. Few reliable measurements of these parameters have been performed and no large body of data has yet been collected which would allow the prediction of solder lifetime. We should remember that the temperature cycling rates depend on the kinds of devices in the chip and can vary from very slow to quite high.

So we are not able to predict solder failure lifetime from models as yet, but we do have some experimental evidence as to what the failure mechanisms are in microelectronic packages. Observations of solder balls after failure show significant cavitation at interphase boundaries in some cases, which strongly implies that interfacial creep is at least partly responsible for the failure (Stone *et al* 1985b). However, it is notoriously difficult to determine the deformation mechanism that causes failure from metallographic observations, especially when a multiplicity of deformation modes are operating simultaneously. It seems likely that there is no one mechanism of failure in solder joints, because the stress and strain levels will depend on the position of the solder joint on the chip and the kind of chip which is being packaged. To make matters worse, we have so far assumed that the stress distribution in solder joints is relatively homogeneous, so that it has been possible to assume that only one deformation mechanism is operating in any particular solder joint. Clatterbaugh and Charles (1985) have used finite element analysis to model local stress distributions in solder balls and found that the stresses are developed in an extremely inhomogeneous manner. In particular, the shape of the solder ball is a very important contributory factor in determining the local stress levels. Thus our assumption that a single deformation mechanism will operate is unlikely to be true and our goal of accurately predicting the lifetime of solder components seems to recede even further. The best we can do is to use the information we do have on deformation and failure in solder materials to propose some design criteria that will limit the damaging effects of the principal deformation modes we have isolated above. Some of these criteria are listed in table 7.5.

In practice, the most common reason for solder joint failure is probably contamination of the surfaces to be joined, which prevents solder wetting and bonding, or incorporation of gross defects like dust particles or voids into the solder balls. These kinds of fabrication problems lie well outside the range of the mechanistic models we have been trying to construct and can only be solved by scrupulous cleanliness, careful processing methods during soldering operations and the choice of suitable materials for the contacts and pins.

Table 7.5 Design criteria for increasing lifetime of solder joints.

Design feature	Effect
Match thermal expansion coefficients of substrate and chip	All these lower the strain rate and stress in the solder and so decrease fatigue and creep rates
Increase the height of the solder ball	
Use smaller chips	
Increase the solder/metallisation contact area	Reduces the shear stress on the solder/contact interface
Seal the package hermetically and avoid contamination from solder flux	Reduces the possibility of corrosion which inhibits the healing of fatigue cracks by rewelding
If creep rupture is primary failure mechanism, use soft solder	Increases fraction of total deformation accommodated by fatigue
If creep failure is primary failure mechanism, use hard solder	Increases fraction of total deformation accommodated by creep

7.7 POTTING MATERIALS

In an earlier section, the use of polymer or glass materials as potting agents for chip assemblies was described. These materials are required to protect the chip surface and wiring interconnects from mechanical damage or corrosive attack during service. The principle of a potting process is to flow a highly fluid thermoplastic or thermosetting material over the chip and substrate to create a solid block of hardened compound around the delicate wire bonds. This section will describe the properties of some of these materials.

Potting compounds must, of course, be insulating materials, which is a common property of a wide range of polymers and glasses. It must also be possible to flow the polymer precursor, or molten glassy, material around the wires connected to a chip surface, leaving no mechanical damage or bubbles which could lead to local corrosion. This flow process is potentially the most difficult part of the potting operation to control, and a low-viscosity material is always needed to limit damage. Glasses are at a major disdavantage here, as they usually have to be at quite high temperatures to have low viscosities. Even when using polymer precursors with very low viscosities, the relatively weak aluminium wire-bond joints can easily be sheared from the contact pads and so gold-wire bonds are normally used for polymer-packaged devices.

The potting material must also bond very strongly to the substrate, wires, pins and chip surface, and so good adhesion to metals, ceramics and glass surfaces is required. This is a most important condition, for unless good seals are formed at these joints corrosive species from the atmosphere can attack the chip metallisation. For the same reason, the diffusion rate of moisture through the potting material must be very slow. It is worth considering what it is that controls the adhesion of polymer materials to silicon, aluminium, Kovar and alumina. We have already shown that the bond across a polymer–ceramic interface cannot be very strong, as only weak van der Waals' forces are involved. At the same time, it was shown that strong chemical bonds (M–O) may be formed at polymer–metal interfaces. The polymer potting compound can form chemical bonds with some metallic components, but not to those covered in an ionically bonded oxide layer. It is thus not surprising that potting compounds decohere easily from glass and aluminium (covered with an Al_2O_3 layer of course) layers at the top surface of chip metallisation. This explains why the hermetic sealing of polymer-potted packages is not usually as reliable as methods using metallic cans or ceramic packages.

The thermal expansion coefficient of the polymer should also be as similar as possible to that of the semiconductor and the substrate material. This is extremely difficult to achieve since the normal thermal expansion coefficients of epoxy resins, for example, are 5 to 10 times higher than of silicon and gallium arsenide. During cooling from the curing temperature, which is usually above 150 °C, potting polymers will shrink much more than the other materials in the package. This mismatch in thermal expansion coefficients will put a significant strain on all the potted components and can result in damage to wires and even distortion of the soft aluminium alloy conduction tracks on the chip surface. A schematic illustration of these effects is shown in figure 7.10. A more detailed consideration of the thermomechanical stresses generated during potting has been presented by Steidel (1983).

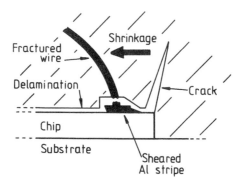

Figure 7.10 A schematic illustration of the damage which can be caused by shrinkage of a potting polymer during curing: wire fracture, distortion of metallisation tracks and decohesion of the polymer from the chip surface.

The mismatch in thermal expansion coefficients between potting agent and chip can be overcome by adding fillers to the polymers, fine silica or alumina particles for instance. The thermal conductivity of the potting material will be improved at the same time, adding another path through which heat can be dissipated from the chip surface. A low elastic modulus is usually an advantage for a potting material to allow the local stresses generated around wires and chip to be relaxed by plastic flow. Table 7.6 gives an indication of how well four families of potting polymers approach the properties of an ideal material. It is clear that selecting an ideal potting polymer is not going to be easy, although polyurethanes and some epoxy resins have reasonable properties.

Table 7.6 Comparison of properties of some important classes of potting polymers.

	Epoxy resins	Silicones	Polyurethanes	Polysulphides
Dielectric strength	Good	Good	Good	Good
Elastic modulus	High	Low	Wide range	Low
Tensile strength	High	Low	Wide range	—
Viscosity in precursor	Low	Low	Low	High
Adhesion to package	Excellent	Poor (to ceramics)	Good	Good
Moisture diffusion rate	High	High	Low	Very low

The high temperatures needed to lower the viscosities of glasses to a level which makes them suitable for potting applications are a major disadvantage in packaging. In addition, hot glasses often give off volatile oxide species like Na_2O and K_2O. Sodium and potassium ions are especially effective at assisting corrosion processes at aluminium contact tracks, and so glassy materials containing these elements are to be avoided in packages. For the same reason, the removal of chloride and sodium ions from the polymer precursors is an extremely important process in successful potting operations.

The long-term reliability of a potted chip or circuit depends mainly on the exclusion of corrosive species like water (see §5.5). The choice of a potting material which adheres well to the packaged materials, and the matching of the thermal expansion coefficients as far as possible, are important factors in ensuring that the hermeticity of the package is preserved in service. If due care is taken over these choices, polymer potting is a simple, cheap and reliable way of making chip packages in a fully automated process. New generations of mixed polymers (silicone plus epoxy mixtures), or fluorocarbon compounds, are now being produced with very low thermal expansion coefficients and elastic moduli. These materials may prove to offer significant improvements in the reliability of potted components and thus extend the application of polymer potting to the more expensive chips and circuits, which are currently packaged mainly in ceramic or metallic containers.

7.8 HYBRID CIRCUITRY

Up to this point in our consideration of packaging technologies we have looked mostly at methods of mounting and encapsulating semiconductor chips. These may be in the form of individual chips, or multiple chip arrays attached by flip chip methods to a substrate. A complete electronic circuit consists of more than logic and memory elements on a chip, however; resistors and capacitors are also usually needed to create a useful array of devices. These additional electrical elements can be attached as separate components to a printed circuit board, PCB, making the required electrical connections with the chips in leadframe or ceramic substrate packages. This is the first method which was commercially developed for the fabrication of complex circuits, but it produces a rather cumbersome assembly because the individually packaged components are relatively large objects and the interconnection paths on a conventional PCB are quite widely spaced. The concept of hybrid circuitry is to mount chips as individual unpackaged components onto a substrate where the interconnection paths, resistor and capacitor components are laid down as an array of patterned films rather similar to those that form the metallisation layers on a chip surface. These features are more widely spaced than the conductors in chip metallisation schemes however, but can be finer than is usually achieved on a PCB (see §7.9). In this way the size of a complete electrical circuit can be considerably reduced and the whole hybrid circuit can be hermetically sealed in a Kovar or ceramic package. This kind of hermetic sealing would be impracticable for a circuit mounted on a relatively massive PCB assembly. An illustration of the reduction in size of a circuit that can easily be achieved when moving from PCB to hybrid packaging is shown in figure 7.11(a). A sketch of a hybrid circuit mounted in a metal flatpack is shown in figure 7.11(b).

The substrates for hybrid circuitry are frequently alumina sheets produced by doctor blade or dry pressing processes. However, glass substrates are also used, although great care has to be exercised in handling these fragile materials. Both AlN and BeO are contenders to replace alumina where high thermal conductivity or low thermal expansion coefficients are required, but these ceramics are, as we have already mentioned, very much more expensive than alumina. Metallic substrates are also used, with aluminium being a very cheap choice and copper-covered Invar being matched in thermal expansion coefficient with the silicon chips. The mounting of chips to hybrid substrates, and the formation of electrical contacts between chip and substrate metallisation, can be carried out by conventional die attachment and wire-bonding processes, although individually packaged components can be easily soldered to contact pads on the substrate surface. The new fabrication processes which have to be introduced here are those concerned with the deposition of conducting, resistor and capacitor films onto the substrates. There are two methods that are used for the deposition of thick and thin film metallisation layers. We should note that even the thick film layers are only a few tens of micrometres thick. The materials used in these two technologies are different and will therefore be discussed separately.

7.8.1 Thick Film Hybrid Circuitry

The feature that distinguishes thick film circuits is that conducting resistor and capacitor features are added to the ceramic surface in a screen-printing process. The electrical properties of the thick films are determined by the composition of the ink, its thickness and the firing conditions. The inks are a mixture of glass particles, metal or metal oxide particles and an organic 'vehicle'. The glass component is often itself a mixture of borosilicate, or lead borosilicate, glass with bismuth oxide additions. The production of a successful ink requires the mixing of the glass frit, metal or metal oxide powders and polymer together to

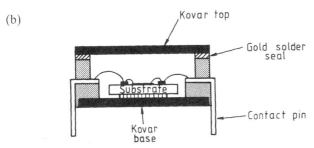

Figure 7.11 (a) An illustration of the reduction in size of a microelectronic system which can be achieved by moving from PCB component mounting to a more compact hybrid system. (Courtesy of Dr G Jordan and Marconi Research Ltd.) (b) A sketch of how a hybrid circuit mounted on a ceramic substrate might be hermetically packaged in a metal flatpack assembly.

achieve just the right composition and viscosity. The viscosity is important because the printing stage of the process involves pushing the ink through photolithographically patterned masks with a polymer blade known as a squeegee. Too high a viscosity will make the flow of the ink through the wire mesh screen supporting the mask uneven, while too low a viscosity will allow the ink to spread out over the substrate surface under the mask with consequent loss of feature resolution. Many commercial inks have the useful property of becoming less viscous when they are being painted onto the substrate surface and then 'setting' once they are no longer moving, i.e. pseudoplastic behaviour. This ensures that the feature resolution is good in the printed pattern. The typical feature separation might be of the order of 100 μm, which is a relatively crude lithography process compared with those we discussed in Chapter 2.

Once the ink is printed onto the substrate surface, it is dried at low temperatures to remove the organic components and then fired at much higher temperatures (700–900°C) to melt the glass phase and sinter the metal and metal oxide particles. The molten glass reacts with the ceramic material to 'glue' the glass/metal (metal oxide) composite to the substrate. We expect glassy materials to form strong chemical bonds with ceramics like alumina as they are both ionic materials. In addition, the glass is observed to react with the glassy phase between the alumina grains allowing mutual interdiffusion of the oxide compounds, the signal of a strongly bonded interface.

Now we must try and understand how the different electrical properties which are required in a thick film circuit are developed in the fired inks. An ink that is filled with pure metal particles—gold, silver or platinum for instance—will form a sintered array of interconnected metal grains when fired and will so create a highly conducting film. The glass phase here provides the adhesion to the substrate and will also adhere to the surface of the metal particles if there is sufficient free oxygen to stimulate the formation of M–O bonds. In this way a strong and adherent conducting layer is formed during firing.

Resistor inks are more complex, as often compositions containing a mixture of metal and metal oxide particles are required to give films with resistor values in the desired ranges. The volume fraction of conducting particles in a resistor ink is usually much lower than in a conductor ink, so the metal particles cannot sinter together during firing to give a continuous metallic conduction path. Conduction in these inks is effected by carrier hopping between metal and metal oxide particles separated by thin glass layers. In addition, the metal oxide particles can sometimes be semiconductors, which will further influence the character of the carrier transport in the films. The precise structure of the fired ink will depend on the extent to which chemical equilibrium has been reached between the metal and metal oxide phases, which itself depends on the firing temperature, oxygen partial pressure and firing time. It is not surprising therefore that the resistivity of fired resistor inks depends heavily on the details of ink composition, firing process and even the shape of the original metal and metal oxide particles. The fabrication of hybrid components with well controlled resistance values is thus extremely difficult.

A process of 'trimming' the resistor values towards the optimum figures has been developed and is routinely used to adjust these values after the firing cycle is complete. This simple process involves altering the area of the resistor film through which current can pass. A convenient way to do this is with a high-power laser beam which vaporises the resistor material leaving narrow cuts across the layer. In this way a component deliberately made with too low a resistance can be adjusted to the correct value. The gaps that result from the laser trimming process can be seen running across many of the dark resistor layers in figure 7.12(a).

The materials that are used in powdered form to make conductor and resistor thick film inks can now be introduced. Table 7.7 lists a range of these materials and outlines some of the advantages and problems associated with particular ink formulations. Conductor inks are often made from a mixture of metallic powders, although the resistivity of the fired films will always be increased by alloying. Binary and ternary inks often have complex phase diagrams as well, Pd/Ag/Pt for instance, and so the fired ink composite may well contain a large number of different phases. Prediction of the metallurgical structure of these inks after firing proves extremely difficult. Many of these noble metal inks are also very sensitive to partial dissolution in molten solder, and care must be taken to limit this effect by choosing inks where the concentration of soluble components like gold and silver is minimised. Palladium is added to silver inks to reduce this dissolution effect and also to stop surface electromigration of silver over the substrate. Plating the ink layers with a metal like nickel, which is insoluble in molten solders, is another way of protecting the conductors from dissolution. Resistor inks contain metal oxide particles because it has been found that the stability and reproducibility of the resistor values in layers made from these inks are better than given by other less complex mixtures. A more complete discussion of the composition and properties of thick film inks can be found in Holmes and Loasby (1976).

Special problems have been found with the adhesion of gold conductor inks since the conditions for achieving good glass/gold bonds are sometimes difficult to obtain in a commercial ink firing process. An elegant solution to this has been to add a few volume per cent of CuO or CdO to the gold powder. Partial reduction of these oxides during firing gives a metallic layer firmly bonded to the gold particles. This layer is itself bonded to the undecomposed oxide, and the oxide to the glass frit. In this way the CuO or CdO acts as a binding agent at the problematic gold/glass interface.

Thick film hybrid technology is used to package some very sophisticated circuits indeed. Figure 7.12(b) shows an example of a mixed

(a)

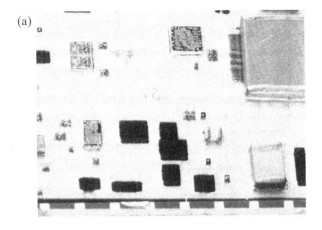

(b)

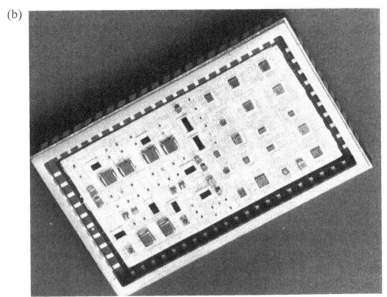

Figure 7.12 (a) A photograph of part of a hybrid circuit showing the laser cuts used to trim the resistors to the desired values. (b) A photograph of a complete hybrid circuit containing unpackaged chips, resistor and capacitor components. (Courtesy of Dr G Jordan and Marconi Research Ltd.)

digital and analogue multilayer hybrid containing 20 bare silicon chips and numerous resistors and capacitor components in the form of thick films. Such is the interconnection complexity needed for such an assembly that six layers of thick film conductor patterns separated by insulating pure-glass thick film layers have been deposited on the substrate. The structure of these conduction layers is remarkably similar to the design of a chip metallisation scheme. Here, however, the feature size is perhaps 100 times larger than would be found on a chip surface, and each layer of conductor or insulator is screen printed, not deposited from the vapour phase. The whole assembly can be hermetically packaged to protect the exposed chip metallisation by sealing a Kovar cover over the gold-plated seal visible around the edge of the substrate. Some aspects of thick film hybrid manufacturing technology are discussed in more detail by Horne (1986).

Table 7.7 Properties of some typical thick film ink compositions.

Material	Advantages	Disadvantages
Conductor inks:		
Gold	High conductivity	Expensive Difficult to solder
Silver	High conductivity Cheap Can be soldered	Surface electromigration
Pd/Ag	Good solderability Reduced electromigration	Lower conductivity
Pt/Au	Good solderability	Expensive Lower conductivity
Resistor ink:		
Ru/RuO (+Bi)	Very stable resistor Very low thermal coefficient of resistivity	Expensive
Dielectric inks:		
Glass ceramics, or ceramic loaded glasses, are used for almost all dielectric layers on thick film hybrids		

7.8.2 Thin Film Hybrid Circuits

The concept of thin film circuitry is similar to that of the thick film assemblies discussed in the previous section, except that the conductor, resistor and capacitor films are now only a fraction of a micrometre thick. The thin layers are usually vacuum sputtered or evaporated onto ceramic or glass substrates, the component features being defined by a photolithographically patterned mask (additive process), or by photolithographic resist patterning over a blanket evaporated layer and wet etching (subtractive process). Krokoszinski *et al* (1985) have recently argued that the latter technique, subtractive processing, offers very much less flexibility in the kind of circuits which can be produced. In particular, the deposition of crossover features where the conduction tracks pass over one another is not possible in a subtractive process, and so the most complex thin film circuits must be made by an additive technology. Flexible substrates can also support thin film circuitry, and polyimide or cellulose acetate sheets can be used for this application. The deposition and patterning processes used in the fabrication of thin film circuits have been described in Chapters 4 and 2 respectively. The materials used to form thin film hybrid features are very numerous because sputtering or evaporation techniques can be used to deposit layers of very many elements and compounds. Some of these materials are listed in table 7.8.

Table 7.8 Materials for thin film hybrid components.

Materials		Properties
Conductors:		
	Cr or Ni/Cr	Adhesion layers under other conductor materials
	Al	Cannot be soldered to, but otherwise very useful conductor
	Au	Sensitive to solder dissolution; resistivity increases with indiffusion of underlying Cr
	Cu	Sensitive to atmospheric attack
Resistors:		
	Ni/Cr	108 $\mu\Omega$ cm and very stable
	Ta	Can be partially anodised to trim
	(β tantalum has a higher resistance than α)	the resistance values
	TaN	Very stable resistors
	SnO_2	About 1400 $\mu\Omega$ cm
	Cr/Si/O cermets	Stable high resistivity
	Al_2O_3/NiCr/Al_2O_3 sandwiches	Stable resistance and zero temperature coefficient of resistivity
Capacitors:		
	SiO	Composition hard to control
	Ta_2O_5	Very high dielectric constant
	Parylene (Poly-para-xylylene)	A remarkable stable polymer with low dielectric constant
	Al_2O_3 or Y_2O_3	Easily sputtered oxides
	Al/SiO_2/Al	Popular stable capacitors

Conductor tracks are often two-layer films with a thin adhesion layer covered by a highly conducting layer. Reactive metals like chromium, titanium and nickel are used for the adhesion layers, as they react readily to form M–O bonds at the interface with the ceramic or glass substrates. Aluminium, copper or gold conduction tracks can then be deposited onto these strongly adherent layers. Problems associated with each of these materials are specified in table 7.8. Effects like the dissolution of noble metals in molten solder are of course especially damaging for thin film conductors, because the volume of the film in contact with the solder is so small.

Resistor layers can be made from a variety of materials depending on the exact values of the resistors required. Metal alloys like Nichrome, Ni–Cr, or β tantalum have rather high resistivities for metals and can be used for the fabrication of low-value resistors. Compounds like TaN and SnO_2 which can be sputtered or evaporated to form stoichiometric layers are used for resistors with higher values. Thin film resistors are relatively simple to prepare, and this is one of the most important advantages of thin film hybrid technology when compared with the thick films described above. Trimming the values of β tantalum resistors can be achieved by partial anodising treatments to reduce the area through which the current can flow.

The most popular materials for dielectric and capacitor films are silicon dioxide, SiO_2, tantalum oxide, Ta_2O_5, and materials like Al_2O_3 and Y_2O_3. Silicon monoxide can also be evaporated fairly easily, but the precise composition of the film usually lies between SiO and SiO_2. This variation in composition does not seem to degrade the capacitive properties of the layer. Ta_2O_5 layers are formed by the complete anodic oxidation of evaporated tantalum films in acid solutions. The very high dielectric constant of this oxide, $\varepsilon = 21.2$, makes it especially useful for fabricating high-value capacitors. An unusually stable polymer, parylene, can be evaporated at low temperatures to give thin films with a very low dielectric constant, $\varepsilon = 2.65$, for low-value capacitors. The problem with this material is that it is hard to control the thickness of the layer which is deposited.

The use of thin film hybrid circuits for the fabrication of complex systems has been somewhat overtaken by thick film technologies, although the thin film processes were the first to be developed. This is due to the greater reproducibility of the properties of the conductors made in a typical thick film process and the relative ease with which wire-bonding and soldering operations can be carried out to the rather thicker films. This has meant that thick film circuits have a larger share of the hybrid market at present, but the closer spacing of the interconnection paths that can be achieved in a thin film evaporated technology has led to a revival of interest in this process. The interested reader is directed to Holland (1965) and Berry *et al* (1968) for a more complete discussion of thin film materials.

7.9 MULTILAYER CERAMIC SUBSTRATES

The most dramatic departure from the conventional single-layer ceramic substrate, which may have several layers of metallisation on the top surface, is the development of the multilayer ceramic (MLC) modules pioneered by IBM in the 1970s. These modules were designed to hold large numbers of chips, thus giving a very large computing power in a relatively small package. Even more significantly, the number of chip-to-substrate joints could be very considerably reduced in such a package, compared with the packaging technologies discussed above. The reliability of a device package has been found to be strongly dependent on the number of joints that have to be fabricated. A more dispersed array of individually packaged chips might have 10 to 12 times the number of chip-to-substrate joints found in a package containing fewer, more complex, chips. Ho (1982) has given a detailed description of the advantages attendant on using MLC modules for fabricating sophisticated electronic systems.

Because of the large number of chips packaged onto a small area, the density of interconnection paths which needs to be contained on the substrate is much too high to be achieved in a few layers of metallic tracks. This is why the multilayer modules were developed. The MLCs are produced by assembly of a series of insulating sheets each covered in a customised pattern of conduction tracks. Connections between the conduction planes are made through vias punched through the sheets before assembly. MLCs are made from continuously cast alumina green sheets, with screen-printed refractory metal conduction tracks. Tungsten, molybdenum or molybdenum alloys are used as the metallic conductors. These sheets are then laminated into a complete assembly under pressure and fired in a carefully controlled atmosphere to remove the organic binding materials from both ceramic and conducting layers. The sintering temperature must be relatively high to allow liquid phase sintering of the alumina sheets; a typical firing temperature for Mo/Al_2O_3 MLCs would be 1500 °C for instance. The remarkable feature of this firing process is that the composite structures shrink by about 20% in all linear dimensions and yet the integrity of the conduction paths in three dimensions throughout the module remains unimpaired. The final stages in the production process involve brazing on external connecting pins, electroplating all the outer terminal pads and pins with nickel and gold layers, and mounting the chips on the top surface by flip chip solder techniques.

The details of the fabrication process are of course carefully guarded commercial secrets, but the successful production of these complex components remains a remarkable technological achievement. In 1983, the largest assembly of chips reported on an MLC was about 120, each with 120 solder joints to the top metallic contact layer (Goldmann and Totta 1983). These modules form part of IBM mainframe computers and each can contain 25 000 logic devices or 300 000 memory bits. A cross section of a typical MLC substrate is shown in figure 7.13, where the top surface contact pads and the wiring lines running through the bulk of the module can be seen.

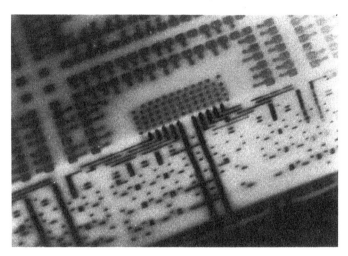

Figure 7.13 A photograph of a section through a multilayer ceramic module, showing the contact pads on the top surface where the chips are attached and the interconnection paths running between the ceramic layers. (Courtesy of Drs L Goldmann and P Totta, and IBM Inc.)

Even though there are a very large number of metallic conduction paths available in the MLC, the heat dissipated by the chips on the top surface is hard to remove efficiently. IBM have made use of the fact that the flip chip technology used in these modules exposes a considerable area of the back surface of the chips, giving an additional path for heat removal. An array of spring-loaded pistons can be bought into contact with the chips and cooled by the flow of water or helium. This assembly is called the thermal conduction module, TCM. When such extreme care is taken in the handling of semiconductor wafers during chip fabrication, it may seem remarkable that a mechanical device such as this should not affect the operation of the chips. Although it is true that the presence of a single dislocation in the active surface of a chip can damage or even destroy the circuit, we must remember that the back surface of the chip is some hundreds of micrometres away from the active chip surface. Mechanical damage created by the pistons on this back surface is unlikely to penetrate through to the active side of the chip. The TCM combines a very high density of powerful chips with excellent thermal dissipation properties—up to 300 W/module, one-third of a single-bar electric fire. More detailed descriptions of these extraordinary composite components can be found in the reports of Rinne and Barbour (1982) and Goldmann and Totta (1983).

We have seen that these modules provide extremely high densities of interconnection paths between individual chips, and this allows many chips to be packaged on a single substrate. However, there are still limitations on the speed with which signals can propagate between the chips. The size of the module is obviously a factor in determining this delay time and so there is a considerable effort being put into reducing the length of individual contacting paths. If the module is made smaller, the cross-sectional area of each conduction track will be reduced and contacting metals with higher conductivities than molybdenum or tungsten will have to be used, gold or copper for instance. These metals have melting points well below the firing temperatures needed to give effective sintering of the alumina sheets. The use of alternative insulating materials has thus been investigated to try and assemble very small modules containing highly conductive metals at low firing temperatures. Various glass compositions can be used to replace the alumina sheets, although they will give a very fragile module. Polyimide sheets are also an attractive alternative because of the low cost and flexibility associated with the processing of polymer materials and the excellent stability and durability of polyimide itself. In fact, polyimide layers supporting a high density of conducting tracks can be used on top of an MLC substrate, giving a high density of signal lines running through a material of very low dielectric constant, $\varepsilon = 3.5$. In this way the delay in the propagation of signals between chips can be reduced considerably. Composite MLC/polyimide/Cu substrates are suitable for the packaging of GaAs devices with very high operating speeds.

The advantage of thin film metallisation is that the spacing of the individual features can be very much finer than is possible in screen-printed layers like those on thick film hybrids or MLCs. Because of this, a great deal of attention needs to be paid to the possibility of fabricating complete multilayer substrate modules from this kind of technology. The spacing of conduction tracks in a screen-printed process is of the order of hundreds of micrometres, while the limitation on the spacing of evaporated thin film features depends only on the resolution with which the mask can be made. If lithography techniques like those used for the fabrication of integrated circuits are chosen, the feature separation in thin film metal layers can easily be reduced to 10 μm or less. This means that a thin film module will have a much higher packing density of conduction tracks and so will need fewer layers of metallisation to give the same complexity of interconnection as an MLC; it will also be cheaper and easier to make. These modules can be made by drawing on the combined experience of thin metallic film deposition processes developed for thin film hybrid-circuit fabrication and the thin dielectric layer deposition techniques used for insulating layers on the chips themselves. Ho (1982) has reviewed the techniques that might be used to make these modules and there seems to be no very good reason why this kind of packaging technology should not challenge wafer scale integration, WSI, for the most compact way of increasing computing power. There are enormous technological problems that have to be solved before these wafer scale extended chip arrays can be produced commercially. Thin film modules supporting very high densities of chips offer about the same computing-power-to-area ratio. In addition they are much easier to test, faulty components can be replaced easily, and they may be produced without the development of any new fabrication technologies.

7.10 PRINTED CIRCUIT BOARDS

Whatever the technology that is chosen to mount the semiconductor chips into a package, these individual packages must be further interconnected to build up any complex array of devices in a large electronic system. The DIP chip holders, more compact hybrid chip carriers, or even the MLCs or thin film substrates, have themselves to be supported on a substrate that functions as a thermal heat sink, as an array of often very complicated electrical connections between the individual packages, and as a mechanical support. Most of electronic circuitry is built up on the printed circuit board, PCB.

The boards may be made by passing woven glass cloth through epoxy resin solutions and then removing the solvent in a careful drying process. Even then, large voids full of solvent vapour are usually formed in the curing process, and reducing the number and size of these voids is an important part of successful board manufacture (Seraphim et al 1985). Individual boards can then be produced by laminating several layers of these 'prepreg' sheets together, giving a composite material with excellent mechanical and dielectric properties.

When the packing density of chips, and so of conduction tracks, is low, the single-sided PCB can adequately contain the required density of interconnecting metal tracks. A typical PCB has one side laminated with a copper sheet about 30 μm thick, which is patterned by defining the required metal conduction paths in photoresist and etching away the remainder of the copper. Thicker copper layers can then be electro-plated onto these starting layers if required. Chip carriers of all kinds can then be soldered to connection sites defined in the copper tracks. Protection of these metallic tracks from atmospheric corrosion is often achieved by electroplating a layer of solder some 10 μm thick over the exposed surfaces. The solder will, of course, only be deposited on the copper and not on the insulating boards. Before proceeding to describe the more complex forms of PCBs which have been developed to allow for a much denser array of chip packages, it should be emphasised that simple single-sided PCBs are by far the most common substrates in most electronics applications.

The first development of the PCB away from the single-sided board was to double-sided structures consisting of two planes of copper conductor tracks connected by copper-plated via holes running through the boards. These were produced in a manner similar to that described for single-sided boards. However, even more interconnection capacity was soon required, and the multilayer PCB was designed to fill this need. We can see a general principle at work here: as a circuit layout becomes more complex, single layers of metallic interconnecting tracks are not sufficient to carry the required density of signal and power lines. In chip metallisation, ceramic substrates and now in PCB design, the necessity of producing multilayer conductors has been demonstrated. Multilayer PCBs are made by laminating a series of thin laminate sheets (about 100 μm thick), each with the required pattern of connecting copper tracks lithographically defined on both surfaces. Lamination of the complete multilayer is achieved by heating at around 170 °C and compressing the stacked sheets with interleaving layers of epoxy-soaked glass cloth (prepreg) to provide the final cured bond. Four, or more, planes of conducting tracks can be bonded into a multilayer PCB, and contact between the planes is provided by plated holes drilled through the whole structure. The drilled holes can be filled with copper by initially using autocatalytic electrodeless plating of the side walls of the holes and then filling the remainder of the holes by ordinary electroplating.

The description of the manufacture of multilayer PCBs may give the impression that the processing stages are relatively straightforward. This conception is reinforced when we recall that these boards are after all rather massive components of an electrical system. Feature sizes are defined in fractions of an inch, rather than the fractions of micrometres within which manufacturing tolerances have to be controlled in the production of integrated circuits. However, to produce multilayer PCBs successfully requires very great control over the positioning of the copper tracks on the board surfaces and the drilling and plating of the via holes. Distortion of the thin sheets before lamination can occur, especially when the temperature fluctuations are large and the humidity high. It is often difficult to design metal patterns where the drilled holes will always intersect the required tracks on each conduction plane because of this distortion of the sheets. Computer-controlled drilling equipment for very accurate placing of the holes is now standard in PCB production. The actual drilling of the holes can also cause considerable damage to the boards, because the heat generated by the friction of the drill in the board degrades the epoxy binder. A more complete description of problems in multilayer PCB manufacture has been given by Payne (1982) and Seraphim et al (1985). The cost of a complex printed circuit board is now a significant fraction of that of a complete electrical system, and considerable efforts are being made by manufacturers to introduce new structures and new materials that will further improve the properties of these supporting and interconnecting components.

7.10.1 New Laminate Structures

The thermal expansion coefficient of the glass fibre/epoxy resin laminated boards was not initially considered to be of great importance when most devices were packaged as DIPs. The leads from these packages could easily absorb the mismatch in thermal expansion between the boards and the packages by flexing slightly. The recent trend to go to leadless chip carriers in a surface-mounted technology removes this freedom of movement. Any mismatch in thermal expansion must now be taken up in the solder joints between package and board. The larger the chip carrier, the greater the relative movement that must be accommodated by the solder balls at the periphery of the carrier. We have seen that the strain excursion in a solder ball is an important parameter in determining lifetime to failure, and the long-term reliability of PCB packages can therefore be improved by matching the thermal expansion coefficient of the laminated board more closely to that of the chip carriers.

Simple glass fibre/epoxy laminates have a thermal expansion coefficient about double that of alumina. It is rather easy to lower this coefficient by substituting quartz or Kevlar (poly p-phenylene tetraphthalamide) fibres for glass ones. These fibres have very low thermal expansion coeffients; indeed that of Kevlar is negative (-2×10^{-6} K^{-1}). Alternatively, the PCBs can be laminated onto a metallic substrate which has the required thermal expansion coefficient, such as Kovar. These metal-cored boards are rather more difficult to produce than ordinary PCBs, but they have the additional advantage that an efficient thermal conduction path has been added to dissipate the waste heat from the chip packages. Speicher and Blackburn (1982) have given values of the thermal expansion coefficients of a variety of laminated boards, and some of these are shown in table 7.9. It is obvious that metal-cored boards offer significantly better properties than the more standard copper/epoxy/glass boards.

Table 7.9 Properties of printed circuit board laminates.

Board materials	Thermal expansion coefficients $(10^{-6}\ K^{-1})$	Thermal conductivity $(W\ cm^{-1}\ K^{-1})$
Epoxy/glass	13.5	0.003
Epoxy/Kevlar	6.7	0.004
Alumina chip carriers	6.0	0.2
Copper-cored polyimide	5.8	~0.9
Kovar-cored/epoxy/glass	~6.0	~0.9

The use of epoxy resins as the bonding polymer in PCBS must be considered something of an industry standard. However, it is also possible to use polyimide as the organic binder and to produce boards with very similar properties to those discussed above. Indeed we can use polyimide to replace the whole epoxy/glass composite. Lamination of polyimide/copper boards is an attractive method of producing PCBS. These boards can have very high thermal conductivities and are often designed especially for surface-mounted components. A more novel approach to PCB materials is to replace the metal/organic board with a wholly organic structure, with polymer conductors running through a polyimide matrix. These conducting layers are deposited by screen-printing techniques. Conducting polymers were introduced in Chapter 1, but for PCB applications simple metal-filled polymers are convenient. (Silver or carbon particles in the polymer matrix are popular.) Alternating screen-printed layers of insulating and conducting polymer can be laid down and the whole package cured at once. The thermal conductivity of such a board will be rather poor, but for mounting arrays of device packages which do not dissipate very much heat this kind of all-polymer board may prove a very cheap alternative to multilayer metal/polymer structures.

REFERENCES

Ashby M F and Verrall R A 1973 *Acta Metall.* **21** 149
Bailey F P and Black K J T 1978 *J. Mater. Sci.* **13** 1045
Berry R W, Hall P M and Harris M T (eds) 1968 *Thin Film Technology* (Princeton, NJ: Van Nostrand)

Burkstrand J 1981 *J. Appl. Phys.* **52** 4795

Clatterbaugh G V and Charles H K 1985 *Proc. Conf. IEEE 35th Electronics Components, Washington, DC* p 60

Coble R C 1963 *J. Appl. Phys.* **34** 1679

Coffin L F 1959 *Trans. ASME* **76** 438

Derby B D and Wallach E R 1982 *Met. Sci.* **16** 49

Flock W 1985 Electronic Packaging Materials Science ed E A Geiss *et al* (Pittsburgh, PA: MRS) p7

Goldmann L S and Totta P A 1983 *Solid State Technol.* 91

Grivas D, Murty K L and Morris J W Jr 1979 *Acta Metall.* **27** 731

Ho C W 1982 *VLSI Electronics, Microstructure Science* ed. N G Einspruch (New York: Academic) p 103

Holland L (ed.) 1965 *Thin Film Microelectronics* (London: Chapman and Hall)

Holmes P J and Loasby R G 1976 *Handbook of Thick Film Technology* (Ayr: Electrochemical Publications)

Horne D F 1986 *Microcircuit Production Technology* (Bristol: Adam Hilger)

Krokoszinski H-J, Oetzmann H, Gernoth H and Schmidt C 1985 *J. Vac. Sci. Technol.* **A3** 2704

Lawley A 1976 *In Situ Composites II* ed. M R Jackson *et al* (Lexington, MA: Xerox) p 451

Mapper D, Bolus D J and Stephen J 1981 *Proc. 11th Int. Conf. Nuclear Track Detectors, Bristol* p 815

Marinis T F and Reinert R C 1985 *Proc. Conf. IEEE 35th Electronics Components, Washington, DC* p 73

May T C and Woods M H 1979 *IEEE Trans. Electron Dev.* **ED-26** 2

Morris P A, Handwerker C A, Coble R C, Gabble D R and Howard R T 1985 *Electronic Packaging Materials Science* ed E A Geiss *et al* (Pittsburgh, PA: MRS) p 89

Norris K C and Landzberg A H 1969 *IBM J. Res. Dev.* **13** 266

Payne S M 1982 *The Marconi Rev.* **XLV** no. 225 65

Rinne R A and Barbour D R 1982 *Electrocomponent Sci. Techn.* 51

Seraphim D P, Lee, L C, Appelt B K and Marsh L L 1985 *Electronic Packaging Materials Science* ed. E A Geiss *et al* (Pittsburgh, PA: MRS) p 21

Speicher P S and Blackburn E C 1982 *Electrocomponent Sci. Techn.* **10** 31

Steidel C A 1983 *VLSI Technology* ed. S M Sze (Singapore: McGraw-Hill) p 551

Stone D, Hannula S P and Li C-Y 1985a *Proc. Conf. IEEE 35th Electronics Components, Washington, DC* p 46

Stone D, Homa T R and Le C-Y 1985b *Electronic Packaging Materials Science* ed. E A Geiss *et al* (Pittsburgh, PA: MRS) p 117

Wu M Y and Sherby O D 1984 *Acta. Metall.* **32** 1561

Ziegler J F and Lanford W A 1979 *Science* **206** 776

8

Materials for Optoelectronic Devices

8.1 INTRODUCTION

In this chapter I shall introduce the materials which are used as the active components of some of the most common optoelectronic devices. These are semiconductor materials which can efficiently radiate light under electrical stimulation, or which can be used to measure the intensity of incident light, and are thus used in the production of lasers, light-emitting diodes (LEDs) and photodetectors. Consideration of the materials technology associated with solar cells will be deferred to the next chapter. Silicon and gallium arsenide have been shown in earlier chapters to be the only semiconductors which are important in integrated circuit technology. We shall see that in optoelectronic devices a much wider range of compound semiconductors have to be used to obtain emission in all the important spectral ranges and to achieve the most sensitive detection efficiencies.

Many optoelectronic devices are manufactured as individually packaged components, but they can also be integrated into larger systems which may contain only optical components, or include a mixture of transistor and optical devices. It is convenient to use the structure of an optical communications system as a framework around which to describe the properties required of semiconductor materials in optoelectronic devices. This is because the choice of materials is often governed not only by the efficiency with which they can be made to operate in an individual component, but also by how well these components can be integrated into a larger system. The materials for light-emitting and light-detecting devices which have been most extensively studied are those which are most obviously compatible with an optical communication system based around silica optical fibres. Before describing the properties of the optically active materials themselves, I shall briefly introduce the structure of an optical communications system and the properties of optical fibres. We shall see that the choice of silica determines that near-infrared wavelengths must be used to obtain the optimum efficiency. In the later sections of this chapter I will also include descriptions of the semiconductors which can be used as the active materials in a number of devices not intended for use in an optical communications system. These will include compounds and alloys which both emit and detect light in the far-infrared and the visible spectral ranges.

8.1.1 Optical Communications Systems

Figure 8.1 shows in schematic form the components which make up an optical communications system based on an optical fibre as the transmission medium. An incoming electrical signal modulates the emission from a light source, laser or LED, and this light is then channelled as efficiently as possible into an optical fibre. At the far end of the fibre, which may be many miles away, a detector converts the light signal back into an electrical signal. Once again, the coupling of the fibre to the detector must be designed to minimise the optical losses. The value of such a system is that the speed of transmission, and also the information density, in a single optical fibre can be far greater than in a conventional electrical conductor.

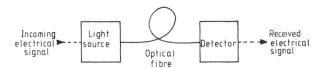

Figure 8.1 A schematic illustration of the components of a communications system based on an optical fibre.

The component which has the most influence on the overall performance of such a system is, not surprisingly, the optical fibre itself, and the successful development of fibre materials in which the attenuation of an optical signal is very low has allowed efficient optical fibre communication systems to be produced. Many optical fibres are made from silica 'doped' with germanium, or more properly germania, GeO_2. The addition of germania alters the refractive index of the silica, and a fibre with a germanium-rich core region of higher refractive index than the outer cladding layers will act as a waveguide along which an optical beam will propagate by total internal reflection, as shown in figure 8.2. Such a fibre can have an extremely high mechanical strength if care is taken to avoid the creation of microflaws during manufacture. The width of the germanium-doped core can vary from around 100 μm in multimode fibres, in which many optical modes can propagate but the signal dispersion is likely to be high, to about 10 μm in single-mode fibres. Fibres with graded germanium concentrations are also produced and B_2O_3 or P_2O_5 can be substituted for GeO_2 in other low-loss fibre materials. The properties and applications of these different kinds of silica fibres have been discussed by Barnoski (1981), Savage (1985) and Izawa and Sudo (1986).

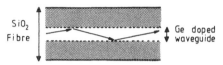

Figure 8.2 A sketch of the structure of a germanium-doped silica fibre, showing the core region along which the light propagates by total internal reflection.

The most important property of the core region of silicon fibres is the optical absorption coefficient, which will vary with the wavelength of the propagating light. Absorption in silica is controlled to a large extent by impurity species like the heavy metals, Fe, Cu, Cr and V, and particularly hydroxyl groups, OH. Even if great care is taken during fibre production, some hydroxyl groups will be incorporated, giving strong absorption at about 0.95, 1.25 and 1.38 μm. There are, however, spectral windows in commercial silica fibres at wavelengths of about 0.85, 1.1 and 1.5 μm, where the absorption is low. Efficient information transfer can be achieved with a longer length of fibre if the source emits at one of these wavelengths. In addition, the material dispersion is minimised at about 1.3 μm in many optical fibres and the waveguide dispersion of the optical signal decreases both with decreasing wavelength and decreasing spectral width of the propagating light. Thus a careful choice of the emission wavelength and the spectral width of the optical source will make a long-distance optical fibre communication system easier and cheaper to construct. (Intensity booster stations along a fibre are especially expensive to install.) We will see that the restrictions in source wavelength imposed by the use of silica fibres limit the choice of emitter and detector materials in optical systems quite considerably. Lasers operating at wavelengths between 1.1 and 1.5 μm, which produce light with a much narrower spectral width than the average LED, are a particularly good choice of source in an optical communications system. This means that detectors have to be designed which show their peak sensitivity at these same wavelengths. The materials used in these devices, and the structures of the devices themselves, are described in the following sections.

The technology of connecting optical fibres to light sources and detectors, and the design of durable fibre cables, are also important in controlling the performance of an optical system. In particular, joining two fibres so as to have negligible losses requires a precise matching of the optical waveguides at the centre of each fibre. The ends of the fibres are then kept together by fusion or a mechanical locking process. Similarly, the efficient connection of a fibre to a light-emitting device requires considerable thought to achieve a good match between the end of the fibre and the shape of the beam emitted from the source. An example of a particularly successful design of a joint between an optical fibre and a high-efficiency LED will be given in §8.4. Further details on the practical aspects of the design of a communications system incorporating optical fibres can be found in Kressel (1980), Barnoski (1981), Senior (1985) and Saul *et al* (1985).

8.2 MATERIALS FOR LIGHT-EMITTING DEVICES

The choice of material from which to fabricate a light-emitting device like an LED or a laser is dictated by several considerations: the efficiency of the radiative process in the material; the wavelength of the radiation which can be emitted; and the ease with which the material can be synthesised and fabricated into a complete device. The last of these is the topic of the bulk of this chapter; LEDs and lasers will be considered separately in §§8.4 and 8.5 respectively. The techniques for crystal growth described in Chapter 3 will frequently be mentioned in these sections when describing how individual devices can be fabricated. The special topic of the properties and applications of p–n junctions and heterojunctions will be discussed in §8.3. We will see that it is at these junctions that the light is generated in almost all semiconductor laser and LED devices. However, the most basic choices of the materials to use in these devices are those associated with the emitting efficiency and wavelength, and we will first consider how these parameters are controlled by the structure and composition of the semiconductor material. We will be interested in the phenomenon of electrical excitation of a semiconductor such that it emits light, i.e. electroluminescence.

8.2.1 Recombination and Luminescence in Semiconductors

In Chapter 1 I introduced the concept of recombination — the mutual annihilation of a hole with an electron which releases an amount of energy equal to the initial energy separation of the carriers. It is possible for the recombination to occur by a number of different mechanisms, and the most important one here is when most of the released energy goes into the creation of a photon with a wavelength which is characteristic of the precise radiative recombination event. There are also recombination processes which do not create a photon of useful wavelength, and we shall call these non-radiative. It is the competition between these two kinds of recombination which determines the radiative efficiency of each semiconductor material. In order to select an efficiently radiating material for LED and laser devices, we must define more clearly what determines whether the overall recombination process in a semiconductor is dominated by radiative or non-radiative electronic transitions.

8.2.1(a) *Radiative transitions*

Figure 8.3 illustrates several kinds of possible recombination processes in a semiconductor material, some of which create a photon with a wavelength which lies in a useful range of about 0.5–30 μm. The first two of these are: (a) electron–hole recombination across the whole of the band gap, which is called a band-to-band transition; and (b) recombination via luminescent states near the valence or conduction bands. In some cases these luminescent transitions can be via simple donor or acceptor states and are called band-to-shallow state, or donor-to-acceptor, transitions. The energy released in these radiative recombination events, and the wavelength of the emitted photons, obviously depend on the width of the band gap, E_g, and the position of the radiative states in the band gap. In heavily doped n-type GaAs, a dilute band of electronic states is formed at the top of the band gap – tailing states. Under these circumstances, the distinction between band-to-band and shallow state-to-band radiative transitions is lost and the energy of the released photons can range from slightly less to slightly greater than E_g depending on the exact position of the Fermi level. The relationship between the energy of the emitted photons, E_g, and their wavelength, λ_p, is given by the familiar expression $\lambda_p = hc/E_g$. An approximate form of this expression, which is very useful, is $\lambda_p = 1.24/E_g$, where λ_p is measured in micrometres and E_g in electron volts.

However, these radiative transitions do not occur with equal likelihood in all semiconductors and the most important cause of this variation in behaviour lies in the form of the band structure of the materials. Semiconductor materials can be roughly divided into two kinds: those with a direct band gap and those where the minimum separation of conduction and valence bands is between points at different positions in momentum space. In a semiconductor material

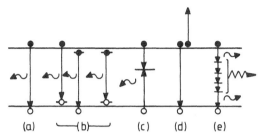

Figure 8.3 A schematic illustration of several possible kinds of recombination processes in semiconductor materials: (a) band-to-band recombination; (b) band-to-shallow state transitions; (c) recombination at a state deep in the semiconductor band gap; (d) an Auger recombination process; (e) a multiphonon process in which the first and last steps are shown as radiative transitions.

with a direct band gap, the transition of an electron from the minimum of the conduction band to the top of the valence band occurs with no change in momentum. Electronic transitions of exactly the same kind in a material with an indirect band gap require a change in both the energy and the momentum of the carrier. Effectively this means that a radiative transition of kind (a) across the whole of the band gap in gallium phosphide, which has an indirect band gap, releases a photon but must also create phonons which propagate away through the crystal lattice. This simultaneous production of a photon and phonons is a much less likely event than the simple release of a photon in the equivalent recombination event in a direct band gap material like gallium arsenide. Thus, although radiative band-to-band recombination events can occur in both direct and indirect band gap semiconductors, they are very much more likely in direct band gap materials even at high temperatures. Equivalently, we can say that the lifetime of a minority carrier in a direct band gap semiconductor is very short, as it can easily recombine with one of the majority carriers to release a photon. The minority carrier lifetime before a radiative recombination event is termed τ_{rad}. Non-radiative recombination events of a kind described below can also occur in a direct band gap semiconductor of course and can lower the minority carrier lifetime even further. In an indirect band-gap semiconductor, τ_{rad} will be relatively long, as is shown by a comparison of the values in direct band gap gallium arsenide, 10^{-9} s, and in very pure germanium, which has an indirect band gap, 10^{-2} s. It is clear from this huge difference that the amount of light released by radiative recombination processes in an indirect band-gap material will usually be a small fraction of that we can expect to obtain from a semiconductor with a direct band gap.

The minority carrier lifetime is also dependent on the majority carrier concentration in the semiconductor, N_m; a high carrier concentration is normally needed for efficient recombination. If shallow state-to-band radiative recombination events contribute significantly to the luminescence, then the capture cross section of the donor or acceptor state for carriers, σ, is another important parameter in determining the luminescent efficiency of the material. In a direct band-gap material, we can estimate the minority carrier lifetime before a recombination event involving a shallow state from

$$\tau_{rad} = (\sigma v_{th} N_m)^{-1} \qquad (8.1)$$

where v_{th} is the thermal carrier velocity in the semiconductor.

From the above we can see that an efficient luminescent material might be expected to have the following features: a direct band gap; a high doping concentration in the recombination region; and possibly a shallow state with a high capture cross section for carriers as well. We would expect heavily tellurium-doped GaAs, which has a donor level only 0.03 eV below the conduction band, to be an efficient luminescent material, while GaP should show almost no electroluminescence even when very heavily doped.

In fact, semiconductor diodes emitting in the visible wavelength range are very often made in GaP, and a reasonable luminescence efficiency is achieved in this indirect band gap material by the introduction of some unusual doping elements. In Chapter 1 the concept of an isoelectronic dopant was introduced and I explained that the disturbance to the local potential around a particularly poorly fitting isoelectronic dopant in a semiconductor lattice could result in the creation of shallow states. Nitrogen can substitute for phosphorus atoms in GaP and will create a neutral state just below the conduction band which is an efficient electron trap. Recombination of an electron trapped at this state with a hole is much more efficient than band-to-band recombination in GaP because it can occur without the production of phonons as well as a photon. This is a result of the localisation of the electron at the nitrogen atom, a bound electron, which then has a diffuse momentum. Recombination of a hole with this bound electron (which can be described as the formation and decay of a bound exciton) can thus conserve momentum and give an efficient radiative transition, one where τ_{rad} is small. Another important isoelectronic centre in GaP is the more complex Zn + O state. Substitution of a zinc atom on a gallium site and an oxygen atom on an adjacent phosphorus site will preserve the local electron density and create a state about 0.3 eV below the conduction band. Once again, electrons are readily trapped by this neutral deep state and radiative recombination can occur by the same mechanism as described for nitrogen-doped GaP. The process of radiative recombination at a deep state is illustrated by transition (c) in figure 8.3. The introduction of N or Zn + O isoelectronic states into GaP does not reduce the value of τ_{rad} to the very low levels found in heavily doped GaAs, but the radiative efficiency is increased sufficiently to allow useful LEDs to be made from this indirect band-gap semiconductor. A particularly authoritative account of the properties of these isoelectronic states in GaP has been given by Bergh and Dean (1976).

8.2.1(b) Non-radiative transitions

Competing with the radiative recombination transitions which I have described above are a number of recombination processes which do not result in the release of a photon in a wavelength range of interest to the device engineer. These transitions often limit the efficiency of a radiative device, and so it is of great importance to understand how they occur and to limit them as much as possible. We can divide these non-radiative processes into two classes: those associated with a new mechanism of recombination and those which occur as a result of the presence of an impurity or lattice defect in the semiconductor crystal.

In the first of these categories, I should briefly describe the mechanism of Auger recombination. In a heavily doped semiconductor it is possible for the energy released by the recombination of an electron and a hole to be transferred to another nearby carrier instead of creating a photon. This process is illustrated in (d) in figure 8.3, where an electron in the conduction band of an n-type semiconductor is promoted to a higher-energy state as a result of a local recombination event. This electron will gradually drift back to its initial energy, a process called thermalisation. The rate of Auger recombination increases with the square of the doping concentration in the semiconductor and also depends inversely on the width of the band gap. This means that an effective limit is set on the doping concentration which can be used in the recombination region of a direct band-gap semiconductor. Any increase in radiative efficiency as a result of increasing the density of free carriers is more than offset by rapid non-radiative Auger recombination.

The second important non-radiative recombination process involves multiphonon transitions, where for instance an electron passes from the conduction band to the valence band in a series of small steps releasing many phonons at each transition. This kind of recombination process can only occur when there are several electronic states closely spaced throughout the band gap. It is possible that these transitions may occur by a combination of radiative and multiphonon steps, as shown in (e) in figure 8.3, but the wavelength of the emitted light in such a situation would probably lie in the far-infrared range which is not exploited in luminescent devices. In a multiphonon process, even a very small decrease in energy of a carrier will require the generation of many phonons because their allowed quanta of energy are very small. The larger the number of phonons required for each transition the less likely the transition becomes. Some metallic elements create more than one electronic state in the band gap when they are present as impurities in a semiconductor lattice (see figure 1.5). The five acceptor states characteristic of copper in GaAs are expected to form a particularly effective 'step-ladder' for multiphonon transitions, and the presence of copper in III–V materials has been linked with rapid non-radiative recombination. Metallic impurities of this kind are called luminescence 'killers', since they can have a dramatic effect on the output from radiant devices. Complexes formed between dopant impurity atoms and native point defects also produce many states in the band gap, and we might expect multiphonon events to be especially likely around these kinds of defects.

Even more damaging multiphonon recombination processes can occur at extended lattice defects. We have seen in Chapter 4 that grain boundaries in semiconductor materials can create a very high density of band-gap states which lead to very rapid non-radiative recombination around the boundaries. Both dislocations and free surfaces in semiconductors have a similar effect, as do precipitates formed as a result of attempting to dope a material to a level in excess of the solid solution limit of the dopant element. It is not unreasonable to consider these kinds of defects as infinitely fast recombination centres. Much of Chapter 3 has been taken up with a description of the techniques which can be used to grow single-crystal layers of semiconductor materials for luminescent devices, and I shall assume here that these materials can be produced without grain boundaries. However, in §3.5 I showed that it can sometimes be difficult to avoid the formation of dislocations near interfaces in layered device structures and there will always be free surfaces on the semiconductor sample in which the device is fabricated. It is thus likely that at least some lattice defects will be present in every luminescent device, and in the following section I shall present some experimental observations which illustrate how the properties of hetero-interfaces in semiconductor materials can be degraded by the presence of dislocation arrays. Later sections of this chapter will describe how dislocations are also important in determining the long-term reliability of laser devices in service.

The luminescent efficiency of a semiconducor material can be thought of as depending on the competition between radiative and non-radiative recombination processes, and so on the relative values of τ_{rad}, which I have defined above, and τ_{nonrad}, the minority carrier lifetime as a result of all the non-radiative recombination mechanisms. An expression like (8.1) can be used to estimate the minority carrier lifetime around lattice defects (point defects, dislocations and precipitates) which create a density of band-gap states, N_d, with capture cross sections, σ_d. The effect of Auger recombination on τ_{nonrad} must be calculated independently. A convenient measure of luminescent efficiency is the quantum efficiency, η_Q, which may be defined as

$$\eta_Q = \frac{\tau_{nonrad}}{\tau_{rad} + \tau_{nonrad}}. \tag{8.2}$$

Obviously we require not only as small a value of τ_{rad} as possible, but also a material which contains no grain boundaries, dislocations, second-phase precipitates or killer impurities, if a high luminescence efficiency is to be obtained.

This section has given a necessarily brief introduction to recombination processes in semiconductor materials. Many more details on these often very complex events can be found in the comprehensive treatments by Bergh and Dean (1976) and Sze (1981).

8.2.1(c) The effect of misfit dislocations on luminescent devices

In the previous section I referred to the deleterious effects which dislocation arrays can have on the performance of luminescent devices. In this section I will give two examples of how the electrical and optoelectronic properties of heterointerfaces can be modified by the presence of a misfit dislocation array. In §1.6 we saw that dislocations in semiconductor materials create deep states in the semiconductor band gap and that the position and density of these states is determined by the core structure of the dislocation and probably by impurity segregation as well. The presence of deep states at the dislocation cores will result in charge trapping and the formation of potential barriers at the defect, just as described in Chapter 4 for grain boundaries in semiconductors.

Woodall et al (1983) carried out a classic experiment on the influence of dislocations on the electrical properties of GaInAs/GaAs heterointerfaces. By controlling the thickness and indium content of GaInAs layers on GaAs substrates it was possible to grow dislocation-free interfaces by keeping the ternary alloy layers thinner than the critical value needed to nucleate a misfit dislocation array (see §3.5 and equation (3.15)). GaInAs layers of thickness exceeding this critical value were also grown and a widely spaced array of dislocations was observed at the heterointerface. Electrical measurements showed that there was no potential barrier at the undislocated interface, an ohmic interface, but that there was a significant barrier at the dislocated interface. The Schottky barrier height at these interfaces, Φ_B, increased with the density of misfit dislocations. A plausible explanation of this observation is that the Fermi level is pinned at the deep states created at the cores of the interfacial dislocations. The presence of a sufficiently dense array of these dislocations can then have the startling effect of changing the electrical properties of the interface from ohmic to strongly rectifying. For efficient operation the series resistance of a luminescent device is usually required to be as low as possible, and clearly this will be hard to achieve in a device which contains heterointerfaces containing high dislocation densities. A model for recombination at grain boundaries in semiconductor materials has already been described in Chapter 4 and a very similar model has been developed for dislocations (Figielski 1978). The dislocations in heterointerfaces are thus expected to be preferential sites for non-radiative recombination as well.

As an indication of how effective a misfit dislocation array can be at reducing the intensity of the light emitted from a luminescent device, figure 8.4 shows a total light cathodoluminescence image from part of a GaInAsP epitaxial layer on an InP substrate. The technique of cathodo-luminescence microscopy will be described in §10.2 and gives a very direct indication of the effect that lattice defects have on the efficiency of device material. Here the dark lines and spots are regions of enhanced non-radiative recombination at the cores of dislocations both in an array at the interface between the GaInAsP layer and the InP substrate and also threading up through the epitaxial layer. Experiments of this kind can be used to reveal that the rate of recombination is not the same at all dislocations, which is usually interpreted as being due to

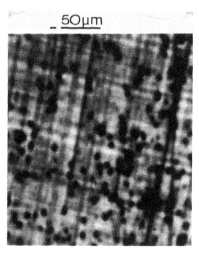

Figure 8.4 A total light cathodoluminescence image from a GaInAsP layer on an InP substrate. The lattice misfit between deposit and substrate is about 5×10^{-3}. The dark lines in the image are regions of enhanced recombination around individual dislocation lines at the heterointerface and the dark spots show recombination around threading dislocations. (Courtesy of Dr M Al-Jassim.)

variations in the atomic structure and segregated impurity concentrations at the cores of the dislocations. Kressel (1975) has proposed an approximate relationship between the misfit strain at a GaAs/GaAlAs heterointerface and the interfacial recombination velocity, S, at the misfit dislocations:

$$S \sim 2 \times 10^{-7} \Delta a/a \qquad (8.3)$$

where a is the lattice parameter of the substrate material and Δa the lattice parameter difference between substrate and epitaxial layer. This expression emphasises the care which must be taken to lattice-match the layers in a heterojunction device if a reasonable efficiency is to be obtained.

8.2.2 Semiconductor Compounds and Alloys

In §8.2.1 I described how some decisions about the range of semiconductor materials in which we might expect to be able to fabricate efficient luminescent devices have already been made. A direct band-gap material is likely to have a much higher quantum efficiency than a semiconductor with an indirect band gap, unless isoelectronic doping techniques are used to encourage radiative recombination processes. For this reason, I shall now discard silicon and germanium as luminescent materials, since these have $\tau_{nonrad} < \tau_{rad}$. Direct band-gap compound semiconductors like GaAs, and isoelectronically doped GaP which behaves like a direct band-gap material, seem good choices for laser and LED materials. However, the wavelength of the light which is emitted from both these semiconductors is controlled by the width of their band gaps, 1.42 eV and 2.24 eV respectively. In GaAs, where the luminescence is dominated by transitions across very nearly the whole band gap, the emitted light can only have a wavelength around 0.9 μm. This is quite close to the absorption window at 0.85 μm in silica fibres. Nitrogen-doped GaP emits green or yellow light at about 2.2 eV (or 0.57 μm), the exact wavelength depending on the nitrogen concentration which determines the position of the isoelectronic state in the band gap. GaP(Zn + O) in which recombination occurs from a trapping state some 0.3 eV below the conduction band, will emit light of longer wavelength than GaP(N) and the energy of the photons will be lower. Emission is in the red or orange spectral range, at about 1.95 eV (or 0.65 μm). We can see that neither GaAs nor GaP can emit light in the violet or blue range of the visible spectrum, 0.4–0.5 μm, or in the infrared region beyond 0.9 μm. I have described how emission at wavelengths between 1.1 and 1.5 μm is needed for compatibility with silica optical fibres, and it is also desirable to be able to produce lasers and LEDs which produce light at any chosen wavelength in the visible spectrum.

If we assume that most direct band-gap semiconductors emit photons with energy very close to the width of the band gap, i.e. that band-to-band transitions dominate the radiative recombination process, then it is possible to select some materials which will produce light in the wavelength regions not covered by GaAs and GaP. Figure 8.5 shows how the band gaps of some potentially useful direct band-gap semiconductor compounds lie relative to the electromagnetic spectrum in the range from the near ultraviolet to the near infrared. I have included in this figure some III–V and II–VI binary semiconductors, a few indirect band-gap compounds like GaP and AlAs, which we will discuss further below, and some more unusual chalcopyrite semiconductor compounds. It is clear that a number of materials can be chosen which will give emission in the visible spectrum, although none of them are binary III–V semiconductor compounds with direct band gaps. We shall see in the next section that it is normally necessary to form a p–n junction in a luminescent device, which can be very difficult in many of the II–VI compounds. This is because they cannot be amphoterically doped for the reasons I have described in §1.4. The heterojunction technology I shall introduce in the following section is a possible way of overcoming this limitation of II–VI materials.

Blue luminescence can be produced in wide band-gap semiconductors like ZnS, GaN and some of the polytypes of SiC, although in all these materials the radiative transitions occur via deep states ((c) in figure 8.3). Nitrogen and aluminium impurities create suitable deep states in SiC, which has an indirect band gap, and so the radiative recombination process must be similar to that described for GaP(N), although these impurities are not isoelectronic dopants. Lithium-doped ZnS emits blue light, while zinc-doped GaN emits light at the violet end of the visible range. These materials are much more difficult to grow as epitaxial films

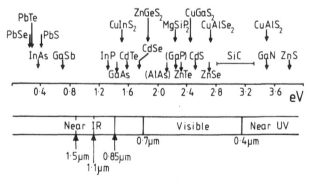

Figure 8.5 An illustration of the relationship between the band gaps of some semiconductor materials and the wavelength of the light which they can emit in band-to-band luminescent transitions.

than GaAs and GaP and can also be hard to dope in a controlled manner. The very high melting point of the SiC polytypes makes the preparation of luminescent devices from this material particularly problematical. GaN films can be deposited by the CVD techniques described in §3.7.

For the chalcopyrite compounds, the growth of stoichiometric crystals, and the control over the conductivity type, is even more difficult. Many of them have extended phase fields at high temperatures (see figure 1.5). This explains why these materials have not been exploited in luminescent devices, even though they have band gaps which are conveniently distributed across the range of the visible spectrum.

8.2.2(a) Ternary and quaternary III–V alloys

I shall now concentrate on how we can produce efficient luminescent semiconductor materials which have band gaps in the range 0.8–1.5 eV, so as to allow emission at 0.85, 1.1 and 1.5 μm, the transmission windows for silica fibres. Extended reviews on the choice of semiconductors for operation in these wavelength ranges have been given by Kressel and Butler (1977), Kressel (1980) and Pearsall (1982). Firstly, let us consider the potential of the two binary III–V compounds which have direct band gaps lying in the range of interest, GaAs and InP. Figure 8.5 shows that neither of these compounds is ideally matched to the absorption windows in silica optical fibres, and so it has proved necessary to use ternary and quaternary III–V alloys to produce material with precisely the required band gap.

The band gaps of ternary III–V alloy semiconductors of the form $A_xB_{1-x}C$ or AC_xD_{1-x}, where A and B are group III and C and D are group V elements, vary monotonically, although not linearly, with the value of x. This means that a ternary alloy composition $A_xB_{1-x}C$ will always have a band gap which lies in between those of the compounds AC and BC. From figure 8.5 we can see that carefully chosen alloy compositions in the ternary systems $InAs_xP_{1-x}$, $Ga_xIn_{1-x}As$ and $GaSb_xAs_{1-x}$ will have band gaps of the required width for emission at 1.1 or 1.5 μm. Similarly, the addition of aluminium to GaAs to form $Ga_xAl_{1-x}As$ will create a semiconductor which can emit at 0.85 μm. GaP_xAs_{1-x} alloys luminesce predominantly in the visible spectrum. If we wish to predict the band gap of a ternary alloy formed between two direct band-gap semiconductors, AC and BC, we must use the fact that the dependence of the band gap, E_g, on x is given by an expression of the form

$$E_g(x) = E_g^{BC} + bx + cx^2 \qquad (8.4)$$

where b and c are constants characteristic of each ternary semiconductor material and E_g^{BC} is the band gap of BC. The variation of E_g with x

in $Ga_xIn_{1-x}Sb$ is shown schematically in figure 8.6(a). Some ternary alloys, $Ga_xIn_{1-x}As$ for example, show an almost linear variation of E_g with x. However, AlAs and GaP are semiconductor compounds with indirect band gaps and so the alloys $Ga_xAl_{1-x}As$ and GaP_xAs_{1-x} show a transition in band structure somewhere in the middle of the composition range. Gallium-rich GaAlAs alloys have a direct band gap like GaAs, but more aluminium-rich alloys have an indirect band gap and so would be expected to show relatively inefficient luminescence. The variation of E_g with x for GaP_xAs_{1-x} alloys is sketched in figure 8.6(b). If we are to select direct band-gap alloys for luminescent applications, it is important to know where the direct–indirect transition occurs. Table 8.1 lists the values of x at which this transition is observed for a number of ternary III–V alloys and the band gap at these compositions. The first four alloys in the table can luminesce in the visible spectrum since the maximum width of the direct band gap falls at the lower end of the range 1.8–3.0 eV in which visible light is radiated.

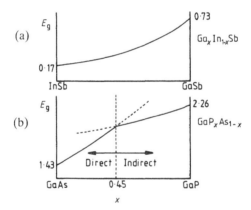

Figure 8.6 The variation of band gap with composition for: (a) the $Ga_xIn_{1-x}Sb$ ternary III–V system which has a direct band gap over the entire composition range; (b) GaP_xAs_{1-x} alloys in which a transition from a direct to an indirect band gap occurs at $x = 0.45$.

The range of band gaps covered by the III–V alloys extends all the way from InSb at 0.18 eV to GaP at 2.24 eV, which means that we can select ternary alloys which will luminesce at any given wavelength from about $7\,\mu m$ down to $0.55\,\mu m$. Data on the variation of E_g with composition for a wider range of ternary III–V alloys have been collected by Casey and Panish (1978). GaAsP alloys are particularly important materials for LEDs because their emission wavelength can extend continuously across much of the visible spectrum, especially if isoelectronic doping with nitrogen is used to give efficient luminescence in alloy compositions which have an indirect band gap.

Table 8.1 The composition and band gap at which the direct–indirect transition occurs in a number of ternary III–V alloys (after Casey and Panish 1978).

III–V alloy	Composition (x)	E_g (eV)
$Ga_xAl_{1-x}Sb$	0.8	0.96
GaP_xAs_{1-x}	0.45	1.98
$Ga_xAl_{1-x}As$	0.6	1.92
$Ga_xIn_{1-x}P$	0.73	2.24
$Al_xIn_{1-x}P$	0.44	2.33
$Al_xIn_{1-x}As$	0.68	2.05

It might seem that the choice of an alloy for emission at any particular wavelength is quite simple given the large number of possible ternary III–V alloy combinations and especially since isoelectronic doping techniques can help to increase the number of efficient luminescent materials. However, in the previous section I described how lattice defects like dislocations can degrade the luminescent efficiency of a semiconductor material by providing a high density of deep states at which non-radiative recombination can occur. Section 3.5 showed how misfit dislocation networks are nucleated at the interface between semiconductor layers whose lattice parameters are not exactly matched. Later in this chapter I will show that laser and LED devices often contain several thin layers of different semiconductor alloys and every interface in such a structure is a potential site for the generation of a misfit dislocation array. From the results shown in §8.2.1(c) we can see that it is crucial that the lattice parameters of adjacent layers be well matched if efficient luminescent devices are to be produced.

Figure 8.7 shows the relation between band gap and lattice parameter for the III–V alloy systems GaInAsP and GaAlAsSb. The ternary alloys which I have been considering so far are to be found on the lines joining individual binary compounds, with full lines indicating a direct band-gap material and broken lines alloys with indirect band gaps. The lattice parameters of these ternary alloys vary approximately, but not exactly, linearly with composition. This diagram makes it quite clear why GaAlAs alloys on GaAs substrates have been widely used to prepare LED and laser devices; the lattice parameters of the two binary compounds GaAs and AlAs are very similar. This means that GaAlAs alloys of any composition can be grown closely lattice matched with GaAs substrates, and there is little danger of misfit dislocation arrays being formed at the heterointerfaces. However, the thermal expansion coefficients of GaAs and GaAlAs alloys differ quite significantly and this can result in some dislocations being produced at the interfaces during cooling, even though there is negligible lattice mismatch at the growth temperature. Nevertheless, this fortuitous lattice matching between AlAs and GaAs has allowed the fabrication by relatively simple

LPE techniques of very efficient luminescent devices operating in the range 0.8–0.9 μm. $Al_xGa_{1-x}P$ alloys are also well lattice matched to GaP substrates, but have an indirect band gap at all values of x and so cannot be used for the production of efficient luminescent devices.

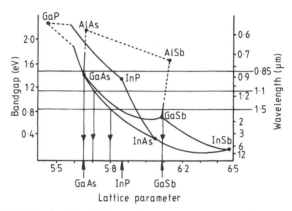

Figure 8.7 A sketch of the relationship between band gap and lattice parameter for the quaternary III–V systems GaInAsP and GaAlAsSb.

The choice of alloy compositions for emission in the 1.1 and 1.5 μm ranges is less simple. The only binary III–V semiconductor compounds which are readily available as substrates for the growth of luminescent layers are GaAs, InP, InAs and GaSb. (Binary compounds which contain aluminium are normally sensitive to chemical attack by moisture and so are difficult to use as substrate materials.) Ideally we must be able to match the lattice parameter of the luminescent alloys to one of these compounds. Alloys of composition $In_{20}Ga_{80}As$ and $GaAs_{85}Sb_{15}$ have band gaps suitable for emission at 1.1 μm, but have lattice parameters which are much greater than that of the closest substrate material, GaAs (figure 8.7). Large lattice mismatches are also generated between InAsP alloys with band gaps in the required ranges and InP substrates. In the 1.5 μm emission range, the lattice mismatches are also large, with ternary alloy compositions in the systems GaInAs, GaAsSb and InPAs all having lattice parameters which are a long way from that of InP. However, gallium-rich GaAlSb alloys can emit at 1.5 μm and are quite well lattice matched with GaSb substrates. In general, it is hard to obtain a perfect lattice match between a substrate and a ternary alloy emitting at either 1.1 or 1.5 μm.

As an indication of the precision with which the lattice parameters have to be matched to avoid the formation of misfit dislocation arrays, figure 8.8 shows some experimental data on the relation between mismatch and the observation of interfacial dislocations for GaInAsP alloy layers with a range of compositions grown on InP substrates (Nakajima 1982). Even mismatchs of 10^{-3} are sufficient to generate misfit dislocation arrays which will severely degrade the luminescent efficiency of any device. When the lattice mismatch exceeds 10^{-3}, the dislocations can only be avoided if rather thin epitaxial layers are grown, less than about 2 μm thick, in accordance with the theory which we developed in §3.5. Figure 8.8 also shows how the thermal expansion coefficient mismatch between substrate and deposit layer shifts the region of dislocation-free growth to one side of the composition at which the lattice parameters are precisely matched. In cases where the thermal expansion mismatch is large, it may prove almost impossible to grow dislocation-free interfaces. Exactly lattice-matched $Ga_{47}In_{53}As$ layers are frequently grown on InP substrates for a wide range of optoelectronic and transistor device applications, since dislocation arrays are not formed in these interfaces.

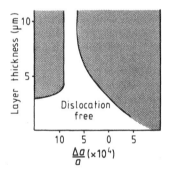

Figure 8.8 Experimental data on the maximum thickness of GaInAsP layers which can be grown on InP substrates without the generation of misfit dislocations. As the lattice parameter mismatch exceeds 10^{-3}, only very thin epitaxial layers can be grown dislocation free (after Nakajima 1982).

Ternary III–V alloys are thus not ideal materials in which to make luminescent devices. Once the width of band gap is chosen in material in a particular alloy system, the lattice parameter is fixed as well. Selecting an alloy for emission at a required wavelength, and then matching the lattice parameter of the alloy to that of a suitable substrate, is only possible in a few systems, the most widely exploited being GaAlAs alloys on GaAs substrates which do not emit in the important range 1.1–1.5 μm. This explains why quaternary III–V alloys are attractive materials for use in luminescent devices; here both the band gap and the lattice parameter can be varied independently throughout the quaternary regions in figure 8.7. GaInAsP alloys which are perfectly lattice matched with InP substrates can luminesce in the range 0.9–1.7 μm and, when lattice matched with GaAs, in the range 0.6–0.7 μm. These alloys are thus suitable for use in optical sources emitting both in the absorption windows of silica fibre material and in the visible spectrum. Similarly, GaAlAsSb alloys containing low concentrations of antimony can be closely, but not perfectly, lattice matched with GaAs substrates. Antimony-rich alloys can be perfectly lattice matched with GaSb. These alloys give emission at wavelengths around 0.85 and 1.5 μm respectively. As well as the advantage of being able to select the emission wavelength and lattice parameter independently, quaternary III–V alloys can be chosen to reduce the stress generated during cooling from the growth temperature. We have already seen that the thermal expansion coefficient of AlAs differs from that of GaAs for instance, and so dislocations can be generated at the GaAlAs–GaAs interface during cooling. The addition of a small amount of phosphorus to the GaAlAs layers will change both the lattice parameter and the thermal expansion coefficient, reducing the stress generated in the epitaxial layers (Rozgonyi et al 1974). A discussion of the properties of a wider range of quaternary III–V alloys has been given by Casey and Panish (1978). It can be quite difficult to grow epitaxial thin films of quaternary alloys which have precisely controlled compositions, and some discussion of the preliminary work which needs to be done before LPE or CVD techniques can be used to deposit satisfactory layers is included in Chapter 3. Tsang (1985) has recently edited a number of papers which describe the techniques that have proved most successful in the preparation of epitaxial layers for devices which operate as sources in optical communications systems – LPE, MOCVD and MBE.

Recently, it has been found that many of the ternary and quaternary III–V alloy semiconductors can both order during growth and spinodally decompose. For a description of these two phenomena, interested readers are directed to Haasen (1978). Figure 8.9(a) shows a diffraction pattern taken from an AlInAs alloy where ternary ordering of the aluminium and indium atoms in the unit cell is clearly identified by the 'extra' diffraction spots over and above those which would be expected from the sphalerite lattice. A suggested arrangement of the atoms in the unit cell is shown in figure 8.9(b). Some experimental and theoretical data on this ordering phenomena can be found in Martins and Zunger (1986) and Kuan et al (1987). Nakajima et al (1977) have determined that the GaAlAsSb quaternary system contains a miscibility gap, and Stringfellow (1982) predicted the same behaviour in many quaternary combinations of Al, Ga and In with P, As and Sb. Spinodal decomposi-

(a)

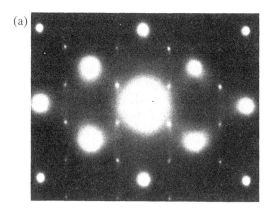

(b)

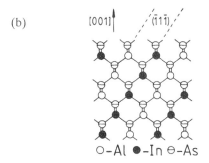

$O-Al$ ● $-In$ ⊖ $-As$

Figure 8.9 (a) A transmission electron diffraction pattern taken from an AlInAs epitaxial layer. The extra weak spots show that ordering of the aluminium and indium atoms to separate {111} planes has occurred (Norman and Booker 1985). (b) A suggested structure for the ordered AlInAs alloy, showing that alternate ($\bar{1}1\bar{1}$) planes consist of pure aluminium and indium. (Courtesy of Dr A Norman.)

tion can thus occur during growth and cooling of alloys with compositions which lie over a considerable region of the quaternary fields. The form of the miscibility gap in the GaInAsP alloy system has been calculated by de Cremoux *et al* (1980) and is sketched in figure 8.10(a). Figure 8.10(b) shows an electron micrograph of part of a layer of $Ga_{41}In_{59}As_{87}P_{13}$, where the alternating light and dark contrast indicates that spinodal decomposition has occurred during growth. Reliable data on the extent of the composition fluctuations in these materials are sparse, but it is becoming clear that both ordering and phase separation may be rather universal features of III–V alloy semiconductor layers grown by the techniques described in Chapter 3. We should not forget that the composition of these alloys may also vary due to random fluctuations in the chemistry of the gas phase or liquid phase during growth. These accidental composition fluctuations will be superimposed on those due to spinodal decomposition, giving semiconductor material which is rather inhomogeneous on the fine scale. How much effect these fine-scale composition variations have on the luminescent and electrical properties of these alloys remains unclear, although several sets of experimental data indicate that electrical properties like carrier mobility are rather low in materials which show spinodal decomposition. As the size of individual luminescent devices is reduced, these variations in composition may become very significant in determining the overall efficiency.

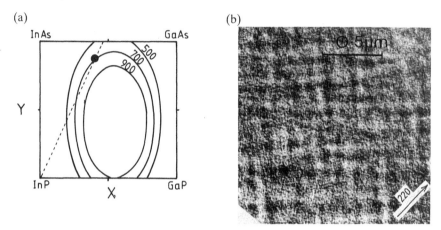

Figure 8.10 (a) The form of the miscibility gap calculated for the GaInAsP quaternary alloy system. The broken line indicates the composition of alloys which are perfectly lattice matched with InP and the circle marks a composition of $Ga_{41}In_{59}As_{87}P_{13}$ (after de Cremoux *et al* 1980). (b) A transmission electron micrograph of an epitaxial layer of the composition marked in (a), showing the contrast variations characteristic of a spinodally decomposed material. (Courtesy Dr A Norman.)

8.2.2(b) Materials for emission at wavelengths longer than 2 μm

In order to achieve efficient luminescence at wavelengths longer than 2 μm, direct band-gap materials with small band gaps are needed. (We are still assuming that band-to-band transitions dominate the radiative recombination process.) Because of the small band gaps, these devices must be operated well below room temperature to avoid the thermal excitation of carriers. $InAs_xSb_{1-x}$ and $Ga_xIn_{1-x}Sb$ are the only ternary III–V alloys which can operate in this wavelength range. GaInAsSb and InAsPSb quaternary alloys, which can be lattice matched to InAs and GaSb substrates respectively, can also give emission in the range between 2 and 4 μm. Both these quaternary alloy systems have miscibility gaps and so the alloys can be difficult to grow as homogeneous layers. A number of II–VI alloys have small band gaps as well.

$Hg_xCd_{1-x}Te$ in particular has been widely investigated for photodetector applications and can be reasonably well lattice matched with CdTe substrates. More attention has been paid to IV–VI alloys based on the lead salts PbTe, PbSe and PbS. These alloys luminesce in the wavelength range from about $5 \mu m$ all the way out to $34 \mu m$, and considerable use has been made of the ternary alloys PbSnTe and PbSSe for lasers. The unusual characteristics of several of the II–VI and IV–VI ternary alloys, where the width of the band gap goes to zero at some points in the composition range and can be sensitive to operating temperature as well, have been mentioned in §1.2.

The II–VI and IV–VI alloys are, in general, much more difficult than the III–V alloys to grow either as bulk single crystals or epitaxial thin films. The problems encountered in the preparation and growth of HgCdTe alloys with well controlled stoichiometries and electrical properties have been described in Chapter 3 and by Micklethwaite (1981). The lead salts can also be hard to deposit, although LPE, MBE and evaporation techniques have all been used to grow epitaxial layers. We have seen in Chaper 1 that the conductivity type and carrier concentrations in most II–VI and IV–VI materials are controlled by native point defect concentrations. This is usually regarded as an inconvenient characteristic of these materials, but it is possible in some cases to change the electrical properties of the surface layers of an epitaxial film by annealing in an atmosphere containing appropriate concentrations of the component elements. This method of forming junctions in these materials will be mentioned again in later sections.

The lattice match of the lead salts on available substrate materials like PbTe is often poor. Fortunately, the high density of interfacial dislocations which must be present at heterointerfaces like that between a PbSnTe layer and a PbTe substrate seems to degrade the luminescent properties to a lesser extent than is common in the III–V alloys. It is not at all clear why this should be the case, but it allows useful luminescent devices to be produced from the lead salts, even when the lattice mismatch between the epitaxial layer and the substrate is as high as 10^{-2}. A more comprehensive list of materials suitable for use in luminescent devices operating at long wavelengths can be found in Casey and Panish (1978) and Horikoshi (1985).

8.3 p–n JUNCTIONS AND HETEROJUNCTIONS IN LUMINESCENT DEVICES

So far in this chapter I have concentrated on the structure and the electronic properties of the semiconductor materials used in the production of efficient luminescent devices. We have seen that it is radiative recombination which is the basic mechanism of luminescence in semi-conductors, but I have not yet explained the manner in which these recombination processes are stimulated in lasers and LEDs. I shall briefly introduce how simple luminescent devices operate, concentrating on the role of p–n junctions in this process. Heterointerfaces, those formed between semiconductor materials of different composition and band gap, also have a number of important uses in these devices and will be considered separately. Detailed treatments of the phenonena of injection, carrier confinement and waveguiding in lasers and LEDs can be found in standard texts like Kressel and Butler (1977) and Sze (1981). Here it is appropriate only to introduce the basic material concepts underlying these processes.

8.3.1 Minority Carrier Injection

One method of exciting rapid recombination in a semiconductor crystal is by the introduction of a concentration of minority carriers far in excess of the equilibrium level. Under these conditions, luminescent recombination will readily occur in materials of the kind that we have selected above and light will be emitted in all directions from the region of recombination. A convenient way of introducing an excess of minority carriers into a semiconductor is by carrier injection across a forward-biased p–n junction. It is also possible to achieve the same effect by avalanche injection at a heavily reverse-biased p–n junction, or at a metal/semiconductor junction, although I shall not consider these excitation modes further in this chapter.

Figure 8.11 illustrates the operation of a p–n junction under forward bias in a semiconductor material where efficient radiative recombination occurs. An excess of electrons is injected into the p-type material, and an excess of holes into the n-type material, and so recombination can occur on both sides of the junction. The $J–V$ characteristics of an ideal p–n junction can be described by the expression for the diffusion current:

$$J = J_0[\exp(eV/kT) - 1].\qquad(8.5)$$

In real junctions, however, this equation is modified by several effects, of which recombination in the space-charge region is the most important in luminescent devices. Non-radiative recombination of the injected holes and electrons can occur at deep states at the heterointerface or in the space-charge region. If we assume that these trap states are roughly at the centre of the band gap, then non-radiative recombination alters the $J–V$ characteristics of the p–n junction such that

$$J \propto \exp(eV/nkT)\qquad(8.6)$$

where $n = 1$ for the ideal junction as above and is approximately equal to 2 when recombination in the space-charge region dominates. Trap states at p–n junctions in luminescent devices can be produced both by misfit dislocation arrays and by the segregation of killer impurities to the interface. The non-radiative recombination current can dominate the $J–V$ characteristics of p–n junctions in luminescent devices when the forward bias is small, but the diffusion current will usually dominate at high applied bias. Thus it is possible to obtain reasonably efficient luminescence from a device containing a highly defective p–n junction simply by increasing the applied voltage. The luminescent output from a simple p–n junction LED, L, will vary with applied voltage in much the same way as the total current:

$$L \propto \exp(eV/nkT) \tag{8.7}$$

where n varies between 1 and 2 as above. Gooch (1973) and Bergh and Dean (1976) have discussed the electrical properties of p–n junctions in luminescent devices in greater detail.

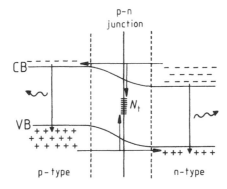

Figure 8.11 A schematic illustration of the operation of a forward-biased p–n junction in a luminescent device. Both electrons and holes are injected across the junction and radiative recombination occurs in n-type and p-type material. Recombination at trapping states at the junction is also shown.

Another important parameter of these p–n junctions is the ratio of the fractions of the total diffusion current carried by injected electrons and holes, J_e and J_h respectively. This ratio depends on the conductivity of the n- and p-type semiconductor materials and the minority carrier diffusion lengths on both sides of the junction and is given roughly by

$$\frac{J_e}{J_h} \sim \left(\frac{D_n L_p N_D}{D_p L_n N_A} \right) \tag{8.8}$$

where D_n and D_p are the minority carrier diffusion coefficients in n- and p-type material, L_p and L_n are the minority carrier diffusion lengths in n- and p-type material and N_D and N_A are the donor and acceptor dopant concentrations on either side of the p–n junction. For a p–n junction in GaAs, where N_A is 5×10^{18} cm^{-3} and N_D is 2×10^{18} cm^{-3} for instance, the electron current is about 80% of the total diffusion current across the junction. Under these circumstances, most of the light will be emitted from the p-type side of the junction. Luminescent devices with the highest efficiencies are made when injection of electrons into the p-type material is the dominant contribution to the junction current. It is clear from equation (8.8) that increasing the doping level in the n-type material will increase the electron injection current, and this is a simple design rule for the production of efficient luminescent devices.

Electroluminescence can be excited in a semiconductor material simply by applying a forward bias across a p–n junction, which is the fundamental mechanism by which most LEDS operate. However, the conditions for achieving lasing action at a p–n junction are a little more complicated. There are two competing processes which can occur when an energetic photon propagates through a semiconductor lattice. Firstly, it can be absorbed with the creation of an electron–hole pair, and carriers produced in this way will then recombine either radiatively or non-radiatively after a time characteristic of the doping concentration and trap density in the material, as I have discussed above. Alternatively, the photon can stimulate a recombination event, releasing a second photon which will be coherent with the first, i.e. stimulated emission. Under normal conditions, where the carrier concentrations in the semiconductor are close to the equilibrium levels, this second mechanism is not important because the population of minority carriers is too low for there to be any significant probability of a stimulated recombination event. However, if the carrier population in the semiconductor can be inverted, efficient stimulated emission can occur. What I mean by population inversion is that the probability of a state near the bottom of the conduction band being occupied is greater than the same probability for a state at the top of the valence band. The formal definition of population inversion is that the two quasi-Fermi levels, one for electrons and one for holes, are separated by more than the width of the band gap, E_g. If this condition is achieved in a semiconductor, rapid stimulated emission can occur by recombination of the excess concentrations of electrons and holes. Figure 8.12(a) shows a diagram of a p–n junction in a heavily doped semiconductor, in which I have sketched the situation of carrier inversion on the p-type side of the junction. The photons generated by radiative recombination in the p-type semiconductor will stimulate further coherent emission. This is the principle of operation of a semiconductor laser. The p–n junction needs to be heavily forward biased before the inverted population is created.

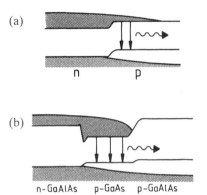

Figure 8.12 A schematic diagram of population inversion in p-type semiconductor material close to a heavily forward-biased p–n junction. Recombination of the excess concentrations of electrons and holes leads to the emission of a beam of coherent light. (b) A sketch of the band structure of a double heterojunction laser in which carriers are confined in a thin GaAs layer by the potential barrier at a second GaAs/GaAlAs heterojunction.

The injected electrons can also recombine non-radiatively at trap states in the junction itself, just as I have described for p–n LEDs, and this lowers the efficiency of the laser diode. It is clearly important to have as low a dislocation density as possible in the active junction region. It is also convenient to constrain the injected electrons to remain close to the p–n junction, because this allows the inverted carrier population to be maintained with a lower junction current. The threshold current at which lasing begins is a particularly important parameter of a laser diode, as it controls the temperature of the p–n junction during operation (§8.5). The development of laser structures in which the injected carriers can be confined to a region close to the p–n junction has been one of the major steps in the fabrication of efficient semiconductor lasers. The principles of carrier confinement have also been applied to the design of high-performance LEDs.

8.3.2 Carrier Confinement and Waveguiding

In a simple p–n junction laser structure like that sketched in figure 8.12(a), the injected electrons can diffuse away from the junction into the bulk of the p-type material. At some distance from the junction, the concentration of electrons in the conduction band will decrease to such a low level that the carrier population is no longer inverted. The electrons which diffuse out of the region of population inversion cannot contribute to stimulated emission, but will continue to recombine giving incoherent emission like that from a LED. Figure 8.12(b) shows the structure of a laser diode in which the injected electrons are confined close to the junction by a potential barrier in the p-type material. I have sketched a double heterojunction (DH) laser structure in this figure where a semiconductor material with a relatively small band gap, GaAs, is surrounded by layers of a semiconductor with a larger band gap, GaAlAs in this case. Electron injection still occurs from the n-type GaAlAs across the p–n heterojunction into the p-type GaAs, but then the electrons are confined to the GaAs layer by the second p-type GaAlAs layer which places a potential barrier at the edge of the conduction band. This second junction is an example of an isotype heterojunction, as the doping type is the same but the band gap and composition of the semiconductors different. The particular value of this kind of laser structure is that the threshold current for stimulated emission is much lower than in a simple p–n junction laser. The diffusion length of electrons in heavily doped p-type GaAs is approximately 1 μm and so this is the ideal thickness of the GaAs active layer where all the stimulated emission occurs. This is a convenient thickness for deposition by many of the thin film epitaxy techniques described in Chapter 3.

The laser structure illustrated in figure 8.12(b) contains two heterojunctions between GaAs and GaAlAs, and these interfaces can have useful properties in optoelectronic devices other than carrier confinement. The first of these concerns the shape of the band edges at p–n heterojunctions (anisotype junctions). Anderson (1962), early in the study of semiconductor heterojunctions, described the form of the band edges in all the possible combinations of materials. He showed that the important parameters in deciding the form of the band edges are the band gaps of the two semiconductors, E_{g1} and E_{g2}, their electron affinities, χ_1 and χ_2, and their workfunctions, Φ_1 and Φ_2. These are the same parameters which we used to model the band structure of metal–semiconductor contacts in §2.2. In figure 8.13 I have indicated one possible result of taking a p-type semiconductor with a small band gap and bringing it into contact with a wide band gap n-type material. The p–n heterojunction has a pronounced spike and notch structure at the edge of the conduction band, and under forward bias this interface will have a band structure similar to that shown at the left-hand side of figure 8.12(b). The difference in the height of the potential step at the valence band edge and that at the conduction band edge increases the ratio of electron injection over hole injection very markedly; for instance, if the band gaps E_{g1} and E_{g2} differ by as little as 0.2 eV then the electron current across the heterojunction will be a factor of 10^3 greater than the hole current. This kind of anisotype heterojunction can therefore be used to improve the efficiency of a laser diode by increasing the injection ratio of electrons to holes. The J–V characteristics of a heterojunction are complicated by the presence of the spikes and notches in the band edges, but to a first approximation they behave like the p–n junctions described above (Milnes and Feucht 1972), Once again, the density of carrier trapping sites at the heterojunction and in the space-charge region will control the efficiency of any heterojunction luminescent device.

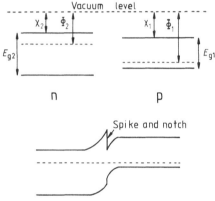

Figure 8.13 A sketch of the band-edge structure at an anisotype heterojunction between a p-type semiconductor with a small band gap and an n-type material with a larger band gap.

However, it is not only the electronic properties of the semiconductor materials which are altered by the presence of a heterojunction; the optical properties of semiconductors change with composition as well. In the case of the GaAs/GaAlAs double heterojunction structure shown in figure 8.12(b), the refractive index of GaAlAs is greater than that of GaAs and increases roughly linearly with aluminium content. Thus in this laser structure not only do the GaAlAs layers allow carrier injection and confinement, but they also act as optical waveguides. The light emitted in the GaAs layer is totally internally reflected from the heterointerfaces in exactly the same manner as illustrated for the optical fibre in figure 8.2. This confines the laser light to well defined paths in the diode, which can be very convenient for the effective coupling of the light into optical fibres.

Finally, heterojunction structures can have two additional uses in luminescent devices. An n-type isotype heterojunction can be used to improve the quality of ohmic contacts to the top and bottom of a light-emitting diode. We have seen in §2.2 that it can be hard to make good ohmic contacts to lightly doped semiconductors, and a degenerately doped capping layer, which plays no active part in the generation of the light, can assist in the formation of better contacts. It is important that the capping layer is very well lattice matched to the underlying layers of course. Secondly, much of the light emitted at the p–n junction in a luminescent device will be absorbed if it passes out of the device through material with a band gap of width just equal to, or less than, the energy of the photons. However, if much of the light path lies through material of larger band gap, very little absorption will occur as the photons will no longer be able to excite electron–hole pairs. An example of a heterojunction LED design which makes use of this effect will be described later in this chapter.

In this brief section I have only introduced the very basic properties of heterojunctions. The efficient operation of many LEDs and laser diodes depends on the use of these junctions and in particular on their crystalline perfection and freedom from killer impurities. I shall discuss how these very perfect heterojunctions and p–n junctions can be produced in the next section. The GaAs/GaAlAs double heterojunction laser structure, which I have used as an example above, is one of the most successful applications of these junctions in devices, but there are numerous other more complex laser and LED structures. Further details on the role of semiconductor heterointerfaces in a wider range of devices can be found in Sharma and Purohit (1974).

8.3.3 The Preparation of p–n Junctions and Heterointerfaces for Lasers and LEDS

I have described above how junctions between heavily doped compound semiconductors (III–V, II–VI and IV–VI materials for emission over a wide range of wavelengths) are needed in laser and LED structures and have emphasised that the density of trapping states at these interfaces is a most important parameter in determining the efficiency of these devices. In addition, the doping levels on both sides of the junction have to be carefully controlled, the interfaces made as flat and as chemically sharp as possible, and the layer thickness in a multilayer structure carefully chosen and accurately grown. I shall now briefly review the processes by which these junctions can be prepared, starting with the most common technique for the deposition of multilayer luminescent device structures, liquid phase epitaxy, LPE.

8.3.3(a) III–V and IV–VI compounds

A section in Chapter 3 was devoted to a discussion of the growth of thin epitaxial films of semiconducting materials by LPE. There it was shown that a wide range of binary III–V and IV–VI compounds, as well as their ternary and quaternary alloys, can easily be deposited from dilute solutions onto suitable substrates. High dopant concentrations can also be included in the layers, although the dopant concentration profile may not be uniform across the layer if its partition coefficient is far from unity. Multilayer structures may be deposited using the sliding substrate technique illustrated in figure 3.17 and both homo- and heterointerfaces of all kinds are simply prepared by selection of a series of melts with suitable compositions. The fact that the growth temperatures in LPE are considerably lower than the melting temperature of the material being deposited can also help in controlling the concentration of electrically active point defects (see figure 3.16(b) for example).

It is possible to use CVD techniques (§3.7) to grow similar multi-layered structures containing homo- and heterointerfaces. These techniques have the additional advantages that many substrates can be processed at once and that non-equilibrium concentrations of some dopants can be incorporated in the growing layers (as has been described for GaP(N) for instance). If really sharp interfaces are required, where no interdiffusion between the layers occurs during growth, MBE becomes an attractive deposition process (§3.8). This is because the epitaxial growth temperature in MBE can be significantly lower than in LPE or CVD. This can be especially important in the preparation of sharp interfaces in IV–VI heterostructure devices where the solid state diffusion rates can be particularly high.

High-quality luminescent device structures can only be grown if a substrate material with a suitable lattice parameter is available. I have shown how ternary and quaternary III–V alloys with a wide range of band-gap widths and so emission wavelengths can be lattice matched with the common substrate materials. Horikoshi (1985) has described a similar advantage when quaternary IV–VI alloys are used for the preparation of lasers for operation at long wavelengths. While the use of ternary and quaternary alloy materials widens the choice of emission wavelengths, it is still important that very perfect substrate crystals are available on which to grow these epitaxial layers. In §3.3 the growth of large crystals of III–V compounds like GaAs and InP was described, and InAs and GaSb crystals can also be grown by the same techniques for use as substrates for III–V heterostructure devices. It is vital that the dislocation density in these substrate crystals be as low as possible, or it is likely that threading dislocations will propagate into the active regions of a device during epitaxial layer growth. These threading dislocations can act as highly effective recombination centres and also play an important part in the degradation of laser diodes in service (§8.6). Bulk substrates of binary and ternary IV–VI materials can be prepared by Czochralski, Bridgman or vapour growth techniques, but often contain substantial composition fluctuations and high dislocation densities.

A simple method has been used to reduce the number of dislocations that propagate up from a defective substrate into a growing epitaxial layer. The introduction of sharp composition changes in the optically inactive material far from the p–n junction will create misfit dislocation arrays at each heterointerface. A threading dislocation can be trapped at the stress fields around these arrays, bending around to lie in the plane of the interface. This effect can occur at the epilayer/substrate interface, or higher up in a multilayered epitaxial structure, but in both cases the propagation of threading dislocations is suppressed (Matthews and Blakeslee 1974, Rozgonyi et al 1974). Thus by using substrate wafers containing few lattice defects, making a careful choice of the composition of the active epilayers, and, if necessary, introducing composition variations in optically inactive material, it is possible to grow heterostructure devices in III–V and IV–VI alloys containing negligible dislocation densities and with low threshold currents and high irradiances.

8.3.3(b) II–VI compounds

The growth of thin films of II–VI materials of device quality is generally much harder than for the III–V compounds and alloys. This is in part a result of the dominant influence that native point defects have on the electrical properties of these semiconductors and this same characteristic makes amphoteric doping of the binary compounds very awkward (see §1.4). These problems have limited the use of materials of this family in homojunction devices to a few ternary alloys which can be amphoterically doped, i.e. HgCdTe and, to a lesser extent, ZnSeTe and related alloys. Evaporation, MOCVD and MBE techniques have all been used to produce epitaxial layers of ternary II–VI alloys.

Heterojunction devices have been prepared in II–VI materials by chemical deposition, evaporation, MBE and CVD techniques, as has been reviewed by Sharma and Purohit (1974). p–n heterojunctions are most easily formed when one of the II–VI materials is naturally a p-type semiconductor and the other an n-type material. For example ZnTe can only be prepared p-type and most other binary II–VI materials are naturally n-type. It is thus relatively simple to make a ZnTe/ZnSe p–n heterojunction by evaporating ZnTe onto a ZnSe substrate. However, the preparation of these device structures requires that substrate wafers with high crystalline quality are available, and these are often difficult to prepare. Reasonably perfect CdTe crystals have been produced on which HgCdTe epitaxial layers can be grown, but substrates of materials like ZnSe or CdS are usually of much poorer quality than those of InP or GaAs. Some of the II–VI materials which are available in single-crystal form are very soft and so dislocation and twin densities can be high. The replacement of CdTe by a material with a higher yield stress, like $Cd_{60}Mn_{40}Te$ for instance, can improve the quality of the epitaxial layers (Bridenbaugh 1985). CdMnTe alloys are readily grown by Bridgman techniques.

The problems of fabricating p–n junctions in II–VI semiconductors has resulted in the development of an alternative strategy for the production of luminescent devices in these materials. MIS diodes, where electrons are injected from a metal contact into a II–VI crystal, can make reasonably efficient devices. Relatively little commercial exploitation of II–VI hetero or homojunction devices has been attempted when compared with the vast range of LEDs and lasers made from III–V or IV–VI materials, and we can understand this because of the rather intractable nature of these semiconductors.

8.3.3(c) Diffusion and implantation

p–n homojunctions can also be prepared in epitaxial layers by diffusion or implantation (see §2.4). Junctions formed by indiffusion of a dopant element have been used to make very efficient devices, especially in III–V and IV–VI materials. The phenomenon of diffusion in compound semiconductors has been described in §1.5. Most donor impurities diffuse too slowly in III–V compounds for the formation of p–n junctions in a reasonable time and so the diffusion of acceptor dopants into n-type material is the only practical way of producing junctions in III–V devices. Careful control of the partial pressure of the volatile group V elements is required before the p–n junction can be created at a chosen depth beneath the surface of the n-type material. Dopant diffusion rates in ternary and quaternary III–V alloys are often not well known, but zinc diffusion rates in $GaAs_xP_{1-x}$ alloys decrease as the GaP fraction increases, bringing the p–n junction position in any given indiffusion anneal closer to the surface. Bergh and Dean (1976) have given a detailed account of the advantages and disadvantages of using diffused p–n junctions to make LED diodes in GaP. Similar diffusion processes can be used to make p–n junctions in IV–VI and II–VI alloys as well, although in these materials it is usually the concentration of one of the native point defects which is altered in the sample surface. For instance, annealing an n-type HgCdTe alloy results in the loss of volatile mercury from the sample surface and the formation of a p-type layer because of the acceptor properties of mercury vacancies. Section 1.5 contains some data on dopant diffusion in III–V compounds and on the diffusion of the elemental components in CdS.

Dopant implantation is not widely used for the fabrication of p–n junctions in luminescent materials because of the lattice damage which is introduced during the implantation process (see §2.4). It is not always possible to anneal out this damage completely, and we have already discussed the deleterious effect which dislocations have on the radiative efficiency of luminescent devices. However, heavy p-type doping of

ZnSe by implantation with Li ions can produce p–n junctions in a material which is normally only an n-type conductor. Implantation of boron as an n-type dopant into HgCdTe alloys can also be used to produce p–n junctions. In both these cases, the use of very light ions, Li^+ and B^+, means that the amount of lattice damage will be small.

Finally, I shall mention a remarkable way of producing a p–n junction in LPE silicon-doped GaAs. Silicon is an amphoteric dopant in GaAs, an n-type dopant when substituting on gallium sites and a p-type dopant on arsenic sites (§1.4.2). The equilibrium distribution of the silicon atoms between these two sites depends on the precise growth temperature and the cooling rate. During growth from a melt containing about 0.1 at. % silicon, the GaAs layer switches from highly n-type to highly p-type at a temperature of about 800 °C (Rosztoczy 1968). This allows the formation of p–n homojunctions in GaAs in a single LPE deposition process, reducing the chances of contamination at the junction and giving a very cheap method of growing material for highly efficient infrared LEDs.

8.3.4 ZnS Phosphor Devices

All of the luminescent devices which I shall describe in §§8.4 and 8.5 are made from semiconductor materials which are very carefully grown in the form of single crystals. However, it is also possible to obtain surprisingly efficient luminescence from materials in powder form. This has been exploited in devices based on some II–VI semiconductor materials, particularly ZnS. It may seem surprising that a semiconductor compound with a band gap of 3.68 eV can emit light in the visible range, but copper impurities in ZnS create an acceptor trap state about 0.95 eV above the valence band edge. Radiative recombination can occur between holes trapped at this state and free electrons from the conduction band to release visible photons; copper is called an 'activator' element. Impurity elements like Cl, Br, Al and In are called 'coactivators' because they increase the solubility of copper in ZnS and so increase the intensity of the emitted light. ZnS(Cu + Cl) powders are reasonably efficient emitters of green light for example. Manganese additions change the dominant emission from green to yellow, but here the recombination at copper atoms excites the core electrons in adjacent manganese atoms, leading to the spontaneous emission of characteristic yellow light.

The doped ZnS powder is usually supported in an insulating binder of plastic or glass and the composite material is a very good electrical insulator. The electronic processes leading to electroluminescence are complex, but one popular theory supposes that the addition of excess copper results in the formation of p-type Cu_xS compounds, either on the outer surface of the ZnS grains or as small precipitates in the grain interiors. This creates a large number of p–n heterojunctions and the injection of carriers across these junctions from the small band-gap Cu_xS into the ZnS stimulates radiative recombination. The luminescent efficiency obtained from such a device is much lower than from a LED or laser diode, but suitably doped ZnS powders can be used to make relatively cheap large-area luminescent displays. Ivey (1963) and Vecht and Werring (1970) have described the production and operation of these devices in more detail.

8.4 THE DESIGN OF LEDs

Figure 8.14 shows a schematic view of a simple LED structure in a GaP(N) crystal. The basic components of the device are the p–n junction at which the light is generated when a forward bias is applied, the GaP substrate in which the homojunction has been prepared and an ohmic contact on the top and bottom of the semiconductor crystal. The metals and alloys which can be used to form good ohmic contacts to the semiconductor materials of interest in the fabrication of LED diodes have been discussed in §2.2. I have sketched the diode with a small ohmic contact on the top surface and this design allows the maximum amount of light to escape from the device. The bottom ohmic contact is patterned through an array of holes in a silica film and this pattern can be produced by a lithographic process similar to those described in Chapter 2. It is advantageous to limit the area of this ohmic contact because the light emitted at the p–n junction is strongly absorbed at metal–semiconductor interfaces where significant interdiffusion and reaction processes have occurred. Ohmic contacts to GaP will normally be of this kind (see table 2.2) and replacing some of the contact area by the reflective GaP–SiO_2 interface will decrease the amount of light absorbed at the ohmic contact. This kind of patterned contact structure is a compromise between the conflicting desire to have large-area contacts, which will have low resistances, and to allow as much light as possible to escape from the device. A large bottom contact also acts as a heat sink, which is important since the efficiency of a LED decreases rapidly with the temperature of the p–n junction. A significant amount of heat is generated in the diodes during operation.

One final observation I shall make about the LED structure illustrated in figure 8.14 relates to the decision to place the p–n junction right at the bottom of the LED crystal so that the escaping light has to pass through a considerable thickness of gallium phosphide. The rate at which photons are absorbed in a semiconductor material depends on the band gap, the doping levels and the doping type (see Chapter 9). The photons generated by the recombination of injected electrons on the p-type side of the junction will have an energy slightly less than the width of the GaP band gap since the isoelectronic nitrogen state lies just below the conduction band. We might expect that a significant proportion of the light generated in the p-type GaP will be absorbed as it passes through the thick layer of n-type material (100–200 μm).

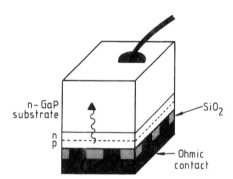

Figure 8.14 A sketch of one possible structure for a p–n junction LED in a GaP crystal.

However, even a relatively low concentration of an n-type dopant in GaP will create an impurity band near the edge of the conduction band and so populate the states at the bottom of the conduction band. This means that the effective width of the band gap is slightly increased, a phenomenon called the Moss–Burstein shift. As a result, photons with energy slightly less than E_g will pass through n-type GaP without being absorbed to any great extent, since they do not possess sufficient energy to promote an electron from the top of the valence band to the unoccupied states in the conduction band. The n-type material thus acts as a window for the emitted light. It is convenient to invert the GaP crystal obtained by the growth of a thin epitaxial p-type layer on an n-type wafer to keep the active p–n junction close to the heat sink at the bottom contact.

Another way of making sure that the light generated at the p–n junction escapes from the device without significant absorption is by growing layers of graded composition in ternary alloy materials. These can serve two useful purposes. Firstly, we have seen that it is often hard to grow an epitaxial layer of a ternary alloy with the required band gap on the available substrate materials without a high density of dislocations being created at the heterointerface. As an example, let us consider the preparation of a device in $Ga_{66}In_{34}As$, which emits at 1.06 μm. This alloy is poorly lattice matched with the most suitable substrate material, GaAs (see figure 8.7). We expect, therefore, that a high density of interfacial dislocations will be formed if the alloy is grown directly on GaAs and that some of these dislocations will propagate up through the growing layer. The luminescent efficiency of this material is likely to be poor because of enhanced non-radiative recombination at these dislocations. However, if a graded layer is grown, starting with pure GaAs directly on the substrate and slowly increasing the In content up to 17 at. %, then the total dislocation density in the layer will be reduced. In particular, the regions of crystal containing the p–n junction can be completely free of these defects. An example of the distribution of dislocations in a graded layer of GaInAs on GaAs is shown in figure 8.15. The second advantage of graded layer device structures is that the light generated at the p–n junction can escape through the much thicker n-type layer and the GaAs substrate not just because of the Moss–Burstein shift, but also because the band gap steadily increases through the graded layer and into the substrate. Graded layers thus allow the device engineer the option of placing a very effective optical window close to the p–n junction, and these structures have been included in the design of some high-performance LEDS.

Figure 8.15 A cross-sectional transmission electron micrograph of a GaInAs epitaxial layer with a graded composition grown on a GaAs substrate. A high density of dislocations can be seen at the original growth interface at the right-hand side of the micrograph. The indium content increases as we proceed away from this interface, and the p–n junction is located near the left-hand side of the micrograph where the dislocation content is fairly low. (Courtesy of Dr M Hockly.)

LED devices which emit in the visible spectrum are usually packaged inside resin lenses. This is because compound semiconductor materials like GaP and GaAsP have high refractive indices, lying in the range 2.3–3.6. Light emitted from p–n junctions in these materials thus has a high probability of being reflected back into the device at the semiconductor/air interface. Potting the LED in a roughly hemispherical lens of a resin material which has a refractive index of around 1.5 decreases the amount of internal reflection at the edges of the cube of semiconductor material and allows more of the emitted light to escape from the package. A description of some of the most popular designs of resin lenses has been given by Sze (1981).

Figure 8.16(a) shows a sketch of one kind of simple LED package. The LED chip containing the p–n junction is sawn from a wafer and attached to one contact pin by a solder joint and has a wire bonded from the top ohmic contact to the second pin on the package. The solder and contact pin act as a sink for excess heat generated by recombination at the p–n junction. The whole of this assembly is then moulded into the resin lens in much the same way as described for plastic potted packages in Chapter 7. Figure 8.16(b) shows a detail from a metallographic cross section through a LED package of this type. The GaP diode is approximately 200 μm across and the ohmic contacts and bonding wire are clearly visible. Great care has to be taken during the assembly of these devices that the sawing, bonding and soldering operations do not introduce dislocations into the GaP. For this reason, thermocompression wire bonding is not often used to join the wire to the top ohmic contact. The wires are normally of gold because they are then compatible with gold-based ohmic contact alloys on the III–V semiconductors. Bergh and Dean (1976) have given many more examples of LED package types, including those where many devices of different kinds are integrated together on a single substrate.

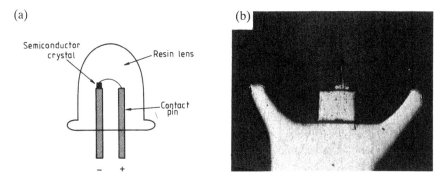

Figure 8.16 (a) A sketch of the structure of a simple LED package. (b) A metallographic cross section through a GaP LED diode showing the bonding wire, contact pin and GaP crystal.

The semiconductor materials which are chosen for LEDs operating in the visible range have already been mentioned in §8.2.2. Isoelectronically doped GaP and GaAsP alloys emit red, yellow and green light and are almost universally used in this range. Some heterostructure GaAs/GaAlAs LEDs can also operate at the very edge of the red end of the visible range. ZnS, SiC and GaN diodes can be used for the emission of blue light.

8.4.1 Infrared LEDs

GaAs, and ternary and quaternary alloys in the GaInAsP and GaAlAsSb composition fields, emit light in the wavelength range from 0.9 to 1.6 μm (see figure 8.7) and so can be used to produce LEDs suitable for direct coupling into optical fibres. GaAs diodes amphoterically doped with germanium or silicon are frequently used for emission at 0.9 μm. Both these dopant elements have rather high solubilities in GaAs, and so very high carrier concentrations can be obtained in the active recombination regions without the danger of precipitate formation. I have already mentioned that the rather wide spectral band width of LED devices (30–40 nm) leads to quite severe dispersion in optical fibres, but LEDs are often simpler to package than laser diodes and are less sensitive to temperature fluctuations during operation. High-performance IR LEDs have thus been developed for use in optical communications systems. Figure 8.17 shows the structure of an IR LED designed for efficient coupling with an optical fibre. Here a GaAlAs/GaAs/GaAlAs double heterostructure LED on a GaAs substrate has a well etched all the way through the substrate. The end of a silica fibre can be introduced into this well and fixed with resin. A limited-area bottom ohmic contact concentrates the generation of light

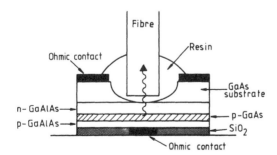

Figure 8.17 The structure of a particularly sophisticated infrared LED containing a limited-area bottom ohmic contact and a well etched into the GaAs substrate where an optical fibre can be attached (after Burrus and Miller 1971).

exactly under the end of the fibre (Burrus and Miller 1971). Kressel and Butler (1977) have given a very complete description of the structure of some sophisticated LEDs, and the principles of stripe geometry devices which I shall present in the following section are also applied to the design of these devices.

8.5 THE DESIGN OF SEMICONDUCTOR LASERS

In this section I shall describe the structures of some of the most successful laser diodes, concentrating on the properties of the materials from which these devices are constructed. Extensive reviews of a wider range of laser structures can be found in Kressel and Butler (1977), Casey and Panish (1978) and Sze (1981).

First let us consider how a laser diode can be made from a wafer substrate supporting the optically active layers of semiconductor material. The p–n junction may be grown-in during the deposition of the epitaxial layers, or may be added by a subsequent diffusion process. Top and bottom metal contacts are then evaporated, using alloys of the kind discussed in §2.2, and may be annealed to create ohmic contacts with low resistances. Composite metallisation structures of the kind discussed at the end of Chapter 5 often form excellent contacts to laser diodes and may include a 'glue' layer of Ti or W to ensure adequate adhesion of the contact to the surface of the semiconductor. One of the two ohmic contacts can be evaporated with a stripe geometry in order to improve the operational characteristics of the devices, as is discussed below. Highly doped semiconductor capping layers can be added to the epitaxial layer structure to assist in the formation of satisfactory ohmic contacts. This can remove the need for annealing the top contact, which may be important as it has been found that the uncontrolled indiffusion of impurity elements during the annealing of a contact can severely damage the operating characteristics of the device.

Small dice are now cut out of the wafer and have rough and damaged sides as a result of the mechanical sawing action. Two of these damaged sides are cleaved to give perfectly flat and undamaged surfaces, and these facets can be polished if required. III–V compounds and alloys cleave preferentially along {110} planes and so two parallel facets can easily be cleaved in a die cut from a (001)-oriented wafer. After these operations the laser die has the form shown in figure 8.18, a simple p–n junction laser configuration. The cleaved facets act as partial reflecting planes for the light emitted at the p–n junction, and so the die itself acts as the lasing cavity, i.e. the Fabry–Perot cavity. Because of the very high density of excitation centres in a semiconductor crystal, compared for instance to the atomic density in a gas laser cavity, the Fabry–Perot cavity can be as short as 100 μm in a semiconductor laser. A typical laser die might be about 300 μm long and 100 μm wide. We can see that the coherent laser light is emitted parallel to the p–n junction in a diode of this kind, while the light is emitted in all directions from the p–n junction of the LED structures sketched above. This sequence of fabricating steps is more or less the same for diodes made with both III–V and IV–VI semiconductor materials.

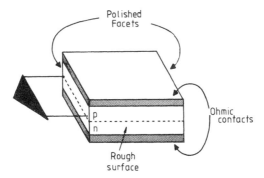

Figure 8.18 A sketch of the structure of a simple p–n junction laser die, showing how the polished surfaces are used to define the lasing cavity.

Laser diodes are normally packaged by soldering the p-type side of the die to a header, often with a low-melting-point indium-based solder. The top contact is wire bonded to the package and, just as I have described for LEDs, it is most important to avoid introducing lattice damage into the semiconductor crystal during these soldering and wire-bonding operations. Lasers operate at high current densities, with often more than 10^4 A cm^{-2} passing across the p–n junction in the active region of the device. Thus a great deal of heat is generated in the p-type material near the junction where most of the radiative recombination occurs. This is why laser dice are soldered with the p-type side down; the diffusion path for heat out of the active region of the device is shortest in this configuration. The temperature of the p–n junction determines the threshold current for the onset of lasing and if it increases too much the laser action can be quenched altogether. The rate of lattice damage at the junction also seems to increase with temperature, as I shall describe in §8.6.

From the above it is clear that the thermal conductivities of the materials used to package semiconductor laser diodes must be chosen especially carefully. Highly conductive indium solders and copper substrates are often used in laser packages and metallised BeO substrates are an excellent choice when the laser is to form part of an integrated optical/transistor hybrid circuit. However, the thermal conductivity of most compound semiconductors is very poor: GaAs a factor of four worse than aluminium; and CdTe and $Ga_{50}Al_{50}As$ a further factor of four worse than GaAs for example. In general, ternary alloys have thermal conductivities which are much lower than those of the binary compounds, and so it is very important to keep the thickness of these alloys between the p–n junction and the bottom metal contact as small as possible. A layer of GaAlAs only 1 μm thick can often give the dominant contribution to the total thermal resistance between the junction and the heat sink. Table 8.2 gives a few examples of the thermal conductivities of the materials used in laser diodes, contacts and packages. Joyce and Dixon (1975) have developed a quantitative model for the heat flow in a laser diode and from this model, and a knowledge of the thickness and thermal conductivities of each layer in the laser structure, a precise calculation can be made of the junction temperature. Even with very thin p-type semiconductor layers, the p–n junction is usually some 10 K above the temperature of the heat sink.

Table 8.2 The thermal conductivities of materials used in the fabrication of laser diodes and laser packages.

Material	Thermal conductivity ($W\ cm^{-1}\ K^{-1}$)
Silver	4.0
Copper	3.9
Gold	3.2
Aluminium	2.0
Silicon	1.5
Platinum	0.73
Indium	0.9
Titanium	0.22
Diamond	9.0
Indium phosphide	0.68
Gallium arsenide	0.46
Gallium phosphide	0.61
Indium arsenide	0.27
Indium antimonide	0.17
$Ga_{50}Al_{50}As$	0.12
$Ga_{50}As_{50}P$	0.1
$Ga_{50}In_{50}As$	0.05

8.5.1 Laser Structures With Stripe Geometries

The laser diode I have sketched in figure 8.18 has ohmic contacts deposited over the whole of the top and bottom surfaces of the semiconductor die. However, there are several advantages in having the active region of the laser in the form of a thin stripe running between the ends of the Fabry–Perot cavity. This limits the area of the p–n junction at which lasing occurs, in a manner similar to that shown for the IR LED in figure 8.17. The threshold current is often significantly decreased, and this, with the additional assistance of thermal dissipation into the surrounding inactive semiconductor material, can lower the operating temperature at the p–n junction. The laser light is also generated far from the rough side walls of the die where rapid non-radiative recombination will always occur, and with a smaller active region the number of interfacial defects which can affect the radiative efficiency, and the rate of device degradation, will be very much reduced. The mechanism by which lattice defects assist in the gradual degradation of the performance of laser diodes will be described in §8.6. Stripe geometry lasers are often very reliable because of this low density of dislocations in the recombination region. Finally, the light from a stripe geometry laser will be emitted from a narrow source which is very suitable for direct coupling into an optical fibre.

Many methods have been developed to limit the area of the active layer in which lasing occurs, including the use of isolating oxide layers to define the shape of one of the ohmic contacts, and the fabrication of buried heterojunction devices where the active region of the diode is itself grown with a stripe geometry. Examples of these two kinds of laser structures are sketched in figures 8.19(a) and (b) respectively. The stripes are typically 10 μm in width and the cross-hatched regions indicate the regions of the p-type GaAs layers where the lasing action occurs. In both these structures I have shown the inclusion of a heavily doped GaAs layer directly under the top contact to improve its ohmic character. It is clear from figure 8.19 that the fabrication of these laser structures requires the growth of quite complex multilayered epitaxial deposits, and at least one lithographic process is needed to define the stripe. The lithographic and etching techniques described in Chapter 2 can be used to pattern the layers of semiconductor, contact metal and oxide. Selective chemical etches are often available to dissolve one semiconductor material in a multilayered structure but not the others. There are several other processes by which devices with stripe geometries can be produced, for example, mesa isolation and proton bombardment and these have been described by Casey and Panish (1978). Figure 2.17 shows an example of a circular mesa prepared by dry etching a GaAs wafer, and stripe geometries can be etched in the same kind of process. Proton bombardment is a particularly simple process for stripe definition, involving the implantation of H^+ ions to create lattice defects which quench the lasing activity in regions around the undamaged stripe.

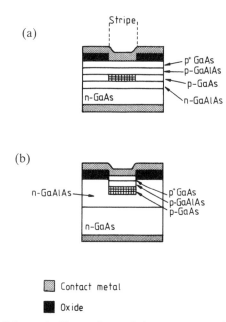

Figure 8.19 Schematic illustrations of the structures of two kinds of stripe geometry lasers: (a) an oxide isolated contact stripe; (b) a buried heterostructure laser diode.

8.5.2 Multiple Quantum Well Lasers

So far in this chapter we have assumed that the wavelength at which light is emitted from a semiconductor diode is governed by the width of the band gap and in some cases the position of trap states in the band gap as well. However, a new generation of microelectronic devices has been developed, based on the electronic properties of carriers in quantum well structures of the kind illustrated in figure 3.30. The original suggestion that exciting new physical effects might be observed if it were possible to produce heterostructures where the individual layers were thinner than the mean free path length of the carriers was made by Esaki and Tsu in 1970. At that time, however, techniques for the growth of very perfect epitaxial layers with well controlled compositions, layer thickness and doping content had not been developed. MBE and MOCVD growth processes now provide exactly the required kind of layered structures, as has been described in Chapter 3. It is thus possible to grow heterostructures where the thicknesses of the individual semi-conductor layers are as small as a few tenths of a nanometre, so that the band gap and the doping concentration can be varied on this scale. Figure 8.20 illustrates the band structure of one kind of superlattice heterostructure, where layers of a wide band-gap semiconductor, A, alternate with material with a smaller band gap, B. For obvious reasons, structures of this kind are often called multi-quantum wells, MQWs. I have shown the A layers to be heavily n-type doped while the B layers contain no dopant at all. The periodicity of this superlattice structure, and the extremely small dimensions of the layers, result in the formation of new superlattice bands which can lie between the offset band edges of the two semiconductors. Electrons in the A layers will tend to fall into the superlattice band in the conduction band of the B layers, where they are free to move and have very high mobilities since there are no lattice defects and no impurity ions in the B layers. The exceptionally high mobilities of the carriers in the thin B layers means that a number of very fast transistor devices can be made in superlattice structures, as described by Ando *et al* (1982), Kelly and Nicholas (1985) and Chang and Giessen (1985).

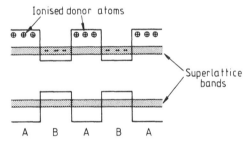

Figure 8.20 A sketch of the band structure of a simple multiple quantum well structure, showing the superlattice bands and the localisation of the free carriers in the undoped B layers.

However, superlattice structures can also have useful optoelectronic properties and, of particular relevance to this chapter, can be used to make unusually efficient lasers. GaAs/GaAlAs heterostructure MQWs have been comprehensively studied and are usually made by replacing the active p-type GaAs layer shown in all the laser diode structures illustrated above by a stack of alternating GaAs and GaAlAs epitaxial layers with thicknesses of about 1 to 5 nm (Holonyak *et al* 1980). The layer thicknesses and the band offsets control the position of the superlattice bands and so determine the emission wavelength. A number of advantages are found in these MQW lasers including: a reduction in the threshold current to as low as 250 A cm^{-2} (at least four times lower than the best conventional heterostructure laser); the ability to select the emission wavelength by changing the thickness of the individual super-lattice layers; and the surprising fact that a superlattice between two indirect band-gap semiconductor materials can have optical characteristics like those of a direct band-gap material. While the GaAs/GaAlAs system is by far the most studied, MQWS have been made in a number of other combinations of materials, such as InAs/GaSb, Ge/GaAs and PbTe/PbSnTe. Some of these superlattice structures have band offsets of a completely different kind to that shown in figure 8.20 and consequently have other interesting electronic properties. The possible applications of MQW structures are still being investigated. The references given above contain further details on these exciting and novel devices, which are examples of how the MBE and MOCVD crystal growth techniques can be used to produce new materials for microelectronic applications.

To conclude this section, it seems worthwhile to remind ourselves of the semiconductor materials which are most often used as the active components of laser diodes of the kinds I have described above. In the visible range, ternary III–V alloys like GaAsP and InGaP can be used, although the lattice match to the available substrate compounds is usually poor (see figure 8.7). Quaternary GaAlAsP alloys emitting in this wavelength range can be matched very much more closely with the substrate materials. GaAs/GaAlAs heterostructure lasers can operate just at the edge of the visible range, but are the most common and successful devices in the range 0.8–0.9 μm. Between 1.1 and 1.6 μm we can choose between GaInAs and GaAsSb alloys lattice matched to GaAs and GaInAsP alloys matched to InP. Beyond this range, the antimonide and IV–VI alloys are available, including a number of lead salts. Mercury cadmium telluride alloys are also useful for emission at long wavelengths. The principle of avoiding lattice defects in the active regions of the devices is important for all these materials, and so the development of techniques for the growth of very perfect heterostruc-ture epitaxial layers has been one of the most significant steps in the production of efficient laser diodes.

8.6 MECHANISMS OF DEGRADATION OF LASERS AND LEDs

The last few sections have concentrated on describing how luminescent devices can be designed and fabricated. However, little attention has been paid to the fact that almost all devices of this kind degrade in service. Both catastrophic failures after short lifetimes, and a more gradual deterioration in light output and an increase in threshold current, have been observed. This degradation is a serious problem when luminescent devices are being designed for use in optical communications systems where lifetimes of 10–20 years may be required. Considerable efforts have been directed at identifying the mechanisms by which the degradation occurs, and in this section I will describe four of these mechanisms. We will see that the optical flux and the operating temperature are both important in controlling the rate of degradation, and so it is not surprising that it is in laser diodes that the most rapid deterioration of properties is observed. leds generally only degrade by the point defect mechanism described in §8.6.3.

8.6.1 Catastrophic Facet Damage

Laser diodes operating at high current densities are often observed to suffer from disintegration of the cleaved mirror facets. Investigation of the morphology of some of these failed diodes shows that gross mechanical damage extends over the whole of the area of the facet. The accepted model for this effect assumes that rapid non-radiative recombination occurs at the surface states created by lattice defects on the facet, or at defects introduced by a careless cleaving procedure. These non-radiative transitions result in the release of considerable amounts of energy, leading to local dissociation of compound semiconductor materials and the creation of further point and extended defects which can then act as recombination centres. Runaway heating of the surface region can occur if the optical flux is sufficiently high, and some local melting in extreme cases. A critical current density of 5×10^6 W cm^{-2} is observed to be the value at which GaAs/GaAlAs laser diodes start to show catastrophic facet damage. The irradiance levels in leds are usually much too low to cause facet melting or dissociation, even if the density of surface states is high.

Coating the laser facets with Al_2O_3 or SiO_2 antireflectant films, of thickness about half the wavelength of the emitted light, seems to limit the facet damage, perhaps by passivating some of the surface states. Many ternary and quaternary alloy semiconductors are also less susceptible to facet damage than GaAs/GaAlAs devices. This may be because the density of surface states is relatively low in some alloys (as it is in GaInAsP near to InP in composition for instance), or because the band gap is small. This last factor will reduce the amount of energy released to the lattice by each non-radiative recombination event at the diode surface. We might expect, therefore, that lasers which operate in the infrared may prove more resistant to facet damage at high current densities than GaAs/GaAlAs diodes.

8.6.2 Dark-line Defects

A second, more gradual, mode of degradation observed in laser diodes is the formation of lattice defects in the active region of the devices. Rapid non-radiative recombination at these defects results in the appearance of dark lines running across the emission region of the laser, which is why they are known as dark-line defects, or DLDs. During operation, these defects gradually increase in size until they cover a significant fraction of the total light-emitting area of the diode and the radiance of the device is decreased. These defects seem to form most readily in diodes which have been damaged during mounting, leading to the suspicion that they are associated with dislocations in the laser crystals. Closer examination of devices containing DLDs has shown that each defect consists of a three-dimensional array of dislocations nucleating from a pre-existing defect like a threading dislocation or one introduced by mechanical damage. Recombination at these dense dislocation arrays causes the observed reduction in emission.

A good deal of research has been directed at understanding why these dislocation arrays nucleate and grow preferentially in the active regions of laser diodes. It has been proposed that the dislocations migrate both by glide and by climb mechanisms, although climb is usually observed only in laser materials with wide band gaps. A link between a wide band-gap semiconductor laser material and enhanced dislocation climb can be made as follows. Dislocation climb requires the creation or annihilation of point defects and this normally means that dislocations can only climb slowly, except at high temperatures. However, if there is non-radiative recombination at dislocations in the active region of the laser, considerable numbers of phonons are released to the lattice immediately around each dislocation core. These lattice vibrations can increase the rates of point defect formation and migration processes and so encourage the rapid climb of dislocations—the so-called 'phonon kick' mechanism. This will only occur in regions of high irradiance and explains why dislocation arrays are not created outside the active region of the devices. The width of the semiconductor band gap determines how much energy is transferred from electronic to vibronic states in multiphonon transitions, which is why dislocation climb is usually observed only in laser crystals which emit in the visible or near-infrared spectrum. The position in the band gap of the defect electronic states is also an important factor in controlling the efficiency of this energy transfer. The mechanisms of multiphonon processes have been reviewed by Stoneham (1981) and Lang (1984), and more complete descriptions of the mechanisms by which DLDs are formed have been given by Newman and Ritchie (1981) and Petroff (1985).

DLD formation can be avoided if care is taken to exclude all dislocations from the active region of a laser diode. The use of stripe geometries can help to improve the reliability of the devices, because by limiting the volume of the active regions the chance of including a dislocation is reduced. Crystal growth and device manufacturing techniques have now been developed to the point where the formation of DLDs is no longer a very serious degradation mechanism in GaAs/GaAlAs lasers. However, a large number of threading dislocations are expected to propagate up into the active regions of GaInAsP devices grown on InP substrates, because these substrates often contain a high dislocation density. Observations of DLDs in these laser materials are surprisingly rare, perhaps because of the smaller amount of energy released in each non-radiative recombination event as a result of the relatively small band gaps of semiconductor materials which emit in the wavelength range between 1.1 and 1.6 μm. Alternatively, the dislocations in these alloys may be intrinsically immobile. Finally it is possible that some phase separation may occur in these materials, as described in §8.2, increasing the yield strength of the semiconductor crystal. Similarly, the addition of ternary or quaternary alloying elements to the active regions of laser diodes may result in a significant solution strengthening and an improvement in the resistance of the material to the nucleation of dislocation arrays.

8.6.3 Gradual Degradation Mechanisms

Even when no dislocations are nucleated in the active regions of a luminescent device, the emitted intensity and the threshold current are both observed to degrade gradually during service. Relatively little is known about the details of these degradation mechanisms, but they are thought to involve the interaction of point defects and the emitted photons. A high photon flux seems to encourage not just dislocation and point defect migration but the dissociation of some point defect clusters and the creation of vacancies as well. The radiation-assisted dissociation of the Zn + O complexes in GaP(Zn + O) will lead to a reduction in the luminescence for instance, and processes of this kind may be important in the degradation of all GaP LEDs. Point defects at which non-radiative recombination is rapid can be created in the active regions of other semiconductor materials and cause a deterioration in the performance of devices. These point defects are much harder to characterise than the extended defects which form DLDs, and the technique of deep level transient spectroscopy (DLTS) has been particularly valuable in determining characteristics like the position of electronic states created by point defects in the band gap, and their capture cross section for carriers (Lang 1974). A range of point defects may be responsible for the gradual degradation of luminescent devices; radiation-induced dopant–vacancy (and impurity–vacancy) complexes and antisite defects have been identified. The migration of these defects into the active region of the diodes seems to be partly controlled by the local stress levels, and so the precise structure of a device may play an important part in determining its operational lifetime. The role of point defects in the gradual degradation of lasers and LEDs has been reviewed by Newman and Ritchie (1981).

A second mode of gradual degradation is the oxidation of the mirror facets on a laser diode. The emitted light intensity will decrease as the oxide layer increases in thickness. It has been suggested that point defects are produced at the oxide–semiconductor interface and that these defects then increase the rate of local recombination. The rate of oxidation depends on the composition of the laser material and is often observed to be slower on GaAlAs than on GaAs. Coating the facets with an antireflectant layer of Al_2O_3 also improves the stability of the diode.

8.6.4 Die-bond Degradation

In §8.4 I described how GaAs/GaAlAs laser diodes are attached to a heat sink by a solder with a high thermal conductivity to try and keep the operating temperature of the p–n junction as low as possible. A typical metallisation scheme to a laser diode might include a Ti/Au contact to the bottom p^+ semiconductor material and a thick indium solder layer to attach the diode to a copper heat sink. The softness of the indium solder will ensure that stresses generated during die bonding, and by thermal expansion mismatches between the copper and semiconductor, do not create lattice defects in the active regions of the device. However, indium and gold react at low temperatures to form intermetallic compounds and Kirkendall voids are often formed during these reactions. This is precisely the same phenomenon as I have already described in Chapter 5 for the notorious gold–aluminium reactions in metallisation systems on silicon devices. Void formation can lead to decohesion of the die from the heat sink in extreme cases, but the thermal resistance of the die bond will be significantly increased even by the presence of a few voids. The operating temperature of the p–n junctions will also be increased, which will degrade the luminescent efficiency of the device and reduce its stability. Metallographic sectioning has proved particularly useful in understanding how this degradation process proceeds (Plumb *et al* 1979). This kind of reaction can be avoided by reducing the thickness of the gold contact layer, or by using a gold/germanium eutectic alloy solder to replace indium. A diamond heat sink, which is quite closely matched in thermal expansion coefficient to GaAs, can also help improve reliability by reducing the strain excursions which the rather brittle eutectic solder joint must undergo.

The use of stable metallisation systems, and stripe geometry devices to reduce the number of defects in active regions, has resulted in the production of lasers with very long operating lifetimes — well in excess of 10^5 h in many cases. LEDs often have even longer lifetimes, because at lower optical flux levels the degradation mechanisms described above are very slow, or do not occur at all. The excellent reliability of these devices has necessitated the design of accelerated testing programmes to estimate their lifetime. Temperature or current 'overstressing' is often used to speed up the deterioration process, but may not be a very accurate method for estimating lifetimes if different degradation mechanisms operate at different temperatures. We have seen that several mechanisms may be operating in each device, each with a characteristic activation energy. The relative importance of these mechanisms in determining the performance of a luminescent device may thus change with the temperature. For this reason the results of overstressing tests must be treated with caution, unless care is taken to ensure that the dominant degradation process is the same at both operating temperature and overstressing temperature.

8.7 LASER DESIGN FOR INTEGRATED OPTICAL SYSTEMS

The basic principle of integrated optics is to fabricate a variety of optical components on the surface of a semiconductor wafer. The luminescent devices, switches, waveguides and detectors must thus be formed wholly in the epitaxial layers grown on the substrate. In §8.3 we saw how the light emitted by a GaAs/GaAlAs laser can be forced to remain within the waveguide formed by surrounding the active GaAs region with epitaxial layers of GaAlAs with a higher refractive index. In this way, the active region of the device acts as the optical waveguide. However, in an integrated optical system it is often necessary to have passive waveguides as well, and a way must be found to couple lasers into these thin film waveguides. In addition, the cleaved mirror facets which form the Fabry–Perot cavity in a discrete laser diode cannot be fabricated in an integrated optical system and a thin film laser structure must be designed in which partial internal reflection takes place at the edge of the active layer.

The distributed feedback (DFB) laser replaces the cleaved mirror facets by an optical grating above the active layer, which acts to reflect some of the emitted light back into the lasing region. Corrugating the surface of the optical confinement layer around the active region makes an excellent optical grating and corrugations of the correct periodicity are readily produced by ion etching, or masking and chemical etching (Casey *et al* 1975). The coupling of a heterojunction laser into a thin film waveguide can be achieved by a design called the 'leaky mode' laser, where the emitted light is permitted to leak out of the active region of the laser by careful control over the refractive index of the confinement layer. Light can also be allowed to escape into a waveguide through a corrugated grating of the correct spacing. Another design is the composition coupler which is illustrated in figure 8.21. Here a heterojunction laser is etched at one end to produce a sloping profile and a waveguide of precisely chosen composition grown over this surface. Some of the light propagating along the active GaAs layer will couple into the waveguide (Aiki *et al* 1976). Further details on the design of a wider range of thin film devices suitable for fabrication in integrated optical systems can be found in the review article by Conwell and Burnham (1978). Most of the devices of this kind have been based on GaAs and GaAlAs because the technologies of crystal growth and selective etching are well developed in these materials.

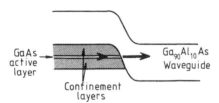

Figure 8.21 A sketch of a composition coupler which links a double heterojunction laser into a thin film waveguide (after Aiki *et al* 1976).

8.8 MATERIALS FOR LIGHT-DETECTING DEVICES

As an introduction to this section I shall describe two of the device types which can be used to measure the intensity of incident light — the photoconductive device and the photodiode. These two kinds of detector do not function in the same way and so different properties are required in the materials which function as the active regions in the two cases. However, as we shall see, several semiconductor materials can be used both as photoconductors and in photodiodes. Firstly, I shall describe the processes by which an incoming photon can create free carriers in a semiconductor material. These processes are the same whether a photoconductor or a photodiode is involved. Figure 8.22

illustrates the two basic processes of carrier photoexcitation: band-to-band excitation which creates an electron–hole pair; and excitation between a mid-gap state and a band where only a single free carrier is produced. Photoexcitation will only occur in a band-to-band process when the incident photon has an energy in excess of the width of the band gap, E_g, and so the value of E_g defines the long-wavelength limit at which photons are absorbed (recall that the relationship between the energy and the wavelength of the photon is $\lambda_{max} = hc/E_g$). In the second process, photoexcitation can occur from the valence band to an unfilled acceptor state, or from an un-ionised donor state to the conduction band. The energy required to stimulate transitions of this kind may be very much smaller than E_g. In particular, semiconductors doped with impurities which create shallow states can be used for the detection of long-wavelength light.

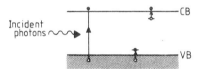

Figure 8.22 A schematic illustration of the two most important mechanisms by which an incident photon can be absorbed in a semiconductor material; (a) band-to-band excitation; (b) shallow-state-to-band transitions.

The basic structure of a photoconductive device is shown in figure 8.23(a). A slab, or thin film, of a semiconductor material, with an ohmic contact at either end, is exposed to a flux of photons. Carriers are generated in the semiconductor material by the photons and this increases the conductivity. This change in conductivity can be measured by an external circuit and so the intensity of the incident light can be determined. A critical parameter in the performance of a photoconductor is the free carrier lifetime; any recombination process within the bulk of the semiconductor material will significantly lower the sensitivity of the device (Bratt 1977, Sze 1981). Materials which are to be used for photoconductors are thus very carefully purified to remove all the species which can create deep trapping states at which recombination will be rapid.

Figure 8.23(b) shows the structure of a simple p–n junction photodiode. Incident photons will create carriers near the reverse-biased junction and the electrons and holes will be separated across the junction by the field in the depletion region. The junction current is then a measure of the intensity of the incident photon flux. If the p- and n-type sides of the junction are heavily doped, the carrier transit time across the junction is small and a fast device is produced. However, the thicker the p- and n-type material, the longer the carriers will take to diffuse to the contacts. Unfortunately, to increase the fraction of the total number of incident photons absorbed in the semiconductor material, and so improve the detection efficiency of the device, the p- and n-type layers often have to be as much as a hundred micrometres thick. Speed of operation and detection efficiency are thus conflicting requirements in a photodiode, and the structure of a device must be chosen to suit each particular material and application. However, there is no need for very pure starting materials in the fabrication of photodiodes, since heavy doping levels are used. This makes a number of complex alloy semiconductors suitable for the production of photodiodes, although they may be hard to prepare in a sufficiently pure form for use as photoconductors.

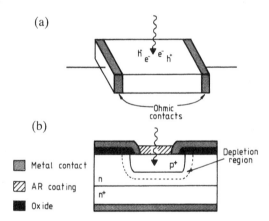

Figure 8.23 A schematic illustration of the structure of: (a) a simple photoconductor device; (b) a p–n junction photodiode.

There are several other kinds of photodiode devices: the p–i–n diode, the metal–semiconductor diode, the avalanche photodiode and various phototransistors. The structure and performance of these devices is described by Stillman and Wolfe (1977), Schinke *et al* (1980) and Sze (1981). The same semiconductor materials can be used as the active elements of all these devices, and silicon photodiodes are especially common because the technology for the fabrication of complex planar structures in silicon is very well developed.

In the following two short sections I shall present some of the semiconductor materials which are used in photodetecting devices. These will be divided into three classes depending on the wavelength range in which they operate: the visible range; the range between 0.8 and 1.6 μm needed in optical communications systems; and the long-wavelength range, 5–100 μm. This last wavelength range is extremely important in a large number of military applications.

8.8.1 Photoconductor Materials

For detection in the visible range, 0.4–0.7 μm, the most important photoconductor material is CdS, which has a very high sensitivity peaking at about 0.6 μm. Free electrons can have exceptionally long lifetimes in this semiconductor if dopants like iodine or chlorine are added to create cadmium vacancies which form very effective trap states for the free holes. This leaves the electrons able to migrate through the photoconductor with little chance of spontaneous recombination with a free hole. The band gap of CdS is 2.46 eV and so only light at the blue end of the visible spectrum will have sufficient energy to excite band-to-band transitions. The inclusion of copper as an impurity creates a trap state which allows absorption of photons with energies as low as 1.7 eV, which means that CdS(Cu + Cl) photoconductors have high detection efficiencies over the whole of the visible spectrum.

In the near-infrared range, germanium crystals doped with either nickel or gold can be used as photoconductors. These dopants create acceptor states at 0.23 and 0.15 eV above the valence band edge respectively, and electrons can be excited from the top of the valence band into these unoccupied states to leave free holes in the valence band. These germanium devices must be cooled to about 77 K to avoid thermal excitation of electrons to fill the acceptor states. In the same wavelength range, 1–4 μm, semiconductor compounds with smaller band gaps can be used as band-to-band photoconductors, e.g. InSb, PbSe and PbS.

For the detection of longer-wavelength light, we must either use a semiconductor with a very small band gap indeed, 0.1 eV or below, or one containing an impurity which introduces a very shallow state. Ternary alloys like $Pb_{80}Sn_{20}Te$ and $Hg_{80}Cd_{20}Te$ are suitable for use as photoconductors in the 10 μm range. However, it can be difficult to grow single crystals of these alloys with well defined stoichiometry and doping levels, and some of the particular problems associated with the preparation of HgCdTe alloys are highlighted in Chapter 3. Even so, this ternary II–VI material has been very widely used in photoconductive devices because it can be prepared with a very low residual carrier concentration, giving high detection efficiencies. A comprehensive account of the production of devices in HgCdTe has been given by Broudy and Mazurczyk (1981).

Finally, for detection of light of even longer wavelengths, germanium doped with copper, gallium or zinc, phosphorus-doped silicon, and GaAs doped with a variety of elements, all have very shallow states, allowing photoexcited transitions across extremely small energy gaps. These materials must, of course, be cooled to very low temperatures to ensure the shallow states are not saturated by thermal excitation. The necessity for cooling long-wavelength photoconductive crystals to liquid helium temperatures can cause problems of bond decohesion because of the thermal expansion mismatches between metal contacts and semiconductor, and between contacts and wires.

Figure 8.24 indicates the wavelength ranges over which the materials discussed above can operate and also gives some indication of their relative sensitivities. As I have explained, it is vitally important that the materials from which photoconductor crystals are made are as pure as possible to avoid enhanced recombination at impurity states. This is especially true in doped photoconductors like gold-doped germanium, where the solid solubility limit of the activating dopant is only about 10^{16} cm^{-3}. Zone-refined germanium can have a residual impurity concentration as low as 10^{13} cm^{-3}, and the dopants may be added by zone-levelling techniques. HgCdTe crystals can be grown with a free carrier density of only 10^{15} cm^{-3}, a remarkably low value for such a complex and intractable alloy. The preparation of these exceptionally pure semiconductor materials is possible only if very pure feedstock elements are available, and the successful production of high-quality HgCdTe is due in no small part to the development of highly efficient techniques for the purification of the component elements (see for instance Hirsch *et al* (1981) and Grovenor (1987)).

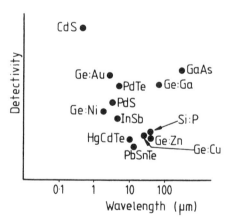

Figure 8.24 A comparison of the wavelength ranges for peak sensitivities in a number of photoconductor materials (after Sze 1981).

8.8.2 Photodiode Materials

A wide range of photodiodes have been fabricated in the semiconductor materials which are commercially available in the form of high-quality wafers and for which the technology for the fabrication of microdevices is well developed. Silicon photodiodes are most efficient around $0.8 \mu m$ and so are conveniently coupled with GaAs/GaAlAs laser sources in optical fibre systems. The main problem with these devices is the low value of the optical absorption coefficient in silicon, which means that the p- and n-type layers have to be at least $50 \mu m$ thick if a significant fraction of the incident photons are to be absorbed. The most efficient diodes are produced in semi-insulating silicon wafers, where care has been taken to remove all the metallic impurities which create deep trapping states. Silicon will not absorb photons with wavelengths over about $1.1 \mu m$, and germanium, GaAs, GaInAsP and CdTe can all be used to make photodiodes suitable for detecting the light from optical sources emitting at $1.1–1.5 \mu m$. For shorter wavelengths of $0.3–0.7 \mu m$, metal–semiconductor diodes, for example, Ag/ZnS and Au/Si are used.

HgCdTe alloys are once again very important materials for photodiodes which operate in the $10 \mu m$ range. Epitaxial thin film material is usually p-type because it is mercury deficient and can easily be type converted to form p–n junctions by the controlled indiffusion of mercury or an n-type dopant like indium (Reine *et al* 1981). Ion implantation of boron can also be used to produce n-type surface layers. Many of the IV–VI alloys can be amphoterically doped, and PbSnTe and related materials may be used to make useful photodiodes in this same wavelength range.

So far I have only considered homojunction photodiodes, but an enormous range of heterojunction devices have been fabricated and tested. Two interesting examples of this kind of photodiode are the p-type GaInAs epitaxial layer lattice matched to an n-type InP substrate, and the heterojunction formed by evaporation of a thin n-type CdS film onto a p-type $CuInSe_2$ crystal. In the first of these structures a suitable dopant has to be added to both the substrate and epitaxial layer, while in the second the conductivity types are intrinsic to the two semiconductor materials. We have seen in §1.4 that CdS can only be prepared as an n-type material, and annealing a $CuInSe_2$ crystal in Se vapour produces Cu vacancies and so p-type material. In both these heterojunction devices the incident light passes through the wide band-gap substrate material, InP and $CuInSe_2$ respectively, and is absorbed in the thin layers of GaInAs or CdS. This ensures that the carrier generation always occurs close to the p–n junction, which improves the operating speed of the photodiodes. Sharma and Purohit (1974) have reviewed the use of heterojunctions of all kinds in photodiodes. For more general reading on the application of some of the materials I have described above in photoconductor devices and photodiodes, the reader is directed to the volumes edited by Willardson and Beer (1977, 1981) and Keyes (1980).

REFERENCES

Aiki K, Nakamura M and Umeda J 1976 *Appl. Phys. Lett.* **29** 506
Anderson R L 1962 *Solid-State Electron.* **5** 341
Ando T, Fowler A B and Stern F 1982 *Rev. Mod. Phys.* **54** 437
Barnoski M K 1981 *Fundamentals of Optical Fibre Communications* (New York: Academic)
Bergh A A and Dean P J 1976 *Light Emitting Diodes* (Oxford: Clarendon)
Bratt P 1977 *Semiconductors and Semimetals* **12** ed R K Willardson and A C Beer (New York: Academic) p 39
Bridenbaugh P M 1985 *Mater. Lett.* **3** 287
Broudy R M and Mazurczyk V J 1981 *Semiconductors and Semimetals* **18** ed. R K Willardson and A C Beer (New York: Academic) p 157
Burrus C A and Miller B I 1971 *Opt. Commun.* **4** 307
Casey H C and Panish M B 1978 *Heterostructure Lasers* (New York; Academic)
Casey H C, Somekh S and Ilegems M 1975 *Appl. Phys. Lett.* **27** 142
Chang L L and Giessen B C (eds) 1985 *Synthetic Modulated Structures* (New York: Academic)
Conwell E M and Burnham R D 1978 *Ann. Rev. Mater. Sci.* **8** 135
de Cremoux B, Hirtz P and Ricciardi J 1980 *Gallium Arsenide and Related Compounds* (Inst. Phys. Conf. Ser. 56) p 115
Esaki L and Tsu R 1970 *IBM J. Res. Dev.* **14** 161
Figielski T 1978 *Solid-State Electron.* **21** 1403
Gooch C H 1973 *Injection Luminescent Devices* (Chichester: Wiley)
Grovenor C R M 1987 *Materials for Microelectronics* (London: Institute of Metals)
Haasen P 1978 *Physical Metallurgy* (Cambridge: Cambridge University Press)
Hirsch H E, Liang S C and White A G 1981 in *Semiconductors and Semimetals* **18** ed. R K Willardson and A C Beer (New York: Academic) p 21
Holonyak N, Kolbas R M, Dupuis R D and Dapkus P D 1980 *IEEE J. Quant. Electron.* **QE16** 170
Horikoshi Y 1985 in *Semiconductors and Semimetals* **22C** ed. W T Tsang (New York: Academic) p 93
Ivey H F 1963 *Advances in Electronics and Electron Physics* suppl. 1 (New York: Academic)
Izawa T and Sudo S 1986 *Optical Fibres* (Dordrecht: Reidel)
Joyce W B and Dixon R W 1975 *J. Appl. Phys.* **46** 855
Kelly M J and Nicholas R J 1985 *Prog. Phys.* **48** 1699
Keyes R J 1980 *Optical and Infrared Detectors* (New York: Springer)
Kressel H 1975 *J. Electron. Mater.* **4** 1081
—— 1980 *Semiconductor Devices for Optical Communication* (Berlin: Springer)

Kressel H and Butler J K 1977 *Semiconductor Lasers and Heterojunction LEDs* (New York: Academic)

Kuan T S, Wang W I and Wilkie E L 1987 *Appl. Phys. Lett.* **51** 51

Lang D V 1974 *J. Appl. Phys.* **45** 3023

—— 1984 *Ann. Rev. Mater. Sci.* **12** 377

Martins J L and Zunger A 1986 *J. Mater. Res.* **1** 523

Matthews J W and Blakeslee A E 1974 *J. Cryst. Growth* **27** 118

Micklethwaite W F H 1981 in *Semiconductors and Semimetals* **18** ed. R K Willardson and A C Beer (New York: Academic) p 47

Milnes A G and Feucht D S 1972 *Heterojunction and Metal–Semiconductor Junctions* (New York: Academic)

Nakajima K 1982 in *GaInAsP Alloy Semiconductors* ed. T P Pearsall (New York: Wiley) p 43

Nakajima K, Osamura K, Yasuda K and Murakami Y 1977 *J. Cryst. Growth* **51** 87

Newman D H and Ritchie S 1981 in *Reliability and Degradation* ed. M J Howes and D V Morgan (New York: Wiley) p 301

Norman A G and Booker G R 1985 *J. Appl. Phys.* **57** 4715

Pearsall T P 1982 *GaInAsP Alloy Semiconductors* (New York: Wiley)

Petroff P M 1985 in *Semiconductors and Semimetals* **22A** ed. W T Tsang (New York: Academic) p 379

Plumb R G, Goodwin A R and Baulcomb R S 1979 *Solid State Electron Dev.* **3** 206

Reine M B, Sood A K and Treadwell T J 1981 in *Semiconductors and Semimetals* **18** ed. R K Willardson and A C Beer (New York: Academic) p 201

Rosztoczy F R 1968 *J. Electrochem. Soc.* **17** 513

Rozgonyi G A, Petroff P M and Panish M B 1974 *J. Cryst. Growth* **27** 106

Saul R H, Plee T and Burrus C A 1985 in *Semiconductors and Semimetals* **22C** ed. W T Tsang (New York: Academic) p 193

Savage J A 1985 *Infrared Optical Materials* (Bristol: Adam Hilger)

Schinke D P, Smith R G and Hartman A R 1980 in *Semiconductor Devices for Optical Communications* ed. H Kressel (Berlin: Springer) p 63

Senior J 1985 *Optical Fibre Communications* (London: Prentice-Hall)

Sharma B L and Purohit R K 1974 *Semiconductor Heterojunctions* (Oxford: Pergamon)

Stillman G E and Wolfe C M 1977 in *Semiconductors and Semimetals* **12** ed. R K Willardson and A C Beer (New York: Academic) p 291

Stoneham A M 1981 *Rep. Prog. Phys.* **44** 1251

Stringfellow G B 1982 *J. Cryst. Growth* **58** 194

Sze S M 1981 *Physics of Semiconductor Devices* (New York: Wiley)

Tsang W T 1985 *Semiconductors and Semimetals* **22** (New York: Academic)

Vecht A and Werring N J 1970 *J. Phys. D: Appl. Phys.* **3** 105

Willardson R K and Beer A C (Series ed.) 1977, 1981, 1985 *Semiconductors and Semimetals* **12, 18, 22** (New York: Academic)

Woodall J M, Pettit G D, Jackson T N, Lanza C, Kavanagh K L and Mayer J W 983 *Phys. Rev. Lett.* **51** 1783

9

Materials for Solar Cells

9.1 INTRODUCTION

Of the semiconductor devices I will describe in this book, by far the widest range of materials, device structures and fabrication techniques are used in the production of solar cells. In Chapter 2 I showed that integrated circuits, whether in silicon or gallium arsenide, are produced by a series of processing stages which are more or less the same whatever the structure of the individual devices in the circuit. Similarly, the laser diodes and LEDs described in the previous chapter are all rather similar in structure and the fabrication techniques basically the same for both sets of devices. However, photovoltaic devices have been produced with an enormous number of different structures and from a wider range of semiconducting materials than we have yet considered for any other microelectronic application. In addition, many of the techniques used to prepare solar cells seem, at first glance, to be closer to the kitchen sink than the clean environment in which integrated circuits are manufactured.

I will start this chapter with an introduction to the fundamental principles which govern the structure and performance of simple p–n homojunction solar cells. We will first decide how to select the materials from which to prepare an efficient solar cell of this kind. Then the structure of some other kinds of solar cells will be mentioned: the heterojunction cell, the Schottky barrier cell and the metal–insulator– semiconductor (MIS) cell. The influence of the various interfaces in a solar cell on the performance of the complete devices will also be described. The next section will consider the specific properties of some of the materials used in successful solar cells, starting with devices based on single-crystalline silicon and gallium arsenide and introducing the structure of some single-crystal heterojunction cells. These devices are quite similar in structure and fabrication technology to the optoelectronic devices described in the previous chapter. Later sections will concentrate on solar cells made from some very non-ideal materials such as amorphous and polycrystalline thin films. Some of the more unusual techniques for depositing thin films of semiconductor materials will be introduced, along with a short discussion on the very complex structure, and highly non-equilibrium chemistry, of thin film cells. In many cases the details of the electrical and optical processes which occur in these devices during operation are poorly understood, but some highly efficient, and relatively cheap, solar cells can be prepared by extremely simple fabrication processes. Excellent extended reviews of the physics and technology of all kinds of solar cells can be found in the volumes by Hovel (1975), Chopra and Das (1983) and Fahrenbruch and Bube (1983).

9.2 THE BASIC FEATURES OF SOLAR CELLS

Figure 9.1 shows a schematic illustration of a simple p–n homojunction solar cell. (As in the previous chapter, a homojunction is one formed between two regions of the same semiconductor, one doped p-type and the other n-type.) This figure will be used to describe the fundamental electronic processes which occur in solar cells. Let us assume that solar light strikes the front surface of the semiconductor material after passing through an ohmic contact. Most of the incident photons pass through the thin n-type layer on this surface and are absorbed in the thicker underlying p-type region, which we shall therefore call the absorber layer. Some of the photons will have sufficient energy to excite

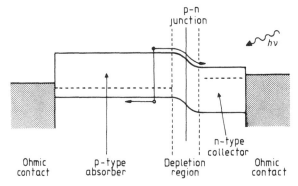

Figure 9.1 A schematic illustration of the operation of a homojunction solar cell where the incident solar light creates free carrier pairs in the absorber layer which are then separated across the p–n junction.

electron–hole pairs in this layer and these carriers will then be separated by the field across the p–n junction, as shown in the figure. The n-type layer collects the minority carriers generated in the absorber layer and is thus called the collector layer. The absorber layer is normally chosen to be p-type material because the mobility of electrons in most semiconductor materials exceeds that of the holes. This means that the electrons (minority carriers in the p-type region of course) diffuse rapidly to the p–n junction, which reduces the chance of recombination in the bulk of the p-type layer. The solar cell in figure 9.1 is completed with an ohmic contact to the p-type semiconductor.

The efficient operation of such a cell depends on: (i) the absorption of a large fraction of the incident solar photons with the creation of as many electron–hole pairs as possible; and (ii) the separation of these carriers across the p–n junction before many of the minority carriers can be lost by recombination in the absorber layer or at the junction itself. Both the front and back surfaces of the semiconductor are likely to be sites for extremely rapid recombination as well, and we might expect surface recombination to be an especially important mechanism for decreasing efficiency in the thin film devices described in later sections.

In order to make an informed choice of the materials which can be used to make efficient solar cells we must look more closely at the mechanisms of photon absorption in semiconductors, carrier transport across p–n junctions, and the density of possible recombination sites.

9.2.1 Photon Absorption in Semiconductors

The fundamental mechanism of photon absorption of interest in solar cells is that which results in the production of an electron–hole pair. I shall assume here that the intrinsic photoexcitation process illustrated in figure 8.22 is the only important carrier generation mechanism in solar cells. Semiconductor materials with direct band gaps will absorb photons with an energy, hv, greater than the band gap, E_g, since the promotion of an electron from the valence band into the conduction band can occur without any change in the momentum of the electron. This process is called 'direct absorption'. We can define an absorption coefficient, α, using the expression

$$I(x) = I_0 \exp(-\alpha x) \tag{9.1}$$

where $I(x)$ is the light intensity a depth x beneath the semiconductor surface and I_0 is the intensity of the light incident on the surface. The absorption coefficient in a direct band-gap semiconductor is given by

$$\alpha = A(hv - E_g)^{1/2} \tag{9.2a}$$

where A is nearly constant in the wavelength range of interest in solar cells and depends strongly on the refractive index of the semiconductor.

A convenient way of thinking of the importance of the value of α in solar cell materials is that about 90% of the incident photons are absorbed in a layer of the semiconductor of thickness $2/\alpha$. Equation (9.2a) shows that α increases sharply as the energy of the photons exceeds E_g, so that we should expect a well defined absorption edge in semiconductor materials with a direct band gap.

By contrast, in a semiconductor with an indirect band gap, the absorption of photons can only occur with the simultaneous creation or absorption of phonons. The likelihood of this second-order process is much smaller than of direct absorption. We can see that this situation is an exact analogue of the difference in light-emission efficiencies in direct and indirect band-gap semiconductors, as described in §8.2. The value of α in semiconductor materials with indirect band gaps depends on the phonon energy, E_p, as well as $h\nu$ and E_g and, when $h\nu$ is approximately equal to E_g,

$$\alpha = \frac{B\ (h\nu - E_g + E_p)^2}{E_g^2 [\exp(E_p/kT) - 1]} \qquad (9.2b)$$

where B is another constant (Smith 1968). The absorption edge in indirect band-gap materials is usually less well defined than in the direct band-gap semiconductors. Figure 9.2 illustrates the variation in absorption coefficient with photon energy for several of the semiconductor materials used in solar cells. Some materials can show both kinds of optical absorption characteristics at different photon energies. Germanium is a good example of a semiconductor which has this behaviour.

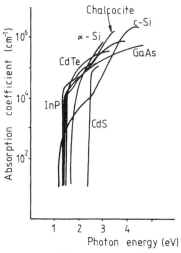

Figure 9.2 A comparison of the variation of absorption coefficients with photon energy for a number of common semiconductors, showing the poorly defined absorption edge in crystalline silicon and that CdS is transparent to photons with energy less than 2.4 eV.

We should now consider the characteristics of the solar light incident on a photovoltaic cell. Figure 9.3 shows the spectrum of solar radiation after passing through the earth's atmosphere in a direction which minimises the thickness of the atmospheric layer. This is known as the solar spectrum for air mass 1, or AM1. A considerable fraction of the incident spectrum is absorbed on passing through this layer of atmosphere, mostly by water and carbon dioxide molecules. The molecules which result in the particularly strong absorption bands are identified in figure 9.3. We can see that semiconductor materials with band gaps around 1 eV will absorb almost all the incident solar photons, but that as E_g is increased the fraction of solar power which can be used to create electron–hole pairs decreases. This gives us our first information on how to choose a semiconductor material for the absorber layer in a solar cell—the band gap must be in the range 1–2 eV. We have also seen that the absorber layer in a solar cell should have a thickness of $2/\alpha$ if over 90% of the incident photons are to be absorbed. Figure 9.2 illustrates the large difference in the values of the absorption coefficient in the energy range 1–2 eV for semiconductors with direct and indirect band gaps. A striking example of this difference can be seen if CdTe and crystalline silicon are compared, with values of α of 3×10^4 and 10^3 cm^{-1} respectively for photons of energy 1.5 eV. A layer of CdTe, which has a direct band gap, only 1 μm thick will absorb most of the solar photons, while thicknesses of 30 μm or more of silicon are required to absorb an equivalent fraction of the incident photons. We shall see below that if the highest efficiency is to be achieved, the ideal thickness of a silicon layer is more than 100 μm.

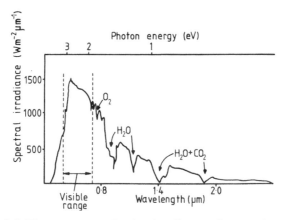

Figure 9.3 The spectrum of solar irradiance after passing through the earth's atmosphere, AM1. Strong absorption bands are created by the O_2, H_2O and CO_2 molecules in the air (after Thekaekara 1974).

This simple introduction to the absorption of photons in semiconductors allows us to understand why solar cells made in CdTe and silicon may have very different structures, but there are several other important effects on the values of the absorption coefficient which I shall only mention here. The potential field which exists at the p–n junction in a solar cell can alter the value of α, i.e. the Franz–Keldysh effect. In addition, the absorption edge is shifted to higher energies in a very heavily doped semiconductor by the Moss-Burnstein effect (see §8.4). Further details on the mechanisms by which photons are absorbed by semiconductors can be found in Moss *et al* (1973).

9.2.2 Carrier Transport Across p–n Junctions in Solar Cells

The second important consideration when selecting a semiconductor for use in a solar cell is how the total amount of power which can be extracted from the device depends on the band gap of the semiconductor. We can gain an understanding of the relation between power and E_g by looking at the current which flows across the p–n junction in a simple solar cell.

We have already seen in Chapter 8 that the J/V characteristics of a p–n junction are described by the expression

$$J = J_0[\exp(qV/nkT) - 1] \qquad (8.6)$$

where $n = 2$ when recombination dominates the current flowing across the junction. This is the form of the diode equation that we shall use in this chapter, emphasising the importance of recombination processes in solar cells, especially in the heterojunction devices I shall describe in §9.2.3. A discussion of the importance of tunnelling via deep states on the performance of p–n junction solar cells has been given by Hovel (1975). We must now look at how equation (8.6) is modified when a cell is exposed to solar light.

If we assume that the photons incident on the device illustrated in figure 9.1 generate a photoexcited current across the p–n junction of J_L, then the diode equation becomes

$$J = J_0[\exp(qV/nkT) - 1] - J_L. \qquad (9.3)$$

This new contribution to the current across the p–n junction shifts the J–V curve of the device in the manner shown in figure 9.4. This diagram can be used to define some of the most important operating parameters of a solar cell. The maximum voltage which can be generated across the device under any given intensity of illumination is called the open circuit voltage, V_{oc}, and the maximum current across the junction is called the space charge current, J_{sc}. These two parameters are given by the intersection of the J–V curve of the illuminated device with the voltage and current axes as shown in figure 9.4. The maximum power generated by the solar cell, P_m, is however only $V_m J_m$, and once again these two parameters are defined in the figure. The relationship between V_{oc} and J_{sc}, which are both relatively easy to measure, and P_m is given by the fill factor, ff:

$$ff = \left(\frac{P_m}{J_{sc}V_{oc}}\right) = \left(\frac{V_m J_m}{J_{sc}V_{oc}}\right). \tag{9.4}$$

We can think of the fill factor as a measure of the 'squareness' of the J–V curve. Finally, we can define the solar efficiency of a cell, η_s, as P_m/P_s, where P_s is the solar power incident on the device. These four parameters, V_{oc}, J_{sc}, ff and η_s are often used to characterise the performance of solar cells.

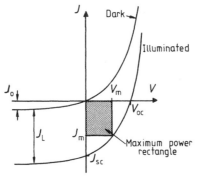

Figure 9.4 A schematic illustration of the J–V curve characteristic of a simple p–n junction solar cell under illumination. The open-circuit voltage, V_{oc}, short-circuit current, J_{sc}, and maximum voltage and current, V_m and J_m, are defined in the figure.

The open circuit voltage, V_{oc}, is usually the parameter which depends most heavily on the crystalline perfection and purity of the materials used in the solar cells. This is important because the power generated by the cell depends directly on the value of V_{oc}. From equation (9.3), V_{oc} is given by

$$V_{oc} \sim \frac{nkT}{q} \ln\left(\frac{J_L}{J_0} + 1\right). \tag{9.5}$$

It is clear that V_{oc} can be increased by decreasing the dark current, J_0, across the p–n junction, or by increasing the diode factor, n. The simplest way of reducing J_0 is to increase E_g, but as E_g is increased the fraction of the total incident solar spectrum which can be used to excite carriers in the absorbing region of the solar cell will decrease. This will decrease J_L, which in turn decreases both V_{oc} and J_{sc}. The conflicting demands of increasing V_{oc} and keeping E_g as low as possible result in the greatest solar efficiency being found in semiconductor materials with band gaps in the range 1.4–1.5 eV. The calculated variation of η_s with E_g is sketched in figure 9.5, in which the band gaps of a few of the semiconductor materials used most frequently in solar cells are also shown. Most calculations of solar efficiency assume rather ideal properties for the semiconductor materials, e.g. Loferski (1956), but the calculated maximum values of η_s are still only about 20–30%, even in a cell containing an ideal p–n junction. Most of the solar photons have either a long wavelength and so are not absorbed, or have more energy than is required to create an electron–hole pair, and this excess energy is lost as waste heat in the cell (Rothwarf and Boer 1975).

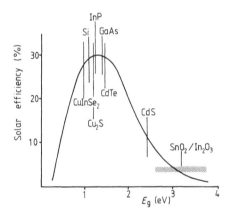

Figure 9.5 A plot of the calculated variation of the maximum efficiency of a solar cell with the band gap of the absorber material. The optimum band gap lies in the range 1.4–1.5 eV, and so we expect the most efficient solar cells to be made in semiconductors like InP, GaAs and CdTe.

The solar efficiency can also be reduced by recombination before the carriers are separated across the p–n junction. Both surface and space-charge region recombination effects are likely to more significant in heterojunction and thin film solar cells than in homojunction cells of the kind shown in figure 9.1, because the density of lattice defects in these devices can be rather high. However, a mechanism for recombination which is common to all solar cells is via deep trapping states in the bulk of the p-type absorber layer. These states can be created by the 'lifetime killer' elements described in §8.2 and will reduce the lifetime of electrons in the p-type material. This effect will obviously be most important in solar cells made in semiconductors with an indirect band gap, where the absorber layers must be thick and the minority carriers must diffuse a considerable distance to the p–n junction. For this reason, silicon material for homojunction cells is usually required to be free of killer impurity elements, although it is not necessary to use materials with the exceptionally high purities needed in photoconductive devices (§8.6.1). The preparation of solar grade silicon will be discussed in §9.3.1.

9.2.3 Heterojunction Solar Cells

There are several kinds of solar cells in which the simple p–n homojunction illustrated in figure 9.1 is replaced by a heterojunction: semiconductor/semiconductor heterojunctions, Schottky barrier cells and metal–insulator–semiconductor devices. While these solar cell configurations operate in much the same way as a homojunction cell, there are sufficiently important differences to make it worthwhile describing them separately.

The principle of many heterojunction solar cells is to replace the n-type collector layer in the simple p–n homojunction cell, which plays little part in absorbing the incident photons, with a layer of a wide band gap semiconductor material through which most of the solar photons can pass to the absorber layer. This window layer can be very thick if its band gap is sufficiently large. However, in an efficient solar cell the recombination rate at the heterointerface must be as low as possible. The mechanisms of dislocation generation at heterointerfaces are considered in §3.5 and the density of misfit dislocations shown to depend on the lattice mismatch between the two semiconductor materials. In §8.2 we saw how the recombination rate at a heterointerface depends on the misfit dislocation density, so it is clear that it is important when choosing materials for the absorber and window layers on a heterojunction solar cell to try and match their lattice parameters and crystal structures. Later in this chapter we will see that the high density of threading dislocations propagating through a thin epitaxial layer from a highly misfitting heterointerface can reduce the efficiency of a solar cell. A particularly successful heterojunction solar cell has been made by the combination of GaAs and GaAlAs layers, which are very well lattice matched (see §9.3.2).

In heterostructure solar cells, the absorbing layer can have either n- or p-type conductivity, although the higher mobility of minority carriers in p-type semiconductor material usually means that the absorber is chosen to be p-type. A wide range of semiconductor materials can be included in heterojunction photovoltaic devices, including some rather unusual compounds with wide band gaps: tin oxide, SnO_2; indium oxide, In_2O_3; and indium tin oxide, or ITO, in which the tin is used as a dopant in the indium oxide to increase its conductivity. These compounds can be conveniently deposited by evaporation, sputtering or chemical techniques and are used as transparent window layers. Copper indium diselenide, $CuInSe_2$, a I–III–VI chalcopyrite semiconductor, can be used as an effective absorber layer. The electrical properties and structure of the semiconductor materials which are used in solar cells are given in table 9.1. We shall need this data when describing the structure of practical heterojunction solar cells in §§9.3.3 and 9.5.

Table 9.1 The structure and properties of semiconductor materials used in solar cells.

Semiconductor	Crystal structure and lattice parameters (nm)		Band gap[†] (eV)	Electron affinity (eV)
Si	Diamond cubic	0.543	1.12(I)	4.05
GaAs	Sphalerite	0.565	1.43(D)	4.07
InP	Sphalerite	0.587	1.35(D)	4.4
CdTe	Sphalerite	0.648	1.56(D)	4.3
$CuInSe_2$	Tetragonal	$a = 0.578$	1.01(D)	4.15
		$c = 1.16$		
Cu_xS	Ortho	$a = 1.18$	1.2 (D)	4.3
Chalcocite	-rhombic	$b = 2.73$		
		$c = 1.35$		
CdS	Wurtzite	$a = 0.416$		
		$c = 0.675$	2.42(D)	4.5
	Sphalerite	0.582		
SnO_2	Tetragonal	$a = 0.474$	∼ 3.7[‡]	∼4.8
		$c = 0.319$		
In_2O_3 and	BCC	∼1.01	between 2.8	∼4.3
indium-rich ITO			and 3.7[‡]	

[†]D = direct, I = indirect.
[‡]The properties of these oxide compounds depend on the deposition conditions and the doping levels.

Heterojunction devices can be operated with the light incident either on the absorber layer or on the collector, as is shown in figure 9.6. The choice of operating mode, front-wall or back-wall illumination, illustrated in figures 9.6(a) and (b) respectively, dictates which materials can be used for the collector layer. In the back-wall configuration, the collector layer must either be a window layer with a large band gap, or a very thin layer such that most of the photons reach the p-type absorber layer. A very wide range of heterojunction solar cell configurations have been fabricated and a more complete discussion of their properties can be found in Sharma and Purohit (1974) and Vanhaotte and Pauwels (1979).

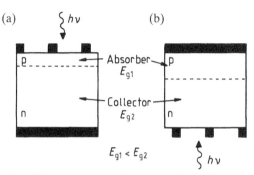

Figure 9.6 An illustration of the operation of a solar cell in (a) front-wall and (b) back-wall illumination configurations. In (b) the collector material must have a very low value of α in order for the solar photons to reach the absorber layer and is called the window layer.

At this point it is worth asking how these particular combinations of semiconductor materials are selected for the production of heterojunction cells. In order to understand this, we have to remember what physical properties must be defined in order to predict the form of the band edges when two semiconductor materials are brought together to form a heterojunction. Figure 8.13 illustrates the model developed by Anderson (1960) for predicting the band structure of a heterojunction. We must remember that interfacial states are ignored in this model, although they probably exist in all practical heterojunctions. From Anderson's model, a number of different band-edge profiles can be predicted, depending on the particular values of parameters like the width of the band gaps, the electron affinities and the workfunctions of the two semiconductors. Vanhaotte and Pauwels (1979) have proposed that a large potential barrier at the heterointerface is needed to decrease J_0 and so increase V_{oc}. A large discontinuity in the conduction band edge at the heterojunction might be expected to give a nearly ideal structure if the absorber is a p-type semiconductor. Under these circumstances, electrons photoexcited in the absorber layer will rapidly diffuse down the sharp potential gradient at the junction and yet the dark current will be extremely low. Figure 9.7 illustrates some of the heterojunction band structures which are predicted to lead to efficient solar cells, and, in principle, these structures can be prepared by a careful choice of the electron affinities, band gaps and doping levels of two semiconductor materials. A similar set of semiconductor combinations can be chosen when the absorber layer is assumed to be an n-type semiconductor.

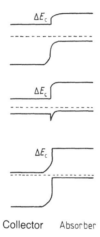

Collector   Absorber

Figure 9.7 Sketches of the band-edge profiles at semiconductor heterojunctions which are predicted to give efficient solar cells when the absorber layer is a p-type semiconductor.

However, spikes and notches can be formed at some heterojunctions (see figure 8.13) and these can interfere with the separation of photoexcited carriers across the junction. It is thus important to avoid the production of these features in the band-edge profiles. The presence of charged interface states will also modify the form of the band-edge discontinuities and enhanced interfacial recombination at these states will change the carrier transport properties of the heterojunction. The states associated with misfit dislocations at the heterointerface might be expected to give a particularly significant contribution to these effects. Fahrenbruch and Bube (1983) have reviewed the various models of carrier transport across heterojunctions and concluded that a model based on a combination of tunnelling, interfacial recombination and thermionic emission gives the best fit to the often very complex form of the J–V curves at heterojunctions in solar cell configurations. The relative importance of each of these mechanisms of carrier transport will depend on the choice of materials at the heterojunction. Ideally, the choice of semiconductors for the absorber and collector layers should be dictated by the precise form of the band-edge structure, the interfacial recombination velocity and the interfacial charge, since all of these control which of the carrier transport mechanisms dominates under any particular set of operating conditions. In practice, rather little is known about some of these parameters, and the combinations of semiconducting materials which have proved successful in heterojunction devices have been chosen more on the basis of an observed high efficiency than by prediction of this efficiency before any cells were fabricated. There are numerous examples of heterojunction cells which should have had high efficiencies on the basis of the simple models of band-edge structure, but which proved very disappointing once fabricated. However, it does seem to be widely accepted that a large band-edge discontinuity at the heterojunction will usually result in a satisfactory performance from the solar cell. Many of the heterojunction configurations which will be described in later sections have band-edge structures of the kind illustrated in figure 9.7.

9.2.3(a) Schottky barrier and MIS solar cells

A second kind of photovoltaic device can be made by placing a Schottky contact on a semiconductor absorber layer. A very thin metallic contact will allow most of the incident solar light to penetrate to the semiconductor, where photoexcitation of carriers occurs exactly as in a p–n homojunction cell. Carrier separation will then occur across the built-in potential at the Schottky barrier, as illustrated in figure 9.8(a). Schottky barrier solar cells are usually very simple to fabricate, requiring only the deposition of one ohmic and one Schottky contact onto opposing sides of a semiconductor layer. The height of the potential barrier at the Schottky contact is a very important parameter in controlling the value of V_{oc} and the efficiency of the device. The carrier transport properties at Schottky barriers are controlled by thermionic emission and tunnelling phenomena similar to those described above for transport across semiconductor heterojunctions. Crowell and Sze (1966) have combined the diffusion and thermionic emission models to generate an expression which describes the performance of a Schottky barrier solar cell (equation (2.2)). In this model, the dark current, J_0, decreases exponentially with the barrier height, Φ_B, which emphasises that the efficiency of these devices can be increased by increasing Φ_B. In §2.2 we saw that the Schottky barrier heights of metal contacts on semiconductors are determined primarily by the position of the interface states associated with defects in the semiconductor surface or states created by the interaction of metal with the semiconductor. This means that we cannot use the values of the workfunctions to predict which metals will form contacts with high barrier heights on a given semiconductor. In fact, on many of the semiconductor materials which have band gaps near the optimum value for an absorber layer (1.4–1.5 eV), there is relatively little variation in barrier height whichever metal is used to form the contact, as can be seen by looking at the values in table 2.1. The practical result of this phenomenon is that values of Φ_B are rarely greater than $\frac{2}{3} E_g$ and that the dark current, J_0, is likely to be quite large in most Schottky barrier solar cells. The values of V_{oc} which are measured in these devices are usually lower than in p–n junction cells in the same semiconductor material, and this is reflected in low solar efficiencies as well.

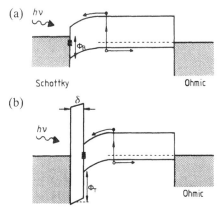

Figure 9.8 Schematic illustrations of the operation of (a) the Schottky contact solar cell and (b) the MIS cell.

It is possible to increase the effective value of Φ_B by including a very thin oxide layer between the metal contact and the semiconductor. Such a structure is known as a metal–insulator–semiconductor, or MIS, solar cell. Figure 9.8(b) illustrates the structure of such a cell. The insulating layer has several effects, the most important of which for our purposes is to block the majority carrier dark current across the junction. This is a major contribution to J_0, and the inclusion of an oxide layer only 1 nm thick results in a significant increase in V_{oc}. If we consider the effect of the inclusion of an oxide layer of width δ to be the addition of a tunnelling barrier of average height $q\Phi_T$, as indicated in figure 9.8(b), then we have to modify equation (2.2) with the inclusion of a new term to take account of the decrease in dark current (Kumar and Dahlke 1977):

$$J_0 \propto \exp[-(q\Phi_T)^{1/2}\delta]. \tag{9.6}$$

As the thickness of the insulating layer increases, we expect J_0 to decrease very sharply. In addition, the trapped charge in the insulator layer, and the changes in surface state density on replacing a metal–semiconductor interface by one between the semiconductor and an insulating layer, can alter both the effective value of Φ_T and the diode factor n. From equation (9.6) we might expect that simply increasing the thickness of the oxide layer would steadily increase the efficiency of MIS solar cells. However, the short-circuit current, J_{sc}, will also decrease as δ is increased above about 3 nm. This is because it becomes difficult for the photoexcited carriers to tunnel across the thicker layers of insulator. The optimum insulator thickness appears to be about 1–3 nm in a MIS cell. The inclusion of a very thin insulator layer between the two semiconductor layers in a heterojunction device will have the same beneficial effect of reducing J_0 and increasing V_{oc}. These are called SIS solar cells. Reviews of the structure and properties of MIS and SIS photovoltaic cells have been given by Ng and Card (1980) and Singh *et al* (1981).

In practice, the distinction between the Schottky barrier and the MIS solar cell is poorly defined. Many Schottky barrier devices will contain a very thin oxide layer unintentionally included at the junction during processing. This layer may be the native oxide layer formed on the surface of the semiconductor before the metal contact is deposited. The insulating layer can also grow at the metal–semiconductor interface in a completed device. Ponpon and Siffert (1978) have shown that the exposure of Au–Si Schottky barrier cells to air results in a gradual decrease in J_0 and increase in V_{oc}. This is presumably due to the indiffusion of oxygen and the formation of an MIS device by reaction at the metal–semiconductor interface. Most MIS solar cells are prepared by deliberately growing the required thickness of an insulator layer on the semiconductor surface before the deposition of the metal contact. Very thin oxide layers can be prepared on silicon-based cells by low-temperature oxidation processes (§6.4) and similar treatments can be used to grow thin oxide layers on GaAs as well. In this second case, the oxide layers are probably a complex mixture of gallium, arsenic and oxygen, and may not have a stoichiometry similar to that of any of the equilibrium phases (see §6.5 and Grovenor *et al* (1987)). Vapour-deposited layers of silicon nitride on silicon, and Sb_2O_3 on GaAs, have also been used to give efficient MIS solar cells. Useful MIS cells can be prepared when the semiconductor is not an expensive single crystal but a cheap polycrystalline thin film. Devices with solar efficiencies greater than 10% have been achieved in MIS cells on large-grained Si and GaAs sheets. Whether similar efficiencies can be obtained in MIS devices fabricated on materials with a smaller grain size has not yet been convincingly demonstrated.

9.2.4 Contacts and Surface Properties

In the last few sections I have described the basic structures of a number of solar cell configurations and shown that a wide range of materials can be used in these devices. However, I have as yet made no mention of the materials used to form the contacts to the cells, or how the surface of a solar cell can be prepared to improve the solar efficiency.

9.2.4(a) Contact structures

Figure 9.1 shows that a homojunction solar cell must have two ohmic contacts—one to the collector and one to the absorber semiconductor layer. Similarly, a Schottky barrier or MIS cell will have one ohmic and one rectifying contact (figure 9.8). It would be convenient if one of the contacts in each case could be transparent to solar light as well as highly conductive, so that the total resistance of the device is kept as low as possible while still allowing most of the solar spectrum to reach the semiconductor material. In §4.3.2 I described the optical and electrical properties of thin polycrystalline metal films and showed that the optical transmission of most metallic films falls off very sharply at about the thickness at which the resistivity approaches that of the bulk metal (figure 4.7). Many metal films grow by the nucleation and agglomeration of island nuclei, so it can prove difficult to deposit thin continuous films for contact applications. Film thicknesses of about 30–50 nm are required to obtain useful sheet resistances in common contact metals like silver, gold, copper and chromium, and at least 50% of the incident solar illumination will be either reflected or absorbed in these films. It is usually regarded as impracticable to use blanket metal contact layers over both surfaces of a practical solar cell device, although I shall show an example of a device which uses a thin platinum front contact in a later section.

In many solar cell designs, the contact on the side of the cell exposed to the solar illumination thus has a discontinuous or grid structure. Figure 9.9 shows the kind of grid contact structure which is commonly used on solar cells. These contact arrays are designed to allow the maximum light transmission, while also giving the cell a small series resistance (which will keep the fill factor high). However, if the grid bars are too widely spaced, the sheet resistance of the semiconductor material itself will contribute significantly to the series resistance of the device. As a result, considerable attention has been paid to designing the most efficient grid shapes by varying the width, A, and the spacing, B, of the contact 'fingers', and by experimenting with rectangular and circular arrays (see Appendix C in Chopra and Das (1983) for example). Grid arrays with fairly coarse features can be deposited by screen-printing techniques like those described for thick film metallisation on ceramic substrates in §7.8. However, when the sheet resistance of the semiconductor material is high (in thin polycrystalline films for example), the contact fingers must be much closer together. Standard lithographic techniques are then used to prepare the contacts (see §2.4).

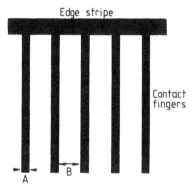

Figure 9.9 A sketch of a simple grid array contact structure, showing the important variables A, the grid bar width, and B, the grid bar spacing.

The ohmic contacts are passive components in solar cells, but any increase in the series resistance of the device will result in a sharp decrease in the efficiency. It is thus important that the correct metallic alloy be chosen to ensure that the contact resistance is as low as possible. Back ohmic contacts to p-type silicon material can be conveniently made with aluminium, as we have seen in Chapter 2. A brief heat treatment after evaporation encourages indiffusion of the aluminium and heavy doping of the underlying silicon. Ti/Pd/Au layered contacts are often used for the front grid contacts to heavily doped n-type silicon, and tunnelling across the contact barrier produces a 'pseudo-ohmic' contact. Here the titanium is a thin glue layer, the palladium a diffusion barrier between the gold and the titanium, and the gold is the thickest layer and carries most of the current. The materials used to make ohmic contacts on compound semiconductors are often quite complex, and a list of some useful ohmic contact alloys is given in table 2.2. Many of these contacts have a very complex microstructure after firing, as has been illustrated in figure 2.5, and degradation of the electrical properties may occur as a result of further diffusion and chemical reactions in service.

Many metals can be used to give high-quality Schottky contacts to the semiconductor materials used in solar cells. Aluminium makes a useful contact to silicon MIS cells, and the presence of a native oxide layer increases the value of Φ_B from 0.4 eV, which is characteristic of the unannealed Al/p-type silicon contact, to a much more useful value of 0.85 eV. Gold and platinum can make satisfactory contacts to Schottky barrier or MIS cells on gallium arsenide or cadmium telluride (table 2.1).

However, not all solar cells need to have a metallic grid contact and heterojunction cells can dispense with metallic contacts altogether if a transparent oxide semiconductor is used as the collector layer. Heavily doped oxides like $SnO_2(F)$, $In_2O_3(Cd)$ and ITO can be degenerate semiconductors with resistivities as low as 10^{-4} Ω cm and can also be almost completely transparent to solar light, with transmittance values of 80–95%. We shall see below that some highly efficient solar cells can be produced with transparent oxide layers functioning as both collector and top contact, but the properties of the films depend heavily on the precise details of the process by which they are deposited. The chemical stability of these oxides in thin film form, and their mechanical hardness, are particularly useful properties in practical solar cells. A more complex oxide material which may prove useful in this application is cadmium stannate, Cd_2SnO_4. Thin films of this material can have both very high conductivities and high transmission coefficients. A review of the properties of these materials has been given by Haake (1977).

9.2.4(b) Antireflectant coatings

The efficiency of solar cells is determined in part by how large a fraction of the incident solar radiation can be used to generate free carriers in the absorbing semiconductor layer. Any reflection of the light from the front surface of the device will obviously reduce the solar efficiency, and the large change in refractive index at air/semiconductor or the air/metal interfaces will result in a significant reflectance. Between 20 and 35% of solar radiation is reflected from the front surface of most kinds of solar cells, unless an antireflectant (AR) coating is included in the design of the device.

If a single-layer AR coating is used on a device in which the semiconductor has a real value of refractive index, n_s, then the optimum refractive index of the coating material, n_c, is given by $\sqrt{n_s}$, and the thickness of the layer should be $\frac{1}{4}$ of the wavelength of the incident light (see for instance Heavens 1965). For silicon, which has a refractive index with only a small unreal component, $n_s = 3.5$ and a coating material with $n_c = 1.87$ is required in a layer of thickness around 140 nm if the incident light has a wavelength of between 0.5 and 0.7 μm. Cu_xS layers have a refractive index $n_s = 3.46 - 0.87i$ and the ideal AR layer has $n_c = 1.94$ and a thickness of about 65 nm (see Appendix B in Chopra and Das 1983).

Some of the materials which have been used for AR coatings on solar cells include: Ta_2O_5 (2.2); SiO_2 (1.4); SiO (1.9); SnO_2 (2.1); Si_3N_4 (2);

In_2O_3 (2.1); Al_2O_3 (1.7); and TiO_2 (2.6). Here the numbers in parentheses give a rough idea of the value of n_c for these materials for light of wavelength 0.55 μm. The precise values of n_c are sensitively dependent on the manner in which the films are deposited, and quite large variations can be expected from film to film of nominally the same material. Sputtering is a particularly effective way of depositing many of these oxide compounds, but the stoichiometry of the layers, and so their optical properties, depend on the details of the sputtering process.

AR coatings on solar cells can not only increase the fraction of the incident light which reaches the absorbing semiconductor, reducing the reflected fraction to less than 5% in most cases, but can also help passivate the top surface of the semiconductor. This will reduce the rate of surface recombination and so increase the efficiency especially in thin film cells. We should not forget that the transparent oxide semiconductors, In_2O_3, SnO_2 and ITO can function as AR layers on heterojunction devices, as well as acting as the collector and the electrical contact.

9.2.4(c) Surface texturing

In addition to an AR coating, it is possible to modify the morphology of the exposed surface of the solar cell in order to increase the fraction of the incident light absorbed in the device. The surface of a silicon (001) wafer can be etched by a weak solution of NaOH in methanol, which preferentially exposes the (111) faces. A series of pyramidal structures will be produced on the wafer surface in a treatment of this kind. Light reflected from such a surface, where the pyramid faces make an angle of 54° with the macroscopic surface normal, will have a high probability of striking the side of another pyramid and being absorbed. The fraction of incident light reflected from a textured silicon surface covered by an AR coating can be as low as 2–3%. Many more complex cell designs incorporating a textured surface, or a carefully shaped pyramidal form for the p–n junction, have been described by Fahrenbruch and Bube (1983).

Thin film solar cells of the kind to be discussed in §9.5 often have rough surfaces as a result of the processes involved in the growth of films from solution or from the vapour phase. These rough surfaces will act in much the same way as the deliberately textured surfaces of single-crystal silicon solar cells and will increase the fraction of the incident solar light absorbed by the device. Chemical treatments can also be used to roughen the surface further if required; the process of roughening CdS film surfaces in HCl solutions will be described in §9.5.4. However, if the surface of the semiconductor is too rough, it may prove difficult to deposit integral thin film contact grid arrays in a reproducible manner. It is thus not always advisable to try and produce the most heavily textured surfaces, in an attempt to maximise the advantage of reduced reflection.

9.2.4(d) Encapsulating solar cells

Solar cells suffer from two kinds of degradation mechanism in service: intrinsic effects due to interdiffusion across heterojunctions and contacts; and extrinsic effects usually associated with the oxidation of the semiconductor surfaces or contact metallisations. In order to try to limit the extrinsic effects, many solar cells designed for terrestrial use are encapsulated to protect them from atmospheric corrosion. This encapsulation can take the form of a thick metallic layer over one side of the device, which has a dual role as back contact and corrosion inhibitor, and transparent SiO or glass layers sputtered over the front surface of the cell. The device can also be completely encapsulated inside a transparent polymer. Polyimide or paralyne can be used as coating layers, while encapsulating adhesives include silicones, acrylics and fluorocarbons. These are exactly the same classes of materials which we have already met when considering the packaging of integrated circuits in Chapter 7.

Solar cells designed for use in space have to be packaged so as to protect them from damage by cosmic particle fluxes, i.e. electrons and protons with a few photons, neutrinos and α particles, in the energy range of 1 keV to 100 MeV. These particles create vacancies and interstitials in the semiconductor layers and form deep states in the band gap which can lower the minority carrier lifetime. Dopant atoms may also play a part in determining the concentration of these recombination centres by combining with the point defects. It has been found that when aluminium is used as a p-type dopant in silicon homojunction cells the rate of recombination is much slower than in boron-doped material. Similarly, the presence of oxygen in silicon seems to reduce the rate at which the minority carrier lifetime is degraded. The addition of lithium to silicon cells also reduces the density of recombination sites—the lithium seems to diffuse to, and combine with, the defect which constitutes the recombination centre (Wysocki *et al* 1966).

GaAs and many other solar cell materials are considerably less sensitive to damage than silicon. The absorber thickness in these direct band-gap materials can be very small, so the chance of a fast particle creating some radiation damage is lower than in a relatively thick silicon cell. However, radiation damage remains a mechanism by which the efficiency of solar cells in space is steadily degraded.

The protection of solar cells from ionising particles is achieved by packaging the cells behind thick transparent sheets, i.e. cover slides. These must be made of a material which is not degraded by the cosmic flux, has a suitable refractive index, and is mechanically strong. Glass and sapphire cover slides have both been used and the cells are often potted in silicone or fluorocarbon resins as well. The cover slides might be anything from 100 μm to several millimetres thick and will absorb the low-energy cosmic particles which are particularly likely to produce serious radiation damage in the devices.

The degradation of solar cells by intrinsic mechanisms is harder to control, since these involve the interdiffusion and chemical reaction of layered structures which are far from equilibrium. The instability of the complex ohmic contact alloys used to prepare low-resistivity contacts to compound semiconductors has been mentioned in §2.2, while the diffusion of elements across heterointerfaces can also alter the electrical and optical properties of semiconductor materials. In the following sections some examples will be given of the modification of solar cell performance during service as a result of these intrinsic effects. It is worth recalling that the more complex the structure and chemistry of the solar cells, the less likely they are to be stable in service.

9.2.5 Design Rules for Solar Cells

We are now able to define some design rules which should assist in the fabrication of useful solar cells. These rules are slightly different for homojunction and heterojunction cells and so are listed separately in tables 9.2 and 9.3 respectively.

Table 9.2 Design rules for homojunction solar cells.

Back contact	Ohmic
	Should reflect light back into cell
Absorber	High mobility of minority carriers needed, therefore p-type
layer	Sufficient thickness to absorb 90% of incident light
Collector layer	n-type, high doping level for low series resistance
	Low surface recombination rate
Top contact	Ohmic
	Grid structure for efficient carrier collection

Note: In addition, all cells should have a textured top surface and an antireflectant coating. The complete devices will be encapsulated to protect them from attack by water and oxygen.

9.3 SOLAR CELLS IN SINGLE-CRYSTALLINE SEMICONDUCTORS

In this section we will consider the structure and performance of some of the more common solar cell configurations prepared in single-crystal, or nearly single-crystalline, semiconductor material. These devices are often fabricated by the introduction of p–n junctions into bulk wafer substrates by diffusion or implantation (§2.4.4), or by the growth of

Table 9.3 Design rules for heterojunction solar cells.

Back contact	Ohmic
	Should reflect light back into cell
Absorber semiconductor (smaller E_g)	High mobility of minority carriers needed, therefore p-type
	Sufficient thickness to absorb 90% of incident light
	E_g between 1.4 and 1.5 eV
	Photoexcited carriers should be generated close to junction
	Low surface recombination rate
	Lattice matched to collector
Collector layer (larger E_g)	n-type, high doping level to reduce series resistance
	E_g as large as possible
	Lattice matched to absorber
	Electron affinity and doping level chosen so no spikes are formed at the band edges
	Low surface recombination rate
Top contact	May also be collector layer
	If metal contact used, should be ohmic and have a grid structure

Note: All cells should have a textured top surface and an antireflectant coating. The complete device should be encapsulated to give protection from attack by water and oxygen.

homo- or heteroepitaxial layers on a single-crystal substrate by the LPE, CVD or MBE techniques described in Chapter 3. Because these processes have been described in some detail elsewhere in this book, here I shall only very briefly show how the solar cells are fabricated. More time will then be spent on a consideration of how the properties of the substrate or epitaxial materials can influence the performance of the photovoltaic devices. Specific examples of materials for single-crystal solar cells will include silicon sheet or ribbon, the GaAs/GaAlAs and GaAs/Si heterojunction cells, and a short discussion of some more unusual heterojunction configurations. A more comprehensive description of all these devices can be found in Hovel (1975) and Fahrenbruch and Bube (1983).

9.3.1 Material for Silicon Solar Cells

The basic structure of a simple p–n homojunction solar cell has already been shown in figure 9.1. The stages in a typical process for the manufacture of such a cell might be as follows:

(1) Cleaning and surface preparation of a commercial p-type silicon wafer (see §2.4).

(2) Diffusion or implantation of a shallow n-type region into the top surface of the wafer. (Epitaxial growth of an n-type layer is also possible.) This n-type region might be about 0.2–0.5 μm deep.

(3) Aluminium evaporation onto the back surface and annealing to cause indiffusion of the aluminium. This creates a p$^+$ layer which ensures that a low-resistance ohmic contact is formed.

(4) A lithography process is used to define a contact grid on the front surface.

(5) Application of antireflectant coating.

Such a cell might have the following operating parameters: $V_{oc} = 0.6$ V, $J_{sc} = 40$ mA cm^{-1}, $ff = 0.8$ and a solar efficiency of about 15% in terrestrial applications. The whole of this fabrication process is based on well established technology for the preparation of silicon integrated devices, so that the yield of satisfactory devices is likely to be high. The only device features requiring lithographic definition, the contact grid patterns, are relatively coarse and are easy to produce with the lithographic tools described in §2.4.

Let us now consider the properties which are required of the silicon wafer material in an ideal solar cell. The p-type absorber layer in a typical silicon cell is at least 100 μm thick, to ensure that a reasonable fraction of the incident solar light is absorbed. This means that the minority carrier diffusion lengths are large, so the recombination rates in the bulk of the silicon wafer must be low if the device is to have a high efficiency. We have already seen in §1.4 that the presence of impurities which create deep levels in the semiconductor band gap will result in a high recombination rate. These impurities are the 'lifetime killers' identified in §8.2. Many of these impurities are also exceptionally rapid diffusers in silicon (see table 1.4) and so can diffuse to the active regions of the solar cell from a contaminated surface if care is not taken during processing to avoid all contact of the silicon wafer with materials containing these elements. Table 9.4 lists some of the most notorious lifetime killer impurities in silicon and gives the concentration of each impurity calculated to be the highest level tolerable in solar cell material which can be used to produce devices with solar efficiencies greater than 10% (Hill *et al* 1976). We can see that many metallic impurities should be excluded from silicon material in concentrations higher than about 10 ppb. By contrast, carbon has a much smaller effect on minority carrier lifetimes and so can be tolerated in higher concentrations.

Also included in table 9.4 are some values of the effective partition coefficient, k_{eff}, of these impurities in silicon (see §3.2). k_{eff} is very small for most of the damaging elements, so we expect that a Czochralski or float zone crystal growth process may be used to grow material in which the concentrations of these impurities are very low, even if the melt is heavily contaminated. These same crystal growth techniques can also produce silicon material with very low dislocation densities. Section 2.3 describes how heavy metal impurities are particularly prone to segregate to dislocations in silicon crystals, where they can form precipitates of silicide phases. The presence of these decorated defects can result in high leakage currents across p–n junctions and high local recombination rates. Both these effects can contribute to degradation of the performance of silicon solar cells. It is thus important to fabricate solar cells in material which is essentially dislocation free. From this discussion, it seems that wafers cut from Czochralski or float zone silicon crystals should be ideally suited for use as the starting material for the production of highly efficient silicon solar cells. However, these wafers are relatively expensive and it would be convenient if a source of 'solar grade' silicon material were available for the production of large-area solar arrays.

Table 9.4 Properties of some killer impurities in silicon solar cells.

Impurity	Deep states	k_{eff}	Calculated maximum concentration for 10% efficiency (cm^{-3})
Cr	1 donor	10^{-5}	5×10^{15}
Cu	3 acceptors	4×10^{-4}	10^{15}
Fe	2 donors	6×10^{-6}	10^{15}
Na	2 donors	2×10^{-3}	2×10^{14}
Ti	1 donor	10^{-5}	4×10^{13}
V	1 donor 1 acceptor	10^{-5}	10^{14}
C	2 donors	7×10^{-2}	10^{17}

There are two ways in which the stringent materials' specifications applied to silicon intended for integrated circuit manufacture can be relaxed when choosing a material for solar cells:

(1) It may be possible to use large-grained polycrystalline material instead of expensive single crystals. We might make as a first approximation a condition that the grain size should be at least as large as the minority carrier diffusion length in the absorber material, L_p.

(2) The only impurity elements which it is absolutely necessary to exclude from the absorber layer are those which create deep trapping states; all other impurities can be tolerated at quite high concentrations even if they are normally considered to be electrically active in silicon.

It is possible, using solar grade silicon with these new specifications, to prepare solar cells which have efficiencies quite similar to those achieved in cells fabricated in Czochralski-grown material. Several methods have been developed to prepare material suitable for the production of cheap silicon solar cells, and we shall now consider to what extent these have been successful in giving high solar efficiencies.

9.3.1(a) The preparation of polycrystalline silicon ribbons and sheets

Two techniques for the growth of thin silicon ribbons were introduced in Chapter 3: dendritic web growth and edge-defined film-fed growth, EFG. Both of these processes look attractive for the production of cheap solar cell material because of the high growth rates that can be achieved and the possibility of the direct growth of ribbons with the optimum thickness of around 100 μm. This second point means that the sawing and polishing operations which must be used to prepare wafers from Czochralski boules are no longer necessary and the 'kerf' loss of material is therefore avoided. The silicon ribbons produced by these growth techniques were shown in §3.3 to be very heavily twinned and often to contain a high dislocation density as well. Because coherent twin boundaries in semiconductor materials are generally electrically inactive, the effective grain size of these ribbons is quite large, certainly of the order of 100 μm. We can, therefore, think of the ribbon as almost single-crystal material. Unfortunately, the very fast rates at which the ribbons are pulled from the melt means that the effective partition coefficients for the impurity elements are much closer to unity than in a quasi-equilibrium process like Czochralski or float zone growth (see equation (3.1)). The result of this is that heavy metal impurities present in the melt will be found in the ribbons as well, with consequent increase in the recombination rate. It is thus important to remove the lifetime killer elements from the starting silicon material, before melting and ribbon growth. A further problem with EFG ribbon is that a high density of SiC precipitate particles are found if a graphite die is used to shape the growing crystal. Dendritic web material often has a lower contamination level, as no die is needed, and the crystalline perfection of the ribbons can be higher than in the EFG material.

Another way of preparing thin sheets of silicon for photovoltaic devices is by a simple CVD process onto any suitable conducting substrate which will form a back ohmic contact to the silicon layer. CVD techniques by which thin epitaxial layers of silicon can be grown are described in §3.7, and the same kind of deposition reactions can be used to grow silicon films on a wide range of substrates. Because these substrates will not generally be lattice matched to the silicon, polycrystalline films will be grown. Chu (1977) has reviewed the structure of films of CVD silicon on steel, steel with a thin tungsten barrier layer, borosilicate glass and graphite. On substrates of this kind the grain size is greater than 10 μm only if the substrate temperature is in excess of 1000 °C, and at these high temperatures the growth rates of the silicon films are extremely rapid. At lower substrate temperatures, a film 100 μm thick will contain numerous narrow columnar grains, like a thicker version of the polysilicon thin film structure shown in figure 4.6. Recombination at the grain boundaries in such a film is expected to limit the efficiency of solar cells and indeed the measured efficiencies rarely exceed 1%. A discussion of the minimum grain size required before a reasonable efficiency is achieved will be given in the next section.

An alternative strategy for preparing silicon sheets for solar cells involves the growth of CVD silicon layers on large-grained silicon sheet or ribbon. Impure silicon ribbon can readily be prepared by EFG or dendritic web growth, as we have seen above. Silicon sheets can also be cast into boron nitride or silicon nitride moulds. These sheets can have grain sizes as large as a few millimetres, but usually contain high levels of boron nitrogen, aluminium and iron as a result of dissolution of the mould by the molten silicon. Similar silicon sheets can be grown on alumina substrates by passing them through a silicon melt. In all these sheet growth methods the intention is to grow large-grained substrates on which purer silicon films can be deposited. The photovoltaic devices can be prepared in this purer material. The melt from which the sheets are prepared can be very contaminated, and even metallurgical grade silicon, MGS, has been suggested as suitable for this application.

It is possible to grow thin epitaxial layers of high-purity silicon containing the p–n junction directly on top of these sheet or ribbon substrates. The layers will take on the large-grained structure of the substrate material. This is a practical example of the granular epitaxy phenomenon described in §4.2. The major problem with this kind of solar cell structure is the diffusion of impurity elements from the substrates into the epitaxial silicon layer during growth. It is probably this effect which limits the efficiency of these homojunction solar cells and will be less severe in devices grown on EFG or dendritic web ribbon than on MGS sheet. The ribbons are rather more expensive than the cast MGS substrates but MGS will always contain a high level of iron, which is a particularly effective lifetime killer. The performance of CVD/EFG silicon devices is very little worse than that of truly single-crystal cells, with $V_{oc} = 0.55$ V, $J_{sc} = 40$ mA cm^{-1} and $ff = 0.75$ giving a solar efficiency of about 11% (Kressel *et al* 1977). A general overview of the relative merits of the various methods of preparing inexpensive silicon material for solar cells has been given by Bachmann (1979).

9.3.1(b) Grain boundaries in silicon solar cells
The properties of grain boundaries in semiconductor materials have been described at some length in Chapter 4, and that information will

now be used to estimate the critical grain size at which the performance of silicon solar cells will be degraded by the presence of these defects. We have already seen that devices grown in material with grains about 100 μm in diameter can have high efficiencies and have made a rough estimate that a grain size equal to about the value of L_p is a reasonable lower limit at which the device manufacturer should aim.

The recombination rate at an isolated grain boundary in a semiconductor material in given by equation (4.11) and it is possible to use this equation in a model of the performance of a solar cell. An ideal model must include variations in the values of the grain boundary barrier height, Φ_B, with illumination and an estimate of the distortion in the form of the band edges where the boundaries intersect a p–n junction. This distortion is a result of the interaction of the potential fields around junction and grain boundary. Figure 9.10 gives an sketch of a possible shape for the conduction band edge where a single grain boundary intersects a p–n junction. For a real situation, where many boundaries with different values of Φ_B cross the p–n junction, the construction of a suitable model becomes extremely complex. Once the model is complete the three-dimensional carrier transport problem has to be solved under conditions in which free carriers are being created by incident photons. Lindholm and Fossum (1981) have given a graphic description of just how difficult it is to use such a model to obtain an estimate of the efficiency of a solar cell. Soclof and Iles (1975) and Card and Yang (1977) amongst many others have developed models which, using many

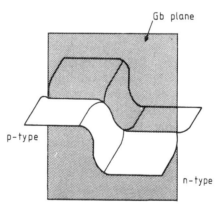

Figure 9.10 A sketch of the shape of the conduction band edge where a grain boundary intersects a p–n junction. The presence of the grain boundary creates a peak in the band edge in the n-type material and a trough in the p-type material, and this results in a local lowering of the band-edge discontinuity across the junction.

simplifying assumptions, permit an estimate to be made of how the efficiency of a polycrystalline silicon solar cell varies with the grain size. Figure 9.11 compares the predictions of these two models with some of the experimental data obtained from cells of the kind described in the previous section. We can see that the trend of increasing efficiency with grain size is both predicted and observed, but that both models are rather pessimistic as to the efficiencies obtained when the grain size exceeds 100 μm. (Some other models err in the other direction, predicting higher efficiencies than are observed.) It is clear that under ordinary illumination conditions single-crystal silicon cells will always have a slightly better solar efficiency than all but the very best polycrystalline cells.

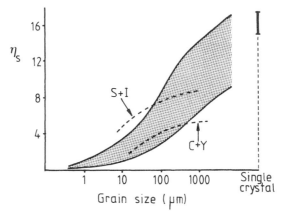

Figure 9.11 A comparison of some experimental data on the solar efficiency, η_s, of polycrystalline silicon solar cells (shaded region) and the predictions of two theoretical models, Soclof and Iles (1975), S + I, and Card and Yang (1977), C + Y. The range of efficiency values for single-crystal silicon solar cells is shown on the far right-hand side of the diagram.

9.3.2 The GaAs/GaAlAs Heteroface Cell

GaAs appears to be a semiconductor material which is almost ideally suited to the preparation of efficient homojunction solar cells. The band gap lies in the range of highest theoretical efficiency (see figure 9.5), the absorption coefficient is high because of the direct band gap so that a 2 μm thick layer will absorb most of the incident solar photons, minority carrier diffusion lengths in material of a reasonable crystalline perfection are longer than 2 μm and, finally, fabrication techniques for GaAs devices are reasonably well developed. The growth of a p-type absorber layer on top of an n-type wafer cut from material grown by Czochralski or Bridgman techniques is readily achieved by LPE or CVD techniques (see Chapter 3). Alternatively, a p-type layer can easily be produced by the diffusion of a p-type dopant into an n-type wafer. The most important lifetime killer impurities in GaAs are iron, nickel, copper, silver and oxygen, and if these elements are excluded from the active regions of the solar cells there seems little reason why very high efficiencies should not be achieved in homojunction devices.

Unfortunately, the recombination rate at the free GaAs surface is extremely high, and because the photon absorption and carrier generation processes occur close to a surface, the solar efficiency of GaAs homojunction cells is surprisingly poor, only about 10%. The efficiency of these cells can be improved by the inclusion of a heavily doped p^+ surface layer on the absorber material, creating a potential gradient which drives the minority carriers away from the free surface region. Another way of overcoming surface recombination problems is by the development of a new structure for the GaAs solar cell, a heteroface device, in which the top surface of the GaAs is passivated by an epitaxial GaAlAs layer. We have seen in Chapter 8 that GaAlAs ternary alloys are exceedingly well lattice matched to GaAs substrates over the whole of the composition range. This means that the misfit dislocation density and the interfacial recombination rate are both expected to be low at the GaAlAs/GaAs interface. By selecting a fairly aluminium-rich composition for the GaAlAs, this layer will have an indirect band gap and a low absorption coefficient for solar photons. Thus the GaAlAs layer can act as a window, allowing incident solar light to reach the GaAs homojunction cell underneath (Woodall and Hovel 1972). The configuration of a simple heteroface cell is shown in figure 9.12, where a p^+-GaAlAs epitaxial layer is used to passivate the surface of the p-type GaAs absorber layer. A device of this structure can have operating parameters $V_{oc} = 0.9$ V, $J_{sc} = 20$ mA cm^{-1}, $ff = 0.75$ and a solar efficiency of 15%.

A single LPE process can be used to produce the p-GaAs and the p^+-GaAlAs layers on an n-type GaAs substrate, since the p-type dopant in the GaAlAs layer will diffuse into the substrate during the high-temperature LPE growth process and compensate for the donor doping in the top surface of the wafer. More precisely controlled band-edge profiles can be prepared by the separate growth of the p-GaAs and GaAlAs layers. The GaAlAs layer is usually kept very thin to ensure that as high a fraction as possible of the incident solar light reaches the absorber layer. A thickness of around 1 μm is convenient for many practical cells and the ternary layer is usually degenerately doped to make it easier to form an ohmic contact on the top surface of the device.

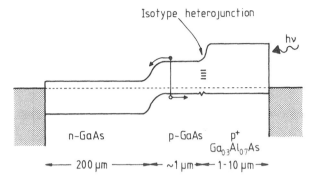

Figure 9.12 A sketch of a simple heteroface GaAs solar cell in which the top surface of the p-type absorber layer is passivated by the addition of a GaAlAs layer in which the solar photons are not absorbed.

GaAs/GaAlAs heteroface solar cells are a good example of the use of relatively simple growth techniques to prepare highly efficient photovoltaic devices. These cells are suitable for space applications because GaAs is more resistant to radiation damage than silicon. However, they are relatively expensive to produce since both the growth of the substrate material and the deposition of the epitaxial layers by LPE are costly operations. These devices are thus not suitable for large-area terrestrial power generation. For further reading on the development and design of these solar cells, including a discussion of the very high efficiencies which can be achieved if these cells are included in concentrator arrays, see Fahrenbruch and Bube (1983).

9.3.3 Heterojunction Cells

In §9.2.5 I gave the basic design rules for the production of reasonably efficient heterojunction solar cells. Here I shall describe the properties of a few of the most common heterojunction cells. This will by no means be a comprehensive review of state-of-the-art heterojunction devices, but will concentrate on a few systems for which some attempt has been made to link the properties of the semiconductor materials to the performance of the devices. All single-crystal solar cells are relatively expensive because of the high cost of the starting material, and we shall see later that many of the successful configurations have also been made in cheaper polycrystalline materials. Figure 9.5 shows that both GaAs and InP have band gaps of almost the optimum width for the absorber layers of heterojunction cells and that CdS, SnO_2 and In_2O_3 are all useful window layers. It is the combination of these materials to make efficient solar cells which will be of the most interest in this section.

9.3.3(a) GaAs on silicon solar cells

GaAs on silicon technology is being investigated for the manufacture of many kinds of GaAs photovoltaic and transistor devices. The driving force for this work is that silicon wafers are very much cheaper than those cut from GaAs crystals. It is thus possible to conceive of devices where a relatively impure silicon material can be used as a substrate for the growth of a high-quality GaAs layer, in exactly the same way as we have already described for low-cost silicon solar cells. Considerable attention has been paid to the deposition of epitaxial GaAs films on silicon substrates by CVD techniques. These epitaxial layers can be grown containing all the junctions necessary for the operation of the devices, and so the GaAs on silicon solar cell is really only a homojunction cell on a foreign substrate. Figure 9.13(a) shows a sketch of the structure of such a cell. The major problem with this kind of solar cell is that there is a 4% mismatch between the lattice parameters of GaAs and silicon and so a very high dislocation density is formed at the heterointerface during the epitaxial growth process. Some of these dislocations will propagate from the interface up into the GaAs layer. Dislocation densities of around 10^8 cm^{-2} are commonly found in GaAs films grown directly on silicon substrates. Figure 9.13(b) shows a transmission electron micrograph of such a layer, and a high density of threading dislocations is clearly seen running up through the GaAs epilayer. In addition, numerous twins and some antiphase domain boundaries between symmetry-related morphological variants are found in these GaAs layers.

In §2.4 I described the deleterious effects that dislocations can have on the performance of devices and circuits in silicon wafers. These properties stem from the segregation of fast-diffusing impurities like iron and copper to the dislocations, creating highly active recombination centres and short-circuit conduction paths across p–n junctions. Dislocations in GaAs act in a similar manner (Ettenberg 1974), so the high density of threading dislocations passing through the GaAs homojunction solar cells on silicon substrates are expected to reduce the solar efficiencies. The dislocations can increase J_0 by providing short-circuit paths across the p–n junction and also decrease V_{oc} by increasing the density of recombination centres in the absorber layer of the device. Yamaguchi et al (1986) have estimated that dislocation densities as low as 10^5 cm^{-2} are required in the GaAs epilayers before an efficiency is achieved which approximates to that of the GaAs/GaAlAs heteroface cell described above.

Thus we have a system in which the GaAs homojunction seems to offer a highly efficient solar cell configuration and silicon a convenient and cheap substrate material. However, the large misfit at the heterojunction results in a solar efficiency of perhaps half of that achieved in homojunction GaAs devices on GaAs substrates. One possible way of improving the performance of these devices is to include another layer between the silicon and the GaAs in an attempt to reduce the density of threading dislocations propagating into the active region of the solar cells. Tsaur *et al* (1982) have demonstrated that an epitaxial germanium layer can readily be grown on the silicon substrate. The lattice parameter of the germanium is almost exactly the same as that of GaAs, and so it is not surprising that the germanium epitaxial layer contains a high density of dislocations. However, when an epitaxial GaAs layer is grown on the germanium, the density of threading dislocations which propagates across the Ge/GaAs interface is relatively low. Using a multi-layered Ge/GaAs structure of the kind sketched in figure 9.13(c), the

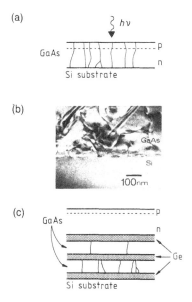

Figure 9.13 (a) A sketch of a GaAs homojunction solar cell in a thin epitaxial film grown on a silicon substrate. A high density of threading dislocations is expected to propagate up from the hetero-interface into the GaAs layer. (b) A cross-sectional transmission electron micrograph of a GaAs/Si heterointerface showing the lattice defects propagating into the GaAs layer. (Courtesy of Dr D Eaglesham.) (c) A sketch of a multilayered Ge/GaAs on silicon structure which can be used to reduce the dislocation content in the top GaAs layer which is where the homojunction solar cell is prepared.

dislocation density in the top GaAs layer can be reduced to $10^6 \, \text{cm}^{-2}$. This is a result of the trapping of the threading dislocations in the Ge/GaAs interfaces in agreement with the model of Matthews and Blakeslee (1974) described in §8.3.

Tsaur *et al* (1982) have also shown that the GaAs on silicon structure can be used in a practical tandem solar cell. Here the principle is that an absorber layer with large band gap is placed on top of a semiconductor with a smaller band gap and both of these layers connected as independent solar cells. Solar photons which pass through the upper semiconductor layer are efficiently absorbed in the lower material with the smaller band gap, thus increasing the fraction of the total incident light which is used to generate free carriers. A description of the theory of tandem solar cells can be found in Sze (1981).

9.3.3(b) Two InP-based solar cells

Table 9.1 shows that InP has a direct band gap of width approximately 1.35 eV. InP absorber layers can thus be used in combination with CdS or In_2O_3 window layers to prepare efficient heterojunction solar cells. There is a relatively good lattice match between the {111} planes of InP and the basal planes of CdS, with a misfit of only 0.32%. This would be considered a rather poor lattice match in heterojunction laser structures of the kind described in Chapter 8, but will give a much lower density of misfit dislocations than the GaAs/Si heterojunction described above. In addition, the electron affinities of InP and CdS are such that no spike is formed in the conduction band edge at a p-InP/n-CdS interface (figure 9.7).

Single-crystal InP/CdS solar cells are usually prepared on InP wafers by the evaporation or CVD growth of epitaxial CdS films. This gives a device with the band structure illustrated in figure 9.14. The CdS layer will have a degenerate n-type character because of the very high density of native defects (S vacancies) produced by the vapour phase growth process. A particular advantage of this structure is that any propagation of interfacial dislocations into the growing epitaxial layer will be in the CdS window layer, not in the InP absorber where enhanced recombination of photoexcited minority carriers would degrade the solar efficiency. Solar photons incident on such a device will pass through the CdS window layer and be absorbed in the top few micrometres of the InP substrate. Typical operating parameters of a high quality InP/CdS solar cell might be $V_{oc} = 0.8 \, \text{V}$, $J_{sc} = 18 \, \text{mA cm}^{-2}$, $ff = 0.74$ and a solar efficiency of 14% (Yoshikawa and Sakai 1977).

The InP/ITO cell seems another favourable combination of absorber and wide band-gap window layer, giving a heterojunction with no conduction band spikes. However, there is a large misfit between the

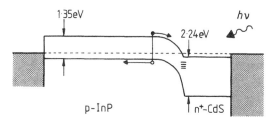

Figure 9.14 A sketch of the structure of a solar cell prepared by depositing degenerate n-type CdS onto a single-crystal InP substrate.

two lattices whichever orientation of InP is used as a substrate. The heterointerface formed when ITO is sputtered onto InP is thus expected to contain a very high density of misfit dislocations and interfacial recombination is expected to be rapid. Even so, the performance of single-crystalline InP/ITO cells is often excellent, with solar efficiencies greater than 14%. It seems possible that indiffusion of the tin from the oxide layer into the InP substrate creates a shallow p–n junction and that the carrier separation actually takes place across this relatively perfect interface rather than the highly defective heterointerface. An alternative explanation of the high efficiency of these cells is that the stoichiometry of the top surface of the InP is disturbed during the sputtering of the ITO layer, creating a buried p–n junction. These devices should thus more properly be thought of as homojunction cells in InP (Bachmann *et al* 1979).

Single-crystal heterojunction solar cells have also been prepared with a variety of other combinations of semiconductor materials, including Si/ITO, CdS/CuInSe$_2$ and CdTe/CdS. However, the combination of silicon and ITO is not one which would be expected to give a band-edge profile of a favourable kind (figure 9.7). The preparation of efficient Si/ITO cells seems to rely on the sputtering process used to deposit the window layer damaging the silicon surface to give a buried homojunction just like the InP case above. CuInSe$_2$ and CdTe are both available as large single crystals grown by Bridgman or vaour techniques and both can be amphoterically doped. The band gaps are 1.01 and 1.56 eV respectively, so both of these compound semiconductors are suitable for use as absorber layers in solar cells. CdS window layers can be quite well lattice matched with CuInSe$_2$ and rather less well with CdTe. Efficiencies of about 12% have been achieved in cells with these two material combinations, although by no means all the devices have performed as well as this. It seems to be generally true that great care has to be taken over the cleanliness of the heterointerfaces if the best efficiencies are to be achieved. The electrical properties of CuInSe$_2$ are particularly sensitive to annealing, which can modify the surface stoichiometry and conduction character. A comprehensive review of the applications of this chalcopyrite compound in photovoltaic cells has been given by Coutts *et al* (1986).

A huge range of other single-crystal heterojunction solar cells have been tested, some approaching the efficiencies of the devices described here, but many giving disappointing results. By no means all the combinations of materials with good lattice matching across the hetero-interface, and where the simple models outlined above predict no spikes or notches in the band-edge profile, give high efficiency cells, and some apparently unattractive combinations can be used to make excellent cells. It is not always clear why the predicted and measured properties differ so widely, but this serves to emphasise the point made in §9.2.3 that the real properties of a heterojunction may depend more on chemical diffusion across the interface, interfacial defect concentrations and lattice damage introduced during fabrication, than the idealised band structure of the junction. A more detailed discussion of the subtle influence of defects and processing conditions on the performance of heterojunction solar cells can be found in Fahrenbruch and Bube (1983).

9.4 AMORPHOUS SILICON SOLAR CELLS

The structure and electrical properties of amorphous silicon are introduced in §1.7, where it is shown that amorphous silicon, α-Si, has an effective band gap of about 1.6 eV, almost the optimum value for a solar cell absorber material. In addition, it has been shown that the optical absorption coefficient of α-Si for solar light is much higher than that of crystalline silicon. The α-Si behaves like a direct band-gap material, with a sharp absorption edge for photons with energies in excess of 1.6 eV. A layer of α-Si 1 μm thick will absorb more than 70% of incident solar photons. However, α-Si also has a very high density of electronic states distributed throughout the band gap, which means that it is not easy to change the conductivity by the addition of group III or V doping elements. Vapour-deposited thin films of α-Si usually contain high concentrations of dangling bonds and a density of mid-gap states in excess of 10^{19} cm^{-3}. The incorporation of hydrogen into amorphous silicon films can help to saturate many of these dangling bonds and reduce the density of band-gap states to a more manageable level of 10^{16} cm^{-3}. A heavily hydrogenated α-Si film can thus be doped in just the same way as a crystalline sample. We can see that hydrogenated silicon should be a useful absorber layer in a photovoltaic device.

Evaporation and sputtering techniques can be used to deposit thin films of α-Si, but films for solar cell applications are more commonly grown by exciting a glow discharge in a silane atmosphere over a substrate heated to 200–400 °C. This is a plasma-enhanced CVD process (see §3.7). The silicon films will be amorphous at low substrate temperatures and will contain 10–50 at.% of atomic hydrogen. The precise hydrogen concentration depends on the conditions in the plasma and the substrate temperature; the electrical properties of the films depend on the hydrogen content. In general, the higher the substrate temperature, the lower the hydrogen content and the lower the resistivity. Heating the films after deposition will also reduce the hydrogen content and the resistivity. If the hydrogen level is high, the density of the α-Si can be as low as 60% of the value for crystalline silicon. We can think of these α-Si films as having the continuous random network structure described in §1.7, but with a microstructure consisting of columns of relatively dense material separated by hydrogen-rich regions (Knights 1977). The most important features of glow discharge α-Si are that it has a high resistivity, is an n-type semiconductor, and can readily be doped by the addition of elements like boron and phosphorus (Spear and LeComber 1975). It is thus possible to prepare thin films of α-Si which contain a p–n junction, just as required in a homojunction solar cell.

A range of glassy and metallic materials have been used as substrates for α-Si solar cells, and the important consideration when selecting these materials is that no diffusion of electrically active impurities should occur from the substrate into the amorphous films at the relatively modest deposition temperatures. Films grown by glow discharge processes usually contain high oxygen, nitrogen and carbon levels, but these impurities do not have a serious influence on the photovoltaic properties of the α-Si. Deposition by glow discharge can be carried out over large areas onto cheap substrates and so α-Si is a popular material for use in inexpensive terrestrial solar arrays.

Most α-Si photovoltaic devices are of the Schottky barrier or MIS type, because these structures are very simple to prepare. Noble metal Schottky contacts to n-type α-Si have higher barrier heights than the same metals on crystalline silicon and so solar cells with high values of V_{oc} can be prepared. For instance, the barrier height of platinum on crystalline n-type silicon is only 0.9 eV, but on α-Si it is 1.1 eV. Interestingly, the Schottky barrier heights on α-Si increase linearly with the workfunction of the metal, which implies that there is only weak pinning of the Fermi level at the α-Si surface (see §2.2). Figure 9.15 illustrates the structure of a thin film α-Si solar cell on a steel substrate. The thick layer of undoped α-Si acts as the absorber and the heavily doped layer at the back surface helps to produce a good ohmic contact. The low conductivity of the glow discharge α-Si layers means that it is impossible to use a grid structure for the top contact. Unfortunately, it has proved difficult to deposit continuous thin metallic Schottky contacts on α-Si devices which also have a high optical transmission. At least some of the incident solar photons will be absorbed or reflected by the thin platinum contact. A device of this kind would typically have $V_{oc} = 0.8$ V, $J_{sc} = 12$ mA cm^{-2}, $ff = 0.6$ and a solar efficiency of 5.5% (Carlson 1977). This is quite a respectable efficiency to obtain from a very cheap device. Alternative cell structures include SnO_2/α-Si hetero-structures and MIS devices like $Pt/TiO_2/\alpha$-Si. These MIS cells have higher values of V_{oc} than the simple Schottky barrier devices, as we would expect from the discussion in §9.2.3. More details on the properties of α-Si thin films, and their use in photovoltaic devices, are given in the review by Fritzche (1980).

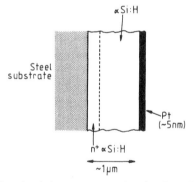

Figure 9.15 A sketch of the structure of a simple α-Si Schottky solar cell. The steel substrate acts as the back contact to the device.

9.5 THIN FILM POLYCRYSTALLINE SOLAR CELLS

A vast range of thin film solar cells have been fabricated and tested, and no attempt will be made here to describe all the materials' combinations and cell configurations. A comprehensive review of the art and technology of thin film solar cell production has been given by Chopra and Das (1983). In this section I shall introduce some of the more unusual techniques by which thin semiconductor films can be deposited for photovoltaic devices and then consider how the high density of grain boundaries in these films will influence the performance of these cells.

Finally, I will describe the preparation and properties of one of the most complex and fascinating thin film devices, the Cu_xS/CdS solar cell.

9.5.1 The Deposition of Thin Polycrystalline Semiconductor Films

Thin films of many semiconductor materials can be deposited by evaporation, sputtering and CVD techniques, and these processes have been described in some detail in Chapters 3 and 4. The drive to develop cheap processes for the fabrication of solar cells has resulted in a range of techniques based on wet chemical deposition being considered for the preparation of thin films of the materials listed in table 9.1. Most of these techniques are selected because they require little or no expensive equipment and can be carried out by relatively unskilled staff. Here I will introduce the application of spray pyrolysis, solution growth, electro-deposition and chemical exchange interactions in the fabrication of thin film photovoltaic devices.

The process of spray pyrolysis involves spraying a solution of salts of the desired elements onto a heated substrate. The salts decompose on the substrate, allowing the elements to react together to form a semiconductor compound. It is very important that the reaction by-products are volatile at the temperature of the substrate, or the growing film will be heavily contaminated. A typical process for the deposition of a film of CdS would involve the spraying of a dilute aqueous solution of $CdCl_2$ and thiourea, $(NH_2)_2CS$, onto a substrate held at 200–500 °C. The overall reaction to form CdS is

$$CdCl_2 + (NH_2)_2CS + 2H_2O \rightarrow CdS + 2NH_4Cl + CO_2 \qquad (9.7)$$

although the actual reaction path will probably be much more complex, with several intermediate stages. Tin-doped highly conducting In_2O_3 films can be grown from solutions of $InCl_3$ and $SnCl_4$ dissolved in a mixture of water and ethanol. The composition of the films depends in a complex manner on the solution chemistry, the spraying rate and the substrate temperature, and in most cases a considerable concentration of the solvent and by-product chemicals will be incorporated in the growing film. Increasing the substrate temperature improves the purity of the films by encouraging the evaporation of the volatile chemical species but also lowers the rate at which the films can be deposited.

It has proved possible to use spray pyrolysis to deposit films of a wide range of semiconductor compounds with properties surprisingly similar to those of the bulk materials. Even a few ternary and quarternary semiconductor alloys have been grown in this way, including $Cd_xZn_{1-x}S$ and the more familiar ITO alloys. Films deposited by spray pyrolysis are generally very adherent and excellent control over the composition can be achieved. However, the grain size in the films is rarely greater than 1 μm, even after a post-deposition anneal. Further details on the use of spray pyrolysis techniques for the preparation of thin films for solar cells have been given by Chopra et al (1982).

Thin films of compound semiconductor materials can also be deposited by solution growth. Here the semiconductor compound is precipitated from a slightly supersaturated solution of weakly soluble salts of the desired elements. The precipitation process has to be controlled very carefully to ensure that the compound grows as a uniform film on the substrate. This is done by releasing precisely determined concentrations of ionic species into solution by the spontaneous decomposition of complex ions. For instance, the concentration of Cd^{++} ions in solution can be controlled by forming a complex with ammonia molecules, $Cd(NH_3)_4^{++}$; the ammonia is called the complexing agent in this case. In the presence of a high concentration of sulphate ions, CdS will slowly precipitate onto a substrate and the concentration of Cd^{++} ions in solution will be replenished by the decomposition of the complex ion:

$$Cd(NH_3)_4^{++} \overset{K}{\rightleftharpoons} Cd^{++} + 4NH_3. \tag{9.8}$$

This decomposition reaction has an equilibrium constant of $K = 8 \times 10^{-8}$ and so the concentration of cadmium ions in solution will be very low and the precipitation reaction slow and controllable. Chopra and Das (1983) have described the conditions under which well characterised films of compounds and alloys like CdS, PbSe, CdZnS and SnO_2 can be deposited from solutions by this process. A convenient feature of solution growth is that there is a terminal thickness to which the semiconductor film will grow and that this thickness can be set to a desired value by choice of the appropriate substrate temperature and concentration of the solution. The choice of complexing agent often has a strong influence on the film thickness, composition and microstructure, and so great care has to be taken to select the complex ion which gives the most attractive combination of properties in the deposited films. The grain size in solution-grown films is usually very small indeed (20–200 nm) and in many cases the semiconductor films consist of a mixture of phases, just as we have seen in films prepared by sputtering or evaporation. For example, CdS films contain grains of both wurtzite and sphalerite structure.

Another technique which can be used in the deposition of thin films of semiconductor materials is screen printing, which we have already described in §7.8 as a method for putting thick film conductor tracks onto ceramic substrates. The deposition of semiconductor films requires the preparation of pastes which contain fine particles of the semiconductor material, and dopant elements as well, in a supporting medium. A paste for the deposition of CdS might contain CdS powder, $CdCl_2$ and $GaCl_2$ in an alcohol medium. The $GaCl_2$ is used to include gallium as a dopant in the CdS film. Such a paste is printed onto a substrate and then fired at 600–700 °C to give a relatively thick CdS film with a grain size of around 10 μm. A very similar process based on the same paste composition might simply paint the paste onto the substrate and use a doctor blade to control the thickness. Alternatively, sedimentation of the CdS particles from a less viscous slurry will also produce a thin film. In all these processes we expect the semiconductor layers to be heavily contaminated with chemicals from the paste, even after firing at high temperatures. Screen-printing techniques can be used to deposit the contact grid arrays as well as the semiconductor layers. The paste compositions for these metallic layers will be of the same kind as described in §7.8. It is thus possible to envisage a complete solar cell device deposited in a series of screen-printing processes.

Electrolysis is a very well known method for depositing metallic contacts and is used for some solar cell contacts, but can also grow thin layers of semiconductor materials. A solution of $CdSO_4$ and H_2TeO_3 in sulphuric acid can be used to deposit CdTe on a titanium cathode. A $CuSO_4$ solution can be used to deposit a thin layer of copper sulphide on a CdS cathode. Saksena *et al* (1982) have shown that these copper sulphide films have a stoichiometry very close to Cu_2S if the temperature of the solution and the current density are chosen correctly. We will see in §9.5.4 that this is the composition of the copper sulphide layer which gives the best efficiency in Cu_xS/CdS solar cells.

An even simpler way of producing a film of Cu_xS on a CdS surface is by dipping a CdS substrate, which can be a single crystal or a fine-grained polycrystalline film, into a CuCl solution. A reaction occurs between the CdS and the solution to grow a thin Cu_xS layer, i.e. an exchange reaction. At a solution temperature of about 100 °C a copper sulphide layer 1 μm thick is grown in a few seconds. However, the Cu_xS layer rarely consists of a single phase and this may have important consequences for the efficiency of the solar cells. The evaporation of solid CuCl onto a CdS surface, followed by a solid phase reaction process to produce copper sulphide, may be a more controlled way of preparing the desired phases. In both these processes, the copper sulphide layer will tend to grow with a grain size approximately the same as that in the CdS substrate. A rather larger grain size is thus achieved than is normally found in films deposited by spray pyrolysis or solution growth.

From this brief introduction to some of the techniques which can be used to deposit thin films of semiconducting materials, it is clear that there are very many possible methods for preparing a thin film solar cell. Before we go on to consider how the presence of grain boundaries influences the performance of these cells, it may prove useful to review how the films prepared by all these methods differ in microstructure and electrical properties. Only with this knowledge can an informed choice be made of the precise deposition process which will result in the fabrication of the most efficient thin film solar cell. Rather than give an exhaustive description of the properties of a range of semiconductor compounds in thin film form, I shall only consider the case of CdS. The variation in the structure and the properties of films of this compound semiconductor deposited by the techniques described above gives a reasonably general view of the way in which the deposition process can effect device efficiency. The rather special case of copper sulphide films will be described in §9.5.4. For a more complete description of the properties of a wider range of thin films the reader is directed to Chopra and Das (1983).

9.5.2 Thin Films of CdS

Table 9.5 summarises some experimental observations on the structure and properties of CdS thin films deposited by evaporation, sputtering, spray pyrolysis and solution growth techniques. The films are usually 10–30 μm thick and are designed for use as window layers in hetero-junction thin film solar cells. We can see that the crystal structures and the grain sizes in the films can be altered by the substrate temperature, T_s, and the choice of deposition process. It is quite common for films to contain a mixture of wurtzite and sphalerite crystals, as has already been described in §4.6. The resistivity of CdS thin films can be varied over about 10 orders of magnitude and depends on the impurity and dopant concentration, the grain size and the height of the potential barriers at the grain boundaries. High levels of oxygen are normally included in chemically deposited films, resulting in highly resistive grain boundaries. This oxygen can be partially removed by annealing the films in a vacuum, but will resegregate to the boundaries if the film is heated in air. The optical properties of the films are quite similar to those of single crystalline CdS, with a slight decrease in refractive index and a broadening of the absorption edge in fine-grained material.

The most important properties of CdS films in solar cells are the position of the absorption edge, which controls whether the layer will be an effective window for solar photons, and the resistivity, which determines the contribution the collector layer makes to the series resistance of the solar cell. The optical properties of CdS films are similar to those of bulk single-crystal material and so we expect CdS thin films to be effective window layers. The presence of a high density of grain boundaries in a window layer can still have some effect on the

Table 9.5 The structure and properties of thin films of cadmium sulphide (after Chopra and Das 1983).

Deposition technique	Crystal structure and grain size	Electrical properties	Optical properties
Evaporation:			
$T_s > 200\,°C$	wurtzite c axis texture columnar grains 1–5 μm	n-type 1–1000 Ω cm (with In doping 10^{-3} Ω cm)	
$T_s < 150\,°C$	sphalerite little texture	$L_p \sim 0.2$ μm	Sharp absorption edge at 0.52 μm
Sputtering:	wurtzite c axis texture columnar grains ~1 μm	n-type 10^8 Ω cm (with In doping ~1 Ω cm)	Refractive index approaches bulk value as grain size increases
Spraying:	wide range of structures depending on T_s, spray chemistry and dopant	n-type $L_p \sim 0.3$ μm high O content at boundaries	
Solution growth:	crystal structure depends on solution some c axis texture small grains, ~100 nm	n-type high O content at boundaries	Fine grain size gives rather diffuse absorption edge

Note: T_s is the substrate temperature and L_p the minority carrier diffusion length.

performance of a solar cell by increasing the sheet resistance of the films. In the next section we shall consider in a little more detail the effect of grain boundaries on the efficiency of thin film solar cells.

9.5.3 Grain Boundaries in Thin Film Solar Cells

In figure 9.11 the optimum grain size in material for polycrystalline silicon solar cells was shown to be greater than 100 μm. We might, therefore, expect the possibility of producing efficient solar cells from thin film semiconductor materials with grain sizes of the order of 1 μm to be very slight. However, we must look rather more closely at what actually happens in a thin film solar cell before jumping to this conclusion. Figure 9.16 will be used to illustrate some of the processes which occur in a photovoltaic device of this kind. Solar photons incident on the front surface of the device will pass through the window layer

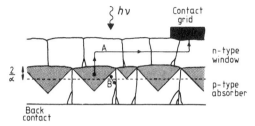

Figure 9.16 A schematic illustration of the process of photon absorption in a thin film solar cell, showing the recombination cones and their relationship to the value of $2/\alpha$.

and be absorbed by the lower semiconductor material which is usually chosen to have a band gap, E_{g1}, in the range 1–1.5 eV. About 90% of the photons with energy greater than E_{g1} will be absorbed in a thickness of $2/\alpha$. This thickness varies from about 100 μm for silicon to 1 μm for semiconductors like CdTe, InP and CuInSe$_2$, which have direct band gaps. The carriers created by the photon absorption must now be separated across the p–n junction if they are to contribute to the current flowing through the cell. This process is illustrated by path A in figure 9.16, where a minority carrier generated in the absorber layer diffuses across the junction and then along the window layer to a grid contact bar. However, the grain boundaries in the absorber layer may be efficient recombination centres for the minority carriers and we must expect some of the photoexcited electrons in the p-type absorber to be lost to these defects. This process is illustrated by path B in the same figure.

It is convenient here to introduce the concept of a 'recombination cone', the shaded regions in figure 9.16. The electrons generated within a cone pass across the p–n junction, while those outside the cones are more likely to recombine at a grain boundary (Rothwarf 1976). The depth within which most of the absorption of the incident photons occurs is indicated by the broken line $2/\alpha$ below the surface of the absorber layer. It is clear that the smaller the grain size, the larger the fraction of the minority carriers which will be generated outside the recombination cones. If $2/\alpha$ is larger than the grain size in the absorber layer, most of the minority carriers will be lost to the grain boundaries and the solar cell will have a very low efficiency. We can now see why the optimum grain size in polycrystalline silicon cells is greater than 100 μm; this is the value of $2/\alpha$. However, if the absorber layer is a direct band-gap semiconductor, the value of $2/\alpha$ is about 1 μm and we expect to be able to make quite satisfactory solar cells in materials with much smaller grains. Of course, the efficiency of the cells will still improve if the grain size is increased. The dramatic difference in the performance of thin film solar cells in direct band-gap compound semiconductors and those in polycrystalline silicon is highlighted in figure 9.17. The same solar efficiency is found in $CdS/CuInSe_2$ cells with grain sizes of around 1 μm and in silicon cells with a grain size of 100 μm. This is precisely what we would expect from the fact that the optical absorption coefficient for solar photons in $CuInSe_2$ is about two orders of magnitude greater than in silicon.

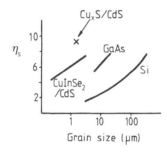

Figure 9.17 A comparison of the variation of solar efficiency with grain size in solar cells fabricated in a number of semiconductor materials.

Another reason for the high efficiency of $CdS/CuInSe_2$ thin film solar cells is that the grain boundary potential barriers in $CuInSe_2$ films are relatively low (Kazmerski *et al* 1976). Equation (4.11) shows that the recombination rate at grain boundaries depends strongly on the barrier height. It is possible that grain boundaries in some compound semiconductor materials can be autopassivated by local changes in stoichiometry, which would explain the low barrier heights measured in these materials. Grain boundaries in silicon have to be passivated by the deliberate introduction of impurity species, as described in §4.3.4.

So far we have followed the electrons generated in the recombination cones across the p–n junction, but they also have to diffuse through the window layer to be collected at the contact grid bar. If the grain size in the window layer is small, this diffusion process involves crossing a large number of grain boundaries. The influence of grain boundaries on the resistivity of a thin film of a semiconductor material was described in §4.3, where it was shown that the sheet resistivity of a film depends exponentially on the barrier height, Φ_B. It is thus important in an efficient solar cell design to choose a material for the window layer which has either a large grain size or a low average value of Φ_B. The grain size is normally considered to depend on the substrate temperature during deposition, the thickness of the film and the details of any annealing treatment, but it is difficult in most materials to obtain a grain diameter greater than the thickness of the film (Maissel and Glang 1983). The average measured values of Φ_B in CdS films lie around 0.15 eV, which is a very low value in a semiconductor with such a large band gap and leads to a rather low sheet resistance in CdS films. However, if oxygen is segregated to the grain boundaries, the value of Φ_B increases dramatically and the sheet resistance of the CdS films also increases. Fortunately, the oxygen can easily be removed from the boundaries by vacuum annealing and the low sheet resistance allied with the ease of depositing CdS films explains why so many thin film solar cell configurations use this compound semiconductor as a window layer.

A final point which should be made in this section concerns the effect that grain boundaries have on the dark current across the p–n junction. We can see from figure 9.16 that the grain boundaries run more or less parallel to the predominant direction of current flow across the p–n junctions. These grain boundaries can be highly effective short-circuit conduction paths, especially when decorated with precipitates of impurity elements which segregate to the boundaries (Matare 1971) and can lead to a dramatic reduction in V_{oc}. Increasing the grain size in the thin film materials is the best way of ensuring that all the deleterious grain boundary effects are minimised. An extended review of the effect that grain boundaries can have on the performance of thin film solar cells has been given by Fahrenbruch and Bube (1983).

9.5.4 The Cu$_x$S/CdS Solar Cell

I have chosen to discuss the Cu$_x$S/CdS solar cell as an example of a practical thin film device because it is one of the most complex cells and yet paradoxically has been shown to give high efficiencies even in crudely made devices. We should remember, however, that this is not the only commercially important thin film solar cell; combinations of absorber materials like GaAs, InP, CdTe and CuInSe$_2$ with CdS window layers also give solar efficiencies in the range 5–8%, and these same absorber films can be combined with the conducting oxide window materials as well.

Let us first consider the structure of a typical Cu$_x$S/CdS solar cell and see how it can be fabricated. The deposition and structure of thin films of CdS have already been described and a convenient first stage in the manufacture of a large-area copper sulphide cell is the evaporation of a layer of CdS some 20 μm thick. This layer will have a columnar grain structure, with a grain size of a few micrometres. Many kinds of substrates can be used for the CdS film, including copper or glass sheets coated in SnO$_2$ or In$_2$O$_3$ to give a good ohmic back contact. Dipping the CdS-covered substrate in HCl for a few seconds will etch the semiconductor to give a highly textured surface.

The substrate carrying the textured CdS layer is now dipped in a cuprous ion solution held at about 90 °C, and a layer of copper sulphide a few hundred nanometres thick grows on the CdS surface by an exchange reaction. The grain boundaries in the CdS substrate are preferential sites for this reaction and so the copper sulphide penetrates quite extensively down the boundaries. A copper grid bar array, and an antireflectant coating, complete the solar cell, and a sketch of the final structure is given in figure 9.18. The highly textured CdS surface gives a structure where photons incident on the device pass obliquely through a thickness of Cu_xS much greater than the nominal layer thickness of 100–300 nm. This ensures that the solar photons are efficiently absorbed in the copper sulphide layer and yet the photoexcited minority carriers are always generated very close to the heterojunction. The process described above is only one way in which Cu_xS/CdS cells can be prepared. Chopra and Das (1983) have given a more complete review of a wider range of production routes.

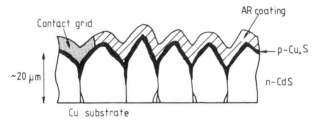

Figure 9.18 A sketch of the structure of a typical thin film Cu_xS/CdS solar cell, with a textured CdS film supporting a thin Cu_xS layer grown by an exchange reaction.

Now we should consider what form the copper sulphide layer takes when prepared in an exchange reaction between CdS and a cuprous ion solution. The exchange reaction proceeds by replacing each cadmium atom by two copper atoms, and the sulphur sublattice may remain relatively undisturbed in this process (TeVelde and Dieleman 1973). There are a number of equilibrium phases in the Cu/S phase diagram in the composition range around Cu_2S: in particular djurleite at $Cu_{1.95}S$; orthorhombic chalcocite, Cu_2S; and digenite at higher temperatures and lower sulphur contents (see figure 9.19). Chalcocite also has a high-temperature hexagonal form, which is visible at the top of this figure. Of these four phases, orthorhombic chalcocite has been found to have the highest optical absorption coefficient for photons of energy 1.5 eV and is a p-type defect semiconductor with copper vacancies producing shallow acceptor states. The resistivity of chemically grown Cu_2S films is quite low, usually around $10^{-2}\ \Omega$ cm, and orthorhombic chalcocite is also the only one of these four phases which has a minority carrier diffusion length greater than 5 nm. This combination of properties means that we expect the chalcocite/CdS solar cell to have a much higher efficiency than those made with any of the other copper sulphides (TeVelde and Dieleman 1973). It has been convincingly demonstrated that the exact composition of the copper sulphide layer does indeed control the performance of the solar cells (Palz *et al* 1972).

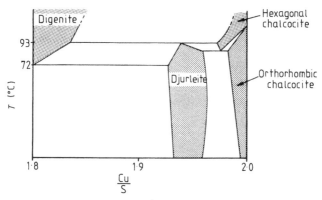

Figure 9.19 A detail of the Cu/S equilibrium phase diagram in the composition range close to Cu_2S. Several phases with very similar compositions are found in this region of the phase diagram (after Potter 1977).

Hadley and Tseng (1977) have shown that after an exchange reaction the copper sulphide layer consists primarily of orthorhombic chalcocite and that this phase is oriented so as to form a low misfit interface with the *c*-axis-textured CdS. However, some of the other copper sulphide phases are usually present in these films as well and it has been found that the relative proportions of the phases can be adjusted after the exchange reaction by low-temperature annealing treatments. The ideal copper sulphide layer is chalcocite with a stoichiometry of $Cu_{1.995}S$, so that the concentration of copper vacancies is high and the resistivity of the film low. One suggested process for obtaining the maximum solar efficiency from a thin film Cu_xS/CdS cell involves annealing the devices at 170 °C for 15 h after the exchange reaction. We can see that while both the Cu/S phase diagram and the microstructure of the thin film solar cells are very complex, it has proved possible to develop 'recipes' which seem to result in the preferential growth of the desired chalcocite phase on the CdS.

Figure 9.20 shows the energy band diagram of a Cu_xS/CdS solar cell, and this combination of materials satisfies many of the requirements listed in table 9.3. This kind of cell can be operated with the light incident on the Cu_xS layer, as shown in figures 9.18 and 9.20, or on the

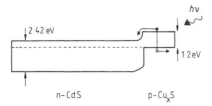

Figure 9.20 A schematic illustration of the band structure of a Cu_xS/CdS solar cell, in this case with the solar light incident on the thin Cu_xS layer.

CdS window layer if the whole device is fabricated on a glass substrate. The electrical transport properties of the p-Cu_xS/n-CdS solar cell are complex and not fully understood, and Fahrenbruch and Bube (1983) have given an excellent review of the relative importance of the processes of diffusion, tunnelling and recombination at the heterointerface. What is of more direct interest here is how the structure and properties of the ideal orthorhombic chalcocite phase influence the operation of these devices. It has been observed that orthorhombic chalcocite has a low surface recombination velocity. This is very important in reducing the recombination losses in a thin film cell, but whether it arisès from an intrinsically low surface state density, or passivation of the surface (perhaps by a thin oxide layer), remains unknown. The precise composition of the chalcocite controls its resistivity and optical absorption coefficient, both of which have a strong influence on the efficiency of the solar devices.

The Cu_xS/CdS interface is, however, highly unstable and can degrade in several ways. The chalcocite can react with air or moisture with the formation of oxide phases, which will eventually result in the disappearance of the chalcocite phase altogether. Copper also diffuses into the CdS collector layer where it creates a deep acceptor state. This state will eventually compensate for the native and donor dopant states and increase the resistivity of the CdS. Both of these effects will increase the series resistance of the cell and reduce its efficiency. Because of these possible degradation mechanisms, a question mark still hangs over the long-term stability of the Cu_xS/CdS solar cell in service. The performance of most of these devices does fall off steadily with time, even when they are hermetically packaged.

From this brief introduction we can see that the Cu_xS/CdS solar cell is a good example of a system where the physics of thc photovoltaic process, and the material reactions during the preparation of the device, are very complex indeed and by no means fully understood. However, it has proved possible to fabricate cells with operating parameters $V_{oc} = 0.52$ V, $J_{sc} = 22$ mA cm^{-1}, $ff = 0.7$ and a solar efficiency of 9%.

More importantly, these cells have been made with very basic equipment and from relatively inexpensive starting materials (Barnett *et al* 1978). This kind of device offers the possibility of cheap power to countries where the sophisticated crystal growth equipment needed to produce single-crystal solar cells has not been developed, and cannot be afforded, and the requirement is for cheap large-area solar arrays.

REFERENCES

Anderson R L 1960 *IBM J. Res. Dev.* **4** 283
Bachmann K J 1979 in *Current Topics in Materials Science* vol. **3** ed. E Kaldis (Amsterdam: North-Holland) p 477
Barnett A M, Bragagndo J A, Hall R B, Phillips J E and Meakin J D 1978 *Proc. 13th IEEE Photovoltaic Spec. Conf.* p 419
Card H C and Yang E S 1977 *IEEE Trans. Electron Dev.* **ED24** 397
Carlson D E 1977 *IEEE Trans. Electron Dev.* **ED-24** 449
Chopra K L and Das S R 1983 *Thin Film Solar Cells* (New York: Plenum)
Chopra K L, Kainthla R C, Pandya D K and Thakoor A P 1982 in *Physics of Thin Films* **12** (New York: Academic)
Chu T L 1977 *J. Cryst. Growth* **39** 45
Coutts T J, Kazmerski L L and Wagner S (eds) 1986 *Copper Indium Diselenide for Photovoltaic Applications* (Amsterdam: Elsevier)
Crowell C R and Sze S M 1966 *J. Appl. Phys.* **37** 2683
Ettenberg M 1974 *J. Appl. Phys.* **45** 901
Fahrenbruch A L and Bube R H 1983 *Fundamentals of Solar Cells* (New York: Academic)
Fritzsche H 1980 *Sol. Energy Mater.* **3** 447
Grovenor C R M, Cerezo A, Liddle J A and Smith G D W 1987 in *Microscopy of Semiconducting Materials, 1987* (Inst. Phys. Conf. Ser. 87) p 665
Haake G 1977 *Ann. Rev. Mater. Sci.* **7** 73
Hadley H C and Tseng 1977 *J. Cryst. Growth* **39** 61
Heavens O S 1965 *Optics of Thin Films* (London: Dover)
Hill D E, Gutsche H W, Wang M S, Gupta K P, Tucker W F, Dowdy J D and Crepin R J 1976 *Proc. 12th IEEE Photovoltaic Spec. Conf.* p 112
Hovel H J 1975 *Semiconductors and Semimetals* **11** ed. R K Willardson and A C Beer (New York: Academic)
Kazmerski L L, White F R and Morgan G K 1976 *Appl. Phys. Lett.* **29** 268
Knights J C 1977 *Solid State Commun.* **21** 983
Kressel H, D'Aiello R V, Lewis E R, Robinson P H and McFarlane S H 1977 *J. Cryst. Growth* **39** 23
Kumar V and Dahlke W E 1977 *Solid-State Electron.* **20** 143
Lindholm F A and Fossum J G 1981 *Proc. 15th IEEE Photovoltaic Spec. Conf.* p 422
Loferski J J 1956 *J. Appl. Phys.* **27** 777
Maissel L and Glang R 1983 *Handbook of Thin Film Technology* (New York: McGraw-Hill)

Matare H F 1971 *Defect Electronics in Semiconductors* (New York: Wiley)

Matthews J W and Blakeslee A E 1974 *J. Cryst. Growth* **27** 118

Moss T S, Burell G J and Ellis B 1973 *Semiconductor Optoelectronics* (London: Butterworths)

Ng K K and Card H C 1980 *IEEE Trans. Electron Dev.* **ED27** 716

Palz W, Besson J, Nguyen Duy T and Vedel J 1972 *Proc. 9th IEEE Photovoltaic Spec. Conf.* 91

Ponpon J P and Siffert P 1978 *Proc. 13th IEEE Photovoltaic Spec. Conf.* p 639

Potter R W 1977 *Econ. Geol.* **72** 1524

Rothwarf A 1976 *Proc. 12th IEEE Photovoltaic Spec. Conf.* p 488

Rothwarf A and Boer K W 1975 *Prog. Solid State Chem.* **10** 71

Saksena S, Pandya D K and Chopra K L 1982 *Thin Solid Films* **49** 223

Sharma B L and Purohit R K 1974 *Semiconductor Heterojunctions* (Oxford: Pergamon)

Singh R, Green M A and Raykanan K 1981 *Sol. Cells* **3** 95

Smith R A 1968 *Semiconductors* (Cambridge: Cambridge University Press)

Soclof S I and Iles P A 1975 *Proc. 14th IEEE Photovoltaic Spec. Conf.* p 56

Spear W E and LeComber P G 1975 *Solid State Commun.* **17** 1193

Sze S M 1981 *Physics of Semiconductor Devices* (New York: Wiley)

TeVelde T S and Dieleman J 1973 *Philips Res. Rep.* **28** 573

Thekaekara M P 1974 *Suppl. Proc. 20th Annu. Meet. Inst. Environmental Science* p 21

Tsaur B-Y, Fan J C C, Turner G W, Davis F M and Gall R P 1982 *Proc. 16th IEEE Photovoltaic Spec. Conf.* p 1143

Vanhaotte G and Pauwels H 1979 *Proc. 2nd Eur. Comm. Photovoltaic Solar Energy Conf.* ed. R van Overstraeten and W Palz p 662

Woodall J M and Hovel H J 1972 *Appl. Phys. Lett.* **21** 379

Wysocki J J, Rappaport P, Davison E, Hard R and Loferski J J 1966 *Appl. Phys. Lett.* **9** 44

Yamaguchi M, Yamamoto A and Itoh Y 1986 *J. Appl. Phys.* **59** 1751

Yoshikawa A and Sakai Y 1977 *Solid State Electron.* **20** 133

10

Failure Analysis and the Investigation of the Structure and Composition of Microelectronic Materials

10.1 INTRODUCTION

In the first part of this chapter we will remind ourselves of some of the mechanisms by which integrated circuits can fail in service and also consider how defects can be introduced during fabrication of the devices. Some of these defects have already been mentioned in Chapters 2 and 6, while the degradation of thin film metallisation structures has been extensively discussed in Chapter 5. The purpose of collecting together these failure modes is to see how a quality-control or failure analyst can set about detecting the site of any particular structural fault or damaging chemical reaction and, if possible, determining its cause. Some of the basic principles of failure analysis will then be described in §10.2.2. It is, however, often necessary to study chemical reactions in multilayered thin film structures in greater detail than is possible in a simple failure analysis programme, and a few of the more commonly used techniques for the detailed investigation of the chemistry and microstructure of integrated circuits will be described in §10.3. Examples of results obtained from these techniques have been included in many places in earlier chapters, but here a comparison will be made of their relative strengths and weaknesses. This will enable us to make an informed choice of which technique to apply to gaining a fuller understanding of any particular kind of failure. Finally, the techniques which have been applied to the study of the crystalline perfection of semiconductor materials will be described in §10.4. These techniques are particularly valuable in the assessment of the quality of bulk single-crystal and epitaxial layer material.

10.2 ANALYSIS OF FAILURE IN INTEGRATED CIRCUITS

10.2.1 Common Modes of Failure

It is convenient to divide modes of failure into two groups: those associated with the manufacturing process itself and those that arise during service. Many of the first group of failures are catastrophic—the device or circuit never works at all—while a large fraction of failures in the second group only become apparent after long periods of normal operation. The manufacturing yield of a device which suffers from failures of the first kind can often be improved by isolating that stage of the manufacturing process in which the defects are introduced and detecting and correcting the faulty process. Many failures of the second kind involve a chemical reaction between components of the microelectronic system and can only be avoided by a radical re-design of the device or metallisation structure. It has become common to describe the lifetime of a microelectronic component in terms of the 'bathtub' curve illustrated in figure 10.1. Quite a significant fraction of devices or systems fail very soon after being put into service, or never function correctly at all, and these are known as 'infant mortality' failures. Most of these components should be detected by quality-control procedures before they leave the manufacturing site. Components are often subjected to a severe short-term current stressing, or thermal annealing cycle, called 'burn-in', specifically to screen out infant mortality failures. The second part of the bathtub curve shows a much lower failure rate over an extended period of time in service. These devices usually fail by gradual degradation mechanisms in the metallisation or within the semiconductor material itself. Finally, the components begin to fail more rapidly as the wearout stage is reached after long service times. The failure mechanisms in wearout are often similar to those characteristic of random service failure.

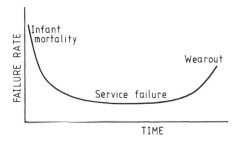

Figure 10.1 A schematic illustration of the 'bathtub' curve, showing the variation in the rate of failure of microelectronic components with time.

Table 10.1 contains a list of some of the common modes of failure in microelectronic devices or components. These have been divided into the two groups introduced above: gross manufacturing defects and material degradation in service. These two groups are of course not entirely independent; for example, a dust particle on a lithographic mask can result in a thinner region than normal in an evaporated metallic conductor track. Here the current density and the temperature will be higher than in the surrounding regions, so that electromigration will be especially rapid and the local flux divergence severe. Likewise, any contamination left in the metallisation structure by a poor cleaning procedure during manufacture can result in a reduced resistance to corrosion and surface electromigration in service. Problems caused by ionic contamination are often very hard to trace back to a particular cleaning step. Reviews of failure modes and mechanisms in silicon- and gallium arsenide-based devices can be found in Howes and Morgan (1981).

Table 10.1 Common failure mechanisms in microelectronic components.

Phenomenon	Effect
(A) Manufacturing defects:	
Mask damage or contamination with dust particles	Incomplete metal patterns
	Pinholes in dielectric layers
Faulty mask alignment	Defective metallisation or device structure in semiconductor
Mechanical damage during handling	Cracking of semiconductor or passivation
	Dislocations in active regions of devices
Contamination due to insufficient cleaning (water, Na^+, Cl^-, etc)	Corrosion of metallisation
	Surface electromigration
	Poor adhesion in metallisation
	High resistance contacts
(B) Degradation mechanisms in service:	
Excessively high temperature	Interdiffusion and chemical reactions
	Degradation of contacts
	Loss of cohesion by void formation
High current density in conductors	Electromigration
Water or ionic species penetrating through passivation or packaging	Corrosion of metallisation
	Surface electromigration
Local high voltages	Dielectric breakdown
Vibration or rapid thermal cycling	Hillock growth on conductors
	Passivation cracking
	Fatigue of metallic components

With such a variety of failure modes in an integrated circuit, it often proves difficult to isolate the precise manner by which a particular component has failed. We shall now consider how to isolate the failure mode and, if possible, the mechanism of failure as well.

10.2.2 A Process for Failure Analysis

This section will introduce a procedure by which faulty microelectronic components can be investigated to determine the cause of failure. A similar kind of process can be used in a quality-control programme, randomly selecting components from a production line for assessment. No description will be given here of the use of electrical assessment techniques to monitor the performance of components, although they are an important part of any quality-control process. Here it is the problems associated with the structure and materials' interactions in a device or circuit which will be described. We will follow the outline of a failure analysis programme given in a series of articles by Richards and Footner (1983, 1987), since these authors give an excellent introduction to the stages needed in an investigation to determine the site and cause of many simple modes of failure. Developing an understanding of failures of a more complex kind may require the use of specialised and sophisticated analysis techniques, and §10.3 will give an introduction to some particular applications of these techniques.

The first stage of any failure analysis programme is the inspection of the component package under an optical microscope. Gross defects such as damage to the packaging material itself, or to the connecting pins, will be readily detectable. A possible next stage is to perform x-ray radiographic analysis on the unopened package. This technique gives a shadow image through the whole package, with the heavy elements absorbing the x-rays strongly just like a medical x-ray radiograph. The package is usually made from ceramic or polymeric materials containing only light elements, which means that metallic wires, contacts and pins within the package can be clearly seen. Any broken or incorrectly connected wires can thus be detected in a radiograph, and cavitation and poor adherence at interfaces may be seen as regions of variable contrast. An example of a radiograph which reveals voids in a solder joint between a chip and a header is shown in figure 10.2, and a back bond of this kind could lead to overheating of the chip in operation. From this example we can see how the internal structure of the component may be revealed in a radiograph, and this technique is a reasonably rapid means of detecting some major bonding and wiring defects without opening the device package.

An alternative means of investigating the integrity of a metallisation structure inside a device package is provided by infrared microscopy. This technique has proved of value in the study of the microstructure of

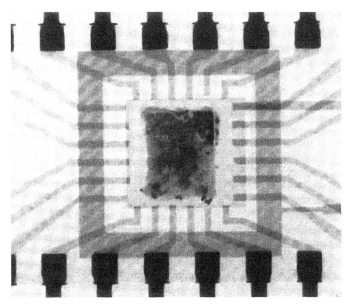

Figure 10.2 A radiograph taken through a complete chip package showing how the package pins and metallic connectors are revealed. The region of uneven contrast under the semiconductor chip shows that there must be voiding in the back bond. (Courtesy of Dr B Richards. © GEC plc. Reproduced with permission of *GEC Journal of Research*.)

Al/Au bonds and a range of defects in metallisation structures. It is hard to evaluate the quality of gold-wire bonds to aluminium contact pads in circuits which are potted in plastic, simply because it is hard to expose the bonds for mechanical testing or microstructural investigation without introducing additional damage. If the plastic package is ground away to expose the bottom of the device chip, infrared illumination can be used to study the undisturbed Al/Au bonds. The silicon chip is transparent to the infrared light and collection of the reflected light allows an image to be formed of individual bonds on the top surface of the chip. Figure 10.3 shows an example of the kind of contrast which is obtained from a bond at which partial reaction has occurred between the gold wire and the aluminium contact to form intermetallic phases by the reactions described in Chapter 5. The grey contrast at the centre of the bonding pad is characteristic of the presence of the intermetallic compounds. Footner *et al* (1986) have been able to show that a clear impression of the mechanical integrity of individual wire bonds can be gained from a careful study of reflection IR images of this kind, and this technique offers another way of investigating the internal structure of a metallisation scheme without the need for complete removal of the package. However, the inclusion of Ti–W barrier layers between the silicon chip and the wire bonds makes a study of the intermetallic formation reaction impossible, since this layer will completely reflect the IR illumination.

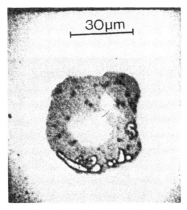

Figure 10.3 An example of an infrared reflection micrograph of an individual joint between an aluminium conductor and a gold wire inside a packaged component. The uneven contrast reveals regions where intermetallic phase formation has occurred. (Courtesy of Drs B Richards and P Footner and GEC Research Ltd.)

If external examination, radiography, and IR microscopy reveal no defects, the surface of the chip itself must be exposed for more detailed study. A series of techniques have been developed for removing the top of packages of the kind described in Chapter 7: single chip cans, ceramic DILs and polymer potted chips. It is most important that these opening procedures should introduce no mechanical or chemical damage to the wiring, metallisation or semiconductor chip, since it is not usually possible to distinguish between the defects which caused the failure of the component and those added inadvertently during the opening of the package. Metal can and ceramic packages are generally easy to open by simple mechanical operations. The can top can be gently sawn or ground off and a knife edge can be used to break ceramic topping sheets free of the glassy cement which bonds them together. If not carried out very carefully, these operations can severely damage the exposed wiring and metallisation on the chip surface, or even crack the semiconductor die.

Opening polymer potted components poses a rather more severe problem, as the polymeric material completely fills all the volume between the bonding wires. For this reason, no mechanical process can be used to remove the plastic and leave the chip and wiring undisturbed. However, most potting polymers can be chemically dissolved in a jet of hot sulphuric acid directed at the encapsulant just over the chip surface. The hot acid will not attack the top passivation layer on the metallisation to any extent and it is unlikely to damage gold or aluminium wires. Figure 10.4 shows an example of a plastic potted component after removal of part of the encapsulating block by this jetting process.

Figure 10.4 An example of a polymer potted component in which the surface of the chip has been exposed by removing the polymer with hot sulphuric acid. (Courtesy of Dr B Richards. © GEC plc. Reproduced with permission of *GEC Journal of Research.*)

At this stage the failure analyst has exposed the top surface of the metallisation on the semiconductor chip and the wires or thick film conductors running from the periphery of the chip to the connecting pins on the package. The interlayer dielectric and glass passivation layers are optically transparent, so that straightforward optical microscopy can reveal some kinds of damage in the metallisation structure. The damaged features must be relatively large (more than 1 μm across) or they cannot be resolved in the optical microscope. Figure 10.5 shows an example of an opened ceramic chip package as seen at low magnification in an optical microscope. The wires connecting the package pins and the contact pads at the outer rim of the chip are clearly visible. At higher magnifications some details of the metallisation layers become visible, and figure 10.6 illustrates how cracks in a glassy passivation layer can be revealed. These cracks may have been produced as a result of the mismatch in thermal expansion coefficients between the glass and the semiconductor chip and can lead to severe local corrosion of the underlying metallic conductors. Mechanical damage to the device can also be seen in the optical microscope, and figure 10.7

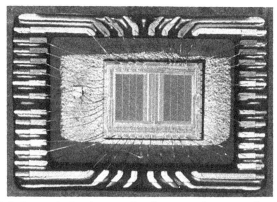

Figure 10.5 An optical micrograph of a ceramic-encapsulated component where the top ceramic sheet has been removed to reveal the chip, wires, package pins and back-bonding solder under the chip.

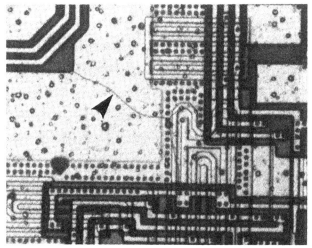

Figure 10.6 An optical micrograph of the top surface of a metallisation structure, showing a fine crack running across a passivation layer (arrowed). (Courtesy of Dr B Richards. © GEC plc. Reproduced with permission of *GEC Journal of Research*.)

gives an example of a crack running across a GaAs die. As a final example of the use of simple optical microscopy in the investigation of damage or potential failure sites in microelectronic devices, figure 10.8 shows a contamination stain left on the surface of a metallisation structure by a careless cleaning procedure during manufacture.

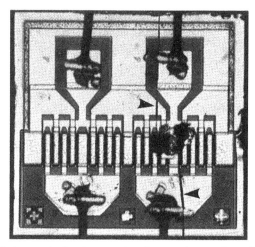

Figure 10.7 An optical micrograph of a crack running all the way across a small GaAs chip. (Courtesy of Dr B Richards. © GEC plc. Reproduced with permission of *GEC Journal of Research.*)

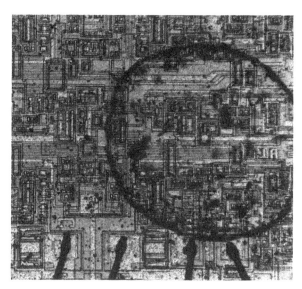

Figure 10.8 An optical micrograph of a gross contamination stain left on the surface of a metallisation structure during manufacture. This kind of contamination can lead to corrosion problems in service. (Courtesy of Dr B Richards. © GEC plc. Reproduced with permission of *GEC Journal of Research.*)

These few examples illustrate the kind of defects which are readily detected in an optical microscope once the package has been opened, i.e. gross mechanical damage, cracks in passivation layers and macroscopic contamination. Misalignment of lithographic masks during the fabrication of complex metallisation schemes can also be seen by virtue of the ability to look through the interlayer dielectric films to the buried levels of metallic conductor tracks. However, not all the defects which can cause failure in metallisation structures can be resolved in the optical microscope. For this reason the scanning electron microscope (SEM) has become a very powerful tool in the analysis of failure in microelectronic devices and circuits. In the following section I shall introduce a few of the ways in which a standard SEM can be used to gain important information on the structure, chemistry and electrical properties of semiconductor materials and devices.

10.2.2(a) The use of scanning electron microscope in failure analysis

The modern SEM has several modes of operation, but here I will concentrate on the use of secondary electron imaging to examine the topography of circuits and devices. More specialised techniques for the examination of the electrical properties of semiconductor materials and devices such as voltage contrast microscopy (VCM), and image formation using the free carriers excited by the electron beam, will be described in §10.2.2(b). Readers seeking an introduction to a wider range of SEM techniques and operating conditions should consult standard textbooks like Wells (1974) and Goldstein and Yakowitz (1975).

The SEM has several major advantages over the optical microscope for the study of microelectronic devices and circuits: a high spatial resolution (about 5 nm in modern machines), a very large depth of field and the ability to obtain excellent topographic contrast. I shall now give two examples to illustrate how the SEM can be used to identify very small defects in metallisation structures. Figure 10.9 shows a high-magnification secondary electron image of a metallic conductor track crossing a step in an underlying dielectric layer. The ability to view the tracks at an oblique angle, and obtain strong topographic contrast, reveals a void where the metal track crosses the step. This defect is too small to be observed by optical microscopy and yet is a site at which excessive Joule heating and highly divergent electromigration is likely. This could lead to the formation of an open circuit, as has been described in §5.4.2. Figure 10.10 is an excellent example of gold dendrites formed by surface electromigration processes, as described in §5.8. These dendrites are too small to be detected in an optical microscope.

Many commercial SEMs are also fitted with facilities for detecting the characteristic x-rays excited from material in the top few micrometres of the sample—energy dispersive x-ray analysis (EDX). This has proved a particularly useful tool in the characterisation of the elements present in

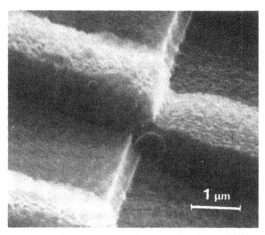

Figure 10.9 A scanning electron micrograph of a void formed where a metal conductor passes over a step in a dielectric layer. (Courtesy of Dr B Richards. © GEC plc. Reproduced with permission of *GEC Journal of Research.*)

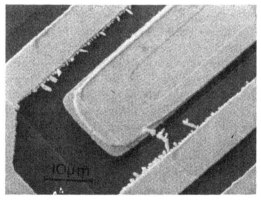

Figure 10.10 A scanning electron micrograph of gold dendrites formed as a result of surface electromigration. (Courtesy of Dr B Richards. © GEC plc. Reproduced with permission of *GEC Journal of Research.*)

contamination stains of the kind shown in figure 10.8. EDX spectra from gross contamination defects often reveal the presence of chlorine, potassium, calcium and sulphur. These are the species which have been linked with surface corrosion and electromigration failure. Regions of a contaminated surface only a few micrometres across can readily be analysed by EDX techniques in SEM. Richards and Footner (1983) have, however, made it clear that the identification of the chemical nature of a contaminating stain may not lead to a precise determination of the stage of the manufacturing process in which it was deposited. In addition, contamination and corrosion problems associated with the presence of carbonaceous species are very hard to study with EDX, since detection efficiencies for light elements are usually very poor. We will see later in this chapter that there are other chemical analysis techniques which are better suited to the study of the distribution of light elements.

Unlike the optical microscope, the SEM cannot form images from layers buried below the surface of a metallisation structure. Defects in metal conductors or dielectric layers below the surface passivating glass layer can only be revealed by a careful process of selective layer removal. As each layer of the metallisation is removed, the component is inspected in the SEM to see if the cause of a particular failure can be isolated. If not, then the next layer is removed and the inspection procedure repeated. This is an enormously time-consuming process and can only be successful if highly selective etches are available for each material in a metallisation scheme. We must for instance be able to strip a phosphosilicate glass layer from a chip surface without damaging the top aluminium conductor layer. A list of some of the most useful selective etches is given in table 10.2, which are similar in some cases to those used for selective wet etching processes during the fabrication of devices. We can see from the comments included in this table that absolute selectivity is hard to obtain and that the removal of a chromium layer, for instance, may cause quite considerable damage in metal layers further down the metallisation scheme. In cases where this damage is not too severe, the SEM can be used to give very detailed topographic information on the structure of defects in very complex

Table 10.2 A list of some selective etches for the removal of layers in metallisation systems (after Richards and Footner 1983).

Layers removed	Etch composition	Comments
Passivation glass	NH_4F/HF/acetic acid	Removes all glassy layers May attack Al, Cr and Ni
Al	HCl	
	$HPO_3/CH_3COOH/HNO_3$	Very rapid etching
Au	Aqua regia	
Cr	Bromine/methanol	Dissolves all metals
W–Ti	H_2O_2 at 50 °C	
	Bromine/methanol	Dissolves all metals
Silicon nitride	Boiling HPO_3	Attacks Al to some extent
Polysilicon	HF/HNO_3/acetic acid	

metallisation systems. It is particularly convenient that good selective etches have been found for the dielectric layers. This allows the conductor tracks, where many defects and failure sites lie in practical metallisation systems, to be exposed for examination.

10.2.2(b) Voltage contrast microscopy

A scanning electron microscope can also be operated so as to detect variations in the local potential on the sample surface, and this can be especially valuable in tracing faulty connections and defects which create open circuits on a chip surface. Variations in the potential on the sample surface result in strong contrast in a standard secondary electron image. This is in part because an SEM detector is biased postively with respect to the sample and so electrons emitted from regions of the sample which are at a negative potential will experience a higher collection field than electrons emitted from regions at a positive potential. This results in an increase in the efficiency with which the secondary electrons are collected by the detector and a significant enhancement in the brightness of the image of parts of the sample at a negative potential. For the same reason, positively biased regions appear dark. The efficiency of secondary electron emission from the surface is strongly affected by local fields as well and so significant contrast changes will be seen around charged conductor tracks. We thus have a very simple way of distinguishing between conductor tracks which are held at different bias levels, i.e. voltage contrast microscopy (VCM). It is quite easy to form electrical connections to a chip in a SEM, so that the devices can be operated normally while the surface of the metallisation is being inspected. Figure 10.11(a) shows an ordinary secondary electron image of the metallisation layers of a simple device with no operating bias applied, while figure 10.11(b) shows the same device with -0.5 V applied to one of the input pins. Some of the metal conductor tracks now appear very bright, indicating that they are connected to this pin. Faults in the metallisation structure, short and open circuits, can easily be detected by a simple experiment of this kind once the pattern of bright conductors has been photographed in a correctly functioning device. Even very complex metallisation structures, where the width of individual conductor tracks may be as small as 1 μm, can be studied, because of the excellent lateral resolution of the SEM. The voltage contrast is normally maximised by using as low an accelerating voltage as possible for the incident electron beam, but if information is required on conductor tracks which lie beneath thin layers of dielectric, an increase in accelerating voltage can allow the electrons to penetrate further into the surface and excite secondary electrons from the buried metal layers.

A slightly more complicated form of VCM chops the SEM beam at exactly the same frequency as the voltage is being modulated in an

(a)

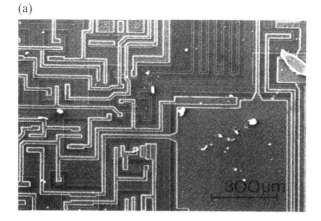

(b)

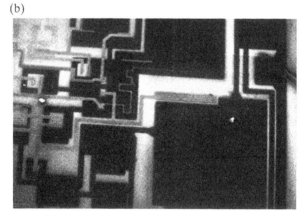

Figure 10.11 (a) A secondary electron SEM image of the top surface of an integrated circuit and (b) a micrograph of the same circuit with a voltage of −0.5 V applied to one of the input pins; (a) is taken with an accelerating voltage of 20 keV, while the strong voltage contrast in (b) is obtained by reducing the accelerating voltage to 2 keV.

operating device and then imposes a small variable phase shift. This technique is exactly analogous to the stroboscopic 'freezing' of a rapidly rotating object when the illumination is pulsed at a frequency equal to the rate of rotation. It is possible to follow the propagation of a signal through an integrated circuit by changing the phase shift and looking for regions of the conductor tracks which appear bright in the SEM image. Signal propagation delays can thus be measured and useful information on the signal waveform in any chosen conductor can also be obtained. Both of these kinds of data can then be compared with those expected from the circuit design of the component to see if it is functioning as expected. Further information on the use of VCM for the study of microelectronic devices can be found in Newbury *et al* (1986).

10.2.2(c) Charge collection microscopy and cathodoluminescence

In this section I will introduce two further modes of operation of an SEM which can be used to obtain useful information on semiconductor devices and also on the properties of defects in the semiconductor materials. Both of these techniques rely on the fact that a high-voltage electron beam (10–30 kV exciting voltage) incident on a semiconductor surface will generate a large number of electron–hole pairs. Each incident electron will generate between 10^3 and 10^4 pairs of free carriers, distributed in a roughly pear-shaped excitation volume which penetrates a few micrometres beneath the sample surface. From a knowledge of the magnitude of the beam current in the SEM it is possible to calculate that the study state concentration of electron–hole pairs under the beam lies in the range $10^{14} - 10^{21}$ cm^{-2}. There are two ways in which the presence of these excess carriers can be detected: (i) by separating them across the built-in field at a p–n junction; or (ii) by observing the light emitted as a result of radiative recombination.

Let us first consider the case where the electron beam is incident on a semiconductor which contains a p–n junction a few micrometres below the surface, as shown in figure 10.12. The electrons generated in the excitation volume will diffuse down the built-in potential gradient at the junction and will be separated from the holes which remain in the p-type material. This looks remarkably like the way in which a p–n junction solar cell operates (figure 9.1) and Holt *et al* (1974) have emphasised the direct correlation between the physical processes which take place in a photovoltaic device and the phenomenon of electron-voltaic charge separation across a junction. We can take the analogy between the solar cell and the electron-voltaic charge collection microscope a stage further by considering what can be measured in the external circuit shown in figure 10.12. If the load resistance, R, is infinitely large, no current flows and an open-circuit voltage will be measured across the specimen, V_{oc}. Alternatively, if R is infinitely small, a short-circuit current, J_{sc}, will flow in the external circuit. These measured parameters are exactly analogous with V_{oc} and J_{sc} in a p–n junction solar cell as defined in figure 9.4. A convenient way of operating the SEM as a charge collection microscope is thus to measure the short-circuit current generated by the electron beam striking the specimen surface. If the beam is scanned over the surface, and the short-circuit current in the external circuit is used to modulate the intensity of a cathode-ray tube, a charge collection image of the sample surface is obtained. The measurement of the short-circuit current, J_{sc}, is by far the most popular way of performing charge collection microscopy on semiconductor samples. We can also obtain charge collection images of semiconductor samples which do not contain a buried p–n junction. Once again, by analogy with solar cell design, it is possible to use a thin Schottky barrier contact to replace the p–n junction. In this way, charge collection images from surfaces of a very wide range of semiconductor materials can be obtained. This kind of imaging technique is called electron beam induced current (EBIC) microscopy. We can also use the external circuit to apply a DC voltage to a specimen which does not contain a p–n junction and to measure the variation of sample current as the electron beam is scanned over the surface. This mode of operation is known as a beta conductivity measurement and is described in Chapter 8 of Holt *et al* (1974).

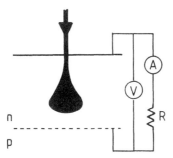

Figure 10.12 A schematic illustration of the excitation of a semiconductor sample by an electron beam. The free carriers will be separated across the buried p–n junction and either the open-circuit voltage or the short-circuit current measured in the external circuit.

Now we should consider the kind of information which can be gained from EBIC microscopy. If a p–n junction intersects the sample surface at right angles, then collection of the current passing across the junction as the electron beam is scanned over the sample can be used to measure the minority carrier diffusion length in both n- and p-type material. The current collected at any position on the sample surface will depend on how rapidly the minority carriers generated under the electron beam can diffuse to the junction, and the minority carrier diffusion length can be accurately estimated from the way that the measured current varies with the distance from the junction. An enormous number of EBIC measurements of the minority carrier diffusion lengths have been reported, and this is one of the most important applications of the technique.

The morphology of buried p–n junctions in complete integrated circuits can also be studied in EBIC microscopy, since the depletion regions around the junctions are regions of especially high charge collection efficiency and so will appear bright in the image. Variations in the concentration of dopant atoms will also affect the efficiency of charge collection, and so the swirl defects in bulk semiconductor crystals described in Chapter 3 can be revealed in an EBIC image. A fourth way of using EBIC microscopy is to identify regions of a sample where electron–hole recombination is particularly rapid. This technique has proved very valuable in the measurement of defect densities in semiconductor crystals, and so will be described in a little more detail.

Figure 10.13(a) shows an electron beam being scanned over a semiconductor crystal which contains a lattice defect, i.e. a grain boundary, dislocation or second-phase precipitate. The figure also shows a thin Schottky barrier contact on the semiconductor surface. If the defect is a preferential site for recombination, by virtue of introducing electronic states at or near the centre of the forbidden band gap (see §§4.3.4 and 8.2), then the short-circuit current will decrease sharply as the electron beam passes over that region of the surface under which the defect lies. This is because some of the carriers excited by the electron beam will recombine at the defect rather than being separated across the Schottky barrier contact. The collected current will have a form like that shown in figure 10.13(b). Many lattice defects are readily detected in an EBIC image if they lie within the excitation volume, and the penetration depth of the incident electrons is about 1–5 μm. This value depends on the beam energy and on which semiconductor is being investigated. Figure 10.14(b) shows an EBIC image from a GaInAs layer grown on a GaAs substrate. The epitaxial layer contains both a planar array of dislocations close to the surface and some threading dislocations, as is shown in the transmission electron micrograph in figure 10.14(a). Both kinds of dislocations are clearly seen in the EBIC image—the planar array as two orthogonal sets of dark lines and the threading dislocations, which run out of the plane of the image, as dark spots. It is the enhanced recombination rate at the dislocations which allows them to be imaged in the EBIC microscopy.

Now let us consider the information which can be extracted from collection of the light emitted from regions of a semiconductor crystal illuminated by the electron beam, i.e. cathodoluminescence. Firstly, we should recall from the discussion in §8.2 that recombination of an electron and a hole can take place by several processes and that not all of these are radiative transitions. Only in semiconductor materials with a direct band gap do we expect the luminescent transitions to dominate over the non-radiative ones. We should thus be able to observe strong cathodoluminescence from direct band-gap materials like GaAs, or from compounds like GaP which can be isoelectronically doped (see §8.2). Most of the light emitted from perfect crystals of these materials under

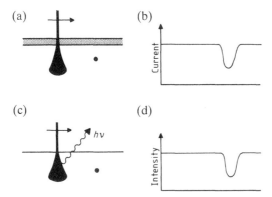

Figure 10.13 Schematic illustrations showing the detection of lattice defects by EBIC and CL techniques in a scanning electron microscope. (a) An electron beam is scanned over a semiconductor sample which contains an extended lattice defect. A Schottky contact is shown on the top surface of the semiconductor. (b) The short-circuit EBIC current which is measured when the beam is scanned over the defect. (c) The electron beam is scanned over a sample of a luminescent semiconductor material. (d) The total emitted light intensity detected as the beam passes over the defect.

an electron beam is released by band-to-band or band-to-shallow state transitions, and so the photons will have an energy very close to the separation between valence and conduction band. Even in highly luminescent crystals, the intensity of the emitted light is very low and care has to be taken to avoid the collection of spurious light signals generated by secondary electrons striking parts of the SEM or light collection and photomultiplication system. The simplest cathodoluminescence (CL) image, a total light image, is formed using all the light emitted from the sample under a scanned electron beam to modulate a cathode-ray tube. Let us once again consider what happens when the beam is scanned over a lattice defect in the semiconductor material. Figure 10.13(c) illustrates the electron beam striking a semiconductor surface (note that no Schottky barrier is needed) and creating electron–hole pairs which spontaneously recombine and release photons. Many kinds of lattice defects will provide preferential sites for non-radiative recombination, and so the collected light intensity will be reduced as the beam passes over one of these defects (figure 10.13(d)). Figure 10.14(c) shows an example of a CL image from the same specimen as in the transmission electron microscope (TEM) and EBIC images described above. The planar dislocation network and the threading dislocations are shown as dark lines and spots; another example of a CL image of dislocations is shown in figure 8.4. In many cases, variations in dopant concentration can also be clearly revealed in CL images, especially if the

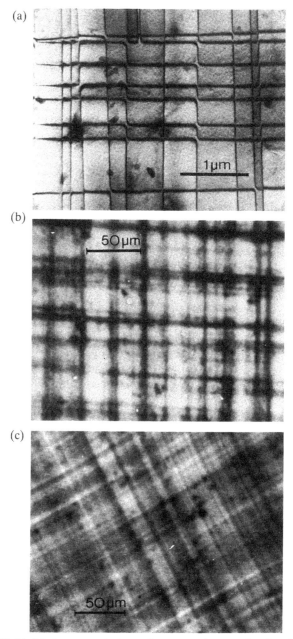

Figure 10.14 (a) A transmission electron micrograph showing a buried dislocation array in a GaInAs epitaxial layer. (b) A scanning EBIC micrograph of the same sample showing how the dislocations are clearly revealed as dark lines. (c) A total light cathodoluminescence image of the same specimen revealing the same dislocation array. (Courtesy of Dr M Al-Jassim.)

dominant mechanism of luminescence is a band-to-shallow state transition. Variations in dopant concentration, or in the stoichiometry of the crystals, are especially clearly imaged in direct band-gap II–VI semiconductor materials.

It is also possible to place a wavelength dispersive spectrometer in the light-collection system in the SEM and to collect a CL spectrum. While much of the emitted light has a wavelength close to that characteristic of a band-to-band transition, many other emission wavelengths may be observed in crystals which contain a high defect density. It is often difficult to identify the character of the defect or contaminant which is responsible for a particular emission band, because there is no simple relationship between the energy of the emitted light and the chemistry or structure of the defect. However, the spatial location of the defect can be determined by forming a CL image using just the intensity in that band to modulate the cathode-ray tube. The lateral resolution of defects in both CL and EBIC images is of the order of $1 \mu m$, so that even quite dense arrays of defects can be studied individually by these techniques. CL spectral peaks are usually rather broad unless the sample is cooled well below room temperature because of phonon scattering effects, and so cooling stages are commonly used for CL experiments.

Some particularly interesting results have been obtained on recombination at individual dislocations in diamond. This recombination process releases photons with energies far below those expected from band-to-band transitions and so must involve a mixture of multiphonon transitions and at least one radiative transition, as sketched in figure 8.3(e). Images of the crystal formed using only the long-wavelength light emitted in these recombination processes show the dislocations as bright lines in a dark background. It is not usually possible to perform the same kind of experiment in semiconductor materials, since the very low energy of the photons released in defect-stimulated radiative transitions of this kind are hard to detect in conventional CL spectrometers. However, Petroff (1981) has demonstrated that certain dislocations in GaAs/GaAlAs samples are non-radiative centres at some wavelengths but radiate intensely at others. A wide range of other kinds of lattice defects emit brightly at characteristic wavelengths, giving well characterised spectral features.

EBIC and CL techniques can be used to study the distribution and electrical properties of defects in a wide range of semiconductor materials. Because of the relative ease with which samples can be prepared for examination by these techniques, and the combination of the identification of the spatial location and electrical activity of defects, CL and EBIC analysis in SEM equipment have become important techniques in the characterisation of bulk crystals and epitaxial layer materials. A combination of TEM, EBIC and CL techniques can allow the crystallography of electrically active lattice defects to be determined, and this has recently been used to obtain useful information on which disloca-tions in semiconductor crystals have high intrinsic recombination rates. On a more macroscopic level, the observation of a high density of recombination centres in epitaxial layers intended for the fabrication of laser diodes is a good indication that the optical efficiency of the diodes will be poor. The role played by dislocations, grain boundaries and precipitates in degrading the electrical and optical properties of micro-electronic devices is described in Chapters 2, 4, 8 and 9. More detailed descriptions of the physics and technology of these techniques can be found in several chapters in Holt *et al* (1974), and in Holt and Datta (1980) and Newbury *et al* (1986). Two recent references which give excellent introductions to the particular strengths of these techniques in the study of defects in semiconductor materials are Leamy (1982) on charge collection microscopy and Yacobi and Holt (1986) on CL micros-copy.

10.3 MICROSTRUCTURAL AND MICROCHEMICAL ANALYSIS TECHNIQUES

We have seen how the optical microscope and the SEM can be used to identify a very wide range of defects and characteristic failure modes in microelectronic circuitry and inside the bulk of the semiconductor material itself. However, there are some kinds of degradation processes which can only be investigated by the use of techniques which provide more detailed information on the chemistry or structure of materials and devices. In particular, the ability to obtain a chemical and microstructur-al depth profile through a complex multilayered device is often very valuable in developing an understanding of degradation mechanisms. The techniques for chemical analysis on which I shall concentrate in this section are those which can provide information on: dopant profiles in implanted or diffusion-doped semiconductor crystals; the distribution of light contaminant elements (C, O, H and F in particular) on a device surface; and intermetallic phase formation in metallic thin film struc-tures. However, we will begin this section with a brief introduction to techniques which allow the microstructure of devices and metallisation systems to be studied. Here it is the combination of resolution of submicrometre device features and characterisation of crystallographic defects in complete device structures, which is required. The more macroscopic techniques which have been developed for the investigation of lattice defects in bulk crystals and epitaxial layers will be described in §10.4

10.3.1 Plan View Transmission Electron Microscopy

The value of the technique of transmission electron microscopy in the study of the microstructure and defect content of microelectronic materials has already been shown by many examples in earlier chapters of this book. The very high spatial resolution available in the current generation of electron microscopes has been illustrated by the lattice images in figures 1.10 and 3.24, and the structure of many semiconductor and metallic materials can now be investigated in atomic detail. In addition, the TEM allows the crystallographic nature of many kinds of lattice defects to be determined, including stacking faults, dislocations, grain boundaries and precipitates. Chemical analysis can be performed by EDX techniques, as described in the last section, or by the use of electron energy loss spectroscopy, EELS. In both these techniques the volume of material from which the precise chemical information is taken can be very small, which allows the identification of very small precipitate particles and the study of processes like grain boundary segregation. EBIC and CL microscopy can also be carried out in a modified TEM, enabling the crystallography and the electrical activity of a defect to be determined in a single experiment. For comprehensive descriptions of the theory and practice of all kinds of transmission electron microscopy experiments the reader is directed to Hirsch *et al* (1977) and Thomas and Goringe (1979).

It is not possible here to illustrate adequately the very wide range of problems in microelectronic materials which have been investigated by TEM techniques, but I shall give one example of an application of this important analytical tool in the study of process-induced defects in device structures. Figure 10.15 shows a pair of TEM micrographs showing polysilicon gate contacts to devices in a single-crystal silicon wafer. The use of polysilicon contacts avoids many of the problems associated with chemical reactions at the metal–semiconductor interfaces. However, in this case the polysilicon is deposited with a high intrinsic stress which causes some plastic flow in the underlying silicon, and the dislocations can be clearly seen under the contact in figure 10.15(b). These defects have a severe effect on the device performance.

TEM studies have also played an important part in developing an understanding of the role of defects in the degradation of laser diodes, as described in §8.6 and in the analysis of the role of oxide precipitates in controlling the properties of Czochralski silicon crystals (§3.3.1). The structure of metallic thin films of all kinds has been investigated in the TEM, including the morphology of silicide contacts.

The disadvantages of TEM lie mainly in the very thin samples which are needed (usually less than 500 nm, and considerably less than 50 nm if high-resolution microscopy is to be attempted) and the small volumes

(a)

(b)

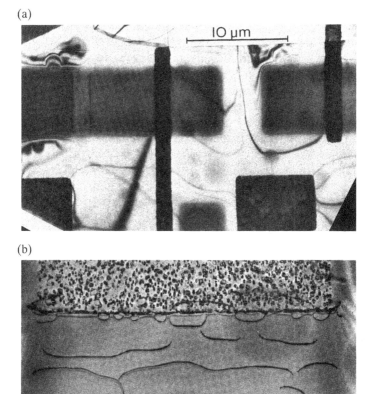

Figure 10.15 (a) A low-magnification transmission electron micrograph of polysilicon contacts deposited onto a silicon wafer. The large dark regions in the wafer are where dopant has been implanted. (b) A higher-magnification image of the same specimen, where the polysilicon gate contact has been removed to reveal dislocations in the silicon wafer between the two implanted regions. These dislocations were produced by the high stress levels generated during the deposition of the polysilicon. (Courtesy of Mr P Augustus and Plessey Research (Caswell) Ltd.)

of material which can be studied in any experiment. The extended specimen preparation time, the destructive nature of the technique and the limited area of the typical sample through which the electrons can penetrate, all conspire to make TEM a technique which is poorly suited to the rapid and routine analysis of the crystalline perfection of large crystals. Only when the unique combination of very high spatial resolution and crystallographic analysis is required will the TEM be perferred to the macroscopic assessment techniques described in §10.4

10.3.2 Cross-sectional Microscopy

I have shown in Chapters 2, 5 and 8 that microelectronic devices and integrated circuits are multilayered structures built up by the sequential deposition of thin films by vapour phase or liquid phase techniques. While metallisation structures can be investigated in some detail in a simple optical microscope, little information on the registry between the features in the metallisation and in the devices buried beneath the semiconductor surface can be obtained in a technique which looks at the devices in plan view. Chemical sectioning techniques allied with SEM investigations can study the structure of a metallisation system in a layer-by-layer manner, but once again cannot be used to investigate the morphology of an entire device. If we wish to look at the structure of a complete device, it is often necessary to make cross-sectional samples in which all the layers can be viewed at once. In many cases it has been found that important information on defect generation during manufacturing processes, and on registration and chemical reaction problems which arise as a result of poorly designed device structures, are only obtained once cross-sectional microscopy techniques have been used.

The simplest form of cross-sectional sample is the metallographic section, where the device or integrated circuit is cut and polished on a plane normal to the top surface of the semiconductor wafer. Figure 8.16(b) is an example of a metallographic cross section through a simple light-emitting diode, and many of the device features are revealed, with the important exception of the p–n junction. Figure 10.16 shows a section through a silicon diode which hs been etched to highlight changes in dopant concentration, so that here the p–n junction is revealed. The aluminium contact on the far right hand side of the figure has penetrated through the top silicon layer and has shorted out the buried p–n junction. (It is quite possible to find polishing procedures which can be used to prepare excellent metallographic samples, even when complete device packages, including printed circuit boards and ceramic substrates, are being sectioned.) This kind of optical microscopy can be used to reveal large processing defects or non-uniform junctions, but will not be able to resolve damage or defects in integrated circuits when the individual metallisation layers and contact vias have dimensions as small as $1 \, \mu m$. Clearly we will have to use microscopic techniques with better resolution to study the structure of these devices in any detail.

Figure 10.16 A metallographic cross section through a silicon diode, showing the penetration of the aluminium contact to short-out a buried p–n junction. (Courtesy of Dr B Richards. © GEC plc. Reproduced with permission of *GEC Journal of Research.*)

Figure 10.17 shows an example of a SEM image of a cross section through one device in an integrated circuit. This kind of sample is conveniently made by cleaving the semiconductor die. Both the individual layers in the metallisation structure and the p–n junctions can be highlighted by chemical etching. Patel and Trigg (1984) have described how cleaving and staining processes can also be used to delineate defects like grain boundaries and dislocations. The morphology of oxide layers can be detected by preferential dissolution to provide topographic contrast which can be seen in the SEM. This simple technique can give an excellent image of the morphology of a whole device, which makes it very valuable for failure analysis and quality control. The design engineer can quickly compare a micrograph like the one shown in figure 10.17 directly with the ideal structure of the devices to see whether all the manufacturing processes have been correctly carried out. The thickness of individual metallisation layers, the depth of p–n junctions and the accuracy of the registration of the lithographic masks can all be seen.

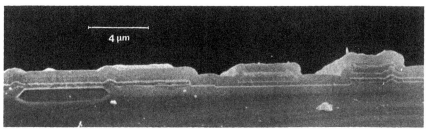

Figure 10.17 An SEM micrograph of a cleaved cross-sectional specimen of a complete device in a silicon wafer. (Courtesy of Dr A Trigg. © GEC plc. Reproduced with permission of *GEC Journal of Research.*)

However, even the SEM may not have sufficient resolution to detect some defects or manufacturing faults, and it is sometimes necessary to prepare cross-sectional specimens thinned for examination in the TEM. Since the optimum thickness of a TEM sample is 20–200 nm, it is a delicate and prolonged procedure to make this kind of specimen from an integrated circuit. The most common method involves sawing the chip into strips which are then mechanically polished to a thickness of about 20 μm, before final ion beam thinning to produce electron transparent regions in precisely the desired place in the circuit. Oppolzer (1985) and Chew and Cullis (1987) have given more complete descriptions of how to carry out this difficult process.

The particular value of the TEM in the study of device structures is that individual features only a few nanometres across can easily be resolved. This allows the morphology of even the thinnest oxide layer to be investigated (see figures 6.6 and 6.8 for example) as well as the density and character of lattice defects, like dislocations introduced by improper handling of the wafer or during poorly designed process steps. Figure 10.18 gives an example of how the structure of an entire device can be imaged in a TEM micrograph. The microstructure of each individual metallisation layer is revealed in such a picture and in some cases the implanted dopant profiles are visible as well. TEM studies have been particularly important in revealing the lattice damage which can be created during the oxidation of silicon wafers through a silicon nitride oxidation mask.

Figure 10.18 A cross-sectional transmission electron micrograph of a complete device structure, showing how lattice and fabrication defects can be clearly revealed. (Courtesy of Dr N Jorgenson.)

Cross-sectional transmission electron microscopy is one of the most commonly used techniques in the study of the morphology and micro-structure of devices and multilayered semiconductor materials. Examples of observations on the crystalline perfection of II–VI and III–V epitaxial layers have been given in figures 3.23 and 8.10(b). Some remarkably successful failure analysis programmes have concentrated on taking large integrated circuits, locating single microscopic defects in the metallisation system or device structures by electrical characterisation or VCM, and identifying the defect in a cross-sectional TEM specimen. In some cases the defects are only a few nonometres in size—a real example of finding a needle in a haystack.

In all the cross-sectional microscopy techniques I have described here, the microelectronic component must be sectioned, and so, of course, they are all destructive processes. There is no question of repairing a faulty device once it has been prepared for observation in an SEM or TEM. As a general principle, it is not possible to repair faults in microelectronic components except by reflow processes during manufacture itself. The purpose of cross-sectional microscopy studies is thus to reveal the character of the damage in a faulty circuit or device and so to identify the mechanism of failure. This will then allow the device engineer to design device structures and manufacturing processes in which these defects do not occur.

10.3.3 Chemical Analysis and Depth Profiling

Microelectronic devices and circuits can consist of a large number of layers of different materials and it is often convenient to be able to obtain chemical composition profiles through all these layers. There are three analytical techniques which are commonly used to study the chemistry of electronic materials and devices: Auger electron spectroscopy (AES), secondary ion mass spectrometry (SIMS) and Rutherford backscattering (RBS). I shall now briefly introduce these techniques and then compare their relative advantages and disadvantages in the investigation of thin film phenomena.

10.3.3(a) Auger electron spectroscopy
The principle of AES analysis is to excite atoms on the surface of a sample with a high-energy (about 5 keV) electron beam. Electrons in the inner shells of the surface atoms are ejected by collisions with the energetic incident electrons, leading to the decay of electrons from higher shells in the same surface atom to fill the vacant states. The energy released in this second process ejects an Auger electron from another higher shell, and these electrons will have an energy characteristic of the separations between the three participating electron levels and so of the atom in which all these processes occur. (A rather similar technique relies on energetic x-rays to eject characteristic electrons from

surface atoms, i.e. x-ray photoelectron spectroscopy or XPS.) The basic features of Auger spectroscopy are:

(1) An Auger spectrum can be used to identify the elements present on the sample surface since it contains peaks at characteristic energies. However, quantitative analysis is often very hard to perform without analysis of a wide range of standards.

(2) The shape and position of some Auger peaks can also contain information on the chemical state of the atom on the sample surface. The Auger peaks for silicon atoms in crystalline silicon and in amorphous silicon dioxide are at different energies for instance.

(3) The escape depth of Auger electrons ranges from about 0.5 to 2 nm, so AES is sensitive only to surface atoms, not to those in the bulk of the specimen.

(4) A map of the distribution of chosen elements in the sample surface can be obtained by scanning a finely focused electron beam over the sample and forming an image where the strength of the Auger signal from any particular element is used to modulate the intensity of a cathode-ray tube. This technique is known as scanning auger microscopy (SAM).

The extreme surface sensitivity of the AES technique is useful in the study of surface reactions, contamination-related corrosion and surface electromigration phenomena. The technique is sensitive to almost all the light elements, with the important exception of hydrogen. If a chemical depth profile is required, then a controlled sputtering process can be used to abrade away the sample. This profiling process is carried out in a series of steps, with AES analysis of the exposed surface at regular intervals. Since sputter depth profiling is also used in SIMS analysis, it is worth considering what effect a sputtering process may have on the morphology of the sample and the quality of the chemical information obtained.

Sputter profiling uses a 0.5–5 keV ion beam, often of inert argon ions, to abrade away the sample surface. The incident ions transfer some of their kinetic energy to the surface atoms, encouraging them to escape into the vapour phase as a mixture of ionic and atomic species. The number of surface atoms removed by each incident ion, the sputter yield, varies with the mass of the incident ion, its chemical reactivity, the angle of incidence of the beam onto the sample surface, the incident energy, and also the mass of the surface atoms and the energy with which they are bound to the surface. There are several unwanted effects which can be produced by sputtering:

(1) The composition of the surface of an alloy sample will rarely be exactly the same as that of the underlying bulk material due to the

preferential sputtering of some species and inhomogeneous mixing and redeposition phenomena.

(2) The ionic bombardment will result in knock-on of surface atoms into the bulk of the sample. With a 3 keV incident ion beam, a mixed layer of thickness approximately 4 nm is produced. This layer will also contain a significant concentration of the bombarding ions. Ion mixing can result in an apparent broadening of the chemical profiles across internal interfaces and may also stimulate phase formation at unreacted interfaces.

(3) Sputtering can roughen the sample surface by random fluctuations in the abrasion efficiency, leading to morphologies containing prominent peaks and troughs. A severe case of surface roughening has already been shown in figure 2.16 for indium phosphide sputtered with argon ions. Most of the other important III–V compounds will show some surface roughening at the bottom of a sputter crater of any significant depth. Different crystal faces can also be abraded at quite different rates, and grain boundaries in polycrystalline samples are preferential sites for rapid sputtering. A peak-to-trough height of about 3–10% of the total sputtered depth is a realistic estimate even in single-crystal samples, although this roughness may be reduced by cooling the sample.

Both ion mixing and surface roughening effects will degrade the depth resolution of a profiling experiment which involves sputtering and AES surface analysis. The deeper the profile extends into the sample surface, the worse the resolution will become. In addition, modifications to the surface chemistry during sputtering can result in the measurement of a very inaccurate chemical depth profile. Because of these limitations, AES depth profiling experiments are often carried out on the sloping surfaces of samples with a bevel polished at a small angle to the original surface. This has the effect of broadening the apparent width of thin layered structures, allowing accurate analysis to be made of the composition of each layer, and avoids the problems associated with ion beam sputtering. A similar experiment can be carried out on the sidewall of a sputter crater, a process called 'sputter edge profiling'. An example of the kind of chemical information which can be obtained in an Auger sputter etch profiling experiment is given in figure 10.19. The sample consists of thin gold and titanium films on a silica substrate, and the presence of a significant concentration of oxygen in the titanium layer is detected. AES depth profiles through thin film metallisation structures have been particularly valuable in the analysis of interdiffusion and chemical reaction processes. Further details on the technology and potential of AES surface analysis can be found in reviews of Chang (1974), Joshi *et al* (1975) and Woodruff and Delchar (1986). The physical mechanisms underlying the sputtering process have been described by Carter and

Colligon (1968), Czanderna (1975) and Auciello and Kelly (1984).

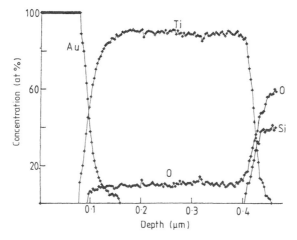

Figure 10.19 A sputter-etched Auger profile through a $Au/Ti/SiO_2$ sample. The oxygen detected in the titanium layer may lead to loss of adhesion between the metal layers in service and increased electrical resistance in the titanium layer. (Courtesy of Dr G Thomas and British Telecom Research Laboratories.)

10.3.3(b) Secondary ion mass spectrometry

The principle of secondary ion mass spectrometry (SIMS) analysis is to use an ion beam to sputter the sample surface, as described for AES depth profiling, but to use mass spectrometry techniques to identify the mass-to-charge ratios of the ionic species in the sputtered material. Sputtering processes remove the surface of a material as a mixture of atomic and ionic species, but the fraction of the sputtered species which are ionised is usually less than 10^{-3}. However, since the total number of sputtered particles is very large indeed, the fact that only a small fraction of them can be detected is usually not important. SIMS analysis can be an extremely sensitive way of obtaining information on the chemical nature of the sample, as we shall see below, but there exists a very serious difficulty in the interpretation of a SIMS mass spectrum—how to relate the number of ions detected in any particular mass peak, Si_{28}^{+} for example, to the concentration of silicon in the sample surface. It has been found that secondary ion yields, i.e. the number of ions of an element ejected from the sample by each incident ion, depend very strongly on the chemical environment in which the element is to be found on the sample surface. Ion yields can vary by as much as a factor of 10^4 for different elements in the same alloy. The crystallography of the surface, as well as the mass and reactivity of the incident ions, can alter sputter yields drastically. If O^{2-} ions are used as the primary sputtering beam, the yield of positive secondary ions from metallic elements is usually very high. Conversely, using O_2^+ or Cs^+ primary ions will result in the detection of strong signals from negative ions produced by the sputtering of non-metallic elements and almost no sign of the metallic elements even if they are present in very high concentrations on the sample surface.

A particularly striking example of the problems which can occur in SIMS analysis is given by the case of sputtering through a Ta_2O_5 layer down to a pure tantalum substrate. The yield of tantalum ions from the oxide is much higher than from the metal substrate, and the SIMS signal from tantalum actually decreases as the sputtering process removes the last of the oxide layer. This shows how strong the influence of chemical environment on the sputter yield can be in some cases. Because of this effect, SIMS is not the ideal analysis technique for obtaining chemical depth profiles through metallisation structures where layers of metal conductors alternate with oxide materials. Some further examples of the difficulties encountered in the profiling of interfaces in multilayered samples have been given by Morabito and Lewis (1975).

However, if the sample is reasonably uniform, without sharp changes in the nature of the chemical bonding, then SIMS can be most useful in giving accurate chemical depth profiles. Two kinds of analysis which can be carried out particularly well in a SIMS facility are the determination of dopant profiles in semiconductor crystals and the study of variations in composition in epitaxial layers of III–V compounds. The detection sensitivity of dopant elements in SIMS is at least two orders of magnitude better than in AES and can be as low as 10^{14} cm^{-3} in some cases, for example boron in silicon. Complex implanted dopant profiles can also be analysed, giving an accurate determination of the depth at which p–n junctions are formed. Similarly, composition variations can be studied in epitaxial structures like the laser crystals described in §8.5. Figure 10.20 is a SIMS profile of the boron concentration in an implanted silicon sample and shows that a combination of high sensitivity and quite good depth resolution can be obtained in this kind of experiment. Figure 10.21 gives an example of SIMS analysis of impurity distributions in MOCVD epitaxial layers. However, the problems associated with ion sputtering for depth profiling apply just as much in SIMS analysis as they do in AES experiments. Surface roughening effects will degrade the depth resolution and will become more severe as the profile depth increases.

In both these kinds of applications, the high sensitivity of the SIMS technique is very important, and in many cases the ability to detect all elements, including hydrogen, is also of value. If quantitative chemical

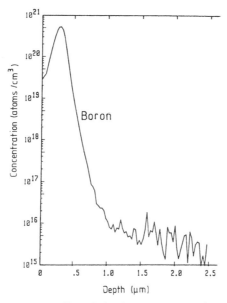

Figure 10.20 A SIMS profile of the boron content in an implanted silicon wafer showing the very high sensitivity of the technique for this dopant. (Courtesy of Dr G Spiller and British Telecom Research Laboratories.)

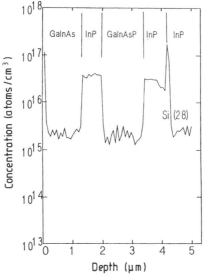

Figure 10.21 A SIMS profile of silicon impurity species in an MOCVD GaInAsP/InP layered structure. The silicon is shown to be concentrated in the InP layers and is introduced in the phosphorus precursor gas. (Courtesy of Dr G Spiller and British Telecom Research Laboratories.)

data are required, it is necessary to calibrate the secondary ion yields against carefully prepared chemical standards. These standards should be as close as possible to the composition and bonding character of the sample being studied. Implantation of primary ions, O^- for example, can also change the chemical nature of the surface, causing a large composition transient at the start of a SIMS analysis before steady state conditions are reached. This last effect is especially important when the abrasion rates are high. It is clear from the above that obtaining quantitative chemical information in a SIMS facility requires considerable effort and a good understanding of the often very complex effects which are occurring on the sample surface.

Another useful way of using SIMS for the analysis of semiconductor materials is by scanning the incident ion beam and using the intensity of the peak from a particular ionic species to form a scanned image on a cathode-ray tube. Imaging SIMS is often used with a very low incident ion beam current so that only a few monolayers of the surface are removed even after an extended period of analysis. This technique is called static SIMS and is valuable in the study of surface contamination problems. Figure 10.22 is an example of a static SIMS experiment on a

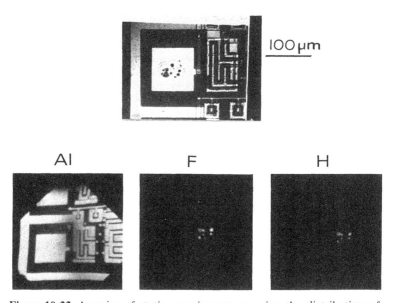

Figure 10.22 A series of static SIMS images mapping the distribution of elements on a blistered aluminium contact pad. Hydrogen and fluorine are clearly present in the damaged region. (Courtesy of Dr G Spiller and British Telecom Research Laboratories. Reprinted by courtesy of The Electrochemical Society, Inc. This paper was originally presented at the Spring 1986 Meeting of the Electrochemical Society, Inc. held in Boston, MA.)

corroded aluminium contact pad. The scanning ion images identify the presence of fluorine and hydrogen in the damaged regions, indicating that this contamination is caused during a dry etching process with gas containing fluorine.

Further details on the use of SIMS analysis in the study of semiconductor materials and devices can be found in McHugh (1975) and Benninghoven (1979).

10.3.3(c) Rutherford backscattering

RBS is the only one of the techniques described here which is non-destructive, but as we shall see the area which is analysed is usually much too large to allow chemical profiling through individual devices in a microelectronic circuit. A high-energy beam of light ions, usually H^+ or He^+ ions with energies of 1 to 4 MeV, is directed at the sample and penetrates deeply below the surface. The high energy of the ions, and their low mass, mean that very little sputtering occurs and there is also almost no lattice damage generated below the surface. A small fraction of the incident ions are backscattered by direct nuclear collisions, and it is the energy of these backscattered ions which is measured in an RBS experiment. The experimental information is contained in the energy distribution of the backscattered ions, and so now we must look at the various mechanisms by which the incident ions lose energy during the backscattering process.

In a head-on elastic collision between two particles, an incident ion of mass m and energy E_0 and a stationary atom with mass M on the sample surface, the incident ion will be backscattered with an energy E given by

$$E = \left(\frac{M - m}{M + m}\right)^2 E_0. \tag{10.1}$$

This means that the ions backscattered by collisions with atoms on the sample surface will have a characteristic energy which depends on the mass of the surface atoms. The higher the mass of the target atom, the greater the energy of the backscattered ions, and so we immediately have a rather crude way of determining the chemical identity of surface atoms from a measurement of the backscattered ion energy. The mass resolution in a system where the incident species are H^+ or He^+ ions is quite good for light elements, but rapidly degrades as the atoms on the sample surface become heavier. For instance, it is hard to separate gallium and arsenic signals from a GaAs surface, which is a great disadvantage when trying to analyse reactions between metal contacts and GaAs substrates.

As an incident ion penetrates into the target, it will gradually lose energy in a series of collisions with the electrons in the atoms of the sample. An ion backscattered by an atom which lies at a depth D under the sample surface will have an energy given by

$$E = \left(\frac{M - m}{M + m}\right)^2 E_0 - 2D \frac{\mathrm{d}E}{\mathrm{d}x}. \qquad (10.2)$$

It is often assumed that the rate of loss of energy with depth, $\mathrm{d}E/\mathrm{d}x$, does not vary with the energy of the ions, although strictly the change in this parameter as the incident ion loses energy to the sample should be included if an accurate determination of depth is to be achieved. $\mathrm{d}E/\mathrm{d}x$ lies between 50 and 200 eV nm^{-1} for most materials and the depth resolution in a RBS experiment is around 25 nm. The information about both the depth of an atom below the sample surface and its chemical identity is contained in the backscattered ion energy, and this can lead to some ambiguities in the interpretation of RBS spectra.

The yield of backscattered ions depends on the atomic number of the sample atoms, Z, and of the incident ion, z:

$$\text{yield} \propto \left(\frac{zZ}{E_0}\right)^2 \qquad (10.3)$$

so that heavy elements scatter many more ions than the lighter elements. As a result, low concentrations of heavy elements are readily detected, but the analysis of light elements in a heavy matrix poses much more of a problem.

Let us now look at the kind of RBS spectrum we expect to detect from a sample which consists of a thin layer of platinum on a silicon substrate. Both the ion yield from the heavy platinum atoms, and the energy of the backscattered ions, will be high. Therefore we should see a prominent peak at the high energy end of the RBS spectrum. The width of this peak will correspond to the thickness of the platinum film. The silicon substrate will give a much lower ion yield, and because it is buried below the platinum layer the highest-energy backscattered ions will have an energy significantly below that expected from silicon atoms on the sample surface. A schematic illustration of the RBS spectrum from a specimen of this kind is shown in figure 10.23. The large difference in mass between the two elements makes this spectrum very simple to interpret. If we now anneal this sample so as to react all the platinum with the silicon substrate to form PtSi, the RBS spectrum will change to that shown in figure 10.24. The platinum peak has broadened, since the PtSi layer is thicker than the initial metal film, and the height of the peak has decreased since the concentration of platinum per unit volume of the surface material is lower. The silicon peak has also broadened, with a significant ion yield now being recorded at the energy characteristic of silicon atoms on the surface of the sample. The stoichiometry of a compound phase can be estimated from an RBS spectrum by using an expression of the form

$$\frac{X_{Pt}}{X_{Si}} \sim \frac{S_{Pt}}{S_{Si}} \left(\frac{Z_{Si}}{Z_{Pt}}\right)^2 \tag{10.4}$$

where S_{Pt} and S_{Si} are the backscattered ion yields detected from a surface of pure platinum and silicon respectively and X_{Pt} and X_{Si} are the atomic fractions of platinum and silicon.

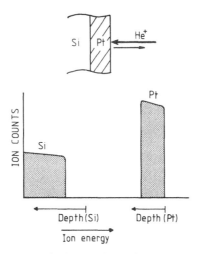

Figure 10.23 A schematic illustration of the RBS spectrum which would be collected from a sample consisting of a thin platinum film on the surface of a silicon wafer.

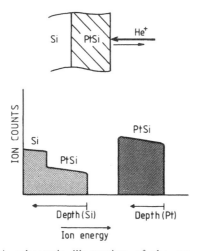

Figure 10.24 A schematic illustration of the RBS spectrum which would be collected from a thin PtSi layer on a silicon wafer.

It is very common to use a combination of x-ray diffraction and RBS analysis to determine the crystallography of the phases which are present and the thickness of each phase. In this way the sequence of phase formation in a thin film reaction can be followed. A great deal of the work on the silicide-forming reactions described in Chapter 4 was carried out with this pair of complementary techniques. In addition, RBS is an extremely valuable tool in the study of interdiffusion between thin films and has been used in many studies of the initial stages of reaction in contact-forming and metallisation systems. Figure 10.25 shows a schematic illustration of the RBS spectrum from a platinum film on a silicon sample after heating to promote interdiffusion, but not compound formation. The trailing edge of the platinum peak has a slope at lower backscattered ion energies, indicating that some platinum has penetrated into the silicon substrate. The front edge of the silicon peak has a similar slope to higher energies. By following the change in these edge positions, an average value for the rate of diffusion over the area of the incident ion beam can be extracted. In interpreting the results of an experiment of this kind, we must note that the silicon diffusion in the platinum will be primarily along grain boundaries in the metal film, but the indiffusion of platinum into the single-crystal silicon substrate will be a bulk phenomenon.

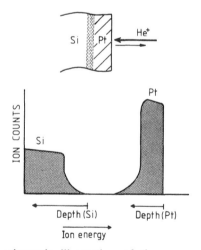

Figure 10.25 A schematic illustration of the RBS spectrum which would be collected from a sample in which interdiffusion has occurred between a platinum film and a silicon wafer.

The non-destructive nature of RBS analysis, and the relatively rapid data collection time, make it particularly suitable for the study of thin film reactions. One limitation of RBS is that the incident beam is usually at least 1 mm in width, and so single microelectronic devices can rarely be investigated. Special samples, which are homogeneous over a large area, can be prepared for RBS analysis, but if the chemical reactions of interest occur in a non-homogeneous manner then the unambiguous interpretation of RBS spectra can be difficult since the chemical and depth information in the spectra will be mixed together in an inseparable way. It is thus important to have determined the morphology of a thin film reaction, perhaps by cross-sectional TEM, before attempting a detailed RBS analysis. A very inhomogeneous reaction between a gold–germanium alloy and gallium arsenide, like that illustrated in figure 2.5, would give an uninterpretable RBS spectrum because a number of phases of different composition are distributed at more or less the same depth in the sample. Howes and Morgan (1981) have given several examples of studies of both interdiffusion and chemical reactions at metal–GaAs interfaces and have illustrated some of the problems associated with the interpretation of RBS spectra of annealed Au–Ge/GaAs contacts.

An enormous number of reviews have been written on the RBS technique and many of these concentrate on the application of RBS for the analysis of the kinetics and phase chemistry of thin film reactions in microelectronic materials. A few of these which give an excellent introduction to this popular and important technique are Morgan (1977), Chu *et al* (1978) and Mayer and Poate (1978).

10.3.3(d) *A comparison of chemical depth profiling techniques*
The principles of AES, SIMS and RBS analysis have now been introduced and here I shall give a brief comparison of the relative advantages of these techniques in the study of interdiffusion and chemical reaction phenomena in microelectronic materials. Table 10.3 lists the main features of all three techniques and shows that SIMS has the best combination of sensitivity and chemical specificity but can be very hard to quantify, especially in multilayered materials. AES analysis is reasonably sensitive, can be roughly quantified, and like SIMS allows maps of surface composition to be formed. RBS analysis is the quickest way of analysing the chemistry of complex layered structures and gives a rough idea of phase compositions, but averages data from a very large area and so samples usually have to be specially prepared if a particular reaction is to be investigated. A full understanding of the chemical reactions which occur in complex multilayered structures can rarely be obtained by the use of only one of these techniques, and it is common to apply AES, SIMS and RBS in combination with x-ray diffraction and TEM/SEM observations to give chemical, morphological and crystallographic information on the same samples. Honig (1976) and Feldman

Table 10.3 A comparison of RBS, AES and SIMS techniques for chemical analysis of microelectronic materials.

	RBS	AES	SIMS
Elements detected	Heavier than Li	Heavier than Li	All, including H
Chemical specificity	Elements with similar mass or atomic number are hard to distinguish	Good	Excellent
Detection limits	0.1 at.%	0.1 at.%	Excellent for most elements, as low as 10^{14} cm^{-3} for some dopants
Spatial resolution	Poor, usually >1 μm No images possible	~1 μm SAM imaging	50 nm–1 μm Secondary ion images
Depth resolution	25 nm Degrades with depth	25 nm Degrades with sputter depth	More than 25 nm Degrades with sputter depth
Ease of obtaining quantitative chemical information	Approximate compositions easily obtained	Approximate composition with standards	Quantification very difficult, even with standards

and Mayer (1986) have given a more comprehensive list of the techniques which can be used to study the chemistry of microelectronic materials and circuits. Details on the valuable surface-sensitive analytical techniques which I have not described here, i.e. low-energy electron diffraction (LEED), ion scattering spectrometry (ISS) and x-ray photoelectron spectroscopy (XPS), can be found in the volumes edited by Czanderna (1975) and Matthews (1975). A useful description of how data from a full range of these techniques can be built up to give a complete picture of a particular thin film reaction can be found in Howes and Morgan (1981).

10.4 THE ASSESSMENT OF CRYSTALLINE PERFECTION

Chapter 3 contains a description of various techniques which can be used to grow semiconductor materials in the form of bulk single crystals and thin epitaxial layers. These materials are normally required to have as low a defect density as possible, and in many earlier chapters we have seen how dislocations, second-phase precipitates and grain boundaries can degrade the performance of microelectronic devices. In this section I will describe a few techniques which have proved particularly useful in assessing the crystalline perfection of semiconductor materials, i.e. defect etching, channelling in an RBS facility and x-ray topography. I have chosen these techniques as examples of how the defect density in a whole wafer or bulk crystal can be studied in a single experiment. We will see that x-ray topography also allows the crystallographic character-isation of extended lattice defects in these samples.

However, we should not forget that there are many other important ways of assessing the quality of a semiconductor wafer or epitaxial layer. These materials are intended for the fabrication of microelectronic devices and so the electrical or optical properties are the most obvious indications of whether they have the desired properties. Depending on the precise application for which a particular material is to be used, the carrier mobility, bulk resistivity and minority carrier lifetime are useful indications of crystalline perfection and the concentration of dopant and other impurity elements. The sharpness of photoluminescence peaks from laser and LED materials also indicates the quality of epitaxial layers. However, these measurements provide rather indirect evidence for the presence of crystalline defects and it may be hard to determine the character and the source of defects. It is especially hard to distinguish the effect of point defects from that of extended lattice defects.

10.4.1 Chemical Defect Etching

Chemical etching techniques have often proved especially revealing in studies of the distribution of dislocations in bulk crystals. A whole range of specialised etching solutions have been developed for delineating the points at which dislocation lines intersect the surfaces of semiconductor wafers. These etching solutions preferentially dissolve the material immediately around the dislocation line, leaving an etch pit on the surface which can easily be observed in an optical microscope. These etches usually work in two stages, first oxidising the semiconductor surface and then dissolving the oxide compounds. An example of the way in which a chemical etching process shows the presence of disloca-tions in a GaAs wafer is given in figure 10.26. Precipitate particles attached to the dislocations are also delineated by this etch. The shape and depth of the etch pits may in some cases depend on the character of the dislocations and it is possible to choose defect etches which give some information on the crystallography of lattice defects.

Table 10.4 contains a list of a few of the most common defect etching solutions used on the semiconductor materials considered in this book.

Figure 10.26 An optical micrograph of a GaAs wafer which has been chemically etched to reveal the dislocations which intersect the wafer surface. The small secondary etch pits delineate precipitate particles on the dislocation lines. (Courtesy of Dr D Stirland and Plessey Research (Caswell) Ltd.)

Table 10.4 Defect etches for silicon and III–V semiconductor materials.

Material etched	Solution
Silicon	HF/CrO_3 (Sirtl)†
	$HF/HNO_3/CH_3COOH$ (Dash)†
	$HF/K_2Cr_2O_7$ (Secco)†
GaAs	Molten KOH ($\sim$350 °C)
	H_2SO_4/H_2O_2
	$AgNO_3/CrO_3/HF$ (Abrahams and Buiocchi, A–B)†
InP	A–B
	HBr/HPO_3
GaInAsP	A–B
	HCl/HNO_3
	HBr/HF

†These solutions are often referred to simply by the names of their inventors.

Some of these solutions are also used to polish the same semiconductors and the concentration of the solution controls whether a particular combination of chemicals polishes or etches. In addition, the dopant type and concentration in the semiconductor may determine the efficiency of the etch in revealing dislocations. General introductions to the procedures which must be followed to obtain a satisfactory delineation of dislocations in silicon and GaAs wafers have been given by Schimmel (1976) and Stirland and Straughan (1976) respectively.

10.4.2 Rutherford Backscattering Channelling Analysis

The technique of RBS analysis has already been described in §10.3, but there is a mode in which the same equipment can be used to study the crystalline perfection of semiconductor samples. If a crystalline sample is tilted such that a set of lattice planes are aligned along the direction of the incident ion beam, then most of the ions will be 'channelled' along the spaces between planes and will not contribute to the backscattered ion signal. The incident ions will be backscattered by the atoms at the top of each lattice plane, but ions which miss these top atoms will pass down through the crystal with only a low probability of meeting an atomic nucleus by which they will be backscattered. In conventional RBS depth profiling experiments of the kind described above, crystalline samples are oriented carefully to avoid the alignment of low index planes with the ion beam. However, the channelling phenomenon can be useful in assessing crystalline perfection. Figure 10.27 shows a schematic comparison of the RBS spectra obtained from a perfect single-crystal sample aligned along a low index direction and a 'random' direction. The backscattered ion yield from the aligned sample is only about 4% of that recorded from the same sample in a random orientation. The aligned spectrum can be used as a 'fingerprint' for a perfect crystal and compared with experimental data from crystalline samples which contain lattice defects. The backscattered ion yield will increase as the sample atoms are shifted off their perfect lattice positions, at the cores of dislocations for instance.

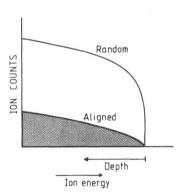

Figure 10.27 A schematic illustration of the difference in backscattered ion yield in a channelled and a random RBS experiment.

One of the most common uses for RBS channelling analysis has been in the study of the structure of wafers which have been implanted with a high dopant concentration (see §2.4). A channelling spectrum of a sample with an amorphous layer on the surface as a result of a heavy implantation treatment might look like figure 10.28(a). The top layer will give an ion yield characteristic of a randomly aligned specimen since the atoms are not arranged on the crystal lattice sites. The undamaged crystal below the implanted depth will give a much lower ion yield characteristic of an aligned spectrum from a perfect crystal. The amorphous region can be removed from the top surface of an implanted wafer by annealing treatments which promote the process of solid phase epitaxy (SPE). During annealing, the crystalline/amorphous interface moves towards the sample surface and this epitaxial regrowth process can be conveniently, and non-destructively, monitored in an RBS facility. Figures 10.28(b) and (c) illustrate two stages in an SPE process as revealed in aligned RBS spectra. In (b) a significant amorphous layer is still left on the wafer surface, although some epitaxial regrowth has occurred, but in (c) a perfect single crystal has been regrown except for a small amount of residual lattice damage right at the wafer surface.

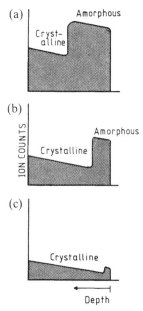

Figure 10.28 A sequence of sketches of channelled RBS spectra from an implanted semiconductor wafer: (a) as-implanted, with an amorphous layer on the top surface; (b) after some solid phase regrowth; (c) after regrowth is complete and only a little residual lattice damage is left near the wafer surface.

Channelled RBS spectra thus allow rapid assessment of the efficiency of SPE processes, but have also proved useful in the study of the quality

of epitaxial layers of the kind described in Chapter 3. Another interesting use of RBS analysis is in the observation of the activation of dopant
atoms. This last process involves the detection of whether implanted
dopant atoms lie on substitutional or interstitial lattice sites in a
semiconductor crystal. Most of the dopant atoms in substitutional sites
will be ionised and will not contribute to an aligned RBS spectrum
because they lie on the atom rows of the perfect crystal. By contrast,
interstitial dopant atoms will increase the backscattered ion yield, since
they lie in the channels between the atom rows.

Channelling analysis in an RBS facility can thus be used to obtain
useful information on the surface layers of semiconductor wafers. The
major shortcoming of the technique is that it gives no information on
the character of the defects which are present in a damaged crystal,
although it can clearly distinguish between an amorphous sample and
one containing various levels of lattice damage. To obtain information
on the kind of defects which are found in bulk crystals and epitaxial
layers we must turn to x-ray topography.

10.4.3 X-ray Topography

The diffraction of x-rays by crystalline materials is a phenomenon which
has been exploited in a number of experimental techniques for many
years. Most of these will be familiar to students of materials science and
have been described in standard references like Cullity (1967) and
Warren (1969). Here I will concentrate on techniques designed for the
study of defects in single-crystal samples. A large number of rather
similar topographic techniques have been developed, but I will only
describe Lang projection topography and double-crystal-rocking curve
analysis. For general reviews on a wider range of x-ray topographic
techniques the reader is directed to Schwuttke (1975), Tanner (1976)
and Meieran (1979).

In principle, an x-ray topograph contains similar information to that in
a transmission electron micrograph, because both techniques are based
on the diffraction of incident radiation by the lattice planes of crystalline
samples. The contrast in an x-ray topograph arises from distortions in
the lattice planes around crystal defects, just like an electron micrograph. Bragg's law is thus the fundamental equation in the theory of
x-ray topographic image formation, as discussed by Tanner (1976).
However, the angular width of the reflecting range around reciprocal
lattice points is much smaller for x-rays than electrons (10^{-5} rad as
opposed to 10^{-2} rad) and this means that x-ray diffraction images are
sensitive to very low levels of strain in the crystal samples, but also that
the images of dislocations are very wide in x-ray topographic images.
The lateral resolution in x-ray images is rarely better than a few

micrometres and so in crystals containing high concentrations of dislocations the defects are not individually resolved. Since the aim of all
semiconductor crystal growth techniques is to reduce the dislocation
content to a minimum, the poor spatial resolution in an x-ray topograph
is not usually a disadvantage since the samples which are of interest
already have low defect densities.

The particular value of x-ray topography as a tool for the assessment
of crystalline perfection lies in the wide range of structural and morphological information which can be obtained from a single image. Localised strains like those around dislocations and second-phase precipitates,
and the abrupt changes in plane orientation at grain boundaries, are
readily observed, but much smaller strains caused by variations in
dopant content can also be detected (see figure 3.7). The Burger's
vectors of dislocations can be found by recording several topographs,
each with a different set of crystal planes set at the Bragg diffraction
condition, and comparing the intensities of the images of the same
dislocation. In addition, x-ray topographs of very large samples can be
recorded, so that the defect content in a large crystal can be assessed in
a single experiment. Another advantage is that very little specimen
preparation is required for most x-ray diffraction experiments. This
contrasts strongly with TEM experiments, which give much the same kind
of information, but each specimen takes a considerable length of time to
prepare and only gives an image of a very small volume of material.

Lang transmission topography is a popular method for studying
defects in semiconductor crystals. Figure 10.29 illustrates the geometry
of the x-ray source, sample and recording film in this kind of experiment. The crystal which can be as thick as a few hundreds of
micrometres since x-rays are only weakly absorbed in most semiconductor materials, is illuminated with a collimated x-ray beam. X-ray tubes
with copper, silver or molybdenum targets are commonly used as the
beam source. The incident x-rays are diffracted by the lattice planes in
the crystal, which is usually oriented so that Bragg diffraction is possible
from only one set of lattice planes. A narrow slit is positioned to allow
only this Bragg reflection through to the recording film. A complete
x-ray topograph of the sample is built up by moving the crystal and the
film together past the slit. In a simple experiment of this kind, extended
lattice defects can easily be seen in the topograph and lattice strains as
small as 10^{-3} are also visible around second-phase precipitates and
compositional inhomogeneities. Figure 10.30 shows an example of a
transmission topograph of dislocations in a silicon wafer on which an
array of devices has already been fabricated. These dislocations seem to
be propagating up to the active surface from sources at the bottom of
the wafer.

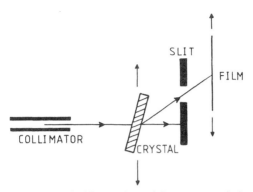

Figure 10.29 A schematic illustration of Lang transmission topography. The equipment needed to perform this kind of experiment is very simple, and exposure times can be very short if a high-intensity x-ray source is used.

Figure 10.30 A transmission x-ray topograph through a silicon wafer on which an integrated circuit has been fabricated. Dense arrays of dislocation loops can be seen propagating up to the top surface of the wafer. (Courtesy of Mr C Dineen and GEC Research Ltd.)

A second way of obtaining useful information on crystalline perfection is to record not a topographic image but the form of the Bragg diffraction peak itself. If the crystal is rocked about an axis which lies in the diffracting planes, the Bragg peak intensity will vary with the rocking angle as the lattice planes are rotated in and out of the precise Bragg condition. The angular width of a Bragg peak will depend on the indices of the diffracting planes, the lattice strain and the defect density in the crystal. The technique of double-crystal-rocking curve analysis has proved particularly sensitive to very small lattice strains, 10^{-5} or less, and the sample configuration in a double-crystal diffraction experiment is illustrated in figure 10.31. A reference crystal is oriented to produce a single Bragg reflection which is then used to illuminate the specimen surface. This beam is diffracted in the second sample only by those planes of the same indices as are excited in the reference crystal. The sample crystal is then rocked about the exact Bragg condition and the intensity of the diffracted peak monitored as a function of rocking angle. If the reference crystal is assumed to have a known lattice parameter and to be defect free, then the measured width and position of the Bragg peak contains information about the lattice parameter and lattice strain in the sample crystal. This kind of technique can be used to study the composition and defect density in thin epitaxial layers. Figure 10.32 shows a comparison of double-crystal-rocking curves from two GaInAs epitaxial layers grown on InP substrates. The epitaxial layers give Bragg peaks at different positions to those from the InP substrates because they have different lattice parameters and the separation of the Bragg peaks can be used to estimate the composition of the layers. The width of the Bragg peak from the epitaxial layers is also an indication of their crystalline perfection, and so in the examples in Figure 10.32 one layer is obviously of much higher quality than the second. It is a reasonable assumption that the dislocation content in the bad layer is much higher than in the other due to poor lattice matching or an uncontrolled growth process. Topographic images can also be recorded in a double-crystal facility and can reveal the small lattice strains associated with in-homogeneous dopant distributions.

X-ray topography thus offers a valuable method for investigating the perfection of semiconductor crystals. The importance of the technique lies in its ability to detect the presence of many kinds of structural and compositional defects and also to analyse their crystallographic charac-ter. This gives the device engineer the chance to understand where and how the defects were generated. In some cases it has been possible to associate observations of high densities of defects in a device material with a particular manufacturing process and then to design a process which does not result in the production of the defects.

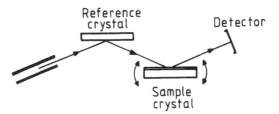

Figure 10.31 A schematic illustration of the principle of double-crystal-rocking curve analysis. The sample crystal is shown with a thin epitaxial layer grown on the surface.

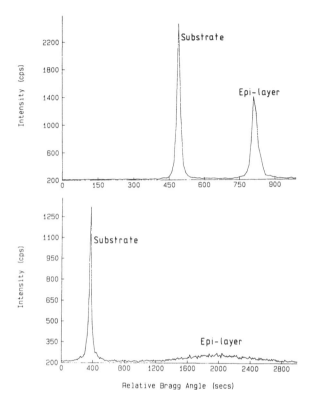

Figure 10.32 A comparison of double-crystal-rocking curves taken from two GaInAs layers grown on InP substrates. In the top case, the lattice parameter of the epitaxial layer is quite closely matched with the substrate and the relative sharpness of the peak from the layer shows that the crystalline quality is high. The lower case shows a large difference in lattic parameter and a very broad diffraction peak from the layer, indicating poor crystalline quality. (Courtesy of Dr M Halliwell and British Telecom Research Laboratories.)

REFERENCES

Auciello O and Kelly R (eds) 1984 *Ion Bombardment Modification of Surfaces* (Amsterdam: Elsevier)

Benninghoven A 1979 *Secondary Ion Mass Spectrometry, SIMS II* (Berlin: Springer)

Carter G and Colligon J S 1968 *Ion Bombardment of Solids* (London: McGraw-Hill)

Chang C C 1974 *Characterisation of Solid Surfaces* ed. P F Kane and G B Larrabee (New York: Plenum) p 670

Chew N G and Cullis A G 1987 *Ultramicroscopy* **23** 175

Chu W K, Mayer J W and Nicolet M-A 1978 *Backscattering Spectrometry* (New York: Academic)

Cullity B D 1967 *Elements of X-ray Diffraction* (Reading, MA: Addison–Wesley)

Czanderna A W 1975 *Methods of Surface Analysis* (Amsterdam: Elsevier)

Feldman L L and Mayer J W 1986 *Fundamentals of Surface and Thin Film Analysis* (New York: North-Holland)

Footner P K, Richards B P, Stephens C E and Amos C T 1986 *24th Annu. Conf. Rel. Phys.* p 102

Goldstein J I and Yakowitz H 1975 *Practical Scanning Electron Microscopy* (New York: Plenum)

Hirsch P B, Howie A, Nicholson R B, Pashley D W and Whelan M J 1977 *Electron Microscopy of Thin Films* (New York: Krieger)

Holt D B and Datta S 1980 *Scanning Electron Microscopy 1980* (Chicago: SEM) p 259

Holt D B, Muir M D, Grant P R and Boswarva I M 1974 *Quantitative Scanning Electron Microscopy* (London: Academic)

Honig R E 1976 *Thin Solid Films* **31** 89

Howes M J and Morgan D V 1981 *Reliability and Degradation* (Chichester: Wiley)

Joshi A, Davis L E and Palmberg P W 1975 *Methods of Surface Analysis* ed. A Czanderna (Amsterdam: Elsevier) p 159

Leamy H J 1982 *J. Appl. Phys.* **53** R51

McHugh J A 1975 *Methods of Surface Analysis* ed. A Czanderna (Amsterdam: Elsevier) p 223

Matthews J W 1975 *Epitaxial Growth A* (New York: Academic)

Mayer J W and Poate J M 1978 *Thin Films, Interdiffusion and Reaction* ed. J M Poate, K N Tu and J W Mayer (New York: Wiley) p 119

Meieran E S 1979 *Characterisation of Crystal Growth Defects by X-ray Methods* ed. B K Tanner and D K Bowen (New York: Plenum) p 1

Morabito J M and Lewis R K 1975 *Methods of Surface Analysis* ed. A Czanderna (Amsterdam: Elsevier) p 279

Morgan D V 1977 *IEE Solid State Electron Dev.* **1** 37

Newbury D C, Joy D C, Echlin P, Fiori C E and Goldstein J I 1986 *Advanced Scanning Electron Microscopy and X-ray Microanalysis* (New York: Plenum)

Oppolzer H 1985 *Microscopy of Semiconducting Materials 1985* Inst. Phys. Conf. Ser 76) p 461

Patel J and Trigg A D 1984 *GEC J. Res.* **2** 240

Petroff P M 1981 *Defects in Semiconductors* ed J Narajan and T Y Tan (Amsterdam: North-Holland) p 457

Richards B P and Footner P K 1983 *GEC J. Res.* **1** 74

—— 1987 *GEC J. Res.* **5** 1

Schimmel D G 1976 *J. Electrochem. Soc.* **123** 734

Schwuttke G H 1975 *Epitaxial Growth* ed. J W Matthews (New York: Academic) p 281

Stirland D J and Straughan B W 1976 *Thin Solid Films* **31** 139

Tanner B K 1976 *X-ray Diffraction Topography* (Oxford: Pergamon)

Thomas G and Goringe M J 1979 *Transmission Electron Microscopy of Materials* (New York: Wiley)

Warren B E 1969 *X-ray Diffraction* (Reading, MA: Addison–Wesley)

Wells O C 1974 *Scanning Electron Microscopy* (New York: McGraw-Hill)

Woodruff D P and Delchar T A 1986 *Modern Techniques of Surface Science* (Cambridge: Cambridge University Press)

Yacobi B G and Holt D B 1986 *J. Appl. Phys.* **59** R1

Index

Milton Keynes UK
Ingram Content Group UK Ltd.
UKHW030901141024
449569UK00025B/1273